TRENDS IN THEORY AND PRACTICE OF NONLINEAR DIFFERENTIAL EQUATIONS

PURE AND APPLIED MATHEMATICS

A Program of Monographs, Textbooks, and Lecture Notes

Contributions to *Lecture Notes in Pure and Applied Mathematics* are reproduced by direct photography of the author's typewritten manuscript. Potential authors are advised to submit preliminary manuscripts for review purposes. After acceptance, the author is responsible for preparing the final manuscript in camera-ready form, suitable for direct reproduction. Marcel Dekker, Inc. will furnish instructions to authors and special typing paper. Sample pages are reviewed and returned with our suggestions to assure quality control and the most attractive rendering of your manuscript. The publisher will also be happy to supervise and assist in all stages of the preparation of your camera-ready manuscript.

LECTURE NOTES
IN PURE AND APPLIED MATHEMATICS

1. *N. Jacobson,* Exceptional Lie Algebras
2. *L.-Å. Lindahl and F. Poulsen,* Thin Sets in Harmonic Analysis
3. *I. Satake,* Classification Theory of Semi-Simple Algebraic Groups
4. *F. Hirzebruch, W. D. Newmann, and S. S. Koh,* Differentiable Manifolds and Quadratic Forms (out of print)
5. *I. Chavel,* Riemannian Symmetric Spaces of Rank One (out of print)
6. *R. B. Burckel,* Characterization of C(X) Among Its Subalgebras
7. *B. R. McDonald, A. R. Magid, and K. C. Smith,* Ring Theory: Proceedings of the Oklahoma Conference
8. *Y.-T. Siu,* Techniques of Extension of Analytic Objects
9. *S. R. Caradus, W. E. Pfaffenberger, and B. Yood,* Calkin Algebras and Algebras of Operators on Banach Spaces
10. *E. O. Roxin, P.-T. Liu, and R. L. Sternberg,* Differential Games and Control Theory
11. *M. Orzech and C. Small,* The Brauer Group of Commutative Rings
12. *S. Thomeier,* Topology and Its Applications
13. *J. M. Lopez and K. A. Ross,* Sidon Sets
14. *W. W. Comfort and S. Negrepontis,* Continuous Pseudometrics
15. *K. McKennon and J. M. Robertson,* Locally Convex Spaces
16. *M. Carmeli and S. Malin,* Representations of the Rotation and Lorentz Groups: An Introduction
17. *G. B. Seligman,* Rational Methods in Lie Algebras
18. *D. G. de Figueiredo,* Functional Analysis: Proceedings of the Brazilian Mathematical Society Symposium
19. *L. Cesari, R. Kannan, and J. D. Schuur,* Nonlinear Functional Analysis and Differential Equations: Proceedings of the Michigan State University Conference
20. *J. J. Schäffer,* Geometry of Spheres in Normed Spaces
21. *K. Yano and M. Kon,* Anti-Invariant Submanifolds
22. *W. V. Vasconcelos,* The Rings of Dimension Two
23. *R. E. Chandler,* Hausdorff Compactifications
24. *S. P. Franklin and B. V. S. Thomas,* Topology: Proceedings of the Memphis State University Conference
25. *S. K. Jain,* Ring Theory: Proceedings of the Ohio University Conference
26. *B. R. McDonald and R. A. Morris,* Ring Theory II: Proceedings of the Second Oklahoma Conference
27. *R. B. Mura and A. Rhemtulla,* Orderable Groups
28. *J. R. Graef,* Stability of Dynamical Systems: Theory and Applications
29. *H.-C. Wang,* Homogeneous Banach Algebras
30. *E. O. Roxin, P.-T. Liu, and R. L. Sternberg,* Differential Games and Control Theory II
31. *R. D. Porter,* Introduction to Fibre Bundles
32. *M. Altman,* Contractors and Contractor Directions Theory and Applications
33. *J. S. Golan,* Decomposition and Dimension in Module Categories
34. *G. Fairweather,* Finite Element Galerkin Methods for Differential Equations
35. *J. D. Sally,* Numbers of Generators of Ideals in Local Rings
36. *S. S. Miller,* Complex Analysis: Proceedings of the S.U.N.Y. Brockport Conference
37. *R. Gordon,* Representation Theory of Algebras: Proceedings of the Philadelphia Conference
38. *M. Goto and F. D. Grosshans,* Semisimple Lie Algebras
39. *A. I. Arruda, N. C. A. da Costa, and R. Chuaqui,* Mathematical Logic: Proceedings of the First Brazilian Conference

40. *F. Van Oystaeyen,* Ring Theory: Proceedings of the 1977 Antwerp Conference
41. *F. Van Oystaeyen and A. Verschoren,* Reflectors and Localization: Application to Sheaf Theory
42. *M. Satyanarayana,* Positively Ordered Semigroups
43. *D. L. Russell,* Mathematics of Finite-Dimensional Control Systems
44. *P.-T. Liu and E. Roxin,* Differential Games and Control Theory III: Proceedings of the Third Kingston Conference, Part A
45. *A. Geramita and J. Seberry,* Orthogonal Designs: Quadratic Forms and Hadamard Matrices
46. *J. Cigler, V. Losert, and P. Michor,* Banach Modules and Functors on Categories of Banach Spaces
47. *P.-T. Liu and J. G. Sutinen,* Control Theory in Mathematical Economics: Proceedings of the Third Kingston Conference, Part B
48. *C. Byrnes,* Partial Differential Equations and Geometry
49. *G. Klambauer;* Problems and Propositions in Analysis
50. *J. Knopfmacher,* Analytic Arithmetic of Algebraic Function Fields
51. *F. Van Oystaeyen,* Ring Theory: Proceedings of the 1978 Antwerp Conference
52. *B. Kedem,* Binary Time Series
53. *J. Barros-Neto and R. A. Artino,* Hypoelliptic Boundary-Value Problems
54. *R. L. Sternberg, A. J. Kalinowski, and J. S. Papadakis,* Nonlinear Partial Differential Equations in Engineering and Applied Science
55. *B. R. McDonald,* Ring Theory and Algebra III: Proceedings of the Third Oklahoma Conference
56. *J. S. Golan,* Structure Sheaves over a Noncommutative Ring
57. *T. V. Narayana, J. G. Williams, and R. M. Mathsen,* Combinatorics, Representation Theory and Statistical Methods in Groups: YOUNG DAY Proceedings
58. *T. A. Burton,* Modeling and Differential Equations in Biology
59. *K. H. Kim and F. W. Roush,* Introduction to Mathematical Consensus Theory
60. *J. Banas and K. Goebel,* Measures of Noncompactness in Banach Spaces
61. *O. A. Nielsen,* Direct Integral Theory
62. *J. E. Smith, G. O. Kenny, and R. N. Ball,* Ordered Groups: Proceedings of the Boise State Conference
63. *J. Cronin,* Mathematics of Cell Electrophysiology
64. *J. W. Brewer,* Power Series Over Commutative Rings
65. *P. K. Kamthan and M. Gupta,* Sequence Spaces and Series
66. *T. G. McLaughlin,* Regressive Sets and the Theory of Isols
67. *T. L. Herdman, S. M. Rankin, III, and H. W. Stech,* Integral and Functional Differential Equations
68. *R. Draper,* Commutative Algebra: Analytic Methods
69. *W. G. McKay and J. Patera,* Tables of Dimensions, Indices, and Branching Rules for Representations of Simple Lie Algebras
70. *R. L. Devaney and Z. H. Nitecki,* Classical Mechanics and Dynamical Systems
71. *J. Van Geel,* Places and Valuations in Noncommutative Ring Theory
72. *C. Faith,* Injective Modules and Injective Quotient Rings
73. *A. Fiacco,* Mathematical Programming with Data Perturbations I
74. *P. Schultz, C. Praeger, and R. Sullivan,* Algebraic Structures and Applications: Proceedings of the First Western Australian Conference on Algebra
75. *L. Bican, T. Kepka, and P. Nemec,* Rings, Modules, and Preradicals
76. *D. C. Kay and M. Breen,* Convexity and Related Combinatorial Geometry: Proceedings of the Second University of Oklahoma Conference
77. *P. Fletcher and W. F. Lindgren,* Quasi-Uniform Spaces
78. *C.-C. Yang,* Factorization Theory of Meromorphic Functions
79. *O. Taussky,* Ternary Quadratic Forms and Norms
80. *S. P. Singh and J. H. Burry,* Nonlinear Analysis and Applications
81. *K. B. Hannsgen, T. L. Herdman, H. W. Stech, and R. L. Wheeler,* Volterra and Functional Differential Equations

82. *N. L. Johnson, M. J. Kallaher, and C. T. Long,* Finite Geometries: Proceedings of a Conference in Honor of T. G. Ostrom
83. *G. I. Zapata,* Functional Analysis, Holomorphy, and Approximation Theory
84. *S. Greco and G. Valla,* Commutative Algebra: Proceedings of the Trento Conference
85. *A. V. Fiacco,* Mathematical Programming with Data Perturbations II
86. *J.-B. Hiriart-Urruty, W. Oettli, and J. Stoer,* Optimization: Theory and Algorithms
87. *A. Figà Talamanca and M. A. Picardello,* Harmonic Analysis on Free Groups
88. *M. Harada,* Factor Categories with Applications to Direct Decomposition of Modules
89. *V. I. Istrăţescu,* Strict Convexity and Complex Strict Convexity: Theory and Applications
90. *V. Lakshmikantham,* Trends in Theory and Practice of Nonlinear Differential Equations

Other Volumes in Preparation

March 21, 1984

TRENDS IN THEORY AND PRACTICE OF NONLINEAR DIFFERENTIAL EQUATIONS

edited by

V. LAKSHMIKANTHAM

Department of Mathematics
University of Texas at Arlington
Arlington, Texas

MARCEL DEKKER, INC. **New York and Basel**

Library of Congress Cataloging in Publication Data
Main entry under title:

Trends in theory and practice of nonlinear differential
equations.

(Lecture notes in pure and applied mathematics;
v. 90)
1. Differential equations, Nonlinear--Congresses.
I. Lakshmikantham, V. II. Series.
QA370.T74 1984 515.3'55 84-15363
ISBN 0-8247-7130-3

MARCEL DEKKER, INC.
270 Madison Avenue, New York, New York 10016

Current printing (last digit):
10 9 8 7 6 5 4 3 2 1

PRINTED IN THE UNITED STATES OF AMERICA

DEDICATION

Dear Fellow Mathematicians:

I have the great privilege to introduce to you Professor E. A. Coddington, a prominent figure of the international mathematical community, and a leading contributor to the theory of ordinary differential operators and equations.

I gladly take the opportunity to present some of the highlights of the outstanding career of Professor Coddington. This presentation is not easy, due to an outstanding record of academic achievements, and due to a very fine personality Professor Coddington embodies. Nevertheless, I will try my best, and at the same time try to comply with what is becoming now a tradition of the mathematical conferences held at The University of Texas at Arlington - to extol the activity of one of our colleagues who has brought an exceptional contribution, a long standing contribution, to the promotion of research in the field of Differential Equations.

Professor Coddington started his mathematical career in 1942--exactly 41 years ago. Until 1946, he was associated with the Navy Department. In 1946 Professor Coddington returned to school for graduate studies and for two consecutive years he was associated with the Johns Hopkins University where he obtained his Ph.D. in 1948. The academic year 1948-1949 for Professor Coddington was spent at Johns Hopkins University where he held an instructorship in mathematics. The next three years, from 1949 to 1952, were spent at the Massachusetts Institute of Technology - again, as an instructor in mathematics.

Since 1952, Professor Coddington was associated with The University of California at Los Angeles where he successively held the positions of Assistant, Associate, and full Professor (the latter, since 1959). From 1968 until 1971, Professor Coddington served as Chairman of the Mathematics Department at The University of California at Los Angeles.

During the period 1955-1981, Professor Coddington held visiting posi-

Professor E. A. Coddington

tions at various universities including: Princeton University (1957-1958), University of Copenhagen (1955-1956 and 1963-1964; as a Fulbright Scholar), University of Paris VI (1972), University of London, Chelsea College (1981) and the Universities of Beijing and Canton (1981).

Besides an impressive number of research papers published in various American and foreign mathematical journals, most of which are dedicated to the theory of differential operators, Professor Coddington has authored a famous treatise *Theory of Ordinary Differential Equations* - jointly with the late MIT professor Norman Levinson (first published by McGraw-Hill

Publishing Co. in 1955), a textbook entitled *An Introduction to Ordinary Differential Equations* (first published by Prentice-Hall in 1958), and recently he has coauthored (with Henrik S. V. de Snoo) the monograph *Regular Boundary Value Problems Associated with Pairs of Ordinary Differential Expressions* - a Springer-Verlag publication in the Lecture Notes in Mathematics Series (1981).

In regard to the treatise *Theory of Ordinary Differential Equations,* one can assert without any doubt that it became the basic reference book in this field for several generations of mathematicians all around the world. Indeed, in 1958, a Russian edition has been published in Moscow. The book has enjoyed a wide circulation not only in the Soviet Union, but also in most East European Countries. A Japanese translation - in two volumes - came out at Yoshioka Publishing House in 1969. In 1972 a reprint of the book came out in India, and according to the testimony of our colleagues in that county, it has been a very successful project. If one thinks that the original English edition has been circulating in the Western World, one can conclude that the *Theory of Ordinary Differential Equations* authored by Coddington and Levinson has indeed enjoyed a world-wide circulation as one of the most authoritative sources in the field. I had the opportunity to use the book in teaching graduate courses in Differential Equations during the period 1958-1962 and I remember the delight of my students in connection with the treatment of most of the classical or less classical problems discussed in the book. The chapter concerning the differential systems in the plane contains so many innovations and is presented in such an attractive manner, that hardly anyone could challenge the presentation of this topic. If nowadays it is very fashionable to deal with integral manifolds, it seems that the first approach to this matter is the one in the book by Coddington and Levinson. There are many more outstanding features of this book and there is no wonder that it still constitutes a basic reference item for researchers in the theory of differential equations. R. Bellman has written a very enthusiastic review of the book in the Bulletin of the AMS, and after almost three decades have passed since its publication, we can ascertain the fact that this is a landmark in the history of differential equations.

The second book Professor Coddington has published constitutes an excellent introductory textbook to the field of Ordinary Differential Equations. In addition to the English version, several foreign publishing houses have produced their versions: Maruzen Co. Ltd. published the

Japanese version in Tokyo in 1964, Prentice-Hall of India reprinted the book in New Delhi in 1968; finally, a Spanish edition was produced by Compaña Editorial Unlimited in Mexico City in 1968.

The third book, authored by Professor Coddington and Henrik S. V. de Snoo, is a monograph dealing with some of the most recent investigations undertaken by the authors. It certainly develops and pursues ideas that appeared in the previous work of the first author, as well as some ideas originated by such famous mathematicians as E. Kamke, E. Hölder, M. G. Krein, and M. A. Naimark. I am not going to present in detail this monograph since Professor Coddington's invited address will be dedicated to the spectral theory of pairs of ordinary differential operators - a topic which seems to be closely related to the ideas developed in the monograph.

Most of Professor Coddington's research work was oriented toward the solution of boundary value problems for systems of ordinary differential equations, involving such core aspects as eigenfunctions expansion, the definition of the adjoint operator, extension of differential operators, and many others.

Identifying the operators with their graphs in the product space, most properties can be formulated in terms of linear manifolds or subspaces of the product space. A theory of linear relations has been built up and successfully developed by Professor Coddington and his co-workers with salient applications to the classical boundary value problems for ordinary differential systems. It is proven that multipoint boundary conditions, and more generally, boundary conditions involving Stieltjes integral, can be dealt with as special cases of the general results obtained in the abstract framework.

In concluding this concise presentation of Professor Coddington's outstanding contribution to the theory of differential equations, I would like to express - and I am sure I can do it on behalf of all the participants - our warmest congratulations to him, and our hope that this conference will emphasize much more than I was able to point out in this brief introductory speech.

Professor C. Corduneanu

PREFACE

An International Conference on *Trends in Theory and Practice of Nonlinear Differential Equations* was held at The University of Texas at Arlington during June 14-18, 1982. This conference was sponsored by the U. S. Army Research Office and The University of Texas at Arlington. This was the fifth in a series of conferences organized by The University of Texas at Arlington. It is a pleasure to acknowledge the financial support received from the various sponsoring angencies which made the conference possible.

The present volume consists of the proceedings of the conference. It includes papers that were delivered as invited survey talks and research reports as well as contributed papers. There were well over eighty talks and sixteen countries were represented.

The aim of the conference was to feature recent trends in theory and practice of nonlinear differential equations. The following topics were discussed: spectral theory for symmetric pairs of differential operators, Lotka-Volterra systems, generalized inverses for linear manifolds, nonlinear problems at resonance, steepest descent methods, reaction-diffusion equations, Lyapunov stability, stochastic differential equations, comparison and frequency domain techniques, delay differential equations, method of upper and lower solutions, Newton-like methods, periodic solutions of nonlinear problems, population biology, effects of harvesting on population systems, models of toxicant population, nonlinear equations of heat flow, inclusion principle for hereditary systems, neuromuscular systems, transform techniques in qualitative theory of systems, vector Lyapunov functions in the analysis of dynamical properties of differential equations, recent topics on nonlinear contraction semi-groups, set valued extensions of integral inequalities, global controllability of nonlinear delay systems, cone-valued Lyapunov functions, large-scale systems, quasi-solutions, almost periodicity of solutions of parabolic equations, generalized Hopf bifurcation and exchange of stability and nonlinear elliptic problems.

At the conference the participants honored Professor Earl A. Coddington of The University of California at Los Angeles for his valuable contributions to differential equations.

I wish to express my special thanks to my colleagues Professors C. Corduneanu, S. Bernfeld, J. Bolen and G. Ladde for helping me in planning and organizing the conference and to my secretaries Mrs. Gloria Brown, Mrs. Cindy Murdock, Mr. Marian Pruiett, and Mrs. Karen Stafford for assisting me during the conference. I am extremely thankful to Mrs. Pruiett and Mrs. Stafford for their excellent typing and finalizing the proceedings of the conference. My immense thanks are due to publishers Marcel Dekker, Inc. for their cooperation.

V. Lakshmikantham

CONTENTS

Honoring Professor E. A. Coddington iii
Preface vii
Contributors xv

Unbounded Perturbations and the Existence of Differentiable Solutions of the Cauchy Problem in Banach Space 1
Nazar Hussein Abdelaziz

Escape Time by Fixed Point Theorem 5
A. R. Aftabizadeh

Asymptotic Behaviour of Solutions of Nth Order Differential Equations with or without Delay 9
Olusola Akinyele and R. S. Dahiya

Nonconstant Periodic Solutions of Nonlinear Differential Equations 13
Shair Ahmad and Francisco Montes de Oca

Evolution Water Coning Problem 27
Y. Amirat, A. El Kolli and A. Ziani

Conditions for Periodic Solutions to Hill's Equation with an even Periodic Coefficient 31
Bernard A. Asner, Jr.

Metrics and Tolerances 39
Prem N. Bajaj

Global Existence for Nonlinear Second Order Differential Equations 43
John V. Baxley

Asymptotically Nonresonant Nonlinear Equations 51
Ronald I. Becker

Boundedness of Oscillatory Solutions of Fourth Order Linear Ordinary Homogeneous Differential Equations 57
Abdelali Benharbit

A Neighborhood of the Hopf Singularity 61
S. R. Bernfeld

Second Order Differential Systems with Generalized Nonlinear Boundary Conditions 69
S. R. Bernfeld

Dynamics and Bifurcation in one Dimension 77
Louis Block and David Hart

The Effect of Harvesting on Population Systems 81
Fred Brauer

Invariants of Vortex Motions in the Plane 91
Jacob Burbea

Instability and Stability in Volterra Equations 99
T. A. Burton and W. E. Mahfoud

Alternating Bounds Converging to the Solution of a Semi-linear Euler-Poisson-Darboux Equation 105
C. Y. Chan

Comparison Theorems and Stochastic Boundary Value Problems 111
Jagdish Chandra and G. S. Ladde

Constrained Global Null Controllability of Nonlinear Delay Systems 121
Ethelbert N. Chukwu

Spectral Theory for Symmetric Pairs of Ordinary Differential Operators 129
Earl A. Coddington

Almost Periodic Solutions to Some Nonlinear Parabolic Equations 139
C. Corduneanu

Conjugate Points and Disconjugacy of Second Order Systems 143
Fozi M. Dannan

Some Results on the Existence of Free Boundaries for Parabolic Reaction-Diffusion Systems 149
Ildefonso Díaz and Jesús Hernández

A Neutral System with State-Dependent Delays 157
R. D. Driver

Asymptotic Properties of Solutions to a Class of Nonlinear Systems of Differential Equations Arising in Pharmacokinetics and Tracer Kinetics 163
J. Eisenfeld

Some Remarks on Semiflows 171
Saber Elaydi

A Two Dimensional Model for Turbulence 177
W. E. Fitzgibbon

Asymptotic Properties of Inhomogeneous Wave Equations 185
Jerome A. Goldstein and James T. Sandefur, Jr.

On a property of Olicz-Sobolev Spaces 197
J.-P. Gossez

An Optimal L^{∞} Error Estimate for Galerkin Approximations to Solutions of Two-Point Nonlinear Boundary Value Problems 201
John Gregory and Marvin Zeman

Quadratic Systems 207
R. D. S. Grisell

Continuation Theorems for Nonlinear Equations 213
Chaitan P. Gupta

Multiplicative Perturbation of m-Accretive Operators and Evolution Equations in $L^{\infty}(\Omega)$ 221
Ki Sik Ha

Energy Bounds for the Slow Capture by a Center 227
Richard Haberman

An Integral Equation for Helminthic Infections: Stability of the Non-Infected Population 231
K. P. Hadeler

Asymptotic Analysis of Rudimentary Models of Toxicant-Population Interactions 241
Thomas G. Hallam

On the Asymptotic Stability by Lyapunov Function with Semidefinite Derivative 247
L. Hatvani

Ergodic Theorems for Nonlinear Volterra Integral Equations 255
Douglas S. Hulbert and Simeon Reich

Some Remarks on Continuous and Discrete Flows 263
S. K. Kaul

Some Stability Notions 269
Ronald A. Knight

Functional Differential Inequalities and Equations with Oscillating Coefficients 277
G. Ladas, Y. G. Sficas and I. P. Stavroulakis

The Present State of the Method of Upper and Lower Solutions 285
V. Lakshmikantham

On a Conjecture on the Number of Solutions of a Nonlinear Dirichlet Problem with Jumping Nonlinearity 301
A. C. Lazer and P. J. McKenna

Operator Parts of Ordinary Differential Subspaces with Application to Nonlinear Problems 315
Sung J. Lee

Large-Scale Systems, Cone-Valued Lyapunov Functions and Quasi-Solutions 323
S. Leela

A Nonlinear Volterra Integrodifferential Equation Occuring in Heat Flow 331
S.-O. Londen and J. A. Nohel

On Exact Controllability in Hilbert Spaces 339
J.-Cl. Louis and D. Wexler

Continuous Dependence for Demicompact Maps 349
Rafiq Makhluf and Zdenek Vorel

Vector Lyapunov Functions Method in the Analysis of Dynamical Properties of Nonlinear Differential Equations 357
V. M. Matrosov

Some Recent Topics on Nonlinear Semi-Groups 375
Isao Miyadera

Set Valued Extensions of Integral Inequalities 383
R. E. Moore

Equations with Nonlinearities Defined only on Subsets 389
R. Kent Nagle

Operator Parts and Generalized Inverses for Linear Manifolds with Applications 395
Z. Nashed

Some Global Steepest Descent Results for Nonlinear Systems 413
John W. Neuberger

Nonlinear Problems at Resonance via Lyapunov-Like Functions 419
J. J. Nieto

Some Problems Concerning Nonlinear Oscillators in Neural Networks 425
M. Namik Oğuztöreli

An Inclusion Principle for Hereditary Systems 435
Y. Ohta and D. D. Šiljak

On Critical Case of Liapunov's Stability of One-Degree of Freedom System 443
Andrzej Olas

Some Computational Procedures for Nonlinear Problems in R^n 449
M. C. Pandian

Monotone Convergence of Time-Dependent Solutions for Coupled Reaction-Diffusion Systems 455
C. V. Pao

A Simple Conversion of Two-Point BVP to One Point BVP 467
Fred R. Payne and Fung-Tai Ko

Constructive Existence of Solution to a Nonlinear Eigenvalue Problem with Application to Nth Order Reaction Diffusion Problems 477
W. L. Perry

A Frequency Domain Approach for Finding the Limit Behavior of Retarded Differential Equations 483
V. M. Popov

Labeled Graphs and a Class of Generalized Volterra Systems 495
Ray Redheffer

The Method of Quasi Upper and Lower Solutions for Extremal Solutions of Volterra Integral Equations in Abstract Cones 505
G. R. Shendge

Perturbed Bifurcation Theory for a Class of Autonomous Ordinary Differential Equations 511
Ratnasingham Shivaji

Bifurcation of Subharmonic Solutions for a Periodic Epidemic Model 517
Hal L. Smith

Some Comparison and Convergence Theorems in λ(n,k)-Parameter Families 523
S. Umamaheswaram

Obstructions to the Existence of Transverse Volume Elements of Foliations 525
I. Vaisman

The Monotone Method for a Second-Order Differential Equation 535
C. Vargas

Existence of Coupled Minimal and Maximal Periodic Quasisolutions for Systems of First Order PBVP via Quasisolutions 541
A. S. Vatsala

Differential Equations with Piecewise Constant Delays 547
Joseph Wiener

Asymptotic Behavior of Solutions in Nonautonomous Systems 553
Taro Yoshizawa

CONTRIBUTORS

Abdelaziz, Nazar Hussein - Kuwait University, Kuwait.

Aftabizadeh, A. R. - Pan American University, Edinburg, Texas.

Ahmad, Shair - University of Miami, Coral Gables, Florida.

Asner, Bernard A. - University of Dallas, Irving Texas.

Akinyele, Olusola - Iowa State University, Ames, Iowa**

Amirat, Y. - Institut de Mathématiques, U.S.T.H.B., Dar El Beida, Algeria.

Bajaj, Prem N. - Wichita State University, Wichita, Kansas.

Baxley, John V. - Wake Forest University, Winston-Salem, North Carolina.

Becker, Ronald I.* - University of Cape Town, Rondebosch, South Africa.

Benharbit, Abdelali - Pennsylvania State University, York, Pennsylvania.

Bernfeld, S. R. - University of Texas at Arlington, Arlington, Texas.

Block, Louis - University of Florida, Gainesville, Florida.

Brauer, Fred - University of Wisconsin, Madison, Wisconsin.

Burbea, Jacob - University of Pittsburgh, Pittsburgh, Pennsylvania.

Burton, T. A. - Southern Illinois University, Carbondale, Illinois.

Chan, C. Y. - Florida State University, Tallahassee, Florida.

Chandra, Jagdish - U. S. Army Research Office, Research Triangle Park, North Carolina.

Chukwu, E. N. - Federal University of Technology, Yola, Nigeria.

Coddington, Earl A. - University of California, Los Angeles, California.

Corduneanu, C. - University of Texas at Arlington, Arlington, Texas.

Dahiya, R. S. - Iowa State University, Ames, Iowa.

Dannan, Fozi M. - Kuwait University, Kuwait.

Diaz, Ildefonso - University of Santander, Santander, Spain.

Driver, R. D. - University of Rhode Island, Kingston, Rhode Island.

Eisenfeld, J. - University of Texas at Arlington, Arlington, Texas.

Elaydi, Saber - Kuwait University, Kuwait.

---

*Present address: University of Tennessee, Knoxville, Tennessee.

**Present address: Department of Mathematics, University of Ibadan, Nigeria.

El kolli, Amar - O.N.R.S. - Institut de Mathématiques, U.S.T.H.B., Dar El Beida, Algeria.

Fitzgibbon, W. E. - University of Houston, Houston, Texas.

Goldstein, Jerome A. - Tulane University, New Orleans, Louisiana.

Gossez, J. P. - Université Libre de Bruxelles, Bruxelles, Belgium.

Gregory, John - Southern Illinois University, Carbondale, Illinois.

Grisell, R. D. S. - Texas Instruments, Dallas, Texas.

Gupta, Chaitan P. - Northern Illinois University, DeKalb, Illinois.

Ha, Ki Sik - Busan National University, Busan, Korea.

Haberman, Richard - Southern Methodist University, Dallas, Texas.

Hadeler, K. P. - Lehrstuhl für Biomathematik, Universität Tübingen, Tübingen, West Germany

Hallam, Thomas G.- Department of Mathematics and Program in Ecology, University of Tennessee, Knoxville, Tennessee.

Hart, David - University of Florida, Gainesville, Florida.

Hatvani, L.- Bolyai Institute, Aradi vértanúk t. 1, Szeged, Hungary.

Hernández, Jesus - Departamento de Matemáticas, Universidad Autónoma, Madrid, Spain

Hulbert, Douglas S. - University of Southern California, Los Angeles, California.

Kaul, S. K. - University of Regina, Regina, Saskatchewan, Canada.

Knight, Ronald A. - Northeast Missouri State University, Kirksville, Missouri.

Ko, Fung-Tai - University of Texas at Arlington, Arlington, Texas.

Ladas, G. - University of Rhode Island, Kingston, Rhode Island.

Ladde, G. S. - University of Texas at Arlington, Arlington, Texas.

Lakshmikantham, V. - University of Texas at Arlington, Arlington, Texas.

Lazer, A. C. - University of Miami, Coral Gables, Florida.

Lee, Sung J. - University of South Florida, Tampa, Florida.

Leela, S. - Department of Mathematics, State University of New York, (SUNY), Genesco, New York.

Londen, S.-O. - Department of Mathematics, Helsinki University of Technology, Espoo 15, Finland.

Louis, J.-Cl. - Facultés Universitaires Notre-Dame de la Paix, Belgium.

McKenna, P. J. - Department of Mathematics, University of Florida, Gainesville, Florida.

Mahfoud, W. E. - Murray State University, Murray, Kentucky.

Makhluf, Rafiq - University of Southern California, Los Angeles, California.

Matrosov, V. M. - Computer Centre, Irkutsk, U.S.S.R.

Miyadera, Isao - Waseda University, Shinjuku-ku, Tokyo, Japan

Montes de Oca, Francisco* - Department of Mathematics, University of Miami, Coral Gables, Florida.

Moore, R. E. - Department of Mathematics, University of Texas at Arlington, Arlington, Texas.

Nagle, R. Kent - Department of Mathematics, University of South Florida, Tampa, Florida.

Nashed, Z. - Department of Mathematics, University of Delaware, Newark, Delaware.

Neuberger, John W. - Department of Mathematics, North Texas State University, Denton, Texas.

Nieto, J. J. - Facultad de Matematicas, Universidad de Santiago, Spain.

Nohel, J. A. - University of Wisconsin, Madison, Wisconsin.

Oğuztöreli, M. Namik - Department of Mathematics, University of Alberta, Edmonton, Alberta, Canada

Ohta, Y. - Fukui University, Fukui-shi, Japan.

Olas, Andrzej† - Institute of Fundamental Technological Research, Warsaw, Poland.

Pandian, M. C. - University of Texas at Arlington, Arlington, Texas.

Pao, C. V. - Department of Mathematics, North Carolina State University, Raleigh, North Carolina.

Payne, Fred R. - Department of Aerospace Engineering, University of Texas at Arlington, Arlington, Texas.

Perry, W. L. - Texas A&M University, College Station, Texas.

Popov, V. M. - University of Florida, Gainesville, Florida.

Redheffer, Ray - University of California, Los Angeles, California.

Reich, Simeon - University of Southern California, University Park, Los Angeles, California.

Sandefur, James T. - Georgetown University, Washington, D.C., and Cornell University, Ithaca, New York.

Sficas, Y. G. - University of Rhode Island, Kingston, Rhode Island.

Shendge, G. R. - University of Iowa, Iowa City, Iowa.

Shivaji, Ratnasingham - Yarmouk University, Jordan.

Šiljak, D. D. - University of Santa Clara, Santa Clara, California.

Smith, Hal L. - Arizona State University, Tempe, Arizona.

Stavroulakis, I. P. - University of Rhode Island, Kingston, Rhode Island.

Umamaheswaram, S. - University of Hyderabad, Hyderabad, India.

Vaisman, I. - University of Haifa, Haifa, Israel.

Vargas, C. - University of Texas at Arlington, Arlington, Texas.

Vatsala, A. S. - Bishop College, Dallas, Texas.

---

*Permanent address: Universitario Politécnico, Barquisimento, Venezuela.

†Temporary address: Tuskegee Institute, Alabama.

Vorel, Zdenek - University of Southern California, Los Angeles, California.

Wexler, D. - Facultes Universitaires de Namur, Belgium.

Wiener, Joseph - Pan American University, Edinburg, Texas.

Yoshizawa, Taro - Tohoku University, Sendai, Japan

Zeman, Marvin - Southern Illinois University, Carbondale, Illinois.

Ziani, A. - Institut de Mathématiques, U.S.T.H.B., Dar El Beida, Algeria.

OTHER PARTICIPANTS

Dragan, I. - University of Texas at Arlington, Arlington, Texas.

Lui, Roger - San Diego State University, San Diego, California.

Moebes, Travis A. - Houston, Texas.

Pandit, Sudhakar G. - Christopher Newport College, Newport News, Virginia.

Parrott, Mary E. - University of South Florida, Tampa, Florida.

Rio, Sheldon T. - School of Science/Mathematics, Ashland, Oregon.

Sambandham, M.- Georgia Institute of Technology, Atlanta, Georgia.

Sarhan, Mahmoud - University of Petroleum and Minerals, Dhahran, Saudi Arabia.

Schröder, Johann - Mathematisches Institut, Universität Köln, Köln, West Germany.

Torres, Euclides F. - Dallas, Texas.

Tsay, Y. T. - Department of Electrical Engineering, University of Houston, Houston, Texas.

Vidyasagar, M. - University of Waterloo, Waterloo, Ontario, Canada.

Waak, Tom - Bell Helicopter, Fort Worth, Texas.

TRENDS IN THEORY AND PRACTICE OF NONLINEAR DIFFERENTIAL EQUATIONS

UNBOUNDED PERTURBATIONS AND THE EXISTENCE DIFFERENTIABLE SOLUTIONS OF THE CAUCHY PROBLEM IN BANACH SPACE

Nazar Hussein Abdelaziz

Department of Mathematics
Kuwait University
Kuwait

ABSTRACT

The present paper deals with the existence of differentiable (strong) solutions of an abstract Cauchy problem in a Banach space that is perturbed by an unbounded factor.

1. INTRODUCTION

Let X be a complex Banach space and consider the Cauchy problem:

$$\frac{du}{dt} = (A + B)u(t) + f(t) \;;\quad u(0) = x \in X, \quad 0 \leq t \leq K \tag{1}$$

where $u(t)$ and $f(t)$ are functions from $[0,K]$ to X, A and B are linear operators which generate strongly continuous semi-groups $T(t)$ and $S(t)$, $t \geq 0$ (class c_o) of bounded linear operators on X. The operator B may be thought of as a perturbating factor of A. In the case of bounded perturbation (i.e. B is a bounded operator of X), the answer to the question of the existence of the strong solution of (1) is known in some cases, for example; (i) $f(t)$ is strongly continuously differentiable and $u(0) \in D(A)$ (c.f. [2]), (ii) $f(t)$ is Hölder continuous of exponent α, $0 < \alpha \leq 1$ and $T(t)$ is of class D^p, for some appropriate value of p and α (c.f. [3]). The cases considered in the present paper dealing with unbounded perturbations extend (i) and (ii) in part.

2. PRELIMINARIES

In what follows $D(A)$, $D(B)$ will denote the domain of A and B

(2) $\| R(\lambda; A + B) \| \leq \| R(\lambda;A) \| (1 - \gamma)^{-1}, \quad 0 < \gamma < 1, \quad \text{Re } \lambda > w$.

Now the conclusion that $\hat{S}(t)$ is analytic, follows directly from (2) and Theorem 1 of [5] .

The following theorem is an immediate consequence of lemma 1 above and Theorem 3.1 of [3] .

Theorem 1. Let $f(t)$ be Hölder continuous of exponent α , $0 < \alpha \leq 1$. Further let A be the generator of an analytic semi-group and let B be the generator of a (c_o) semi-group. If

(a) $D(A) \subseteq D(B)$,

(b) $\int_0^1 \| BT(t) \| dt < \infty$

Then the abstract Cauchy problem (1), has a (unique) strong solution.

The next result follows directly from Trotter's product Theorem [4] and Philips [2] .

Theorem 2. Suppose that $f(t)$ is strongly continuously differentiable. Further, suppose that A and B generate the semi-groups $T(t)$, $S(t)$ of class (c_o) and that $\| T(t) \| \leq e^{wt}$, $\| S(t) \| \leq e^{w't}$, w and w' are constants. If Range $(\lambda_o I - (A + B)) = X$ for some $\lambda_o > w + w'$ and $D(A) \cap D(B)$ is dense in X, then equation (1) has a (unique) strong solution, for each $u(0) \in D(A) \cap D(B)$.

Case 2. (Commutative semi-groups). This case covers yet a different set of conditions on A and B under which the strong solutions of (1) exists.

Theorem 3. Let $f(t)$ be Hölder continuous of exponent α , $0 < \alpha \leq 1$. Let $T(t)$ and $S(t)$ be semi-groups of class D^p, $p < 1 + \alpha$. If $T(t)$ and $S(t)$ commute, then the abstract Cauchy problem (1) has a (unique) strong solution.

Proof. Again $A + B$ (or its closure) is the generator of a (c_o) semi-group, namely $\hat{S}(t) = T(t)S(t)$ (c.f. [4]). In this case, it is readily verified that $\hat{S}(t)X \subseteq D(A) \cap D(B)$, for all $t > 0$ and that $\overline{\lim_{t \to \infty}} \, t^p \| \hat{S}(t) \| < \infty$, showing that $\hat{S}(t)$ is again of class D^p. Now the conclusion follows from theorem 3.1 of [3] .

respectively, $\rho(A)$ and $R(\lambda;A)$ the resolvent set and operator of A (similarly for the resolvent set and operator of B). The semi-group $T(t)$ is said to be of class D^p (see [2]) if (1) $T(t)$ is differentiable for all $t > 0$, that is to say $T'(t) = AT(t)$ exists as a bounded linear operator on X for all $t > 0$ (equivalently; $T(t)X \subseteq D(A)$, $t > 0$), and (2) $\overline{\lim}_{t \to \infty} t^p \| AT(t) \| < \infty$. We recall that $f : [0,K] \to X$ is called Hölder continuous of exponent α, $0 < \alpha \leq 1$ if $\| f(t) - f(t_o) \| < c \mid t - t_o \mid^\alpha$, c is a constant. A function $u(t) : [0,K] \to X$ is a strong solution of equation (1) if $u(t)$ is strongly continuously differentiable, $u(t) \in D(A) \cap D(B)$, $t \in [0,K]$ and it satisfies the differential equation (1), and $\lim_{t \to 0+} u(t) = x$.

3. THE MAIN RESULTS

We consider two cases depending on whether the semi-groups commute or not.

Case 1. (non-commutative semi-groups). The existence of strong solutions of equation (1) is discussed here under two different sets of assumptions on $f(t)$, A and B. In the first $f(t)$ is Hölder continuous of exponent α, $0 < \alpha \leq 1$, $T(t)$ is analytic (equivalently of class D^1) and $S(t)$ is of class c_o. While in the second $f(t)$ is strongly continuously differentiable, $T(t)$ and $S(t)$ are c_o semi-groups satisfying the Trotter Product theorem [4]. We begin by a lemma on perturbation of analytic semi-groups by the generator of a c_o semi-group. This will partly extend the result of [3] on bounded perturbations.

Lemma 1. Let $T(t)$ be a semi-group of class D^1 with generator A and let $S(t)$ be a (c_o) semi-group with generator B. If

(a) $D(A) \subseteq D(B)$ and

(b) $\int_0^1 \| BT(t) \| \, dt < \infty$,

then $A + B$ is the generator of a semi-group of class D^1.

Proof. In view of the differentiability of $T(t)$ and conditions (a),(b) one can easily verify that the hypotheses of cor. 1, of Theorem 13.4.1, [1] are fulfilled. Accordingly, $A + B$ generates a (c_o) semi-group $\hat{S}(t)$ and,

REFERENCES

[1] E. Hille and R.S. Philips, Functional analysis and semi-groups, Amer. Math.Soc. Colloquium Publications, Vol. 31 (1957).

[2] R.S. Philips, Perturbation Theory for semi-group of linear operators, Trans.Amer.Math.Soc., 74 (1953), 199-221.

[3] M.G. Grandall & A. Pazy, On the differentiability of weak solutions of a differential equation in Banach space, Jour. of Math. and Mechs., vol. 18, No. 10 (1969), 1007-1016.

[4] H.F. Trotter, On the product of semi-group of operators, Proc.Amer. Math.Soc. 10 (1959), 545-551.

[5] K. Yosida, On the differentiability of semi-groups of linear operators, Proc. Japan Acad., 34 (1958), 337-340.

ESCAPE TIME BY FIXED POINT THEOREM

A. R. Aftabizadeh

Department of Mathematics
Pan American University
Edinburg, Texas

The solutions of the differential system

$$(*) \qquad x' = A(t,x)x + b(t,x)$$

have been studied by several authors such as C. Corduneanu [1] and A. Kartsatos [3,4]. In this paper we shall prove the existence of solutions of (*) under some adequate conditions. It is likely that the solutions might not exist on whole $R^+ = [0,\infty)$, but we can estimate the escape time.

Consider the linear differential system

$$(1) \qquad x' = A(t)x + b(t),$$

and assume that $A(t)$ is an $n \times n$ matrix whose elements are continuous functions on the interval $R^+ = [0,\infty)$, and $b(t)$ is an n-dimensional vector whose components are continuous functions on R^+. T. Ważewski [5] (see also Problem 4, page 81 of [2]), proved every solution $x(t)$ of (1) satisfies the following inequality:

$$(2) \quad \|x(t)\| \le \|x(0)\| \exp\left(\int_0^t \lambda(s)ds\right) + \int_0^t \|b(u)\| \exp\left(\int_u^t \lambda(s)ds\right)du, \quad t \in R^+,$$

where $\lambda(t)$ is the largest eigenvalue of the symmetric matrix $H = \frac{1}{2}(A+A^*)$.

In [1], the following result has been proved:
Consider the differential system

$$(**) \qquad x' = A(t,x)x + b(t,x), \quad x(0) = \eta.$$

Assume that $0 \le t < \beta \le +\infty$, and:
i) the matrix $A(t,x)$ is $n \times n$, and its entries are continuous complex

functions defined in

$$\Delta:\quad 0 \le t < \beta,\ x \in \mathbb{C}^n = \text{the n-dimensional complex space;}$$

ii) there exists a nonnegative function $\phi(t)$, continuous on $[0,\beta)$, such that

$$\Lambda(t,x) \le \phi(t),\ (t,x) \in \Delta,$$

where Λ is the largest eigenvalue of the matrix $\frac{1}{2}(A+A^*)$;

iii) the vector b is continuous in Δ, and there exists a nonnegative function ψ, continuous on $[0,\beta)$, such that

$$\|b(t,x)\| \le \psi(t),\ (t,x) \in \Delta.$$

Then, the system (**) has at least one solution $x(t)$, defined on the whole interval $[0,\beta)$, satisfying $x(0) = \eta$. This solution satisfies the inequality

$$\|x(t)\| \le \|\eta\| \exp\left(\int_0^t \phi(s)ds\right) + \int_0^t \psi(u)\exp\left(\int_u^t \phi(s)ds\right)du,\ t \in [0,\beta).$$

The following result holds true:

Theorem: Consider the system

$$x' = A(t,x)x + b(t,x),\ x(0) = \eta,\ t \ge 0,\ x \in \mathbb{C}^n, \tag{3}$$

where $A(t,x)$ is an $n \times n$ matrix whose entries are continuous functions, and the vector b is continuous. Assume that:

i) there exist continuous functions $\lambda(t) > 0$, and $\mu(t) \ge 0$ on R^+, such that

$$\Lambda(t,x) \le \lambda(t)\|x\| + \mu(t) \quad \text{for all} \quad t \ge 0,\ x \in \mathbb{C}^n,$$

where Λ is the largest eigenvalue of the symmetric matrix $H = \frac{1}{2}(A+A^*)$;

ii) there exists a nonnegative function $\beta(t)$, continuous on R^+, such that

$$\|b(t,x)\| \le \beta(t)\|x\|.$$

Then system (3) has at least one solution $x(t)$ on some interval $[0,T)$, $T \le +\infty$, where T can be determined from

$$\|\eta\| \int_0^T \lambda(s)\exp\left(\int_0^s [\beta(r) + \mu(r)]dr\right)ds = 1, \tag{4}$$

or is $+\infty$, when the left hand side of (4) is ≤ 1 for any $T > 0$.

Proof: A solution of our problem, which exists on a certain interval $[0,T)$, $T \le +\infty$, is an element of the space $C([0,T),\mathbb{C}^n)$, i.e., the space of all continuous maps from $[0,T)$ in $\mathbb{C}^n$, with the topology of uniform convergence on every compact interval in $[0,T)$.

In order to use the fixed point theorem of Schauder-Tychonoff, we wish to choose a convex set M and an operator T, such that the fixed element of T is a solution of the system (3). Define

$$D = \left\{(t,x):\ 0 \le t < T \text{ and } \|x\| \le \alpha(t),\ \alpha(t) > 0 \text{ and continuous}\right\}.$$

Let $M = \{u(t)\}$ be the set of functions of $C([0,T),\mathbb{C}^n)$ satisfying

$$\|u(t)\| \le \alpha(t),\ 0 \le t < T.$$

First we notice that set M is convex and closed in $C([0,T),\mathbb{C}^n)$. Define the operator T on M in the following way:

If $u(t) \in M$, then $x = Tu$ will be the unique solution of

$$x' = A(t,u)x + b(t,u),\ x(0) = \eta. \tag{5}$$

$A(t,u)$ and $b(t,u)$ are continuous functions on $[0,T)$, then system (5) has a unique solution $x(t) \in C([0,T),\mathbb{C}^n)$. Equation (5) is a form of (1), then using T. Ważewski's inequality we have

$$\|x(t)\| \le \|\eta\| \exp\left[\int_0^t \Lambda(s,u(s))ds\right] + \int_0^t \|b(s,u(s))\| \exp\left[\int_s^t \Lambda(r,u(r))dr\right] ds. \tag{6}$$

In order to show that $TM \subset M$, we like to find a function $\alpha(t)$ such that the right hand side of (6) be less than or equal to $\alpha(t)$, or using conditions (i) and (ii) we need to find $\alpha(t)$ such that

$$\|\eta\| \exp\left[\int_0^t [\lambda(s)\alpha(s) + \mu(s)]ds\right] + \int_0^t \beta(u)\alpha(u) \exp\left(\int_u^t [\lambda(s)\alpha(s) + \mu(s)]ds\right)du \le \alpha(t),\ t \in [0,T) \tag{7}$$

This is a nonlinear integral inequality of Voltera type we can solve. Let

$$L(t) = \alpha(t) \exp\left[-\int_0^t [\lambda(s)\alpha(s) + \mu(s)]ds\right], \tag{8}$$

then from (7) we have

$$\|\eta\| + \int_0^t \beta(u)L(u)du \le L(t). \tag{9}$$

Let us determine $L(t)$ such that equality holds true in (9), i.e.,

$$\|\eta\| + \int_0^t \beta(u)L(u)du = L(t)$$

then $\qquad L'(t) = \beta(t)L(t),\ L(0) = \|\eta\|$

which implies that

$$L(t) = \|\eta\| \exp\left(\int_0^t \beta(s)ds\right).$$

This and (8) lead to

$$-\frac{d}{dt} \exp\left(-\int_0^t \lambda(s)\alpha(s)ds\right) = \|\eta\|\lambda(t)\exp\left(\int_0^t [\beta(s) + \mu(s)]ds\right).$$

Taking integral from 0 to t, we obtain

$$\exp\left(-\int_0^t \lambda(s)\alpha(s)ds\right) = 1 - \|\eta\| \int_0^t \lambda(s)\exp\left(\int_0^s [\beta(r) + \mu(r)]dr\right)ds,$$

or

$$\alpha(t) = \frac{\|\eta\| \exp\left(\int_0^t [\beta(r)+\mu(r)]dr\right)}{1-\|\eta\| \int_0^t \lambda(s)\exp\left(\int_0^s [\beta(r)+\mu(r)]dr\right)ds}, \quad \text{for } t \in [0,T)$$

as long as denominator does not vanish. Actually T can be determined from

$$\|\eta\| \int_0^T \lambda(s)\exp\left(\int_0^s [\beta(r) + \mu(r)]dr\right)ds = 1$$

Therefore, $TM \subset M$. Moreover, the set $\{x'(t)\}$, $x \in TM$ is uniformly bounded in every compact subinterval of $[0,T)$. Consequently, TM is compact in the sense of the uniform convergence on subinterval of $[0,T)$, that is, it is compact in the space $C([0,T),\mathbb{C}^n)$.

All these considerations lead to the existence of a fixed point, i.e., $Tx = x$. Proof is complete.

REFERENCES

[1] C. Corduneanu, Sur les systèmes différentiels de la forme y' = A(x,y)y + b(x,y), Anl. Stiint. Univ. "Al.I. Cuza" Iasi, (N.S.) 4 (1958), 45-52.

[2] C. Corduneanu, Principles of Differential and Integral Equations, Chelsea Publishing Company, 1977.

[3] A. G. Kartsatos, Nonzero Solutions to Boundary Value Problems for Nonlinear Systems, Pacific J. of Math., Vol. 53, No. 2 (1974), 425-433.

[4] A. G. Kartsatos, A Boundary Value Problem on an Infinite Interval, Proc. Edinburgh Math. Soc. Series II, 19 (1974) 245-252.

[5] T. Ważewski, Sur la limitation des intégrales des systèmes d'équations différentielles linéaires ordinaires, Studia Mathematica, T.X., 1948.

ASYMPTOTIC BEHAVIOUR OF SOLUTIONS OF Nth ORDER DIFFERENTIAL EQUATIONS WITH OR WITHOUT DELAY

Olusola Akinyele†

Departments of Mathematics, Iowa State University
Ames, Iowa and University of Ibadan, Nigeria

R. S. Dahiya

Department of Mathematics
Iowa State University
Ames, Iowa

ABSTRACT

We study the asymptotic behaviour of solutions of the differential equation $u^{(n)}(t) + f(t,u(r(t))) = h(t)$ as $t \to \infty$ for certain classes of functions $f(t,u)$ and $h(t)$.

1. INTRODUCTION

Consider the differential equation

$$u^{(n)}(t) + f(t,u(r(t))) = h(t) \tag{1}$$

where

$$(2)\begin{cases} \text{(a)} & f(t,u) \text{ is continuous on } [0,\infty) \times R, \\ \text{(b)} & h(t) \text{ is continuous and non-negative on } [0,\infty), \\ \text{(c)} & r(t) \text{ is continuous, } r(t) \to \infty \text{ as } t \to \infty \text{ and} \\ & r(t) \le t. \end{cases}$$

For second order equations the asymptotic behaviour has been given by Bellman [1] for $f(t,u) = \pm t^{\sigma}u^{m}$, $r(t) = t$, $h(t) = 0$; Bellman [1], Moore and Nehari [4], Trench [7], Waltman [8] and a few others for $r(t) = t$, $f(t,u) = a(t)u^{2m+1}$, $h(t) = 0$ and Cohen [2], Tong [6] for $r(t) = t$ and $h(t) = 0$. We are interested in the asymptotic behaviour of solutions of (1) as $t \to \infty$ for certain classes of functions $f(t,u)$. The results here

---

† Current Address: Department of Mathematics, University of Ibadan, Nigeria

are extensions of results obtained earlier by Cohen [2], Tong [6] and Singh [5]. Our Theorem 1 generalizes the result of Cohen [2] and Singh [5] while Theorems 2 and 3 extend the result of Tong [6]. For other related results the reader is referred to [3]. We only state our main results here, the details will appear elsewhere.

2. MAIN RESULTS

Theorem 1. Assume that the following hold:

(i) $r(t) \equiv t$,

(ii) the partial derivatives $\frac{\partial^{n-1}}{\partial u^{n-1}} f(t,u) = A(t,u)$ exist on $[0,\infty) \times R$ and $A(t,0) > 0$,

(iii) $|f(t,u)| \leq A(t,0)|u|$ on $[0,\infty) \times R$,

(iv) $\frac{1}{(n-1)!} \int^{\infty} t^{n-1} A(t,0)dt < \infty$,

(v) $\int^{\infty} h(t)dt < \infty$.

Then the equation (1) has solutions which are asymptotic to the solutions of $u^{(n)}(t) \equiv 0$ as $t \to \infty$.

We further extend Theorem 1 to the following result:

Theorem 2. Assume that there exist two non-negative functions $v(t)$, $\phi(t)$, for $t \geq 0$ and a continuous function $g(u)$ for $u \geq 0$ such that the following hold:

(i) $\frac{1}{(n-1)!} \int_{1}^{\infty} v(t)\phi(t)dt < \infty$,

(ii) $r(t) = t$,

(iii) for $u > 0$, $g(u) > 0$ and non-decreasing on $[0,\infty)$,

(iv) $|f(t,u(t))| \leq v(t)\phi(t)g\left(\frac{|u(t)|}{t^{n-1}}\right)$ for $t \geq 1$, $n \geq 1$,

(v) $\int^{\infty} h(t)dt < \infty$.

Then equation (1) has solutions which are asymptotic to the solutions of $u^{(n)}(t) = 0$ as $t \to \infty$.

The following result also extends Tong's result [6].

Corollary 1. Assume all the hypotheses of Theorem 2. Assume in addition that $h(t) = 0$. Then the equation (1) has solutions which are asymptotic to $\alpha_0 + \alpha_1 t + \ldots + \alpha_{n-1} t^{n-1}$, with $\alpha_{n-1} \neq 0$ as $t \to \infty$.

We now consider an application of Theorem 2.

Example. Consider the fourth-order non-linear equation

$$u^{(iv)} + t^{-8} u^2 \sin^2 u = e^{-ct} \qquad c > 0, \quad t \geq 1. \tag{3}$$

$f(t,u) = t^{-8} u^2 \sin^2 u$ and $\frac{\partial^3 f}{\partial u^3}(t,0) = A(t,0) = 0$. Theorem 1 therefore is not applicable. However, setting $v(t) = t^{-8}$, $\phi(t) = t^6$, $g(u) = u^2$, $h(t) = e^{-ct}$ and $r(t) = t$, it can be easily checked that all the hypotheses of Theorem 2 are satisfied and so (3) has solutions which are asymptotic to $\alpha_0 + \alpha_1 t + \alpha_2 t^2 + \alpha_3 t^3$ with $\alpha_3 \neq 0$ as $t \to \infty$.

The following result for the equation (1) with delay further extend our Theorems 1 and 2.

Theorem 3. Assume that there exist two non-negative functions $v(t)$, $\phi(t)$ for $t \geq 0$ and a continuous function $g(u)$ for $u \geq 0$ such that the following hold:

(i) $\frac{1}{(n-1)!} \int_1^\infty v(t)\phi(t)\,dt < \infty$,

(ii) for $u > 0$, $g(u) > 0$ and non-decreasing on $[0,\infty)$,

(iii) $|f(t,u(r(t)))| \leq v(t)\phi(t) g\left(\frac{|u(r(t))|}{t^{n-1}}\right)$,

(iv) $\int^\infty h(t)\,dt < \infty$.

Then equation (1) has solutions which are asymptotic to the solutions of $u^{(n)}(t) = 0$ as $t \to \infty$.

REFERENCES

[1] R. BELLMAN, Stability theory of differential equations, McGraw Hill, New York 1953.

[2] D. S. COHEN, The asymptotic behaviour of a class of nonlinear differential equations, Proc. Amer. Math. Soc. 18(1967), 607-609.

[3] R. S. DAHIYA and B. SINGH, Certain results on nonoscillation and asymptotic nature of delay equations, Hiroshima Mathematical Journal, 5(1975), no. 1, 7-15.

[4] R. A. MOORE and Z. NEHARI, Non-oscillation theorems for a class of non-linear differential equations, Trans. Amer. Math. Soc. 93(1957), 30-52.

[5] Y. P. SINGH, The asymptotic behaviour of solutions of a nonlinear n^{th}-order differential equations, Math. J. Okayama University, 15 (1971), no. 1, 71-73.

[6] JING-CHENG, TONG, The asymptotic behaviour of a class of nonlinear differential equations of second order, Proc. Amer. Math. Soc. 84 (1982), no. 2, 235-236.

[7] W. F. TRENCH, On the asymptotic behaviour of solutions of second order linear equations. Proc. Amer. Math. Soc. 14(1963), 12-14.

[8] P. WALTMAN, On the asymptotic behaviour of solutions of a nonlinear equation, Proc. Amer. Math. Soc. 15(1964), 918-923.

NONCONSTANT PERIODIC SOLUTIONS OF NONLINEAR DIFFERENTIAL EQUATIONS

Shair Ahmad and Francisco Montes de Oca*

Department of Mathematics and Computer Science
University of Miami
Coral Gables, Florida

ABSTRACT

Let $g: R^n \to R^n$ be of class C' and satisfy $g(0) = 0$. Assume that g is asymptotically linear at infinity and satisfies a certain weak monotonicity condition. Sufficient conditions for the existence of nonconstant periodic solutions of the second order differential system $x''(t) + g(x(t)) = 0$ are given.

INTRODUCTION

Let $g: R^n \to R^n$ be of class C' and satisfy $g(0) = 0$. Assuming g to be asymptotically linear at infinity - this will be made more precise in what follows - we consider the problem of existence of nonconstant periodic solutions of the second order differential system $x''(t) + g(x(t)) = 0$ with prescribed period. For simplicity we take this prescribed period to be 2π. Conditions for the existence of nonconstant periodic solutions having a different prescribed period may be derived from the conditions we give by a linear change of the time scale. In [7] and [8] D.C. Clark considered this problem with the additional assumption that $g = \text{grad } V$, where $V: R^n \to R$ is an even scalar function. If B is the $n \times n$ matrix such that $g(x)$ is asymptotic to Bx at infinity and A is the Jacobian matrix of g at 0 then Clark's conditions can be simply stated solely in terms of the eigenvalues of the symmetric matrices A and B: Assume that B has no eigenvalues of the form m^2, m an integer. Let $\lambda_1, \ldots, \lambda_k$ be the distinct eigenvalues of A with $\lambda_i \geq 1$ for $i = 1, \ldots, k$ and $\mu_1, \ldots, \mu_\ell$ be the distinct eigenvalues of B with $\mu_j > 1$ for $j = 1, \ldots, \ell$. For $i = 1, \ldots, k$ let $\tilde{m}_i$ be the largest integer

---

Present address: Instituto Universitario Politécnico, Barquisimeto, Venezuela.

with $\tilde{m}_i^2 \leq \lambda_i$; let $m_i = 0$ if $\lambda_i = 1$ and let it be the largest integer such that $m_i^2 < \lambda_i$, otherwise. Let p_j be the largest integer with $p_j^2 < \mu_j$ for $j = 1,\ldots,\ell$. Let $i(A) = \sum_{i=1}^{k} m_i$ (multiplicity of λ_i), $\tilde{i}(A) = \sum_{i=1}^{k} \tilde{m}_i$ (multiplicity of λ_i), and $i(B) = \sum_{i=1}^{\ell} p_j$ (multiplicity of μ_j). Using variational methods, Clark showed in [7] that if $i(B) < i(A)$ there are at least $2(i(A) - i(B))$ odd 2π-periodic solutions. In [8] he showed that there are at least $2(i(B) - i(A))$ odd 2π-periodic solutions if $\tilde{i}(A) < i(B)$. In the second paper Clark lets g depend on t, but in this case his conditions seem difficult to verify. Amann [1] and Amann and Zehnder [2] have generalized Clark's results to asymptotically linear Hamiltonian systems. In another paper [3], Amann and Zehnder proved an abstract result on asymptotically linear operators which implies, among other things, that if $g = \operatorname{grad} V$, $g(0) = 0$ and either $i(A) > i(B)$ or $i(B) > \tilde{i}(A)$ then there exists at least one nontrivial 2π-periodic solution. Amann and Zehnder's proof of the abstract result is based on C.C. Conley's generalized Morse theory. A much shorter proof based on classical Morse theory was recently given by K.C. Chang [6].

In this paper we consider the existence of nonconstant 2π-periodic solutions of $x''(t) + g(x(t)) = 0$ where $g(0) = 0$, but where g is neither assumed to be a gradient nor to be odd. We do assume that g satisfies a certain weak monotonicity condition. If A and B are defined as before, if A has no eigenvalues of the form m^2, $m = 1,2,\ldots$, and if $i(B) \neq i(A) \pmod 2$ then we show the existence of at least one nonconstant 2π-periodic solution. In this case, the word "multiplicity" appearing in the definition of $i(A)$ and $i(B)$ is understood to be algebraic multiplicity, which is the same as geometric multiplicity in Clark's case since A and B are symmetric. We give an example to show that there need not be a nonconstant 2π-periodic solution if $i(A) - i(B)$ is a nonzero even integer.

Our treatment of the problem is given in two steps: first, we use an alternative-method-type argument to reduce the problem to looking for zeros of a vector field for the Leray-Schauder type defined on a certain Sobolev space of even periodic functions having mean zero. Second, we show that the Leray-Schauder degrees are different on a small ball centered at zero and on a larger ball centered at zero, implying the existence of a zero in between.

The bulk of our proof consists of establishing spectral properties of certain nonselfadjoint compact linear operators.

SOME PRELIMINARY RESULTS

We consider

(S) $$x''(t) + g(x(t)) = 0,$$

$g \in C'(R^n, R^n)$, $g(0) = 0$. Assume

(1) $$\langle g(x_1) - g(x_2),\ x_1 - x_2\rangle \leq \delta |x_1 - x_2|^2,$$

where $\langle , \rangle$ denotes the usual inner product in R^n. This implies that $Dg(x)$ is never singular for $x \in R^n$ ($Dg(x) \equiv$ Jacobian matrix of g at x.) We assume the existence of an $n \times n$ matrix B such that

(2) $$\frac{|g(x) - B\,x|}{|x|} \to 0 \text{ as } |x| \to +\infty \quad (|\ | \equiv \text{Euclidean norm}).$$

This will certainly be the case if $Dg(x) \to B$ as $|x| \to \infty$.

Let H consist of the set of all even 2π-periodic functions u from $(-\infty, +\infty)$ into R^n such that u is absolutely continuous and $\int_0^{2\pi} |u'(t)|^2 dt < \infty$. H is a Hilbert space with inner product (,) defined by

$$(u,v) = \int_0^{2\pi} (\langle u'(t),\ v'(t)\rangle + \langle u(t),\ v(t)\rangle)dt.$$

We have $H = R^n \oplus \tilde{H}$, where we identify the constant functions in H with R^n, and $\tilde{H}$ consists of all $w \in H$ such that

(3) $$\int_0^{2\pi} w(t)dt = 0.$$

In other words, $H = (R^n)^{\perp}$ in H. By Wirtinger's inequality

(4) $$\int_0^{2\pi} |w(t)|^2 dt \leq \int_0^{2\pi} |w'(t)|^2 dt \quad \text{for all } w \in \tilde{H}.$$

Therefore, if we set

(5) $$(v,w)_1 = \int_0^{2\pi} \langle v'(t),\ w'(t)\rangle dt,$$

then $(,)_1$ is an inner product on $\tilde{H}$ which generates the same topology on $\tilde{H}$ as (,).

We investigate the existence of nonconstant 2π-periodic solutions of (S) which are even. Given $u \in H$, u will be an even 2π-periodic solution of (S) iff

(6) $$\int_0^{2\pi} (\langle u', \phi'\rangle - \langle g(u(t)), \phi(t)\rangle)dt = 0 \text{ for all } \phi \in H.$$

In fact, from a standard argument from the calculus of variations, if $u \in H$ then $u \in C^2$ and u is a solution of (S) if and only if

(6') $$\int_0^{2\pi} (\langle u', v'\rangle - \langle g(u), v\rangle)dt = 0$$ for all absolutely continuous 2π-periodic functions $v: (-\infty, +\infty) \to R^n$ with $|v'| \in L^2[0,2]$ (see, for example [4, p. 154]). Every such v can be expressed uniquely in the form $v = \phi + \theta$ where $\phi \in H$ and θ is odd. Obviously, $u \in H$ implies that

$\int_0^{2\pi}(\langle u',\theta'\rangle - \langle g(u(t)), \theta(t)\rangle)dt = 0$ since the integrand is an odd 2π-periodic function. Therefore (6') will hold for every v in the aforementioned class if and only if (6) holds for every $\phi \in H$. Writing $u = u_o + \tilde{u}(t)$, where $u_o \in R^n$, $\tilde{u} \in \tilde{H}$, and taking into account the decomposition $\tilde{H} = H \oplus R^n$, it follows that u is a 2π-periodic solution of (S) if and only if

$$\int_0^{2\pi}(\langle \tilde{u}',v'\rangle - \langle g(u_o + \tilde{u}(t)), v(t)\rangle)dt = 0 \text{ for all } v \in \tilde{H} \tag{7}$$

and

$$\int_0^{2\pi} g(u_o + \tilde{u}(t))dt = 0. \tag{8}$$

We shall attempt to find $u \in H$ solving (S) by first showing that (8) determines $u_o \in R^n$ as a function of $\tilde{u} \in \tilde{H}$, then substituting this function in (7), thereby reducing the problem to finding the zeros a certain function from $\tilde{H}$ into itself defined implicitly by (7). This is an example of the usual alternative method [5] but differs from most known examples in that we first solve for the projection of a sought-after solution on a finite-dimensional space as a function of the complementary projection onto an infinite dimensional space, instead of the other way around. This reduction will be accomplished by using (1) and some well-known properties of monotone mappings.

<u>Lemma 1.</u> There exists a function $\theta \in C'(\tilde{H},R^n)$ such that if $x \in R^n$ and $v \in \tilde{H}$, $\int_0^{2\pi} g(x + v(t))dt = 0$ iff $x = \theta(v)$. Moreover, $\theta(0) = 0$ and there exists a number k such that

$$|\theta(v)| \le k\|v\|_1 \tag{9}$$

where $\| \ \|_1$ is the norm defined by the inner product $(\ , \ )_1$.

<u>Proof.</u> Given $v \in \tilde{H}$, we define $T_v: R^n \to R^n$ by $T_v(x) = \int_0^{2\pi} g(x + v(t))dt$. Using (1), we see that

$$\langle T_v(x_1) - T_v(x_2), x_1 - x_2\rangle \ge 2\pi\delta|x_1 - x_2|^2. \tag{9'}$$

By elementary results on monotone mapping or by degree theory (see, for example, [4, pp. 106-107]) this imples the existence of a unique element $\theta(v) \in R^n$ such that

$$0 = T_v(\theta(v)) = \int_0^{2\pi} g(\theta(v) + v(t))dt. \tag{10}$$

Since $g(0) = 0$, it is clear that $\theta(0) = 0$. The fact that $g(0) = 0$ and condition (2) imply the existence of a number $M > 0$ such that $|g(x)| \le M|x|$ for all $x \in R^n$. Therefore, from (1), (9'), (10), and (4), it follows that if $v \in \tilde{H}$, then $2\pi\delta|\theta(v)|^2 \le M|\theta(v)|\sqrt{2\pi}\|v\|_1$. This proves (9). To show that $\theta \in C'(\tilde{H},R^n)$ we first note that if $h \in R^n$,

(11) $$\langle Dg(y)h,h\rangle = \lim_{t\to 0}\langle \frac{g(y+th)-g(y)}{t},h\rangle \geq \delta|h|^2.$$

Let $G:R^n \times \tilde{H} \to R^n$ be defined by $G(x,v) = \int_0^{2\pi} g(x+v(t))dt$. If for $x \in R^n$, $v \in \tilde{H}$ and $h \in R^n$, $D_1G(x,v)(h)$ denotes the Frechet derivative of the mapping from R^n into R^n defined by $y \to G(y,v)$, evaluated at x and then applied to h, a routine calculation shows that $D_1G(x,v)(h) = \int_0^{2\pi} Dg(x+v(t))hdt$, so by (11) $\langle DG(x,v)h,h\rangle \geq 2\pi\delta|h|^2$. Thus, the linear transformation $D_1G(x,v):R^n \to R^n$ is one-to-one and hence onto for all $(x,v) \in R^n \times \tilde{H}$. Let $v \in \tilde{H}$. Since $G \in C'(R^n \times \tilde{H}, R^n)$, $G(\theta(v),v) = 0$, and $D_1G(\theta(v),v):R^n \to R^n$ is invertible, by the implicit function theorem, there exists a neighborhood U of v and a function $\phi \in C'(U,R^n)$ such that $G(\phi(w),w) = \int_0^{2\pi} g(\phi(w)+w(t))dt = 0$ for all $w \in U$. By uniqueness of θ, $\phi = \theta|_U$. Hence $\theta \in C'(\tilde{H},R^n)$. This proves the lemma.

Let $N:\tilde{H} \to \tilde{H}$ be defined implicitly (by means of the Riesz-Frechet representation theorem) as follows:

(12) $$(N(v),w)_1 = \int_0^{2\pi} \langle g(\theta(v)+v(t)),\ w(t)\rangle\, dt.$$

The fact that the right-hand side is a continuous linear functional of w follows from the Schwarz inequality and (4). N is compact and continuous. Moreover N maps weakly convergent sequences in $\tilde{H}$ into strongly convergent sequences.

Let B be as in (2). We define a linear mapping $T:\tilde{H} \to \tilde{H}$ implicitly as follows:

(13) $$(Tv,w)_1 = \int_0^{2\pi} \langle Bv(t),\ w(t)\rangle\, dt.$$

T is a linear compact operator. Also N is "asymptotic to T at ∞," i.e.

(14) $$\frac{\|N(v) - Tv\|_1}{\|v\|_1} \to 0 \quad \text{as} \quad \|v\|_1 \to \infty.$$

We next investigate the behavior of N near 0. Since $\theta(0) = 0$ and $g(0) = 0$, it follows from (12) that $N(0) = 0$. Since $g \in C'(R^n,R^n)$ and $\theta \in C'(\tilde{H},R^n)$ it follows by standard calculations that $N \in C'(\tilde{H},\tilde{H})$ and for $v,h \in \tilde{H}$, $N'(v)(h)$ is defined implicitly by

$$(N'(v)h,w)_1 = \int_0^{2\pi} \langle Dg(v+\theta(v))(h(t)+\theta'(v)(h)),w(t)\rangle dt.$$

Since $\theta'(v)$ is a linear map from $\tilde{H}$ into R^n, $\theta'(v)(h) \in R^n$ for $h,v \in \tilde{H}$. Thus, since $\theta(0) = 0$ it follows, by (3), that if $h,\ w \in \tilde{H}$,

(15) $$(N'(0)h,\ w)_1 = \int_0^{2\pi} \langle Dg(0)h(t),w(t)\rangle dt + \langle Dg(0)\theta'(0)h, \int_0^{2\pi} w(t)dt\rangle =$$

$$\int_0^{2\pi} \langle Ah(t), w(t)\rangle dt,$$

where for brevity we set

(16) $$Dg(0) = A.$$

To investigate the existence of nontrivial solutions of $N(v) = v$, which correspond to even 2π-periodic solutions of (S), we investigate the spectral properties of compact linear operators $M:\tilde{H} \to \tilde{H}$ defined by

(17) $$(Mv, w)_1 = \int_0^{2\pi} \langle Cv(t), w(t)\rangle dt$$

where C is a real $n \times n$ matrix. Both $N'(0)$ and T have this form.

Lemma 2. Let M and C be as in (17) and let $\lambda \neq 0$ be a real scalar. There exists $w \in \tilde{H}$ with $w \neq 0$ such that $(\lambda I - M)^k w = 0$ for some integer $k \geq 1$ if and only if $(\lambda m^2 I_n - C)^k b = 0$ for some integer $m \geq 1$ and some $b \in R^n$ with $b \neq 0$. Here I and I_n denote the identity operator and the $n \times n$ identity matrix respectively. In particular the Fredholm mapping $I - M$ is nonsingular if and only if C has no eigenvalues of the form m^2, $m \geq 1, 2, \cdots$.

Proof. Suppose $\lambda \neq 0$ and $(\lambda I - M)^k w = 0$ for some $w \neq 0$, $w \in \tilde{H}$ and some integer $k \geq 1$. Define $w_1 = w$ and $w_j = (\lambda I - M)w_{j-1} = (\lambda I - M)^{j-1} w$ for $j = 2, \ldots, k$ for $k \geq 2$. Since $(\lambda I - M)w_k = 0$, by (17), $0 = ((\lambda I - M)w_k, z)_1 = \int_0^{2\pi} (\langle \lambda w_k', z'\rangle - \langle Cw_k, z\rangle)dt$ for all $z \in \tilde{H}$. So by the discussion following equation (6), $w_k \in C^2$ and

(18) $$\lambda w_k'' + Cw_k = 0.$$

If $k \geq 2$ then $0 = ((\lambda I - M)w_{j-1} - w_j, z)_1 = \int_0^{2\pi} (\langle \lambda w_{j-1}' - w_j', z'\rangle - \langle Cw_{j-1}, z\rangle)dt$ for all $z \in \tilde{H}$ and $j = 2, \ldots, k$. Thus $w_{j-1} \in C^2$ for $j = 2, \ldots, k$ and

(19) $$\lambda w_{j-1}'' + Cw_{j-1} = w_j'', \quad j = 2, \ldots, k.$$

It follows from (18) and (19) that if D^2 denotes differentiation twice with respect to t then $(\lambda D^2 + C)^k w = 0$. Now assuming that $w \neq 0$ it follows from the definition of $\tilde{H}$ that we may write $w(t) = \sum_{p=1}^{\infty} a_p \cos pt$, where $a_p \in R^n$ and $a_m \neq 0$ for some $m \geq 1$. Since w is a solution of a linear differential system with constant coefficients, w has derivatives of all orders. Hence the series obtained by differentiating its Fourier series termwise any number of time converges.

Since for any $p \geq 1$, $(\lambda D^2 + C)^k a_p \cos pt = (C - p^2\lambda I_n)^k a_p \cos pt$, it follows that $0 = (\lambda D^2 + C)^k w(t) = \sum_{p=1}^{\infty} (C - p^2\lambda I_n)^k a_p \cos pt$, from which it follows that $(C - p^2\lambda I_n)^k a_p = 0$ for all $p = 1, 2, \cdots$. Since $a_m \neq 0$, this proves the Lemma.

Lemma 3. If $A = Dg(0)$ has no eigenvalues of the form m^2, $m = 1,2,\ldots$, then 0 is an isolated zero of $I - N$ and there exists $\varepsilon_o > 0$ such that for $0 < \varepsilon \leq \varepsilon_o$,

(20) $$d(I - N, D_\varepsilon, 0) = d(I - N'(0), D_\varepsilon, 0)$$

where d denotes the Leray-Schauder degree and D_ε denotes the open ball of radius ε centered at 0. If B has no eigenvalues of the form m^2, $m = 1,2,\ldots$, then there exists a number R_o such that for $R \geq R_o$

(21) $$d(I - N, D_R, 0) = d(I - T, D_R, 0).$$

Proof. From (15) and Lemma 2, it follows that if A has no eigenvalues of the form m^2, $m = 1,2,\ldots$, then $I - N'(0)$ is invertible. Since $F'(0) = I - N'(0)$, where $F(u) = u - N(u)$, it follows by the inverse function theorem that 0 is an isolated zero of $u - N(u)$ and (20) is a well-known result of degree theory (see [10, p.136]). Assuming that B has no eigenvalues of the form m^2, $m = 1,2,\ldots$, it follows from (13) and Lemma 2 that $I - T$ is invertible. Thus there exists a number $\alpha > 0$ such that $\|(I - T)v\|_1 \geq \alpha\|v\|_1$ for all $v \in \tilde{H}$. By (14) we may choose $R_o > 0$ such that $\|N(v) - Tv\|_1 < \alpha\|v\|_1$ if $\|v\|_1 \geq R_o$. From this (21) follows by a well-known result of degree theory [10, p.108].

Lemma 4. As in Lemma 2, let the $n \times n$ matrix C and the compact linear operator $M: \tilde{H} \to \tilde{H}$ be related by (17). Assume that C has no eigenvalues of the form m^2, $m = 1,2,\cdots$. If C has no real eigenvalues greater than 1, neither does M. If C has eigenvalues greater than 1, let the distinct ones be denoted by $\mu_1,\ldots,\mu_s$. Let $r_1,\ldots,r_s$ be minimal integers such that the null space of $(C - \mu_i I_n)^{r_i + 1}$ is equal to the null space of $(C - \mu_i I_n)^{r_i}$ for $i = 1,\ldots,s$. Let m_i denote the largest integer such that $m_i^2 < \mu_i$. We define

(22) $$i(C) = \sum_{i=1}^{s} m_i \dim \operatorname{Ker}(C - \mu_i I_n)^{r_i}.$$

Let $\lambda_1,\ldots,\lambda_q$ be the distinct eigenvalues of M greater than 1. Let $s_1,\ldots,s_q$ be minimal integers such that $\operatorname{Ker}(\lambda_i I - M)^{s_i+1} = \operatorname{Ker}(\lambda_i I - M)^{s_i}$ for $i = 1,2,\ldots,q$. If $F = \operatorname{Ker}(\lambda_1 I - M)^{s_1} \oplus \operatorname{Ker}(\lambda_2 I - M)^{s_2} \oplus \ldots \oplus \operatorname{Ker}(\lambda_q I - M)^{s_q}$ then

(23) $$\dim F = i(C).$$

Proof. First let us assume that there exist eigenvalues of C greater than 1 and let $\mu_1,\ldots,\mu_s$ be as in the statement of the Lemma. By standard results of linear algebra (see for example [11, p.398] or [12, p.209]), there exists a subspace G of R^n such that

(24) $$R^n = \operatorname{Ker}(\mu_1 I_n - C)^{r_1} \oplus \operatorname{Ker}(\mu_2 I_n - C)^{r_2} \oplus \ldots \oplus \operatorname{Ker}(\mu_s I_n - C)^{r_s} \oplus G$$

and such that G is invariant under multiplication by C.

Let V be the subspace of $\tilde{H}$ such that $v \in V$ if and only if

(25) $$v(t) = \sum_{i=1}^{s} \sum_{j=1}^{m_i} a_{ij} \cos jt,$$

where $a_{ij} \in \operatorname{Ker}(\mu_i I_n - C)^{r_i}$ for $1 \leq j \leq m_i$, $i = 1,\ldots,s$. We claim

(26) $$F = V.$$

To establish (26), we first show that $F \subseteq V$. By (23), it is sufficient to show that $\operatorname{Ker}(\lambda_k I - M)^{s_k} \subseteq V$ for $k = 1,\ldots,q$. Referring to the proof of Lemma 2 note that if $w \in \operatorname{Ker}(\lambda_k I - M)^{s_k}$ then $(\lambda_k D^2 + C)^{s_k} w = 0$. Again referring to the proof of Lemma 2, if $w(t) = \sum_{p=1}^{\infty} a_p \cos pt$, then $\sum_{p=1}^{\infty} (C - p^2\lambda_k I_n)^{s_k} a_p \cos pt \equiv 0$. Hence $(C - p^2\lambda_k I_n)^{s_k} a_p = 0$ for all p. According to (24), we may write $a_p = x_p + y_p$, where $x_p \in \operatorname{Ker}(\mu_1 I_n - C)^{r_1} \oplus \ldots \oplus \operatorname{Ker}(\mu_s I_n - C)^{r_s}$ and $y_p \in G$. Since both G and $\operatorname{Ker}(\mu_i I_n - C)^{r_i}$ are invariant under multiplication by C, $(C - p^2\lambda_k I_n)^{s_k} x_p = (C - p^2\lambda_k I_n)^{s_k} y_p = 0$. Since, according to (24), there are no eigenvalues greater than 1 of the restriction of the linear transformation defined by C to G, and since $p^2\lambda_k > 1$, it follows that $y_p = 0$ for all $p = 1,2,\cdots$. Let $x_p = \sum_{i=1}^{s} x_{ip}$ where $x_{ip} \in \operatorname{Ker}(\mu_i I_n - C)^{r_i}$ for $i = 1,\ldots,s$. Since $\operatorname{Ker}(\mu_i I_n - C)^{r_i}$ is invariant under multiplication by C, it follows, by (24), that for $i = 1,\ldots,s$, $(C - p^2\lambda_k I_n)^{s_k} x_{ip} = 0$. Since the linear transformation defined by multiplication by C, restricted to $\operatorname{Ker}(\mu_i I_n - C)^{r_i}$, has only the eigenvalue μ_i, it follows that $\lambda_k = \frac{\mu_i}{p^2}$. Since, by assumption, $\lambda_k > 1$ and, since m_i is the largest integer m such that $m^2 < \mu_i$, it follows that $x_{ip} = 0$ if $p > m_i$. Therefore, $w(t) = \sum_{i=1}^{s} \sum_{p=1}^{m_i} x_{ip} \cos pt$. This shows $F \subseteq V$. To obtain the reverse inclusion $V \subseteq F$, it is sufficient to show that if for some i with $1 \leq i \leq s$ and some m with $1 \leq m \leq m_i$, $a \in \operatorname{Ker}(\mu_i I_n - C)^{r_i}$, $v(t) = a \cos mt$, then $v \in F$. To this end, we note, by definition of m_i, if $\lambda = \frac{\mu_i}{m^2}$ then $\lambda > 1$. Moreover $(\lambda D^2 + C)^{r_i} v(t) = (C - \lambda m^2 I_n)^{r_i} a \cos mt = (C - \lambda_i I_n)^{r_i} a \cos mt = 0$. By reversing the argument given in the proof of Lemma 2, $(\lambda I - M)^{r_i} v = 0$. Since $\lambda > 1$, $\lambda = \lambda_j$ for some j with $1 \leq j \leq q$ and by definition of s_j, $v \in \operatorname{Ker}(\lambda_j I - M)^{s_j} \subseteq F$. This shows that $V \subseteq F$. Referring to (22) and (25) it is clear that $\dim F = i(C)$. The first statement in the assertion of Lemma 4 follows from Lemma 2. Indeed, since the eigenvalues of M have the form μ/m^2, $m = 1,2,\ldots$, where μ is an eigenvalue of C, if C has no eigenvalues greater than 1, then neither does M.

<u>Remark</u>. Let μ_i, r_i, $1 \leq i \leq s$, be as in Lemma 4. Then $\dim \operatorname{Ker}(\mu_i I_n - C)^{r_i}$ is equal to the algebraic multiplicity of μ_i as an eigenvalue of C.

We now give a weak form of the main result pertaining to the system (S).

Proposition. Assume that (1) and (2) hold and that neither $A = DG(0)$ nor B have any eigenvalues of the form m^2, $m = 1,2,\cdots$. If $i(B) - i(A)$ is an odd integer, then (S) has a nonconstant even 2π-periodic solution.

Proof. Given $u \in H$, u will be a solution of (S) iff (6) holds for all $\phi \in H$. Writing $u = u_o + \tilde{u}$ it follows that (7) and (8) must hold. Lemma 1 implies that $u_o = \theta(\tilde{u})$ and (5), (7) and (12) imply that $(\tilde{u},v)_1 = (N(\tilde{u}),v)_1$ for all $v \in \tilde{H}$. Hence $\tilde{u} = N(\tilde{u})$. Conversely by reversing the steps in this argument one sees that $\tilde{u} = N(\tilde{u})$ implies that $u = \tilde{u} + \theta(\tilde{u})$ satisfies (6) for all $\phi \in H$, and by standard regularity argument u is a solution of (S). Hence, finding nonconstant even 2π-periodic solutions of (S) is equivalent to finding nonzero solutions of the equation $v = N(v)$ on $\tilde{H}$. According to the hypotheses of the proposition and Lemma 3, $v = 0$ is an isolated zero of $I - N$ and there exists $\varepsilon_o > 0$ such that for $0 \le \varepsilon \le \varepsilon_o$, $d(I - N, D_\varepsilon, 0) = d(I - N'(0), D_\varepsilon, 0)$. Moreover there exists $R_o > 0$ such that $d(I - N, D_R, 0) = d(I - T, D_R, 0)$ for $R \ge R_o$. According to a standard result of degree theory (see [10, p.133] or [13, p.66]) if M is a compact linear operator on $\tilde{H}$, if $\lambda_1, \ldots, \lambda_q$ are the distinct eigenvalues of M greater than 1 and $s_1, \ldots, s_q$ are the minimal integer such that $\mathrm{Ker}(\lambda_i I - M)^{s_i} = \mathrm{Ker}(\lambda_i I - M)^{s_i+1}$ and if $\ell = \sum_{i=1}^{q} \dim \mathrm{Ker}(\lambda_i I - M)^{s_i}$, then for any $r > 0$, $d(I - M, D_r, 0) = (-1)^\ell$. From (13), (15) and Lemma 4 it follows that $d(I - T, D_{R_o}, 0) = (-1)^{i(B)}$, and $d(I - N'(0), D_{\varepsilon_o}, 0) = (-1)^{i(A)}$. If $v = 0$ is the only solution of $v = N(v)$ then by the excision property of degree it follows that $d(I - N, D_{R_o}, 0) = d(I - N, D_{\varepsilon_o}, 0)$. Therefore, by the above, $(-1)^{i(B)} = d(I - T, D_{R_o}, 0) = d(I - N, D_{R_o}, 0) = d(I - N, D_{\varepsilon_o}, 0) = d(I - N'(0), D_{\varepsilon_o}, 0) = (-1)^{i(A)}$. From this we see at once that if $i(B) \not\equiv i(A) \pmod 2$ then there exists a solution $\tilde{u} \neq 0$ of $v = N(v)$. By previous remark $u = \tilde{u} + \theta(\tilde{u})$ is a solution of (S). This proves the proposition.

We now give our main result.

Theorem. Assume that g satisfies the condition

$$\langle g(x_1) - g(x_2),\ x_1 - x_2 \rangle > 0,\ x_1, x_2 \in R^n. \tag{27}$$

Assume that $g(0) = 0$ and that (2) holds. Let A be defined as before. If neither A nor B is singular, if neither A nor B has any eigenvalues of the form m^2, $m = 1,2,\ldots$, and if $i(B) - i(A)$ is odd then (S) has a nonconstant even 2π-periodic solution.

Proof. Assuming the conditions of the theorem, there exists a number

$Q_o > 0$ such that if q is an integer and $q > Q_o$, then the function g_q defined by $g_q(x) = g(x) + \frac{1}{q}x$ will satisfy the conditions of the preceding proposition. Let $\theta_q(t)$ be a nonconstant even 2π-periodic solution of $x'' + g_q(x) = 0$ for each $q \geq Q_o$. Define $|\theta_q|_o = \max\limits_{[0,2\pi]} |\theta_q(t)| = \max\limits_{(-\infty,+\infty)} |\theta_q(t)|$. It will be shown that $|\theta_q|_o \leq L$ for some constant L independent of $q \geq Q_o$. Defining $h(x) = g(x) - Bx$ it follows from (2) that $h(x)/|x| \to 0$ as $|x| \to +\infty$. Assume that the claim is false and let $\{q_k\}_{k=1}^{\infty}$ be a sequence of integers such that $|\theta_{q_k}|_o \to \infty$ as $k \to \infty$. Let $\varepsilon > 0$ and choose $r(\varepsilon) > 0$ so that $|h(x)| \leq \varepsilon|X|$ if $|x| \geq r(\varepsilon)$. Letting $M(\varepsilon) = \max\{|h(x)|;\ |x| \leq r(\varepsilon)$ we see that $\frac{|h(\theta_{q_k}(t))|}{|\theta_{q_k}|_o} \leq \varepsilon$ for all $t \in (-\infty,+\infty)$ if $|\theta_{q_k}|_o \geq M(\varepsilon)/\varepsilon$. Therefore, $\frac{h(\theta_{q_k}(t))}{|\theta_{q_k}|_o} \to 0$ as $k \to \infty$ uniformly on $(-\infty, +\infty)$; and if we set $u_k(t) = \theta_{q_k}(t)/|\theta_{q_k}|_o$, the equation $u_k'' + Bu_k + \frac{1}{|\theta_{q_k}|_o} h(\theta_{q_k}) + \frac{1}{q_k} u_k = 0$ implies that the sequence $\{u_k''\}_{k=1}^{\infty}$ is uniformly bounded on $(-\infty,+\infty)$. Hence, from elementary analysis (see [14], p.115), the sequence $\{u_k'\}_{k=1}^{\infty}$ is uniformly bounded on $(-\infty,+\infty)$. It follows from Ascoli's theorem that we may assume without loss of generality that $\{u_k\}_{k=1}^{\infty}$ converges uniformly on $(-\infty,+\infty)$ to a 2π-periodic, even function u satisfying $\max\limits_{(-\infty,+\infty)} |u(t)| = 1$. Since the differential equation satisfied by u_k implies that $\{u_k''\}_{k=1}^{\infty}$ converges uniformly on $(-\infty,+\infty)$, $u \in C^2$ and $u'' + Bu = 0$. As B is nonsingular, the relation $B\int_0^{2\pi} u(t)dt = 0$ implies $u \in \tilde{H}$. Clearly, $(u,v)_1 = \int_0^{2\pi} \langle Bu,v \rangle dt$ for all $v \in \tilde{H}$, which contradicts Lemma 2 and the assumption that B has no eigenvalues of the form m^2, $m = 1,2,\cdots$. This contradiction shows that the sequence $\{\theta_q(t)\}_{q=1}^{\infty}$ is uniformly bounded on $(-\infty,+\infty)$. Since $\theta_q'' + g(\theta_q) + \frac{1}{q}\theta_q = 0$, the same reasoning as was used above shows that we may assume that $\{\theta_q(t)\}_{q=1}^{\infty}$ converges uniformly to a 2π-periodic, even function θ of class C^2 such that $\theta'' + g(\theta) = 0$. We assert that $\theta(t) \not\equiv 0$. Assuming the contrary, we have $|\theta_q|_0 \to 0$ as $q \to \infty$. Let $v_q(t) = \theta_q(t)/|\theta_q|_0$ for $q \geq 1$. We have $v_q'' + Av_q + r(\theta_q)/|\theta_q|_0 = 0$, where $r(x) = g(x) - Ax$. Since $\lim_{|x|\to 0} \frac{|r(x)|}{|x|} = 0$ and since $\max_{[0,2\pi]} |v_q(t)| = 1$, by the same type of reasoning used before, we may assume that $\{v_q\}$ converges uniformly to a 2π-periodic, even solution v of class C^2 such that $\max_{[0,2\pi]} |v(t)| = 1$ and $v'' + Av = 0$. Since $A\int_0^{2\pi} v(t)dt = 0$, the nonsingularity of A implies that $v \in \tilde{H}$. This contradicts Lemma 2 and the assumption that A has no eigenvalues of the form m^2, $m = 1,2,\cdots$. Thus $\theta(t) \not\equiv 0$. If θ were a constant then we would have $g(\theta) = 0$, and hence $\langle g(\theta),\theta\rangle = 0$, contradicting (27). This proves the theorem.

Finally, we give an example of a system of the form (S) where conditions (1) and (2) are satisfied but for which there exists no periodic solution other than the trivial solution. In this case, $i(B) - i(A)$ must be an even integer. In the example we give, $n = 2$ and $i(B) - i(A) = -2$. The construction we give may easily be extended to give an example where $i(B) - i(A)$ is any even integer.

Let an element $x \in R^2$ be denoted by $x = \begin{bmatrix} u \\ v \end{bmatrix}$, and define $g: R^2 \to R^2$ by

$$g(x) = \begin{bmatrix} w^2 u - \varepsilon v \\ \left(\frac{2\varepsilon}{\pi} \tan^{-1} r^2\right) u + w^2 v \end{bmatrix}$$

where $1 < w^2 < 4$, $\varepsilon > 0$, and $r^2 = u^2 + v^2$. We have $g(0) = 0$ and

$$Dg(x) = \begin{bmatrix} w^2 & -\varepsilon \\ 0 & w^2 \end{bmatrix} + \frac{2\varepsilon}{\pi}\begin{bmatrix} 0 & 0 \\ \frac{2u^2}{1+r^4} + \tan^{-1} r^2 & \frac{2uv}{1+r^4} \end{bmatrix}.$$

Therefore, we have $Dg(0) = \begin{bmatrix} w^2 & -\varepsilon \\ 0 & w^2 \end{bmatrix}$ and $i(A) = 2$. Moreover, it is easy to see that $Dg(x) = w^2 I_2 + \varepsilon M(x)$, where $M(x)$ denotes a 2×2 matrix whose elements are bounded on R^2. Since for $h \in R^2$, $\langle Dg(x)h,h\rangle = w^2|h|^2 + \varepsilon\langle M(x)h,h\rangle$, we may choose $\varepsilon > 0$ so small that

$$\text{(28)} \qquad \langle Dg(x)h,h\rangle \geq \frac{w^2}{2}|h|^2 \text{ for all } x \in R^2 \text{ and all } h \in R.$$

It follows from (28) that if $x_1, x_2 \in R^2$, then

$\langle g(x_2)-g(x_1),\ x_2-x_1\rangle = \int_0^{2\pi} \langle Dg(x_1+t(x_2-x_1))(x_2-x_1),\ x_2-x_1\rangle dt \geq \frac{w^2}{2}|x_2-x_1|^2$ for all $x_1, x_2 \in R^2$, and hence (1) holds. If $B = \begin{bmatrix} w^2 & -\varepsilon \\ \varepsilon & w^2 \end{bmatrix}$ then $\frac{|g(x) - Bx|}{|x|} \to 0$ as $|x| \to \infty$. The matrix B has the complex conjugate eigenvalues $w^2 \pm i\varepsilon$; thus $i(B) = 0$. Hence $i(B) - i(A) = -2$. It is easy to show that there exists no nontrivial periodic solution of $x'' + g(x) = 0$.

Remark. In case g is an odd function of x and $i(A) \neq i(B)$ (mod 2) then the condition (1) is not needed to show the existence of at least two nontrivial odd 2π-periodic solutions of (S). In fact if $\hat{H}$ is the subspace of H consisting of odd 2π-periodic functions then since $\int_0^{2\pi} |u(t)|^2 dt \leq \int_0^{2\pi} |u'(t)|^2 dt$ for all $u \in \hat{H}$, $(u,v)_1 = \int_0^{2\pi} \langle u',v'\rangle dt$ defines an inner product on $\hat{H}$ equivalent to (,). If $\hat{N}:\hat{H} \to \hat{H}$ by $(\hat{N}(u),v)_1 = \int_0^{2\pi} \langle g(u),v\rangle dt$ then $\hat{N}$ is completely continuous and standard regularity theory implies that u is an odd 2π-periodic solution of S iff $\hat{u} = N(\hat{u})$. If $i(A) \neq i(B)$ (mod 2) then the exact same reasoning that was used before gives the existence of $u_\circ \neq 0$ such that $u_\circ = \hat{N}(u_\circ)$. Since $\hat{N}$ is odd $-u_\circ$ will also be a nontrivial solution of $u = \hat{N}(u)$.

REFERENCES

[1] H. Amann "Saddle points and multiple solutions of differential equations", Math. Z. 169, (1979), pp. 127-166.

[2] H. Amann and E. Zehnder "Periodic solutions of asymptotically linear Hamiltonian systems", Manuscripta Math. 32, (1980), pp. 149-489.

[3] H. Amann and E. Zehnder "Nontrivial solutions for a class of nonresonance problems and applications to nonlinear differential equations", Annali Della Superiore DiPisa, 7, (1980), pp. 543-603.

[4] M. Berger and M. Berger, Perspectives in Nonlinearity, W.A. Benjamin Inc., Massachusetts, 1968.

[5] L. Cesari "Functional analysis and Galerkin's method", Michigan Math J., 11, (1964), pp. 385-414.

[6] C. Kung Chang "Solutions of asymptotically linear operator equations via Morse theory", Pure and Applied Mathematics, 34, (1981), pp. 693-712.

[7] D. Clark "Periodic solutions of autonomous Hamiltonian systems of ordinary differential equations", Proc. Amer. Math. Soc., 39, (1973), pp. 579-584.

[8] D. Clark "Periodic solutions of variational systems of ordinary differential equations", J. Differential Equations, 28, (1978), pp. 354-368.

[9] G. Fichera "Linear elliptic differential systems and eigenvalue problems", Lecture Notes in Mathematics, No. 8, Springer-Verlag, 1965.

[10] M. Krosnoselski, Topological Methods in the Theory of Nonlinear Integral Equations. Macmillan, New York, 1964.

[11] S. Lang, Algebra. Addison-Wesley, Massachusetts, 1965.

[12] S. Lang, Analysis II. Ibid, 1969.

[13] L. Nirenberg, Topics in Nonlinear Functional Analysis, Courant Institute Math. Sci. Lecture Notes, 1974.

EVOLUTION WATER CONING PROBLEM

Y. Amirat , A. El kolli , A. Ziani
Institut De Mathématiques , U.S.T.H.B
Dar El Beida, Algeria

ABSTRACT

We make a numerical study of the nonstationary water coning problem using a mixed finite element method.

I. INTRODUCTION

We consider a saturated porous medium Ω with boundary $\Gamma = \partial\Omega$ which contains two fluids (oil and water) in the domain Ω_1 (resp. Ω_2) simply connected , such that the interface is $\bar{\Omega}_1 \cap \bar{\Omega}_2 = \Gamma_0$, and $\Omega_1 \cup \Omega_2 \cup \Gamma_0 = \Omega$.

The boundary Γ is decomposed in two parts Γ_d (where the pressure is given) and Γ_γ (where there are two possibilities: a. the flow is given. b. the boundary is water-proof).

We suppose that the two fluids are incompressible and the medium homogeneous and isotropic.

Our purpose is to determine the motion of the fluids (oil and water) in function of the conditions given on Γ_d and Γ_γ .

In every domain Ω_i , i=1,2 we can write the equation of continuity and DARCY law. On the boundary which separates the two fluids the displacement speed is the (normal) filtration speed of the particle (particle of oil or water). See [1] , [2] .

We determine the evolution of Γ_0 by an iterative method: Γ_0 is given at the initial time.We determine the pressure and the pressure gradients in the domain Ω by solving a transmission problem by a finite, mixed element method of degree.

1. Then we solve the non linear hyperbolic equation given on the inter face. By this way we obtain the new boundary Γ_0, apply the next iteration and this until we obtain the position of the interface at time t.

II. EQUATIONS OF MOTION.

We take 4 numbers r_w, r_e, H_w, H such that $o < r_w < r_e$, $o < H_w < H$ and define :

$\Omega =]r_w, r_e[\times]o, H[$

$\Gamma_1 = \{(r_e, z) \; ; \; o \leq z \leq H\}$

$\Gamma_2 = \{(r, o) \; ; \; r_w \leq r \leq r_e\}$

$\Gamma_3 = \{(r_w, z) \; ; \; o \leq z \leq H_w\}$

$\Gamma_4 = \{(r_w, z) \; ; \; H_w \leq z \leq H\}$

$\Gamma_5 = \{(r, H) \; ; \; r_w \leq r \leq r_e\}$

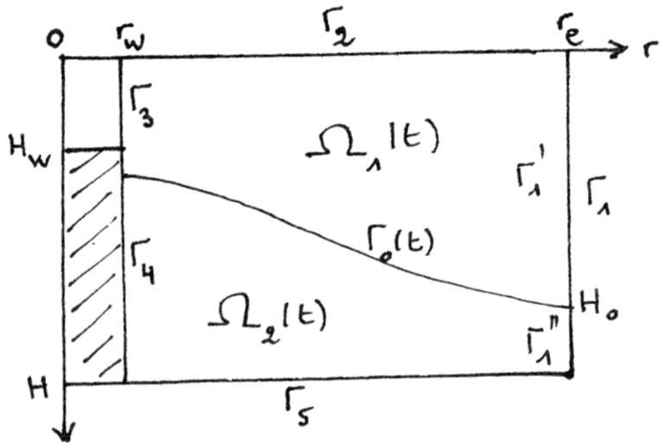

$T > o$ is given. Our purpose is to find:

a. $\xi(r,t)$ "regular" from $]r_w, r_e[\times]o, T[$ to $]H_w, H[$

increasing in r for each t.

b. $P(r,z,t)$ "regular in $\Omega \times]o, T[$ such that :

1. For each t in $]o, T[$

(1) $- C_i . \Delta_r P_i = o$ in $\Omega_i(t)$ $i = 1,2$

Where $\Omega_1(t) = \{(r,z) \in \Omega ; \; z < \xi(r,t)\}$

$\Omega_2(t) = \{(r,z) \in \Omega ; \; z > \xi(r,t)\}$

and P_i is the restriction of P to $\Omega_i(t)$, i=1,2.

(2) $P = P_e + \rho_1 gz$ on $\Gamma_1'(t) = \{(r_e, z) \; ; \; o \leq z \leq \xi(r_e, t)\}$

(3) $\partial P / \partial z = \rho_1 g$ on Γ_2

(4) $P = P_w(t) + \rho_1 gz$ on Γ_3

(5) $\partial P / \partial r = o$ on Γ_4

(6) $\partial P / \partial z = \rho_2 g$ on Γ_5

(7) $P = P_e + (\rho_1 - \rho_2) g H_o + \rho_2 g z$

on $\Gamma_1'' = \{(r_e, z) \;;\; \xi(r_e, t) \leq z \leq H\}$

(8) $P_1 = P_2$ on $\Gamma_o(t)$ the graph of $\xi(r,t)$.

(9) $C_1(\partial P_1/\partial n) = C_2(\partial P_2/\partial n) + C_3$ on $\Gamma_o(t)$.

n is the normal to $\Gamma_o(t)$ oriented from $\Omega_1(t)$ to $\Omega_2(t)$.

2. The evolution of $\Gamma_o(t)$ is given by :

$$\begin{cases} \partial\xi/\partial t = -C_i\left(\sqrt{1 + (\partial\xi/\partial r)^2}\,.\mathrm{grad}(P_i).n - \rho_i g\right) \;;\; i=1,2 \\ \xi(r,o) = \xi_o(r) \quad \text{the initial data , a regular , increasing function with values in } [H_w, H]. \end{cases}$$

This is the formulation of the water coning problem in a petroleum reservoir with two phases (oil and water). The stationary case is studied in [3] , [4] , [5]

P_1 is the oil pressure in $\Omega_1(t)$.
P_2 is the water pressure in $\Omega_2(t)$.
P_e is the pressure at the top , far the weel , constant.
$P_w(t)$ is the pressure at the weel , depending in time t.
ρ_1 (respectively. ρ_2) is the oil (respectively water) density, constant. g is the constant of gravity.
C_1 , C_2 , C_3 , C_4 are constants.

$C_i = k / \mu_i \quad i=1,2$

$C_3 = g(\rho_1 C_1 - \rho_2 C_2)$

$C_{3+i} = \rho_i g C_i \quad i=1,2$

k is the permeability constant of the porous medium.
μ_1 (resp. μ_2) is the viscosity contant of oil (respectively water). $\Gamma_o(t)$ is the interface of the fluids.

ACKNOWLEDGMENT

This work is sponsored by ONRS Contrat AA 75 02 01

REFERENCES

[1] M.Muskat."Principle of oil production" McGraw Hill, New York, 1949.

[2].J.Bear."Dynamics of porous media" Elsevier 1972.

[3].Y.Amirat."Une méthode d'optimisation de domaine dans le problème du cône d'eau en régime stationnaire" Thèse ALGER 1978.

[4].Y.Atik."Sur une formulation variationnelle et approxima-tion numérique du problème du cône d'eau en ré-gime stationnaire" Thèse ALGER 1978.

[5].A.Ziani."Méthode variationnelle et approximation de Muskat dans le problème du cône d'eau en régime station-naire" Thèse ALGER 1976.

CONDITIONS FOR PERIODIC SOLUTIONS TO HILL'S EQUATION WITH AN EVEN PERIODIC COEFFICIENT

Bernard A. Asner, Jr.

Department of Mathematics
University of Dallas
Irving, Texas

ABSTRACT

A n.a.s.c. for periodic solutions to Hill's equation with an even term is derived. These conditions are developed at the half-period and make it possible to determine the behavior of solutions by restricting the analysis for $t \in [0.T/2]$.

INTRODUCTION

In this paper we consider Hill's equation with a damping term and assume the periodic term of period T to be an even function of its argument:

$$y'' + 2cy' + p(t)y = 0$$

$$p(t+T) = p(t), \quad p(t) = p(-t) . \tag{1}$$

Conditions for periodic solutions are of particular interest. The classical theorem of Hill [1],[3],[4], or [5] shows that (1) will have a periodic solution iff

$$|u(T) + v'(T)| = 1 + \exp(-2cT) \tag{2}$$

where u and v are fundamental solutions. If p is even then complete information on the solutions can be found by restricting the time interval to [0,T/2]. This can lead to a considerable reduction in computer calculations. The case of no damping yields trigonometric like results, e.g. the fundamental solutions are even and odd and satisfy the "double-angle formulas" at t=T. These results are used to transform (2) into simple formulas which involve T/2 and the fundamental solutions with no damping.

RESULTS FROM HILL AND FLOQUET THEORY

Abusing notation, we introduce the vector-matrix symbols as

$$y = (y_1 , y_2) = (y , y') , \quad z = (z_1 , z_2) ,$$

$$A = \begin{bmatrix} 0 & 1 \\ -p(t) & -2c \end{bmatrix}, \qquad B = \begin{bmatrix} 1 & 0 \\ -c & 1 \end{bmatrix}, \tag{3}$$

$$C = \begin{bmatrix} 0 & 1 \\ -p+c^2 & 0 \end{bmatrix}, \qquad P = \begin{bmatrix} g(t) & h(t) \\ g'(t) & h'(t) \end{bmatrix}.$$

The canonical form of (1) is

$$y' = A(t)y \tag{4}$$

and the substitution

$$y = \exp(-ct)B\ z \tag{5}$$

transforms (4) into

$$z' = C(t)\ z\ . \tag{6}$$

We note that this substitution has transformed the A matrix to the more familiar form of C where the damping term has been eliminated. P(t) is the fundamental solution to the matrix differential equation

$$P'(t) = C(t)P(t), \quad P(0) = I \;, \tag{7}$$

and the monodromy matrix P(T) , by virtue of C being periodic, is known to satisfy the identity

$$P(t+T) = P(t)P(T). \tag{8}$$

PROPERTIES OF g AND h

In 1939, Brainerd and Weygandt [3] discovered many interesting relations for Mathieu's equation which allowed them to calculate solutions on a reduced time period. Their relations are extended to Hill's equation in the following two lemmas. The technique of proof is also less cumbersome and should make their results more accessible.

Lemma 1. g(t) is even and h(t) is odd.

Proof: g and h are independent solutions to the equation

$$z'' + (p(t) - c^2)z = 0 \tag{9}$$

where $g(0) = 1$, $g'(0) = 0$, $h(0) = 0$, and $h'(0) = 1$. With p even and periodic g(-t) and h(-t) also satisfy (9) and can be expressed as linear combinations of the fundamental solutions:

$$g(-t) = ag(t) + bh(t) \tag{10a}$$

$$h(-t) = cg(t) + dh(t) \tag{10b}$$

Since $g(0) = 1$ and $h(0) = 0$ we have from (10a) $a = 1$. Differentiating (10a) with respect to t gives

$$-g'(-t) = g'(t) + bh'(t) \quad .$$

Evaluating this last expression at $t = 0$ yields $b = 0$ and shows that g is even. Similar arguments for (10b) give $c = 0$ and $d = -1$ so that h is odd.

Lemma 2. The elements of the monodromy matrix satisfy the following identities:

$$g(T) = h'(T/2)g(T/2) + h(T/2)g'(T/2) \tag{11a}$$

$$= 1 + 2g'(T/2)h(T/2) \tag{11b}$$

$$g'(T) = 2g(T/2)g'(T/2) \tag{12}$$

$$h(T) = 2h(T/2)h'(T/2) \tag{13}$$

$$h'(T) = g(T) \quad . \tag{14}$$

Proof: In (8), let $t = -T/2$, to give

$$P(T) = P^{-1}(-T/2)P(T/2) \quad . \tag{15}$$

It is apparent that the inverse of P(t) exists when the Jacobi identity is used. Explicitly,

$$\det[P(t)] = g(t)h'(t) - g'(t)h(t)$$

$$= \exp \int_{0}^{t} \text{trace}[C(s)]\, ds \tag{16}$$

$$= 1.$$

Using the results of Lemma 1, we have in component form,

$$P^{-1}(-T/2) =$$

$$\begin{bmatrix} h'(-T/2) & -h(-T/2) \\ -g'(-T/2) & g(-T/2) \end{bmatrix} = \begin{bmatrix} h'(T/2) & h(T/2) \\ g'(T/2) & g(T/2) \end{bmatrix}.$$

Substituting this expression in (15) and using (16) with $t = T/2$ completes the proof.

Remark. If $g = \cos$ and $h = \sin$ then (14) is $D\sin = \cos$, (13) is $\sin 2x = 2\sin x \cos x$, etc.

STABILITY BOUNDARIES

Returning to the original equation (4) with damping, and denoting its fundamental solution as $Y(t)$, we see that

$$Y(t) = \exp(-ct)BZ(t)$$

is the solution to $Y' = CY$ with $Z(0) = B^{-1}$. Thus

$$Z(t) = P(t)B^{-1}$$

satisfies $Z' = CZ$, $Z(0) = B^{-1}$ and

$$Y(t) = \exp(-2ct)BP(t)B^{-1} \quad .$$

Thus

$$\operatorname{trace}[Y(t)] = \exp(ct) \operatorname{trace}[P(t)]. \tag{17}$$

From the general results of Hill and (2) there exist a point on the stability boundary and a solution of period T of (1), if and only if,

$$\operatorname{trace}[Y(T)] = 1 + \exp(-2cT).$$

Let $t = T$ in (17), use this last expression, and the identities of Lemma 2, to conclude that

$$g'(T/2)h(T/2) = \sinh^2(cT/2) \quad .$$

Similar arguments for a periodic solution of period 2T lead to the following theorem.

<u>Theorem 1. A necessary and sufficient condition for (1) to have a periodic solution is that</u>

(a) $g'(T/2)h(T/2) = \sinh^2 (cT/2)$, <u>if the period</u> = T,

or

(b) $g'(T/2)h(T/2) = -\cosh^2 (cT/2)$, <u>if the period</u> = 2T.

Assuming we have been able to satisfy the conditions of the last theorem it then becomes of interest to determine a vector of initial values that would generate the periodic solution. If the period equals T, then the monodromy matrix Y(T) will have an eigenvalue to one. Hence, there exists an eigenvector y such that $Y(T)y = y$.

Let $x = B^{-1} y$ and the previous defined transformations to get $P(T)\,x = \exp(cT)\,x$. The identities of Lemma 2 can be used to solve for x and then y. These calculations are summarized in the following theorem where the results are presented in a format that is applicable to machine calculations.

<u>Theorem 2. If there exists a periodic solution to (1), then an initial condition y to start the periodic solution is given by:</u>

PERIOD	$g(T/2)g'(T/2) \approx 0$	$h(T/2)h'(T/2) \approx 0$
T	YES	...

$$y = [\ 1\ ,\quad \sinh(cT)/h(T) \quad - c\ ]$$

PERIOD	$g(T/2)g'(T/2) \approx 0$	$h(T/2)h'(T/2) \approx 0$
T	...	YES

$$y = [\ \sinh(cT)/g'(T),\quad 1 - c\ \sinh(cT)/g'(T)\ ]$$

2T	YES	...

y = [1 , -sinh(cT)/h(T) - c]

2T	...	YES

y = [sinh(cT)/g'(T), -1 - c sinh(cT)/g'(T)] .

ACKNOWLEDGMENT

Much of this research was accomplished while the author held an NRC Resident Research Associateship at George C. Marshall Space Flight Center, Systems Dynamics Laboratory, National Aeronautics and Space Administration, Huntsville, AL., U.S.A.

REFERENCES

[1] M. Abramowitz and I. Stegun, Handbook of Mathematical Functions with Formulas, Graphs, and Mathematical Tables. National Bureau of Standards, Applied Mathematics Series. Vol. 55, Washington, D. C. 1964.

[2] B. Asner, Stability Charts for Hill's Equation. Int. J. Non-Linear Mechanics. Vol. 15, 1980, pp. 127-131.

[3] J.G. Brainerd and C.N. Weygandt, Solutions of Mathieu's Equation. Philosophical Magazine, Vol. 30, 1940, pp. 458-477.

[4] P. Pedersen, Stability of the Solutions to Mathieu-Hill Equations with Damping. Ingenieur-Archiv, Vol. 49, 1980, pp. 15-29.

[5] V. A. Yakubovich and V. M. Starzhinskii, Linear Differential Equations with Periodic Coefficients. (English translation) Wiley, New York 1975.

On paper on RHS

TH. 3: <u>G is open</u>. Let $z \in G$. Let $u \in S(z, \epsilon/2)$ be arbitrary. Then $u \, T_\epsilon \, z$. Since $z \, T^b x$ for some integer $b \geq 0$, therefore $u \, T^c x$ for some integer $c \leq b+1$; thus $u \in G$. Hence $S(z, \epsilon/2) \subset G$.

<u>G is closed</u>: better show that $A - G$ is open (exactly as above).

METRICS AND TOLERANCES

Prem N. Bajaj

Wichita State University
Wichita, Kansas

ABSTRACT

In this paper, the correspondence between metric spaces and tolerance spaces is emphasized. In a discrete metric space, a new definition of connectedness is introduced and some results obtained along these lines.

1. INTRODUCTION

Some results on metrics and tolerances have, recently, been given by Chajda [2], Chajda and Zelinka [3]. The results of Chajda and Zelinka needed connectedness of a metric space (even though the hypothesis of connectedness was not explicitly stated); the question arose as to what would be the meaning of connectedness in an integer valued metric. It lead us to introduce a definition of connectedness in an integer-valued metric. Situation appears to be like some problems in a discrete semi-dynamical system [4]; the relation of Metrics and Tolerances with Discrete Semi-dynamical systems is under investigation.

For simplicity, we discuss only metric spaces and leave the extensions to pseudometric and quasimetric spaces ([5], [6]) to the reader.

2. DEFINITIONS

A reflexive and symmetric binary relation T on a non-empty set A is said to be a tolerance relation ([1], [2]) (or simply tolerance) on A. The ordered pair (A,T) is called a tolerance space. Denoting the identity relation by T^0, we define $T^{n+1} = T \cdot T^n$ inductively for any positive integer n.

A tolerance space (A,T) is said to be connected if given x,y in A, there exists a non-negative integer b such that $x\, T^b y$.

Let (A,e) be a metric space and ε a positive real number. Then the relation T_ε defined by $x\, T_\varepsilon y$ whenever $e(x,y) \leq \varepsilon$ gives rise to a tolerance T_ε.

Let (A,T) = connected tolerance space. Define an integer-valued map on A x A by $\delta_T(x,y)$ = least non-negative integer b such that $x\, T^b y$.

Clearly (A, δ_T) is a metric space.

Throughout J denotes the set of non-negative integers.

3. THEOREM

(Preserving the notation). Let (A,e) be a connected metric space and ε a positive real number. Then (A, T_ε) is a connected tolerance space. Moreover if $0 \leq \varepsilon \leq 1$, then $\delta_{T\varepsilon}(x,y) \geq e(x,y)$ for all x,y in A.

Indication of proof: (A, T_ε) being a tolerance is clear. If (A, T_ε) is not connected, then for some x,y in A, $x \neq y$, relation $x\, T^b y$ does not hold for any b in J. Let $G = \{z \in A: x\, T^b z$ for some b in J$\}$. Then G is a nonempty proper subset of A. Moreover G is open and closed which contradicts that the metric space (A,e) is connected.

4. REMARKS

The above theorem does not hold without the hypothesis of connectedness on the metric space (A,e). To see this let $A = \{x: 1 \leq |x| \leq 2\}$ with the usual metric. Let $\varepsilon = \frac{1}{2}$, then the tolerance (A, T_ε) is not connected (and, so, $\delta_{T\varepsilon}$ is not even defined in this case).

Further if $\varepsilon > 1$, for any x,y in A, either of $\delta_T(x,y)$ and e(x,y) can be less.

5. DEFINITION

Let (A,d) be an integer-valued metric. Then (A,d) is said to be connected if there do not exist nonempty disjoint subsets G and H of A such that $G \cup H = A$ and $\min\{d(x,y): x \in G,\ y \in H\} > 1$.

The following theorem follows, almost directly, from the definition.

6. THEOREM

Let (A,d) be an integer-valued metric. The following are equivalent:

(i) (A,d) is connected

(ii) Given a nonempty proper subset G of A, $\min\{d(x,y): x \in G, y \notin G\} = 1$.

(iii) Given a nonempty proper subset G of A, there exist $x \in G$, $y \notin G$ such that $d(x,y) = 1$.

(iv) Given x,y in A, $x \neq y$ there exists a finite sequence $x = x_1, x_2, \dots x_b = y$ such that $d(x_i, x_{i+1}) = 1$, $i = 1, 2, \dots, b-1$.

7. REMARK

By (iv) of Theorem 6, it follows that in a connected integer-valued metric, every pair of points can be joined to each other by a sequence of points. How will these ideas translate to the case of limit sets etc. in a discrete dynamical systems is still under investigation by the author.

8. THEOREM

Let (A,d) be an integer-valued metric. Let (A,d) be connected (in the sense of definition 5 above). Define a relation T on A by $x\ T\ y$ if and only $d(x,y) = 1$. Let $\delta_T: A \times A \to J$ be defined by $\delta_T(x,y) =$ least $b \in J$ such that $x\ T^b y$. Then (A, δ_T) is an integer-valued metric. Moreover $\delta_T(x,y) \geq d(x,y)$ for all x,y in A.

Indication of proof: Since (A,d) is a connected integer-valued metric, therefore the tolerance space (A,T) obtained is also connected.

9. REMARK

In the above theorem, a connected integer-valued metric d gave rise to a connected tolerance T which, in turn, produced an integer-valued connected metric δ_T. However $\delta_T \neq d$ in general. To see this let $A = \{x, y, z, w\}$. Let $d(x,y) = d(x,z) = d(y,w) = 2$, $d(x,w) = d(w,z) = d(z,y) = 1$ and $d(y,y) = d(z,z) = d(w,w) = d(x,x) = 0$.

Then (A,d) is a metric space. Moreover $\delta_T(x,y) = 3$ whereas $d(x,y) = 2$.

REFERENCES

[1] Bajaj, Prem N.: A Note on Metrics and Tolerances, Arch. Math. 17 (1981), 3-6.

[2] Chajda, Ivan: Systems of Equations and Tolerance Relations, Czech. Math. J. 25 (1975), 302-308.

[3] Chajda, Ivan and Zelinka, Bodhan: Metrics and Tolerances, Arch. Math. 14 (1978), 193-200.

[4] LaSalle, J. P.: The Stability of Dynamical Systems, Lecture notes of Talks (1975) given at Mississippi State University.

[5] Pervin, William J.: Foundations of General Topology, Academic Press, 1965.

[6] Willard, Stephen: General Topology, Addison-Wesley, 1970.

GLOBAL EXISTENCE FOR NONLINEAR SECOND ORDER DIFFERENTIAL EQUATIONS

John V. Baxley

Department of Mathematics
Wake Forest University
Winston-Salem, North Carolina

ABSTRACT

In order to solve a two point boundary value problem associated with the differential equation $y'' = f(x,y,y')$ on the interval $[a,b]$, it is first necessary that solutions exist on $[a,b]$. After discussing some previous results and giving two uncooperative examples, we provide rather general sufficient conditions for existence of solutions on the interval $[a,b]$.

INTRODUCTION

What are the minimal hypotheses on a second order differential equation

$$y'' = f(x,y,y') \tag{1}$$

which guarantee the existence of a "reasonably large" number of solutions on the interval $a \leq x \leq b$? To make this question precise, we must have some problem whose solution we hope to find among these global solutions.

My interest in this question originated in the 1966 paper of Keller [13], in which was announced an existence and uniqueness theorem for the boundary value problem consisting of (1) and linear boundary conditions

$$a_0 y(a) - a_1 y'(a) = A,$$
$$b_0 y(b) + b_1 y'(b) = B.$$

His proof used the shooting method and thus sought the solution of the boundary value problem among the global solutions of (1) which satisfy also the boundary condition at $x = a$. Keller's hypotheses on f were not sufficient for solutions of all initial value problems to exist on $[a,b]$; however his proof overlooked that difficulty. Later, Bebernes and Gaines [7] proved a more general result than Keller's using the method of subfunctions.

In [3], essentially best possible conditions on f were formulated in order that every initial value problem consisting of (1) and

$$y(a) = \gamma_1, \; y'(a) = \gamma_2$$

has a unique solution on [a,b] and the result was used with the shooting method to prove an existence-uniqueness theorem for boundary value problems associated with (1). But this theorem is in some important ways weaker than the Keller-Bebernes-Gaines result, whose hypotheses included equations of the form (1) with some solutions which could not be continued to the whole interval [a,b]. Thus, as has been observed also by Jackson [11], it is clearly not necessary that all initial value problems have global solutions in order to solve boundary value problems, only that a "sufficient supply" of such solutions exist.

Therefore, the following assumptions [4] were placed on the function $f(x,y,z)$ in (1):

(a) f is continuous on $R = \{(x,y,z): \; a \le x \le b, \; |y| + |z| < \infty\}$;

(b) f satisfies a uniform Lipschitz condition on each compact set with respect to z, i.e., for each $\rho > o$, there exists $M(\rho) < \infty$ such that

$$|f(x,y,z_2) - f(x,y,z_1)| \le M(\rho) \; |z_2 - z_1|$$

whenever the points (x,y,z_1), (x,y,z_2) belong to

$$R_\rho = \{(x,y,z): \; a \le x \le b, \; |y| + |z| \le \rho\} .$$

Moreover the Lipschitz constant $M(\rho)$ satisfies

$$M(\rho) = o(\log \rho), \text{ as } \rho \to \infty;$$

(c) $f(x,y,z)$ is nondecreasing in y for fixed x, z.

It was proved in [4] that the set of global solutions of (1) is not empty and in fact is sufficiently large that the shooting method can be used to solve a variety of boundary value problems; the boundary conditions may be nonlinear and the results in [4] contain those of Keller, Bebernes, and Gaines.

It is well-known from the work of Bailey, Shampine, and Waltman (see for example [2]) that relaxing the hypothesis (c) requires restrictions on the length of the interval [a,b]. Applications of the results in [4] have led to extensions [5], [6] of their work to a somewhat more general context.

There is an old theorem of S. N. Bernstein [9], quoted in the books [1], [10], which has been largely ignored in the West (for example, it is not mentioned in the survey book [8]) but is of interest here. Bernstein's theorem is that the boundary value problem consisting of (1) and fixed end conditions

$$y(a) = A, \quad y(b) = B,$$

has a unique solution if f is smooth (continuous partial derivatives) and

(i) $\partial f/\partial y > k > 0$

(ii) $|f(x,y,z)| \leq a(x,y)\, z^2 + b(x,y)$

where a(x,y), b(x,y) are bounded on any compact subset of the plane $|x| + |y| < \infty$. This theorem shows that some trade-off is possible between the dependence of f on y and z. In this theorem, the assumption (c) has been strengthened to (i), but the assumption (b) has been weakened to (ii).

Recently, Jackson and Palamides [12] have proved an existence result for boundary value problems associated with (1) which, among other things, assume (c) and that the behavior of f on y = 0 is essentially that of (b) while for $y \neq 0$, f behaves essentially as in (ii); however the precise result does not allow some of the simplest boundary conditions, e.g., the fixed end conditions, but since it does allow conditions such as $y(0) - y'(0) = 0$, $y(1) + y'(1) = 0$, these hypotheses certainly imply the existence of global solutions of (1), sufficient in quantity to solve some boundary value problems. After consideration of some counter-examples, we shall state some recent results which clarify and extend the theorems discussed above.

TWO EXAMPLES

The first example shows that the quadratic growth in z allowed by the Bernstein theorem is not possible without some strengthening of assumption (c). In fact, if f is constant in y, the logarithmic growth in (b) cannot be relaxed at all.

Choose $\varepsilon > 0$ fixed and then choose $K \geq \max\{6, 3/\varepsilon\}$ fixed. Let

$$f(x,y,z) = \begin{cases} K|z|\,(\log|z|)^{1+\varepsilon}, & \text{if } |z| \geq e, \\ Ke, & \text{if } |z| < e, \end{cases}$$

where e is the base for the natural logarithm. Then f satisfies (a), (c) (constant in y), but fails by ε to satisfy (b). For this f, (1) has no solution existing for $0 \leq x \leq 1$, and therefore no boundary value problem, with boundary conditions at x = 0 and x = 1 can be solved.

We verify these assertions as follows. Suppose $\phi(x)$ is any local solution of (1) existing in some neighborhood of x = 0 and which has been continued to its maximum interval of existence.

1) There exists x_1, $0 \leq x_1 \leq 1/3$, at which $\Phi'(x_1) \geq -e$. For if not,

then $\phi'(0) < -e$. Solving (1) by separating variables and using this initial inequality gives, for $0 \leq x \leq 1/3$,

$$\phi'(x) = -\exp\left(\frac{1/\varepsilon}{Kx + c}\right)^{1/\varepsilon}$$

where $0 < c < 1/\varepsilon$ and c is constant. Since $\phi'(x) = -e$ for $x = (1/\varepsilon - c)/K < 1/(\varepsilon K) \leq 1/3$, we have a contradiction.

2) There exists x_2, $x_1 \leq x_2 \leq 2/3$, at which $\phi'(x_2) \geq e$. For if not, then (1) and the initial inequality $-e \leq \phi'(x_1) < e$ gives, for $x_1 \leq x \leq 2/3$,

$$\phi'(x) = \phi'(x_1) + Ke(x - x_1)$$

and $\phi'(x) = e$ for $x - x_1 = (e - \phi'(x_1))/Ke \leq 2/K \leq 1/3$, a contradiction.

3) There exists x_3, $x_2 < x_3 \leq 1$, for which $\phi'(x) \to +\infty$ as $x \to x_3$. For we may solve (1) using the initial inequality $\phi'(x_2) \geq e$ to get, for $x \geq x_2$

$$\Phi'(x) = \exp\left(\frac{1/\varepsilon}{d - Kx}\right)$$

where $Kx_2 \leq d < Kx_2 + 1/\varepsilon$ and d is constant. Thus $\phi'(x) \to +\infty$ as $x \to x_3$ where $x_3 = d/K < x_2 + 1/(\varepsilon K) \leq 1$.

The second example shows that the slightest relaxing of the hypothesis (ii) of the Bernstein theorem can lead to a situation in which the supply of global solutions is too meager to solve some boundary value problems.

Choose $\varepsilon > 0$ fixed and then choose $K > \max\{4, 1/(2\varepsilon)\}$ fixed. Let

$$f(x,y,z) = \begin{cases} Ky\,|z|^{2+\varepsilon}, & \text{if } |z| \geq 1, \\ Ky, & \text{if } |z| < 1. \end{cases}$$

Then f satisfies (i) and is smooth, except for the corners at $|z| = 1$, which can be easily "rounded" if desired. However, f fails by ε to satisfy (ii). For this f, there is no solution of (1) on $0 \leq x \leq 1$ satisfying $y(0) \geq 2 + \varepsilon$. Thus, if the boundary condition at $x = 0$ requires that $y(0) \geq 2 + \varepsilon$, there is no boundary condition at $x = 1$ for which the resulting boundary value problem has a solution.

We now verify these assertions. Suppose $\phi(x)$ is any local solution of (1) satisfying $\phi(0) \geq 2 + \varepsilon$ and suppose $\phi(x)$ has been continued to its maximum interval of existence. As before, we proceed in three stages.

(1) There exists x_1 $(0 \leq x_1 < 1/4)$ for which $\phi'(x_1) \geq -1$ and $\phi(x_1) > 1 + \varepsilon$. If $\phi'(0) \geq -1$ we may let $x_1 = 0$. Otherwise, put

$S = \{t \in [0,1]:\ \phi$ exists on $0 \le x \le t$, $\phi(x) > 1 + \varepsilon$ and $\phi'(x) < -1$ on $[0,t]\}$.
Since $0 \in S$, then $S \ne \emptyset$. Let $x_1 = \sup S \le 1$. Then for $0 \le x < x_1$, we have $\phi(x) > 1 + \varepsilon$, $\phi'(x) < -1$. The definition of $f(x,y,z)$ then gives

$$\phi''(x) > K\,|\phi'(x)|^{2+\varepsilon} \text{ for } 0 \le x < x_1 .$$

Dividing by $|\phi'(x)|^{2+\varepsilon}$ and integrating from 0 to x gives

$$\phi'(x) \ge -\left[(1 + \varepsilon)Kx + 1/|\phi'(0)|^{1+\varepsilon}\right]^{-1/(1+\varepsilon)}, \text{ for } 0 \le x < x_1 . \tag{2}$$

Since the right side of (2) is equal to -1 if

$$(1 + \varepsilon)Kx + 1/|\phi'(0)|^{1+\varepsilon} = 1,$$

it follows that

$$(1 + \varepsilon)Kx_1 + 1/|\phi'(0)|^{1+\varepsilon} \le 1 \tag{3}$$

and since $K \ge 4$ that

$$0 < x_1 < 1/4.$$

The definition of S and x_1 imply that ϕ exists on $0 \le x < x_1$ and ϕ is bounded, ϕ' is bounded above on $[0,x_1)$. The differential equation guarantees that ϕ'' is positive and hence ϕ' is increasing on $[0,x_1)$. Thus ϕ, ϕ' are both bounded on $[0,x_1)$ and the standard theorem on continuation of solutions implies that ϕ exists for $0 \le x \le x_1$. Using (2) and (3), we obtain

$$\phi(x_1) = \phi(0) + \int_0^{x_1} \phi'(x)dx > \phi(0) - 1 = 1 + \varepsilon$$

Since $x_1 = \sup S$ it follows that $\phi'(x_1) = -1$ and we are done.

(2) There exists x_2 ($x_1 \le x_2 < 3/4$) for which $\phi'(x_2) \ge 1$ and $\phi(x) > 1$ for $x_1 \le x \le x_2$. If $\phi'(x_1) \ge 1$ we may let $x_2 = x_1$. Otherwise, put

$S = \{t \in [x_1,1]:\ \phi$ exists on $0 \le x \le t$, $\phi(x) > 1$ and $\phi'(x) < 1$ on $[0,t]\}$.

Since $[0,x_1] \subset S$, then $S \ne \emptyset$. Let $x_2 = \sup S \le 1$. Then for $x_1 \le x < x_2$, we have $\phi(x) > 1$, $\phi'(x) < 1$. The definition of $f(x,y,z)$ then gives

$$\phi'(x) = \phi'(x_1) + \int_{x_1}^{x} \phi'' > -1 + K(x-x_1), \text{ for } x_1 \le x < x_2. \tag{4}$$

The right side of this inequality equals 1 if $K(x-x_1) = 2$; hence

$$x_1 < x_2 \le x_1 + 2/K < 3/4$$

since $K > 4$ and $x_1 < 1/4$. Arguing as before, we see that ϕ exists for $0 \le x \le x_2$. Further, using (4) we have

$$\phi(x_2) = \phi(x_1) + \int_{x_1}^{x_2} \phi' > 1 + \varepsilon + K(x_2-x_1)^2/2-(x_2-x_1) \ge 1 + \varepsilon - 1/(2K) > 1$$

since $0 \le x_2 - x_1 \le 2/K$ and $K > 1/(2\varepsilon)$. Since $x_2 = \sup S$ it follows that $\phi'(x_2) = 1$.

(3) There exists x_3, $x_2 \le x_3 < 1$, for which $\phi'(x) \to +\infty$ as $x \to x_3$. For using (1) for $x \ge x_2$, we have

$$\phi''(x) \ge K|\phi'(x)|^{2+\varepsilon}.$$

Dividing by $|\phi'(x)|^{2+\varepsilon}$ as before and integrating from x_2 to x gives, for $x > x_2$

$$\phi'(x) > \left[1 - K(1+\varepsilon)(x - x_2)\right]^{-1/(1+\varepsilon)}.$$

The right side tends to $+\infty$ as $x \to x_2 + \frac{1}{K(1+\varepsilon)}$ and so there exists x_3 with $x_2 < x_3 < x_2 + \frac{1}{K(1+\varepsilon)} < 1$ for which $\phi'(x) \to +\infty$ as $x \to x_3$.

GLOBAL EXISTENCE

We state below sufficient conditions for equation (1) to have solutions on $a \le x \le b$ which satisfy a subsidiary condition at $x = a$. The following restrictions on the behavior of $f(x,y,z)$ will be used:

α: f is continuous on $R = \{(x,y,z): a \le x \le b, |y| + |z| < \infty\}$.

β_1: f is strictly increasing in y for fixed x, z.

β_2: f is nondecreasing in y for fixed x, z and f satisfies a uniform Lipschitz condition on each compact subset of R with respect to z; that is, given a compact subset W of R, there exists a constant

$M(W)$ so that

$$|f(x,y,z_2) - f(x,y,z_1)| \leq M(W)\ |z_2-z_1|$$

whenever $(x,y,z_i) \in W$ for $i = 1, 2$.

γ: there exists $z_o \geq 1$, $K > 0$, y_1, y_2 (real) so that for every $x \in [a,b]$

$$f(x,y_1,z) \geq -Kz\ \ell og\ z, \text{ for } z \geq z_o,$$
$$f(x,y_2,z) \leq K\ |z|\ \ell og\ |z|, \text{ for } z \leq -z_o.$$

$\delta(A)$: f satisfies γ and further there exist $M_1 > 0$, $M_2 > 0$ so that for every $x \in [a,b]$

$$f(x,y,z) \geq -M_1 z^2, \text{ for } A \leq y < y_1,\ z \geq z_o,$$
$$f(x,y,z) \leq M_2 z^2, \text{ for } y_2 < y \leq A,\ z \leq -z_o\ .$$

Note that if $A \geq y_1$ or $A \leq y_2$ one of the inequalities in $\delta(A)$ is vacuously true.

Theorem 1. Suppose f satisfies α, one of β_1 and β_2, and γ. Suppose that $a_o > 0$, $a_1 > 0$. Then for each real A, the problem

$$y'' = f(x,y,y')$$
$$a_o y(a) - a_1 y'(a) = A$$

has at least one solution existing on the entire interval $a \leq x \leq b$.

If $a_1 = 0$, the hypotheses of Theorem 1 are not enough.

Theorem 2. Suppose f satisfies α and one of β_1 and β_2. For some fixed A, suppose f satisfies $\delta(A)$. Then the problem

$$y'' = f(x,y,y')$$
$$y(a) = A$$

has at least one solution existing on the entire interval $a \leq x \leq b$.

Proofs of somewhat more general theorems (in which the condition $a_o y(a) - a_1 y'(a) = A$ is replaced by a possibly nonlinear condition) will appear elsewhere. The techniques are modified from those used in [4] and involve comparing solutions of $y'' = f(x,y,y')$ with solutions of

$$y'' = \pm K\ |y'|\ \ell og\ |y'|$$

or of

$$y = \pm M(y')^2$$

each of which can be solved in closed form.

These theorems can be applied to solve two point boundary value problems such as those mentioned in the introduction and in which the boundary conditions are allowed to be nonlinear. These applications will also appear elsewhere.

REFERENCES

[1] N. I. Akhiezer, The Calculus of Variations, Blaisdell, New York, 1962.

[2] P. Bailey, L. F. Shampine, and P. Waltman, Nonlinear Two Point Boundary Value Problems, Academic Press, New York, 1968.

[3] J. V. Baxley, Global Existence and uniqueness for second-order ordinary differential equations, J. Differential Equations 23 (1977), 315-334.

[4] J. V. Baxley and S. E. Brown, Existence and uniqueness for two-point boundary value problems, Proc. Roy. Soc. Edinburgh 88A (1981), 219-234.

[5] J. V. Baxley, Nonlinear second order boundary value problems: continuous dependence and periodic boundary conditions, Rend. Circ. Mat. Palermo, 31(II) (1982), to appear.

[6] J. V. Baxley, Nonlinear second order boundary value problems: intervals of existence, uniqueness, and continuous dependence, J. Differential Equations, 45 (1982), 389-407.

[7] J. Bebernes and R. Gaines, A generalized two-point boundary value problem, Proc. Amer. Math. Soc. 19 (1968), 749-754.

[8] S. R. Bernfield and V. Lakshmikantham, An Introduction to Nonlinear Boundary Value Problems, Academic Press, New York, 1974.

[9] S. N. Bernstein, Sur les equations du calcul des variations, Ann. Sci. Ecole Norm. Sup. 29 (1912), 431-485.

[10] I. M. Gelfand and S. V. Fomin, Calculus of Variations, Prentice-Hall, Englewood Cliffs, N.J., 1963.

[11] L. K. Jackson, Boundary value problems for ordinary differential equations, in MAA studies in Mathematics, Vol. 14 (J. Hale, Ed.), 1977.

[12] L. K. Jackson and P. K. Palamides, An existence theorem for a nonlinear two point boundary value problem, to appear.

[13] H. B. Keller, Existence theory for two-point boundary value problems, Bull. Amer. Math. Soc. 72 (1966), 728-731.

ASYMPTOTICALLY NONRESONANT NONLINEAR EQUATIONS*

Ronald I Becker**

Department of Mathematics, University of Cape Town
Rondebosch, South Africa

1. INTRODUCTION

We consider in this paper existence theory for equations of the form

$$(L - Q(x))x = g(x) \tag{1.1}$$

where L is a linear operator from a Banach space B to a Banach space H Existence is studied under the condition that L^{-1} (or more generally $(L - \lambda I)^{-1}$ for some real λ) is compact; $Q' = Q(x)$ lies, for each sufficiently "large" x , in a closed set in the weak operator topology such that $(L-Q')x = 0$ has unique solution $x = 0$; and $g(x)$ is sub-linear in the sense that

$$\lim_{\|x\|_H \to \infty} \inf \frac{\|g(x)\|_H}{\|x\|_H} = 0 . \tag{1.2}$$

We obtain, in the case L and Q are self-adjoint, generalizations to abstract equations of the results for differential equations of the type that assume

$$\lambda_n < \alpha \leqslant Q(x) \leqslant \beta < \lambda_{n+1}$$

where λ_n and λ_{n+1} are successive eigenvalues of L . From the abstract results, we deduce concrete theorems for differential operators, the main task in applying the abstract result being to show that certain conditions on g given by describing pointwise limits of g in function space, actually imply norm limits of the type (1.2) for g . No assumption need be made that L be semi-bounded here. The results for the case

$$Q(x) \leqslant \beta < \lambda_1$$

where λ_1 is the first eigenvalue of L in the case where L is semi-bounded below, require also that $Q(x) > \beta$ which is undesirable. To cope

* Supported in part by CSIR, Pretoria, South Africa.

** Present address: Department of Mathematics, University of Tennessee, Knoxville, Tennessee.

with this, and some other defects, it is desirable to weaken the condition on g to the condition

$$\limsup_{\|x\|_H \to \infty} \frac{(x,g(x))_H}{\|x\|_H^2} = 0 \tag{1.3}$$

(H being here a Hilbert space and $(.,.)_H$ the inner product). Existence proofs in this case require that L be "elliptic" and that a strengthened form of uniqueness hold. With these assumptions similar abstract theorems may be proved, and this will be done in another paper.

The concrete applications we obtain generalize several known results. The results of Mawhin and Ward [8], in which the α and β above are functions differing from λ_n, λ_{n+1} on sets of positive measure, are generalized to a large class of operators, not necessarily semi-bounded, irrespective of the weak solution mechanism used there (see their remark on p.683) but using compactness properties of the resolvent. Further, Mawhin and Ward [9] theorem 2 improve a result of Amarel and Pera [1], and prove existence for second-order periodic equations.

$$x'' + g(t,x) = 0$$

under the conditions that if $\gamma(t) = \liminf_{|x| \to \infty} x^{-1} g(t,x)$ and $\Gamma(t) = \limsup_{|x| \to \infty} x^{-1} g(t,x)$ we have $\int_0^{2\pi} \gamma(t)dt \geqslant 0$, $\int_0^{2\pi} \max(\gamma(t),0)dt > 0$, $\Gamma(t) \leqslant 1$ with inequality on a set of positive measure. Note that 0 and 1 are the first two eigenvalues of $-x''$. (A term of the form cx' is also included in their equation, and this can also be handled by our technique). We generalize this here to cover fairly arbitrary semi-bounded self-adjoint differential operators with compact resolvent, and we also give conditions of the same type between <u>any</u> two successive eigenvalues. In particular, the results also apply to elliptic partial differential equations. The results of Mawhin and Ward [9] stated in their theorem 1 place a nonlinearity of the form $F(x)x'$ in the equation. This requires results using condition (1.3), and generalizations of this will be dealt with elsewhere.

The technique of proof involves multiplying (1.1) by L^{-1} and using Leray-Schauder degree theory. The uniqueness hypotheses mentioned above imply a priori bounds for solutions. This argument is in the spirit of Opial [10]. The introduction of weakly compact sets in the weak operator topology seems natural in the context of the problems considered and has

$$\liminf_{\|x\|_H \to \infty} \frac{\|g(x)\|_H}{\|x\|_H} = 0 . \qquad (2.5)$$

Then the equation

$$(I - PQ(x))x = Pg(x) \qquad (2.6)$$

has a solution in B .

3. PERTURBATIONS OF SELF-ADJOINT OPERATORS

In this section, the operator L and Q in (1.1) will be assumed self-adjoint, L having compact resolvent. The main theorem proves existence under (asymptotic) uniqueness conditions on $(L-Q)x = 0$ given in Lemma 2.

Throughout this section, L will denote a self-adjoint operator having compact resolvent on a Hilbert space H . Then L has discrete spectrum $\{\lambda_i\}$ (where λ_i is repeated according to multiplicity) with no finite limit point, each eigenvalue beig of finite multiplicity. We denote the corresponding normalized eigenfunctions by $\{\phi_i\}$. We assume $\lambda_i \le \lambda_{i+1}$ for all i . We denote by E_n the span of the eigenvectors corresponding to eigenvalues $\lambda \le \lambda_n$, and by F_n the span of these eigenvectors corresponding to $\lambda \ge \lambda_n$. We will also denote the span of all eigenvectors belonging to eigenvalue λ by span $\{\lambda\}$.

LEMMA 1

Let Q be a symmetric operator whose domain contains $\mathrm{dom}(L)$, and let $\lambda_n < \lambda_{n+1}$.

Suppose that one of the following holds :

(a) $Q_1 \le Q \le Q_2$ and

(i) $(\lambda_{n+1}I - Q) \ge 0$, and is > 0 on span $\{\lambda_{n+1}\}$; $(Q_1 - \lambda_n I) \ge 0$

(ii) $(Q - \lambda_n I)x = 0$ has no non-zero solution $x \in \mathrm{span}\{\lambda_n\}$.

(b) $Q_1 \le Q \le Q_2$, L is semi-bounded below and

(i) $(\lambda_{n+1}I - Q_2) \ge 0$, and is > 0 on span $\{\lambda_{n+1}\}$; the matrix $m_{ij} = ((Q_1 - \lambda_n I)\phi_i, \phi_j)$ $(1 \le i,j \le n)$ is non-negative.

(ii) $(Q - \lambda_n I)x = 0$ has no non-zero solution $x \in$ span $\{\lambda_n\}$.

(c) $Q \le Q_2$, L is semi-bounded below and if λ_1 is the smallest eigenvalue of L , then been used in Becker [2], [3] and [4], the first two of which study differential equations on Hilbert spaces. The papers of Mawhin and Ward [8], [9] also use similar techniques.

For lack of space, we give statements in the following sections only of some of the abstract results.

2. NONLINEAR PERTURBATIONS OF NON-SELF-ADJOINT OPERATORS

Let B,H denote Banach spaces satisfying

$$B \subseteq H$$

where the inclusion is with continuous injection. Let $L(S_1,S_2)$ denote the bounded linear operators from space S_1 to space S_2. The weak operator topology in $L(S,S)$ will be denoted by $WOT(S)$ (see Dunford and Schwartz [6], VI.1). Note that $L(H,H) \subseteq L(B,H)$. If $T \in L(H,B)$ we use $\|T\|_{H\to B}$ for its norm or just $\|T\|_B$ if H is understood.

The following hypotheses will be made on our operators :

(A) Let $P,R \in L(H,B)$ be compact.

(B) Let $Q(x)\colon B \to L(H,H)$ be continuous from the strong topology in B to $WOT(H)$.

(C) Let $g(x)\colon B \to H$ be continuous from the strong topology in B to the weak topology in H.

A contracting family for a set $\mathcal{S}$ is a family $\{\mathcal{S}_\varepsilon\}_{0<\varepsilon\leq 1}$ satisfying

(i) $\varepsilon < \varepsilon'$ implies $\mathcal{S}_\varepsilon \subset \mathcal{S}_{\varepsilon'}$ properly

(ii) $\bigcap_{0<\varepsilon\leq 1} \mathcal{S}_\varepsilon = \mathcal{S}$.

THEOREM 1

Let (A), (B) and (C) hold. Let $\mathcal{S}$ be a set with a convex contracting family $\{\mathcal{S}_\varepsilon\}$. Suppose that

(i) $\mathcal{S}_\varepsilon, \mathcal{S}$ are sequentially compact in $WOT(H)$.

(ii) $Q \in \mathcal{S}$ implies $(I - PQ)$ invertible

(iii) $Q(x)$ maps bounded sets in B into bounded sets in $L(H,H)$. Given ε there is a $Q'(x)$ satisfying (B) such that $Q'(x) \in \mathcal{S}_\varepsilon$ for each x and

$$\liminf_{\|x\|_B\to\infty} \frac{\|(Q(x) - Q'(x))x\|_H}{\|x\|_H} = 0 \qquad (2.4)$$

(iv) g maps bounded sets in B into bounded sets in H and

(i) $(\lambda_1 I - Q_2) \geq 0$

(ii) $(Q - \lambda_1 I)x = 0$ has no non-zero solution $x \in \text{span}\ \{\lambda_1\}$.

Then $(L-Q)x = 0$ has unique solution $x = 0$.

We now state a result on the weak sequential compactness of the operator sets we will use.

LEMMA 2

Let $P_1 \leq P_2$ be bounded self-adjoint operators on H. Then the set $\mathcal{S} = \{Q \mid P_1 \leq Q \leq P_2$, Q bounded self-adjoint is sequentially compact in WOT(H).

THEOREM 2

Let H be a Hilbert space. Let (A), (B) and (C) hold. Suppose that $Q(x)$ is self-adjoint for each $x \in H$. Let $\lambda_n < \lambda_{n+1}$.

(i) Let Q_1 and Q_2 be such that one of hypotheses (a), (b) or (c) of lemma 1 holds for all Q such that $Q_1 \leq Q \leq Q_2$.

(ii) $Q(x)$ maps bounded sets in H into bounded sets in $L(H,H)$. Given $\varepsilon > 0$ there is a $Q'(x)$ satisfying (B) such that

$$Q_1 - \varepsilon I \leq Q'(x) \leq Q_2 + \varepsilon I \qquad (\text{all } x) \tag{3.2}$$

and

$$\liminf_{\|x\|_H \to \infty} \frac{\|(Q(x) - Q'(x))x\|_H}{\|x\|_H} = 0 . \tag{3.3}$$

(iii) $g(x)$ maps bounded sets in H into bounded sets in H, and

$$\liminf_{\|x\|_H \to \infty} \frac{\|g(x)\|_H}{\|x\|_H} = 0 . \tag{3.4}$$

Then the equation

$$(L - Q(x))x = g(x) \tag{3.5}$$

has a solution in B.

REMARK

Let us strengthen the hypothesis (ii) on Q in theorem 2 to read

$$\lambda_n I < Q_1 \leqslant Q(x) \leqslant Q_2 \leqslant \lambda_{n+1} I \qquad (x \in H) \tag{3.9}$$

or

$$\lambda_1 I > Q_2 \geq Q(x) \geq Q_1 \qquad (x \in H) \tag{3.10}$$

In the proof of sets $\mathcal{S}_\varepsilon$ may be dispensed with and (3.3) is satisfied automatically. Hence we have

COROLLARY 2.1

Let H be a Hilbert space and let (A), (B), (C) and let part (iii) of theorem 2 hold. Let (3.9) or (3.10) hold. Then

$$(L - Q)x = g(x)$$

has a solution.

REFERENCES

[1] L Amaral, M P Pera, A note on periodic solutions at resonance, Boll. Un. Mat. Ital., 18 (1981) 107-117.

[2] R I Becker, Periodic solutions of semilinear functional differential equations in a Hilbert space, In Functional Equations and Bifurcations, Springer Verlag, Lecture Notes LNM 799 (1980), 23-44.

[3] R I Becker, Periodic solutions of semilinear equations of evolution of compact type, J. Math. Anal. Appl., 82 (1981) 33-48.

[4] R I Becker, Existence of solutions of Hammerstein equations of compact type, Math. Proc. Camb. Phil. Soc. 89 (1981), 149-158.

[5] D G de Figueiredo, The Dirichlet problem for nonlinear elliptic equations: A Hilbert space approach, Springer Verlag Lecture Notes LNM 446 (1975), 144-165.

[6] N Dunford and J T Schwartz, Linear Operators, Vol I, Interscience, Wiley, New York, 1964.

[7] N G Lloyd, Degree Theory, Cambridge University Press, Cambridge, 1978.

[8] J Mawhin and J R Ward, Jr., Nonresonance and existence for nonlinear elliptic boundary value problems, Nonlinear Analysis, 6(1981), 677-684.

[9] J Mawhin and J R Ward, Jr., Nonuniform nonresonance conditions at the two first eigenvalues for periodic solutions of forced Lienard and Duffing equations, To appear, Rocky Mountain J. Math.

[10] Z Opial, Linear problems for systems of nonlinear differential equations, J. Differential Equations 3 (1967), 580-594.

BOUNDEDNESS OF OSCILLATORY SOLUTIONS OF FOURTH ORDER LINEAR ORDINARY HOMOGENEOUS DIFFERENTIAL EQUATIONS

Abdelali Benharbit

The Pennsylvania State University
York, Pennsylvania

ABSTRACT

This paper examines the asymptotic behavior of oscillatory solutions of fourth order linear ordinary homogeneous differential equations. Given are conditions which insure that all oscillatory solutions are bounded.

Consider the differential equation

$$y^{IV} = p(x)y'' + q(x)y' + r(x)y \qquad (1)$$

where p, q, r ε C $[a, \infty)$ for some real number a. A nontrivial solution of (1) is said to be oscillatory if its set of zeros is unbounded above. Nontrivial solutions which are not oscillatory are said to be nonoscillatory. The problem of the asymptotic behavior of solutions of (1) has been investigated by many authors [1-5]. The results and proofs are motivated by [2] and [3]. S. Ahmad [3] gave a very fine classification for the space of all solutions of (1) in the case of bounded coefficients.

First we will prove the following theorem.

Theorem 1 Let p, p', q, r, r' ε C $[a, \infty]$, $r(x) \geq d > 0$, $r'(x) \leq 0$ $p'(x) - 2q(x) < 0$ for all x ε $[a, \infty)$. Then all oscillatory solutions of (1), if they are any, are bounded.

Proof

Consider the function

$$F[y(x)] = r(x)\, y^2(x) + p(x)\, y'^2(x) + y''^2(x) - 2y'(x)\, y'''(x)$$

It can be verified through differentiation that

$$F[y(x)] = F[y(a)] + \int_a^x r'(t)y^2(t)dt + \int_a^x [p'(t) - 2q(t)]y'^2(t)\, dt.$$

Since $r'(x) \leq 0$ and $p'(x)-2q(x) < 0$, it follows that F[y(x)] is strictly decreasing. Now suppose that y(x) is an oscillatory solution of (1), and let x_1, x_2 be zeros of y(x).

If $\max_{x\varepsilon[x_1, x_2]} y^2(x) = y^2(\bar{x})$, $\bar{x} \varepsilon [x_1, x_2]$, then $y'(\bar{x}) = 0$.

If $\bar{x} = x_1$ then $y^2(\bar{x}) = y^2(x_1)$.

If $\bar{x} > x_1$, then we obtain from the fact that $F[y(\bar{x})] < F[y(x_1)]$ and

$$y'(\bar{x}) = 0,$$

$$y''^2(\bar{x}) + r(\bar{x})\ y^2(\bar{x}) < F[y(x_1)].$$

Hence

$$r(\bar{x})\ y^2(\bar{x}) < F[y(x_1)]$$

Thus

$$y^2(\bar{x}) < \frac{F[y(x_1)]}{d}$$

Therefore, in both cases

$$y^2(\bar{x}) \leq y^2(x_1) + \frac{F[y(x_1)]}{d} .$$

Hence $y(x)$ is bounded.

Theorem 2 Let $p, q, r, p', r' \varepsilon C[a, \infty)$, $r(x) > 0$, $r'(x) \geq 0$, $p'(x) - 2\,q(x) \leq 0$ for all $x \varepsilon [a, \infty)$. Then all oscillatory solutions of (1), if there are any, are bounded.

Proof

Consider the function

$$R\ [y(x)] = r(x)\ y^2(x) + p(x)\ y'^2(x) + y''^2(x) - 2y'(x)\ y'''(x) - \int_{x_1}^{x} r'(t)y^2(t)dt$$

where $y(x)$ is an oscillatory solution of (1) and x_1 is a zero of $y(x)$. It can be verified through differentiation that

$$R[y(x)] = R[y(x_1)] + \int_{x_1}^{x} [p'(t) - 2\,q(t)]\ y'^2(t)\ dt$$

If x_2 is a zero of $y(x)$ such that $x_2 > x_1$, then $\max_{x\varepsilon[x_1,x_2]} y^2(x) = y^2(\bar{x})$ where $\bar{x}\varepsilon\ [x_1, x_2]$, and $y'(\bar{x}) = 0$.

If $\bar{x} = x_1$, then $y^2(\bar{x}) = y^2(x_1)$.

If $\bar{x} > x_1$, then $r(\bar{x})\ y^2(\bar{x}) - \int_{x_1}^{\bar{x}} r'(t)\ y^2(t)\ dt \leq R[y(x_1)]$ since $R[y(x)]$ is strictly decreasing and $y'(\bar{x}) = 0$.

But $r' \geq 0$ implies that

$$\int_{x_1}^{\bar{x}} r'(t)\, y^2(t)\, dt \leq y^2(\bar{x})[r(\bar{x}) - r(x_1)] .$$

Hence

$$r(x_1)\, y^2(\bar{x}) \leq R[y(x_1)]$$

or $y^2(\bar{x}) \leq \dfrac{R[y(x_1)]}{r(x_1)}$.

Therefore, in both cases

$$y^2(\bar{x}) \leq y^2(x_1) + \frac{R[y(x_1)]}{r(x_1)} .$$

Hence, it follows that every oscillatory solution is bounded. Combining Theorem 1 and Theorem 2, we obtain the following corollary:

Corollary Let $p, p', q, r, r' \in C\,[a, \infty)$, $r(x) \geq d > 0$, $p'(x) - 2\,q(x) < 0$. If $r'(x)$ does not change sign in $[a, \infty)$, then every oscillatory solution is bounded.

REFERENCES

[1] W. Leighton, Z. Nehari, On the oscillation of solutions of self-adjoint linear differential equations of the fourth order, Trans. Amer. Math. Soc., 89 (1958), 325-377.

[2] S. P. Hasting, A. C. Lazer, On the asymptotic behaviour of solutions of the differential equation $y^{(4)} = p(t)\, y$, Czechoslovak Math J., 18 (93) (1968), 224-229.

[3] S. Ahmad, Asymptotic properties of linear fourth order differential equations, Proc. Amer. Math. Soc., 59 (1976), 45-51.

[4] M. S. Keener, On oscillatory solutions of certain fourth order linear differential equations, Siam J. Math. Anal., 3 (1972), 599-605.

[5] W. E. Taylor, On the solutions of certain non self-adjoint differential equations of fourth order, Atti Accad. naz. Lincei, VIII. Ser., Rend., Cl. Sci. fis. mat. natur., 58 (1975), 851-860.

A NEIGHBORHOOD OF THE HOPF SINGULARITY*

S. R. Bernfeld

Department of Mathematics
The University of Texas at Arlington
Arlington, Texas

1. INTRODUCTION

We give an exposition on the Hopf singularity applied to ordinary differential equations. In particular we look at perturbation of a Hopf differential equation in order to find the maximum number of periodic orbits of the perturbed equation. We discuss the relationship between the number of periodic solutions and the strength of the exchange of stability of the equilibrium as we perturb the Hopf differential system. Much of the discussion in this note come from [2] and [5].

2. PRELIMINARIES

Consider the space $f \in C^{\infty}[R^n,R^n]$ with $f(0) = 0$. For each f consider the differential equation generated by f

(1) $$\dot{p} = f(p), \ p \in R^n.$$

We say f_0 is a Hopf singularity if $J\ f_0(0)$ (Jacobian of f_0 at 0) has 2 purely imaginary eigenvalues, say $\pm i$, and the remaining eigenvalues $\{\lambda_j\}_{j=1}^{n-2}$ satisfy $\lambda_j \neq mi$, $m = 0, \pm 1, \pm 2 \ldots$.

We analyze the Hopf singularity f_0 by analyzing the periodic trajectories of (1) near the origin for those f near f_0 (using the standard C^{∞} topology). We find that the following properties of a neighborhood of f_0 useful:

Property (a): Let k be any finite positive integer.

(i) there exists a neighborhood N of f_0 and a neighborhood U of the origin and a number $p_2 > 0$ such that for every $f \in N$ there are at most k nontrivial periodic orbits of (1) lying in U whose period

* Research partially supported by U. S. Army Research Grant #DAAG 29-80-C0060

is in $[2\pi - \delta_1, 2\pi + \delta_1]$.

(ii) for each integer j, $0 \leq j \leq k$, for each neighborhood N of f_0, $N \subseteq \mathcal{N}$, for each neighborhood U_1 of the origin, $U_1 \subseteq U$, and for each $\delta_2 \in (0,\delta_1)$ there exists $f \in N$ such that (1) has exactly j nontrivial periodic orbits lying in U_1 with period in $[2\pi - \delta_2, 2\pi + \delta_2]$.

In contrast to (a) we consider the case $k = \infty$:

(b) for any neighborhood N of f_0, for any integer $j \geq 0$, for any neighborhood U_1 of the origin, $U_1 \subset U$, and for any $\delta \in (0,\delta_1)$ there exists $f \in N$ such that (1) has j nontrivial periodic orbits lying in U_1 whose period is in $[2\pi - \delta, 2\pi + \delta]$.

Essentially property (a) says that a neighborhood of f_0 looks like a pie with k slices with f_0 at the "center." Each slice consists of those f whose generated differential equation has the same number of periodic orbits. The slices have a nonempty interior but are in general neither open nor closed. Property (b) implies that there are an infinite number of slices with f_0 at the "center."

In order to determine the value k for a particular Hopf singularity f_0 we introduce a stability notion. We say the origin of (f_0) ((f_0) designates the differential equation $\dot{p} = f_0(p)$ generated by f_0) is $2k + 1$ - asymptotically stable (completely unstable) if the origin of (f_0) is asymptotically stable (completely unstable) for those f for which $f - f_0 = o(\|p\|^{2k+1})$.

3. NEIGHBORHOOD OF HOPF SINGULARITY IN R^2

In R^2 Andronov et. al. [1] and Bernfeld and Salvadori [3] obtained the following characterization (we only state the result for stability but a similar result holds for instability).

Theorem 1. In R^2 a neighborhood of (f_0) has property (a) if and only if the origin of (f_0) is h-asymptotically stable (and h is necessarily odd) where $k = \frac{h-1}{2}$. Otherwise a neighborhood of (f_0) has property (b) if and only if the origin of (f_0) is not h-asymptotically stable for any h.

The determination of h can be found using a classical procedure of Poincaré [8] and involves the solution of an algebraic system of equations arising from the construction of a Lyapunov function.

In the case of a vague attractor, $h = 3$, Theorem 1 can be strengthened to show that the pie consists of two slices with the boundary between them contained completely within one of the slices.

The h-asymptotic stability of the origin of (f_0) is equivalent to the existence of a displacement function (the Poincaré map at the origin) $V(f,c)$ having the form

$$V(f_0,c) = g\ c^h + (c^h)$$

$g < 0$, c is a scalar. It follows that for each $f \in N$ (a fixed neighborhood at N) that

$$V(f,c) = a_0c + a_1c^2 + \ldots + g_1c^h + o(c^h).$$

The roots of $V(f,\cdot) = 0$ are those values of c for which a period trajectory (near the origin) intersects the x axis. Thus for each positive root c_1 such that $V(f,c_1) = 0$ there exists $c_2 < 0$ such that $V(f,c_2) = 0$. Thus we are led to the roots of an equation of the form (which we refer to as the bifurcation equation)

$$D(f,c) \equiv b_0 + b_1c + \ldots g_1c^{\frac{h-1}{2}} = 0.$$

Consequently in R^2 we have shown that a neighborhood of a Hopf singularity can be characterized by the unfolding of the polynomial $V(f_0,c)$

$$V(f_0,c) = gc^h$$

where $g < 0$ $(g > 0)$ is equivalent to the origin of (f_0) being h-asymptotically stable (h-completely unstable). Elementary algebraic techniques yield the values of g and h for a given (f_0). For any $f \in N$ (the neighborhood of f_0 given in property (a)(i)) we may easily construct the polynomial whose positive roots will determine the periodic orbits of (1). Thus when analyzing the periodic solutions of (f) it is sufficient to consider the c^h jet of f.

4. NEIGHBORHOOD OF HOPF SINGULARITY IN R^n

In R^n the characterization of a neighborhood of a Hopf singularity is much more difficult. Some progress has been achieved in [5] but there still remain open questions. We elaborate on some of these now.

Rewrite (1) as

$$\begin{aligned} \dot{x} &= \alpha x - \beta y + X(x,y,z,f), \\ \dot{y} &= \alpha y + \beta x + Y(x,y,z,f), \qquad (4.1) \\ \dot{z} &= Az + Z(x,y,z,f), \end{aligned}$$

and in the case $f = f_0$ then $\alpha(f_0) = 0$, $\beta(f_0) = 1$ $A(f_0) \equiv A_0$,

$X(x,y,z,f_0) = X_0(x,y,z)$, $Y(x,y,z,f_0) = Y_0(x,y,z)$, $Z(x,y,z,f_0) = Z_0(x,y,z)$. Recall that the eigenvalues of A_0 satisfy the non-resonance condition, that is its eigenvalues $\{\lambda_j\}_{j=1}^{n-2}$ satisfy $\lambda_j \neq m_i$ $m = 0, \pm 1, \pm 2 \ldots$. Moreover $Z,X,Y \in C^\infty$ for (x,y,z) in a neighborhood of the origin.

We first show that we can construct a scalar bifurcation equation, that is a polynomial, whose roots determine the periodic solutions of (4.1) with period close to 2π. To do this we reduce the problem to a related two dimensional system whose bifurcation properties are the same as those of the original problem. Such a technique has often been used when the existence of a center manifold is known. Since an invariant manifold may not exist we adopt a procedure given in [5]. Consider an $(n-2)$ dimensional polynomial of some degree h, $h > 0$, given by

$$\phi^{(h)}(x,y) = \phi_1(x,y) + \ldots \phi_h(x,y),$$

where $\phi_j(x,y)$ is homogeneous of degree j. Because of the assumptions on A_0 we then are able to determine $\phi_1 \ldots \phi_h$ in order to obtain along the solutions of (4.1) for $f = f_0$

$$\left[\frac{d}{dt}(z-\phi^{(h)}(x,y))\right]_{z=\phi^{(h)}(x,y)} = (x^2+y^2)^{h/2} .$$

Now for any $h > 0$ define the following two dimensional system

$$(S_h) \qquad \begin{aligned} \dot{x} &= -y + X_0(x,y,\phi^{(h)}(x,y)) \\ \dot{y} &= x + Y_0(x,y,\phi^{(h)}(x,y)), \end{aligned}$$

where $X_0(x,y,z) \equiv X(x,y,z,f_0)$, $Y_0(x,y,z) \equiv Y(x,y,z,f_0)$. We then have the following extension of Theorem 1 to R^n.

<u>Theorem 2</u>. If there exists $h > 0$ (and then h must be odd and > 1) such that the origin of (S_h) is h-asymptotically stable or h-completely unstable then property (a) holds with $k = \frac{h-1}{2}$. If for every $h > 0$ the origin of (S_h) is neither h-asymptotically stable nor h-completely unstable, then property (b) holds.

Once again we are able to construct a displacement function for (4.1) $V(c,f)$ having the property that

$$V(c,f_0) = g\, c^h + o(c^h),$$

whenever (S_h) is h-asymptotically stable $(g < 0)$ or h-completely unstable $(g > 0)$. Then for each f near f_0 the periodic orbits of (4.1) can be determined by the positive roots of a polynomial of degree h.

This polynomial can be completely determined by linear algebraic techniques.

One of the open question is to completely describe the invariant compact sets in a neighborhood of a Hopf singularity. Assuming a strong nonresonance condition on the eigenvalues of A_0; that is, $\frac{\lambda_j}{\lambda_k} \neq \frac{p}{q}$, $k, j = 1,\dots,n$ our previous analysis indicates that there exists a neighborhood of the origin containing up to k_i periodic orbits whose period are close to $\frac{2\pi}{\lambda_i}$. Moreover, there exist j dimensional tori, $j = 1 \dots \frac{n}{2}$ (assuming all the eigenvalues of A_0 are purely imaginary). It remains to do an analysis of the number of tori and to relate this to the h-asymptotic stability of the origin of the unperturbed system.

5. EXCHANGE OF STABILITY AND NUMBER OF PERIODIC ORBITS

We now consider a one parameter perturbation of the Hopf differential system in R^n given by

$$(5)_\mu \qquad \begin{aligned} \dot{x} &= \alpha(\mu)x - \beta(\mu)y + X(\mu,x,y,z), \\ \dot{y} &= \alpha(\mu)y + \beta(\mu)x + Y(\mu,x,y,z), \\ \dot{z} &= A_\mu z + Z(\mu,x,y,z), \end{aligned}$$

where $\alpha(0) = 0$, $\beta(0) = 1$. The conditions on X,Y,Z and A_0 are precisely the same as those given in (4.1). We provide a relationship between the maximum number of bifurcating periodic orbits of $(5)_\mu$ in terms of the strength of attraction of the origin of $(5)_0$ coupled with the rate at which the eigenvalues $\alpha(\mu) \pm i\,\beta(\mu)$ crosses the imaginary axis at $\mu = 0$. The results of this section extend the work of Flockerzi [6] and Kielhofer [7] and may be found in [2]. In a future paper with L. Salvadori [4] further results will be given.

By using the same reduction principle as that in the previous section we obtain the two dimensional system

$$(S_h^\mu) \qquad \begin{aligned} \dot{x} &= \alpha(\mu)x - \beta(\mu)y + X(\mu,x,y,\phi^{(h)}(\mu,x,y)) \\ \dot{y} &= \alpha(\mu)y + \beta(\mu)x + Y(\mu,x,y,\phi^{(h)}(\mu,x,y)) \end{aligned}$$

Let us assume the following condition on the eigenvalues $\alpha(\mu) \pm i\,\beta(\mu)$ as they cross the imaginary axis at $\mu = 0$. Namely there exists an integer $s \geq 1$ such that

$$(GT) \qquad \alpha^{i}(0) = 0 \quad i = 0, \dots s-1, \ \alpha^{(s)}(0) > 0.$$

When $S = 1$ we say α satisfies the transversality condition. We then obtain the following result concerning the maximum number of bifurcating orbits of (S_μ).

Theorem 3. Assume that there exists an $h > 1$ such that the origin of (S_h^0) is h-asymptotically stable and that $\alpha(\mu)$ satisfies (GT). Let $k = \frac{h-1}{2}$ and define

$$\ell = \min(s,k).$$

Then for each $\mu > 0$ and sufficiently small there exists at least one and at most ℓ bifurcating periodic orbits whose period is close to 2π.

Here we may think of ℓ as a "measure" of the exchange of stability of the origin which in turn provides an indication of the complexity of the neighborhood of the origin of (5_μ) (with regard to periodic orbits with period near 2π).

In the case in which $\alpha(\mu)$ is analytic and there exists no s such that (GT) is satisfied then $\alpha(\mu) \equiv 0$ in a neighborhood of $\mu = 0$. As before we shall assume the origin of (S_h^0) is h-asymptotically stable and that the origin of (S_h^μ) is completely unstable for $\mu > 0$. Because $\alpha(\mu) \equiv 0$ this instability is due to higher order terms. Thus we assume for each $\mu > 0$ there exists $\bar{h}(\mu)$ such that the origin of (S_h^μ) is $\bar{h}(\mu)$-completely unstable. The special case $\bar{h}(\mu) \equiv 1$ (or equivalently $\alpha(\mu) > 0$ for $\mu > 0$) is the usual assumption in Hopf bifurcation.

Under the assumptions above on the h-asymptotic stability of (S_h^0) and $\bar{h}(\mu)$-complete instability of (S_h^μ) we have the following:

Lemma 1. $\bar{h}(\mu) \equiv \bar{h}$ for all μ small; that is $\bar{h}(\mu)$ is a fixed integer.

Lemma 2. $h > \bar{h}$.

Defining

$$m = \frac{h-\bar{h}}{2},$$

we obtain the following result concerning the exchange of stability of the origin and the maximum number of bifurcating orbits of (5_μ).

Theorem 3. There exists an integer j, $1 \leq j \leq m$, such that the system (5_μ) for μ small and positive, has exactly j periodic orbits in a small neighborhood of the origin with period close to 2π.

REFERENCES

[1] A. Andronov, E. Leontovich, I. Gordon, and A. Maier, Theory of bifurcations of dynamical systems in the plane, Israel Program of Scientific Translations, Jerusalem, 1971.

[2] S. Bernfeld, Generalized transversality, exchange of stability, and Hopf bifurcation, Differential Equations and Applications in Ecology, Epidemics, and Population Problems, Academic Press (1982) p. 279-287.

[3] S. Bernfeld, and L. Salvadori, Generalized Hopf bifurcation and h-asymptotic stability, J. Nonl. Anal. T.M.A. 4 (1980), p. 109-117.

[4] S. Bernfeld and L. Salvadori, Generalized transversality and Hopf bifurcation in R^n, to appear.

[5] S. Bernfeld, P. Negrini, and L. Salvadori, Quasi-invariant manifolds, stability, and generalized Hopf bifurcation, Annali di Mat. Pura ed. Appl. 80 (1982), 105-119.

[6] D. Flockerzi, Bifurcation of periodic solutions from an equilibrium point, Dissertation, Wurzberg, Germany (1979).

[7] H. Kielhöfer, Generalized Hopf bifurcation in Hilbert Space, Math. Meth. in Appl. Sci 1 (1979), p. 498-513.

[8] G. Sansone and R. Conti, Nonlinear Differential Equations, Macmillan, New York, (1964).

SECOND ORDER DIFFERENTIAL SYSTEMS WITH GENERALIZED NONLINEAR BOUNDARY CONDITIONS *

S. R. Bernfeld

Department of Mathematics
The University of Texas at Arlington
Arlington, Texas

Topological methods have proven to be quite useful in obtaining existence results for nonlinear boundary value problems of the form

(E) $$x'' = f(t,x,x')$$

(G) $$g(x(0),x'(0)) = 0 \qquad h(x(0),x'(0),x(1),x'(1)) = 0,$$

where $f\colon I \times R^n \times R^n \to R^n$, $g\colon R^{2n} \to R^n$, and $h\colon R^{4n} \to R^n$ are all continuous, (here $I = [0,1]$), (See [4], [5], [6] for the case $n = 1$ and [2] for the vector-valued case). The prevalent approaches for the case $n = 1$ have been the topological method of Wazewski [6], Kneser-continuum theorem [3], and properties of the consequent mapping [5] i.e. the map of ω onto its boundary, $\partial\omega$, under the action of solutions of (E) (ω is a subset of the domain of f).

In this note we attempt to solve BVP (E), (G) in R^n by combining some elements of degree theory with the Wazewski method and theorem. In particular we prescribe an initial set of points in R^{2n}, call it T, which is assumed to be a continuum. We consider the solution funnel of (E) thru T on $[0,1]$. Kneser's theorem tells us that the funnel is a continuum as well as each of its cross sections. Conditions on f and ω are given which imply that the Leray-Schauder degree of a map related to the solution funnel at $t = 1$ is non zero. For the case $n = 1$, [4], these properties of the solution funnel allow us to conclude the existence of a solution in the funnel, $z(t)$, such that $h(z(0), z'(0), z(1), z'(1)) = 0$. An intermediate value theorem is used for this purpose. One generalization of [4], for the case R^n, has been done in [2] in which our initial set, T, is a connected subset of a

*Research partially supported by U. S. Army Research Grant #DAAG 29-80-C-0060.

parallelopiped in R^n and we studied the solution funnel through each face separately. In [2] the cross-section at $t = 1$ does not allow for a natural extension of an intermediate-value type theorem to R^n so we needed to impose more restrictive conditions on h, such as h is independent of $x'(1)$.

We now attempt to remove some of the restrictions on h by using the complete parallelopiped for the initial set of the solution funnel. By using a degree theory argument we can employ a "generalized" intermediate value theorem in R^n. Of course, we have to pay a price in order to use degree theory; namely we need the projection of the cross section in R^n at $t = 1$ to be convex.

PRELIMINARIES

Consider a point $P = (\tau,\xi,\eta)' \in \Omega = [0,\infty) \times R^{2n}$ and let $\chi(P)$ be the family of all solutions of (E) emanating from the point P. Let $\chi(P)(t)$ be the cross section of $\chi(P)$ at time t. For a solution $x \in \chi(P)$, we denote by

$$G(x|\text{Dom } x) = \{(t,x(t),x'(t)):\ t \in \text{Dom } x\}$$

its graph and by $G(x|I)$ the graph of its restriction $x|I$, $I \subseteq \text{Dom } x$. For a subset $K \subseteq \Omega$ define

$$\begin{aligned} \chi(K) &= \cup\{\chi(P):\ P \in K\}, \\ \chi(K)(t) &= \cup\{\chi(P)(t):\ P \in K\}, \quad \text{and} \\ \chi(I,K) &= \{x|I:\ x \in \chi(K)\}. \end{aligned}$$

Consider a subset ω of Ω such that $\omega^0 \neq \phi$ and $\Omega - \bar{\omega} \neq \phi$. A point $P = (\tau,\xi,\eta) \in \Omega \cap \partial\omega$ is a point of semi-egress of ω, denoted by ω^s, if and only if for each solution $x \in \chi(P)$ there exists an $\varepsilon > 0$ such that $G(x|[\tau-\varepsilon,\tau)) \leq \omega^0$ and $G(x|[\tau,t_2]) \leq \partial\omega$ for some $t_2 \geq \tau$. If moreover for each solution $x \in \chi(P)$ there exists an $\varepsilon > 0$ and $t_3 \geq \tau$ such that $G(x[\tau,t_3]) \subseteq \partial\omega$, $G(x|(t_3,t_3+\varepsilon]) \leq \Omega - \bar{\omega}$, then P will be called a point of strict semi-egress denoted by ω^{ss}. A point $P \in \omega^s$ or $P \in \omega^{ss}$ is a point of egress or a point of strict egress, if in the above, we choose $t_2 = \tau$ and $t_3 = t_1 = \tau$. The set of all points of egress or strict egress will be denoted by ω^s or ω^{ss}.

A point $P \in \omega^s$, $P = (\tau,\xi,\eta)$, is a consequent of $P_0 \in \omega$, $P_0 = (t_0,x_0,y_0)$, with respect to the set ω if there exists a solution $x \in \chi(P_0) \cap \chi(P)$ and a point $t_1 \in [t_0,\tau]$ such that $G(x|[t_1,\tau]) \in \partial\omega$

and if $t_0 < t_1$, then $G(x|(t_0,t_1)) \subseteq \omega^0$. We will say such a solution egresses from ω. The set of all consequent points of P_0 will be denoted by $C(P_0)$ and the so defined mapping is known as the consequent mapping. Similarly if $K \in \omega$ then $C(K) = \cup\{C(P_0): P_0 \in K\}$. Under certain conditions, the mapping C is upper semi-continuous; namely, if we define $S(\omega) = \{Q \in \omega: C(Q) \neq \phi\}$ then if $P \in S(\omega)$ and every solution $x \in \chi(P)$ semi-egresses strictly from ω, then C is upper seim-continuous at the point P and the image $C(P)$ is a continuum in $\partial\Omega$. Moreover, if K is a connected subset of ω then $C(K)$ is connected (see [5]).

For any points $x = (x_1 \ldots x_n) \in R^n$ and $y = (y_1, y_2 \ldots y_n) \in R^n$ then $x \leq y \iff x_i \leq y_i$ $i = 1,\ldots,n$, and, in addition, $[x,y]_i$ will denote the element in R^n given by

$$[x,y]_i = (y_1,y_2,\ldots,y_{i-1},x_i,y_{i+1},\ldots,y_n).$$

If S is any subset of $R^n \times R^n$ then $P_{r_1}S$ and $P_{r_2}S$ are respectively the projections of S onto its first n and second n components.

Let $\alpha(t)$ and $\beta(t)$ be two C^1, R^n-valued functions defined on $[0,1]$ such that

$$\alpha(t) < \beta(t) \qquad 0 \leq t \leq 1.$$

Define the set

$$\omega = \{(t,x,y): \ 0 < t \leq 1, \ \alpha(t) \leq x \leq \beta(t), \ y \in R^n\}$$

Let $\omega(\tau) = \{(x,y): (\tau,x,y) \in \omega)$ and for $I_1 \in [0,1]$ let $\omega(I_1) = \cup\{\omega(\tau), \tau \in I)$. Moreover, for arbitrary subsets $J \subseteq [0,1]$ and $\Lambda \subseteq \omega$

$$\Lambda(J) = \{(t,x,y) \in \Lambda: \ t \in J\}$$

will denote the cross-section of Λ over the set I.

Consider the faces of ω,

$$Q_{\alpha_i} = \{(t,x,y) \in \omega: x_i = \alpha_i(t)\} \quad \text{and}$$

$$Q_{\beta_i} = \{(t,x,y) \in \omega: x_i = \beta_i(t)\}$$

as well as their subsets

$$Q'_{\alpha_i} = \{(t,x,y) \in Q_{\alpha_i}: y_i < \alpha'_i(t)\} \quad \text{and}$$

$$Q'_{\beta_i} = \{(t,x,y) \in Q_{\beta_i}: y_i > \beta'_i(t)\}.$$

Let $Q_{\alpha_i}(\tau)$, $Q_{\beta_i}(\tau)$, $Q'_{\alpha_i}(\tau)$, $Q'_{\beta_i}(\tau)$ be the corresponding cross-sections at $t = \tau$.

Define $[\alpha(\tau),\beta(\tau)] = \prod_{i=1}^{n} [\alpha_i(\tau), \beta_i(\tau)]$ $\tau \in [0,1]$, and $[\alpha,\beta] = \cup\{[\alpha(\tau), \beta(\tau)]$ $\tau \in [0,1]\}$ Let $F_{\alpha_i}(t) = \{x \in R^n: x_i = \alpha_i(t)$ and $\alpha_j(t) \leq x_j \leq \beta_j(t), j \neq i\}$ $F_{\beta_i}(t) = \{x \in R^n: x_i = \beta_i(t)$ and $\alpha_j(t) \leq x_j \leq \beta_j(t), j \neq i\}$ with $F_\alpha(t) = \bigcup_{i=1}^{n} F_{\alpha_i}(t)$, $F_\beta(t) = \bigcup_{i=1}^{n} F_{\beta_i}(t)$ $F_\alpha = \cup\{F_\alpha(t): t \in [0,1]\}$ and $F_\beta = \cup\{F_\beta(t): t \in [0,1]\}$.

Let X,Y be two spaces and $I = [0,1]$. Two continuous maps $f,g: X \to Y$ are called homotopic if there exists a continuous map $\phi: X \times I \to Y$ such that $\phi(x,0) = f(x)$ $\phi(x,1) = g(x)$ and for each $x \in X$ $\phi(x,\cdot)$ never vanishes.

We shall use the following consequence of the antipodal mapping theorem (see [1]).

Lemma 1. Consider the set $R(\tau) = [\alpha(\tau),\beta(\tau)]$ for any fixed τ, and let z be the point of symmetry of this set. Let $\hat{F}$ be any continuous mapping from $R(\tau)$ into R^n. Let A be the continuous mapping of $\partial R(\tau)$ onto itself which maps $y \in \partial R(\tau)$ onto the point which is symmetric to y with respect to z. Assume that for each $y \in \partial R(\tau)$, $\hat{F}(y)$ and $\hat{F}(Ay)$ do not have the same direction. Then $\deg(\hat{F},0,R(\tau)) \neq 0$.

RESULTS

We now obtain existence results under general conditions and then provide applications of this result when we, for example, assume α and β are lower and upper solutions and f satisfies a Nagumo condition.

(1) Let K be a connected component of $Z_0(g) = \{(x,y) \in R^{2n}: g(x,y) = 0\}$ such that for all $i = 1 \ldots n$

$$K \cap Q'_{\alpha_i}(0) \neq \phi \quad \text{and} \quad K \cap Q'_{\beta_i}(0) \neq \phi$$

(2) Every solution in $\chi(K)$ semi-egresses strictly from ω.

(3) Let $K_1 = P_{r_1} K$ and assume

$$K_1 \supseteq [\alpha(0),\beta(0)].$$

(4) For each point z in the n-dimensional parallelopiped $[\alpha(1),\beta(1)]$ there exists an $\bar{x} \in K_1$ and a solution of (E) satisfying $x(0) = \bar{x}$, $x(1) = z$. We now obtain a generalization of the one dimensional

topological principle given in [4].

Proposition 1. Under hypotheses 1-4, there exists a connected set $S[0,1] \subseteq \{x \in \chi(K): G(x|[0,1] \subseteq \omega\}$ and subsets of $S[0,1]$, $S_{\alpha_i}, S_{\beta_i}$, $i = 1 \ldots n$ consisting of solutions x_{α_i}, x_{β_i} such that $(1, x_{\alpha_i}(1), x'_{\alpha_i}(1)) \in Q'_{\alpha_i}(1)$ and $(1, x_{\beta_i}(1), x'_{\beta_i}(1)) \in Q'_{\beta_i}(1)$. Finally

$$P_{r_1}(S_{\alpha_i}(1)) \supseteq F_{\alpha_i}(1) \quad \text{and}$$

$$P_{r_1}(S_{\beta_i}(1)) \supseteq F_{\beta_i}(1).$$

Remark. We may think of $P_{r_1}(S_{\alpha_i})$ and $P_{r_1}(S_{\beta_i})$ to be surfaces of the form $x_i = \ell(x_j)$ $j \neq i$ where $\ell: R^{n-1} \to R$.

In order to use Proposition 1 to obtain the existence of solution of (E) and (G) we impose conditions on α, β, and f to guarantee the hypotheses 2 and 4 of Proposition 1 are satisfied.

We shall essentially assume that g satisfies appropriate conditions so that (1) and (3) are satisfied (see [2] and [5] for details). Finally we need to impose conditions on h that make it compatible with the conclusion of Proposition 1. It is here that we need to use degree theory (or a generalized intermediate value theorem in R^n). We thus need to assume:

(5) f satisfies a Nagumo condition relative to $[\alpha,\beta]$, that is if $x(t)$ is defined on $[0,\tau)$, $\tau \leq 1$ and $\alpha(t) < x(t) < \beta(t)$ for $t \in [0,\tau)$ then $|x'(t)|$ is bounded on $[0,\tau]$ and the bound depends only on α, β, f, and τ.

(6) Every solution of (E) with initial point in K $(x(t,K), x'(t,K)$ has the property that whenever $x(t,K) \in F_{\alpha_i}(t)$ $(x(t,K) \in F_{\beta_i}(t))$, $0 < t < 1$, then $x'_i(t,K) < \alpha'_i(t)$ $(x'_i(t,K) > \beta'_i(t))$.

This can be accomplished by assuming α and β are strict lower and upper solutions respectively. Conditions 5 and 6 imply condition (2) is satisfied. Moreover, they also imply the existence of two point boundary value problems for (E) with boundary conditions $x(0) = x_0 \in [\alpha(0),\beta(0)]$ and $x(1) = x_1 \in [\alpha(1),\beta(1)]$. Thus we also have that condition (4) is satisfied (see [1]).

This allows us to consider the following sets:
For each $z \in [\alpha(1),\beta(1)]$ define

$$S_z = \{ (x_0, y_0) \in K \text{ such that } x(t,x_0,y_0) \in S[0,1]$$

and $x(1,x_0,y_0) = z\}$

and define $S'_z = \{x'(1,x_0,y_0),\ (x_0,y_0) \in S_z\}$. Notice S_z and S'_z are multi-valued and upper semi continuous. At each point on $\partial[\alpha(1),\beta(1)]$ the vector field S'_z points away from the interior of $[\alpha(1),\beta(1)]$ according to Proposition 1.

(7) Assume we can make a continuous single-valued selection from $S'(z)$.

We thus have that $S'(z)$ is a continuous vector field on $[\alpha(1),\beta(1)]$ and according to Proposition 1 satisfies the hypotheses of Lemma 1 on $[\alpha(1),\beta(1)]$. Hence $\deg(S'(z),0,[\alpha(1),\beta(1)]) \neq 0$.

It remains to impose conditions on $h(x(0),x'(0),\ x(1),x'(1))$ to satisfy our boundary value problem. To do this let us assume

(8) for each $(x(0),x'(0)) \in \{ \bigcup_{i=1}^{n} (S_{\alpha_i}(0) \cup S_{\beta_i}(0))\}$

$h(x(0),x'(0),x(1,0,x(0),x'(0),x'(1,0,x(0),x'(0)))$ restricted to its third component is homotopic to S'_z on $[\alpha(1),\beta(1)]$.

Thus $\deg(h(..,u,.),0,[\alpha(1),\beta(1)]) \neq$) which implies that there exists a solution $\bar{x}(t)$ such that $h(\bar{x}(0),\bar{x}'(0),\bar{x}(1,0,\bar{x}(0),\bar{x}'(0)),\bar{x}'(1,0,\bar{x}(0),\bar{x}'(0))=0$ with $(\bar{x}(0),\bar{x}'(0)) \in K$. We thus obtain the following result.

Theorem 1. Under conditions (1), (3), (5), (6), (7), and (8) the BVP (E), (G) has a solution.

Remark. Condition (1) can be weakened in order to allow for the case $g(x(0),x'(0)) = x(0)$ by selecting the initial slopes sufficiently large and sufficiently small. Then one can show that there exists a τ, $0 < \tau < 1$ and an $N > 0$ such that for all i

$$\chi([\alpha(0),\beta(0)] \times \{x' :\ x'_i < -N,\ x'_j = 0\ j \neq i\} \cap Q'_{\alpha_i}(\tau) \neq 0$$

and

$$\chi([\alpha(0),\beta(0)] \times \{x' :\ x'_i > N, x'_i = 0\ j \neq i\} \cap Q'_{\beta_i}(\tau) \neq 0$$

This allows for assumption (1) to be satisfied and we may proceed as before.

Condition (7) is satisfied under the rather restrictive condition of the uniqueness of two point boundary value problems. Much milder conditions in terms of the behavior of the solution set at points of nonuniqueness can be given but we shall not pursue this here.

One particular set of sufficient conditions on h in order to insure (8) is to require that h is a function of $x(1)$ only and for each i

$$h_i(F_{\alpha_i}(1)) \subseteq 0 \quad \text{and} \quad h_i(F_{\beta_i}(1)) \supseteq 0.$$

This assumption can be regarded as sufficient conditions for the application of a generalized intermediate value theorem for they imply that h is homotopic to S_z'.

REFERENCES

[1] S. R. Bernfeld and V. Lakshmikantham, An introduction to nonlinear boundary value problems, Academic Press, New York, 1974.

[2] S. R. Bernfeld and P. Palamides, A topological method for vector-valued and N^{th} order nonlinear boundary value problems, to appear.

[3] P. Hartman, Ordinary Differential Equations, John Wiley, New York, 1964.

[4] L. Jackson and P. Palamides, An existence theorem for a nonlinear two point boundary value problem, to appear.

[5] L. Jackson and G. Klassen, A variation of the topological method of Wazewski, SIAM J. App. Math. 20 (1971), 124-130.

[6] P. Palamides, Y. Sficas, and V. Staikos, Wazewski's topological method for Caratheodory systems, Technical Report 36 (1980), Univ. of Ioannina, Ioannina, Greece.

DYNAMICS AND BIFURCATION IN ONE DIMENSION

Louis Block and David Hart

Department of Mathematics
University of Florida
Gainesville, Florida

The study of discrete dynamical systems arises in connection with both difference and differential equations. It has been found that the general behavior of such systems is organized largely by their periodic orbits. We wish to report recent progress in understanding this behavior, and applications to bifurcation theory. Detailed proofs will appear in [4], [5].

Sarkovskii showed ([8], see also [3] and [6]) that whenever a continuous function $f\colon I \to I$, $I = [0,1]$, has a point of least period k, it also has one of any period n such that n precedes k in the order

$$\begin{array}{l} 1 < 2 < 4 < 8 < \cdots \\ \cdots\ 4\cdot7 < 4\cdot5 < 4\cdot3 < \\ \cdots\ 2\cdot7 < 2\cdot5 < 2\cdot3 < \\ \cdots\ 7 < 5 < 3. \end{array}$$

For example, the existence of a periodic point of period six implies the existence of points of every other even period. Denoting those maps $f \in C^1(I,I)$ which have a point of period k by $F(k)$, n precedes k in Sarkovskii's order just in case $F(k) \subseteq F(n)$. In fact, $\mathrm{Cl}(F(k)) \subseteq \mathrm{int}(F(n))$, and if k is not a power of 2, $F(k)$ is closed (in the C^1 topology). Let the "annuli" of this nested sequence be denoted by $G(k) = F(k) \setminus \bigcup_{m>k} F(m)$.

If f_s is a C^1-continuous family of mappings with $f_0 \in G(1)$ and $f_1 \in G(3)$, then there is a subinterval $a_k < s < b_k$ for which $f_s \in G(k)$. If we introduce $F(\infty)$ as the maps for which all powers of 2 are periods, and $G(\infty)$ as the maps in $F(\infty)$ for which only powers of 2 are periods, the family must also pass through $G(\infty)$.

Recall that the solution set of $f_s^k(x) = x$ can only change if $(f_s^k)' = 1$, by the implicit function theorem. This is of course closely connected to asymptotic stability of such an orbit, and to period-doubling: if $(f_s^k)'(x) = -1$, then $(f_s^{2k})'(x) = +1$, and if in addition $\frac{d}{ds}(f_s^k)' < 0$, a (stable) 2k-periodic orbit branches from the k-periodic orbit (which becomes unstable). This is the only possible "subharmonic bifurcation" in one dimension (for C^1-continuous families): period-tripling or quadrupling cannot occur, due to the topology of the Sarkovskii stratification.

Note that the only <u>new</u> periods which may be created by period-doubling are the powers of two. In any family going from G(1) to G(3) periodic orbits for the powers of 2 are first created, and for some s, $f_s \in G(\infty)$. A sort of explosion occurs as other periods appear; this is associated with the appearance of homoclinic points, "horseshoes", and nonzero topological entropy ([2], [7]). Recall that a "homoclinic" periodic point is one whose stable and unstable sets intersect; in our setting, this means a k-periodic point $p \in f^k(W^u(p,f^k)\backslash\{p\})$, where the unstable set of p, $W^u(p,f^k)$, consists of all $x \in I$ such that for any neighborhood U of p, $x \in f^{kn}(U)$ for some positive integer k. Denote the maps with a k-periodic homoclinic point by H(k), and let H'(k) denote those maps such that f^k has a homoclinic fixed point. Then $H'(k) = \cup\{H(n)$, for n dividing k$\}$, and $H(2^r) \subseteq \text{int}(H(2^{r+1}) \cup H(2^{r+2}))$. In fact, for any odd $m > 1$, $H(m\cdot 2^r) \subseteq F(m\cdot 2^r) \subseteq H'(2^{r+1})$ (hence we concentrate on powers of 2), and conversely $H(2^r) \subseteq F(3\cdot 2^{r+1})$. Also, the closure of $H'(2^r)$ is contained in $H'(2^{r+1})$.

We summarize these results in the following chart.

$$\begin{array}{l}
F(1) \supset F(2) \supset F(4) \supset \cdots F(\infty) \supset \\
\qquad\quad \cdots \\
\qquad\quad \cdots \\
H'(8) \supseteq \cdots F(4\cdot 7) \supset F(4\cdot 5) \supset F(4\cdot 3) \supset H(2) \\
H'(4) \supseteq \cdots F(2\cdot 7) \supset F(2\cdot 5) \supset F(2\cdot 3) \supset H(1) \\
H'(2) \supseteq \cdots \quad F(7) \supset \quad F(5) \supset \quad F(3).
\end{array}$$

A family f_s from G(1) to F(3) must have intervals for which it lies in $H'(2^{r+1})\backslash H'(2^{r-1})$, for every $r \geq 0$. Homoclinic points must be created, whose periods are successively smaller powers of 2 (never skipping more than a single power, i.e. between H(16) and H(2) the family f_s must pass through either H(8) or H(4)). These homoclinic bifurcations must occur via non-transverse homoclinic points, where a transverse homoclinic periodic point x_0 satisfies (1) f^k meets the diagonal transversely, i.e. $|(f^k)'(x_0)| \neq 1$, and (2) W^u and W^s meet "transversely", i.e. $x_0 \in \text{int}(f^k(\text{int}(W^u(x_0,f^k)\backslash\{x_0\})))$, for example if x_0 lies in a noncritical orbit $x_i = f^k(x_{i+1})$ where (x_i) approaches x_0. The maps with a transverse homoclinic point of period k form an open subset of $C^1(I,I)$.

Finally, we remark that for specific families or with additional assumptions, these results may be simplified somewhat, and much more can be said.

REFERENCES

[1] L. Block, Stability of periodic orbits in the theorem of Sarkovskii, Proc. Amer. Math. Soc. 81 (1981), 333-336.

[2] ________, Homoclinic points of mappings of the interval, Proc. Amer. Math. Soc. 72 (1978), 576-580.

[3] L. Block, J. Guckenheimer, M. Misiurewicz, and L. -S. Young, Periodic points and topological entropy of one dimensional maps, Lecture Notes in Math. no. 819 (Springer, Berlin, 1980), pp. 18-34.

[4] L. Block and D. Hart, The bifurcation of periodic orbits of one-dimensional maps, to appear, Ergod. Th. & Dynam. Sys.

[5] L. Block and D. Hart, The bifurcation of homoclinic orbits of maps of the interval, to appear, Ergod. Th. & Dynam. Sys.

[6] T. Li and J. Yorke, Period three implies chaos, Amer. Math. Monthly 82 (1975), 985-992.

[7] M. Misiurewicz, Horseshoes for mappings of the interval, Bull. Acad. Polon. Sci. 27 (1979), 167-169.

[8] A. Sarkovskii, Coexistence of cycles of a continuous map of the line into itself, (Russian) Ukr. Mat. Z. 16 (1964), 61-71.

THE EFFECT OF HARVESTING ON POPULATION SYSTEMS

Fred Brauer

Department of Mathematics
University of Wisconsin
Madison, Wisconsin

ABSTRACT

The size of a population may be modelled in a variety of different ways and the behaviour of the population may be consistent with several different types of model. Under constant-rate harvesting, different models may give quite different behaviour; this is demonstrated by a comparison of first order difference equation models and first order differential-difference equation models.

1. INTRODUCTION

There are many ways to model the size of a population. Even if we confine our attention to deterministic models, we may use differential equations or difference equations to describe populations whose growth rate depends only on population size. To include the effects of time-delay, we may use difference equations or differential-difference equations. The dependence of a population on its food supply may be described by the use of a two-dimensional predator-prey model, while interactions with other species may be indicated by the use of systems of first order equations. If we wish to study the size of a given population, we may fit experimental data to a variety of different models, and as our understanding of the biological mechanisms governing population growth is very incomplete, there may be little reason to prefer one model to another. However, under constant-yield harvesting there may be radical differences in behaviour of the various models. This suggests that harvesting experiments may indicate the unsuitability of some models in certain situations.

In this paper we examine the differences in behaviour between difference equation and differential-difference equation models. Further contrasts could be pointed out by studying other models. For example, enough is known about predator-prey systems modelled by a pair of first order differential equations to categorize the possibilities. When more is understood about the effects of harvesting on populations with age structure or spatial dependence, it may be possible to give a more complete description of the relations between biological mechanisms and the effect of harvesting.

2. EQUILIBRIUM AND STABILITY FOR UNHARVESTED MODELS

Let x represent the size of a population whose growth rate depends only on the population size. For populations considered to have overlapping generations and continuous growth, we will write $x(t)$ to describe population size as a function of time. For populations considered to have non-overlapping generations and discrete growth, we will write x_n to describe population size in the nth generation, and we will assume a fixed time interval T between generations.

The size of a population undergoing discrete growth whose population in the $(n+1)^{st}$ generation is a function $G(x_n)$ of the population size x_n in the nth generation is described by the first-order difference equation

$$x_{n+1} = G(x_n) \qquad (1)$$

The size of a continuously growing population whose growth rate at time t is a function $F(x(t))$ of the population size $x(t)$ at time t is described by the first-order differential equation

$$x' = F(x). \qquad (2)$$

Often, it is convenient to write $F(x) = xf(x)$, so that $f(x(t))$ is the rate of growth per member; then the equation (2) takes the form

$$x' = xf(x).$$

The size of a population undergoing continuous growth for which the growth rate at time t depends on the population size at time $(t-T)$ is described by a first-order differential-difference equation of the form

$$x'(t) = x(t)f(x(t-T)). \qquad (3)$$

Such a description might be appropriate as a description of a population with discrete population changes but overlapping generations, or a population in which we count the number of adults and T is the maturation time.

Models like (3) may arise from considerations of generation time and also from other biological mechanisms such as rate of replenishment of a resource. More generally, the effects of delays may be distributed over the past rather than concentrated, leading to integro-differential equation models of the form

$$x'(t) = x(t)\int_0^\infty f(x(t-s))a(s)ds.$$

We shall not explore models of this type here; rather we shall concentrate on difference models (1) and differential-difference models (3).

An equilibrium of the difference equation (1) is a solution x_∞ of the equation $G(x_\infty) = x_\infty$ or, alternately, the abscissa of an intersection of the curve $y = G(x) - x$ and the horizontal line $y = 0$. An equilibrium x_∞ is asymptotically stable in the sense that if $\{x_n\}$ is a solution of (1) with x_0 sufficiently close to x_∞ then $\lim x_n = x_\infty$ if $|G'(x_\infty)| < 1$ and is unstable if $|G'(x_\infty)| > 1$ [7]. Alternately, an equilibrium x_∞ is asymptotically stable if the slope of the curve $y = G(x) - x$ at x_∞ is negative but sufficiently close to zero,

$$-2 < g'(x_\infty) - 1 < 0. \qquad (4)$$

An equilibrium of the differential-difference equation (3) is a solution x_∞ of the equation $f(x_\infty) = 0$ or, alternately, the abscissa of an intersection of the curve $y = xf(x)$ and the horizontal line $y = 0$. An equilibrium x_∞ is asymptotically stable if and only if

$$0 < -x_\infty f'(x_\infty)T < \frac{\pi}{2}.$$

[3].

Difference models of the form (1) and differential-difference models of the form (3) have been used almost interchangeably by biologists to describe populations with time-lag effects. Although it is clearly understood that different biological mechanisms lead to different types of models, imperfect knowledge of the biological mechanisms operating in specific populations and the considerable similarity in the mathematical behaviour of the two types of models have led to this interchangeability. As a result, difference models, being somewhat easier to analyze mathematically and to solve numerically, have been used to describe biological models with delays. A correspondence has been set up between models of the form (1) and models of the form (3) [6,7] with

$$f(x) = \frac{1}{T}\log\frac{G(x)}{x}, \quad G(x) = e^{Tf(x)}, \qquad (5)$$

Under this correspondence, an equilibrium x_∞ of (1), given by $G(x_\infty) = x_\infty$, is also an equilibrium of (3), given by

$$f(x_\infty) = \frac{1}{T} \log \frac{G(x_\infty)}{x_\infty} .$$

Since

$$x_\infty f'(x_\infty) = \frac{1}{T} [\frac{x_\infty G'(x_\infty)}{G(x_\infty)} - 1] = \frac{1}{T} [G'(x_\infty) - 1] ,$$

the stability conditions for equilibria of (1) and (3) are similar.

3. EQUILIBRIUM AND STABIITY FOR HARVESTED MODELS

In this section we consider the effect of a constant-yield harvest on the models (1) and (3). The size of a population modelled by the difference equation (1) which is subjected to a constant-rate harvest of H members per unit time satisfies the difference equation

$$x_{n+1} = G(x_n) - HT . \qquad (6)$$

An equilibrium of (6) is the abscissa of an intersection of the curve $y = G(x) - x$ and the horizontal line $y = HT$, and is asymptotically stable if the slope $G'(x) - 1$ of the curve $y = G(x) - x$ and the horizontal line $y = HT$, and is asymptotically stable if the slope $G'(x) - 1$ of the curve $y = G(x) - x$ at x_∞ is negative but sufficiently close to zero, just as in the unharvested case. The maximum harvest rate H_c for which there can be an equilibrium is given by

$$H_c = \frac{1}{T} \max \left[G(x) - x\right].$$

If $G''(x) \leq 0$ where the curve $y = G(x) - x$ has negative slope then the slope of this curve decreases from zero as x increases from the point where this maximum is achieved. Thus there is an interval on which (4) is satisfied and this implies that for all harvest rates sufficiently close to H_c there is an asymptotically stable equilibrium.

The size of a population modelled by the differential-difference equation (3) which is subjected to a constant-rate harvest of H members per unit time satisfies the differential-difference equation

$$x'(t) = x(t)f(x(t-t)) - H. \qquad (7)$$

Note that if f and G are related by (5), so that (1) and (3) have been viewed as equivalent, the analogue of (6) is

$$x'(t) = \frac{1}{T} x(t) \left[\log\{G(x(t-T) - HT\} - \log x(t-T)\right]$$

and not (7). Thus it should not be surprising that the solutions of (6) and (7) may behave quite differently.

An equilibrium of (7) is a solution x_∞ of the equation

$$x_\infty f(x_\infty) = H .$$

To describe the asymptotic stability of an equilibrium x_∞ of (7), we define

$$p = f(x_\infty)T, \quad q = x_\infty f'(x_\infty)T .$$

Then the equilibrium x_∞ of (7) is asymptotically stable if and only if

$$p < 1, \quad p < -q < \frac{a}{\sin a},$$

where a is defined as the solution of the equation

$$a = p \tan a ,$$

with $a = \pi/2$ if $p = 0$ [5;3, p.444]. Since $p = a/\tan a < a/\sin a$ implies $\cos a < 1$, which gives $a > 0$, or $p < 1$, the condition $p < 1$ is a consequence of the other two conditions. If $H = 0$, then $f(x_\infty) = 0$, so that $p = 0$ and this result reduces to the special case stated in the preceding section.

4. EXAMPLES

The difference equation analogue of the logistic differential equation is

$$G(x) = xe^{rT(1-\frac{x}{K})} .$$

For $H = 0$, equilibrium is given by $x_\infty = K$, and the condition for asymptotic stability is $0 < rT < 2$. If $rT > 2$, the equilibrium is unstable, but for sufficiently large H, the corresponding equilibrium becomes asymptotically stable.

The analogous differential-difference model is

$$x'(x) = rx(t)[1 - \frac{x(t-T)}{K}] - H ,$$

with

$$f(x) = r(1 - \frac{x}{K}) ,$$

so that

$$p = rT(1 - \frac{x_\infty}{K}) , \quad q = -rT\frac{x_\infty}{K} .$$

It is easy to verify that $H_c = rK/4$ and that $(K/2) < x_\infty \leq K$ for $0 \leq H < H_c$. Thus $p < -q$, and the other condition for asymptotic stability is

$$rT\frac{x_\infty}{K} < \frac{a}{\sin a} .$$

Since

$$\frac{x_\infty}{K} = 1 - \frac{p}{rT} ,$$

this becomes

$$rT < \frac{a}{\sin a} + p = \frac{a}{\sin a} + \frac{a}{\tan a} .$$

The function

$$\frac{a}{\sin a} + \frac{a}{\tan a} .$$

decreases monotonically from 2 at $a = 0$ to $\pi/a2$ at $a = \pi/2$. Thus if $rT < \pi/2$, the equilibrium is asymptotically stable for $0 \leq H < H_c$, while if $rT > 2$, the equilibirum is unstable for $0 \leq H < H_c$. If $\pi/2 < rT < 2$, the equilibrium is asymptotically stable for sufficiently small a, or sufficiently large H, and unstable for sufficiently small H . Thus the behaviour is similar to that of the harvested difference analogue, at least for $rT < 2$.

The difference equation

$$x_{n+1} = x_n \left(\frac{K}{x_n}\right)^{rT} \tag{8}$$

is frequently used for analyzing field data [4], possibly with an alternate form for x_n small which would not affect our analysis in any way; its continuous analogue is the Gompertz equation, with

$$f(x) = r \log \frac{K}{x} = r(\log K - \log x) .$$

For the difference equation (8), $G(x) = K^{rT}x^{1-rT}$, and $G'(x) = (1-rT)(K/x)^{rT}$. The equilibrium is at $x_\infty = K$ if there is no harvest, and the condition for asymptotic stability is $0 < rT < 2$. Since $G''(x) < 0$ if $rT > 1$, under harvesting the equilibrium is asymptotically stable for all H if $0 < rT < 2$ and if $rT > 2$ is unstable for small harvest rates and asymptotically stable for sufficiently large harvest rates.

For the differential-difference model

$$x'(t) = rx(t)[\log K - \log x(t-t)] - H,$$

we have $p = rT \log K/x_\infty$, $q = -rT$. The maximum harvest is

$$H_c = \max_{0 \leq x \leq K} rx \log \frac{K}{x} = \frac{rK}{e} ,$$

achieved for $x = K/e$, and the conditions for asymptotic stability of equilibrium are

$$rT \log \frac{K}{x_\infty} < rT < \frac{a}{\sin a} .$$

Since

$$\frac{K}{e} < x_\infty \leq K, \quad 0 \leq \log \frac{K}{x_\infty} < 1 ,$$

and the first inequality is satisfied. Since $a/\sin a$ increases monotonically from 1 for $a = 0$ to $\pi/2$ for $a = \pi/2$ we see that if $rT < 1$, the equilibrium is asymptotically stable for $0 \le H < H_c$, if $rT > \pi/2$ the equilibrium is unstable for $0 \le H < H_c$, while if $1 < rT < \pi/2$ the equilibrium is asymptotically stable for sufficiently large a, or for sufficiently small H, and unstable for sufficiently large H. This behaviour is not at all analogous to the behaviour of the corresponding difference model.

The difference equation

$$x_{n+1} = x_n e^{rT[1-(x_n/K)\theta]} \tag{9}$$

with

$$G(x) = xe^{rT[1-(x/K)^{\theta}]}$$

is the analogue under the correspondence (5) of the θ model of Ayala, Gilpin, and Ehrenfeld [2]. If there is no harvest, the equilibrium of (9) is at $x_\infty = K$ and since $G'(K) = 1 - \theta rT$, thus equilibrium is asymptotically stable for $0 < \theta rT < 2$. Thus if $0 < \theta rT < 2$, the equilibrium of (9) under harvesting is asymptotically stable for all harvest rates while if $\theta rT > 2$ the equilibrium is asymptotically stable for all sufficiently large harvest rates.

For the differential-difference equation model corresponding to (9), namely

$$x'(t) = rx(t)\left[1 - \left(\frac{x(t-T)}{K}\right)^{\theta}\right] - H,$$

we have

$$p = rT[1 - (x_\infty/K)^{\theta}], \quad q = -\theta rT(x_\infty/K)^{\theta}.$$

The maximum harvest rate is

$$H_c = \frac{rK\theta}{(1+\theta)^{1+(1/\theta)}},$$

achieved for $(x_\infty/K)^{\theta} = 1/(1+\theta)$. The conditions for asymptotic stability are

$$rT[1 - (x_\infty/K)^{\theta}] < \theta rT(x_\infty/K)^{\theta} < \frac{a}{\sin a},$$

the first is satisfied since $(x_\infty/K)^{\theta} > 1/(1+\theta)$. Since $\theta rT(x_\infty/K)^{\theta} = rT - P$, the remaining stability condition is

$$rT < \frac{1}{\theta}\frac{a}{\sin a} + p = \frac{1}{\theta}\frac{a}{\sin a} + \frac{a}{\tan a} = g_{\theta}(a).$$

It has been shown [1] that the function $g_\theta(a)$ is monotone increasing on $0 \leq a \leq \pi/2$ if $0 \leq \theta \leq 1/2$, decreases for small a and then increases if $1/2 < \theta < \pi/4$, with $g(0) < g(\pi/2)$ for $1/2 < \theta < 2/\pi$ and $g(0) > g(\pi/2)$ for $2/\pi < \theta < \pi/4$, and is monotone decreasing on $0 \leq a \leq \pi/2$ if $\theta > \pi/4$. Since $g_\theta(0) = 1 + 1/\theta$, $g_\theta(\pi/2) = \pi/2\theta$, we deduce the following pattern of behaviour:

(i) If $0 \leq \theta \leq 1/2$, the equilibrium is asymptotically stable for all H if $rT < 1 + 1/\theta$, asymptotically stable for small H if $1 + 1/\theta < rT < \pi/2\theta$, and unstable for all H if $rT > \pi/2\theta$.

(ii) If $1/2 < \theta < 2/\pi$, the equilibrium is asymptotically stable for all H if $0 < rT < \min g_\theta(a)$, asymptotically stable for both small and large H but unstable in an intermediate range if $\min g_\theta(a) < rT < 1 + 1/\theta$, asymptotically stable for small H if $1 + 1/\theta < rT < \pi/2\theta$ and unstable for all H if $rT > \pi/2\theta$.

(iii) If $2/\pi < \theta < \pi/4$, the equilibrium is asymptotically stable for all H if $0 \leq rT < \min g_\theta(0)$, asymptotically stable for both small and large H but unstable in an intermediate range if $\min g_\theta(a) < rT < \pi/2\theta$, asymptotically stable for large H if $\pi/2\theta < rT < 1 + 1/\theta$, and unstable for all H if $rT > 1 + 1/\theta$.

(iv) If $\theta > \pi/4$, the equilibrium is asymptotically stable for all H if $0 \leq rT < \pi/2\theta$, asymptotically stable for large H if $\pi/2\theta < rT < 1 + 1/\theta$, and unstable for all H if $rT > 1 + 1/\theta$.

This wide range of behaviours is completely different from the much simpler response of the corresponding difference model to harvesting.

5. CONCLUSIONS

While difference equation models and differential-difference equation models exhibit the same qualitative behaviour without harvesting, their responses to harvesting may be quite different. This suggests that harvesting experiments may indicate which type of model is more appropriate for a given population by showing how harvesting affects the stability of equilibrium. Another, more quantitative, measure of stability of an equilibrium is the characteristic return time. Careful study of the relation between harvest rate and characteristic return time for various types of model may give further insights.

REFERENCES

[1] M. Astaburuaga, Population models with harvesting and delay: Stability of equilibria, Ph.D. Thesis, University of Wisconsin (1978).

[2] F. J. Ayala, M. E. Gilpin, and J. G. Ehrenfeld, Competition between species: Theoretical models and experimental tests, Theor. Pop. Biology 4(1973), 331-356.

[3] R. Bellman and K. L. Cooke, Differential-Difference Equations, Academic Press, New York, 1963.

[4] J. B. S. Haldane, Animal populations and their regulation, New Biol. 15 (1953), 9-24.

[5] N. D. Hayes, Roots of the transcendental equation associated with a certain differential-difference equation, J. London Math. Soc. 25 (1950), 226-232.

[6] R. M. May, On relationships among various types of population models, Am. Naturalist 107 (1973), 46-57.

[7] R. M. May, G. R. Conway, M. P. Hassell, anmd T. R. E. Southwood, Time delays, density-dependence and single-species oscillations, J. Animal Ecology 43 (1974), 747-770.

INVARIANTS OF VORTEX MOTIONS IN THE PLANE

Jacob Burbea

University of Pittsburgh
Pittsburgh, Pennsylvania

ABSTRACT

An extension of the classical invariants of vortex motions in the plane is derived. Also, a sharp inequality for the energy enclosed in a region is established.

1. INTRODUCTION

Methods of complex analysis, as well known, constitute an effective tool in dealing with problems of hydrodynamics in the plane. In this note we shall use methods of complex analysis to derive an extension of the classical invariants of vortex motions in the plane and establish a sharp inequality for the energy enclosed in a plane set whose area-Lebesgue measure is finite. In the literature, as for example in the book of Batchelor [2, pp. 527–530], the classical invariants are derived on the assumption that the flow is at rest at infinity. However, in most applications the flow is not at rest at infinity and in spite of the fact that the latter case may be reduced to the previous one by a suitable change of the frame of reference, it proves more convenient, as we shall do here, to allow the flow to be not at rest at infinity.

Apart from the above mentioned inequality for the energy, the present analysis is based on elementary methods of (complex) function theory.

2. VORTICITY FUNCTIONS

We consider a two-dimensional incompressible inviscid fluid with unit density. In order to carry out such an analysis it proves to be extremely convenient to use complex notation and to adopt the Lagrangian point of view of fluid particles. Accordingly, the paths of the particles of the fluid are given by $z=z(z_o,t)$ where $z=x+iy\in\mathbb{C}$ is the particle of the fluid at time t and which, at the initial time $t=t_o$, it was at the position $z_o=z(z_o,t_o)$. The velocity of the flow is given by $\dot{z}=dz/dt$ and because of incompressibility there exists a (real-valued) stream function $\psi=\psi(z,t)$ so that

$$\dot{z} = 2i\bar{\partial}\psi, \tag{2.1}$$

where $\partial=(\partial_x-i\partial_y)/2$ and $\bar{\partial}=(\partial_x+i\partial_y)/2$. The area-Lebesgue measure is denoted by $d\sigma(z)=(i/2)dz\wedge d\bar{z}$ and we note the relationship $d\sigma(z)=J(t)d\sigma(z_o)$ where $J(t)$ is the Jacobian of the transition $z\to z_o$. It follows from (2.1) that $\dot{d\sigma}(z)=0$.

The vorticity of the flow is $\nu=\Delta\psi=4\partial\bar{\partial}\psi$ and it is convected with the flow, i.e.,

$$\dot{\nu} = \partial_t\nu + 2\mathrm{Re}(\dot{z}\partial\nu) = 0 \quad , \quad (\nu=\Delta\psi) \tag{2.2}$$

which is the well-known Euler equation.

Let $\omega=\omega(z,t)$ be a real-valued piecewise C^1-function with compact support $D(t)$ in the plane, and, which is convected with the flow,

$$\dot{\omega} \equiv \partial_t\omega + 2\mathrm{Re}(\dot{z}\partial\omega) = 0. \tag{2.3}$$

We write

$$\psi_o(z,t) = \frac{1}{2\pi}\int\omega(\zeta,t)\log|z-\zeta|d\sigma(\zeta) \tag{2.4}$$

where the integration is carried over the entire complex plane $\mathbb{C}$, and thus over only the support $D(t)$ of ω. We introduce the "correction-term" $R=R(z,t)$, given by

$$R(z,t) = \psi(z,t) - \psi_o(z,t). \tag{2.5}$$

From (2.4) and (2.5) follows that

$$\nu = \Delta\psi = \omega + \Delta R \tag{2.6}$$

while (2.2) and (2.3) imply that

$$\dot{(\Delta R)} = \partial_t(\Delta R) + 2\mathrm{Re}[\dot{z}\partial(\Delta R)] = 0. \tag{2.7}$$

We shall label ω as the "<u>vorticity of the core</u>" while ΔR will be labeled as the "<u>vorticity of the frame</u>". It follows from (2.1)-(2.5) that the velocity of the flow at infinity is asymptotically $2i\bar{\partial}R$. In fact, $-2i\bar{\partial}R$ may be interpreted as the velocity of the frame of reference for which the flow is at rest at infinity. For this reason, $-2i\bar{\partial}R$ will be labeled as the "<u>velocity of the frame</u>".

3. CLASSICAL INVARIANTS

For a smooth function $f=f(z,t)$, we consider the quantity

$$K_f(t) = \int f(z)\omega(z)d\sigma(z)$$

where the dependence on the time t of the functions in the integrand is suppressed. One may obtain invariants of the motions by solving

$\dot{K}_f(t)=0$ for f. From (2.1) and (2.3) we deduce that $\dot{K}_f(t)=\int\dot{f}(z)\omega(z)d\sigma(z)$ and therefore $K_f(t)$ is invariant whenever $\dot{f}=0$ or, equivalently, when f satisfies the evolution equation $\partial_t f=4\mathrm{Im}[\partial f\bar{\partial}\psi]$. In particular, this will be true, in view of (2.3), if f is functionally dependent on ω. Thus all the quantities

$$K_{\omega^n}(t) = \int[\omega(z)]^{n+1}d\sigma(z) \quad ; \quad n=0,1,\ldots,$$

are invariant. The invariant

$$\kappa = (2\pi)^{-1}K_1(t) = (2\pi)^{-1}\int\omega(z)d\sigma(z) \tag{3.1}$$

is known as the "circulation" of the flow.

Another set of invariants may be obtained by considering f which are not dependent on t explicitly, f=f(z) or $\partial_t f=0$. In this case, by using (2.1) and (2.4)-(2.5),

$$\begin{aligned}\dot{K}_f(t) &= \int[\dot{z}\partial f + \dot{\bar{z}}\bar{\partial}f]\omega(z)d\sigma(z)\\ &= 2i\int[\bar{\partial}R\partial f - \partial R\bar{\partial}f]\omega(z)d\sigma(z)+2i\int[\bar{\partial}\psi_o\partial f-\partial\psi_o\bar{\partial}f]\omega(z)d\sigma(z)\end{aligned}$$

and

$$\int[\bar{\partial}\psi_o\partial f-\partial\psi_o\bar{\partial}f]\omega(z)d\sigma(z) = \frac{1}{4\pi}\int\int|z-\zeta|^{-2}[(z-t)\partial f-(\bar{z}-\bar{\zeta})\bar{\partial}f]\omega(z)\omega(\zeta)d\sigma(z)d\sigma(\zeta).$$

An antisymmetry argument shows that this integral is zero if

$$(z-\zeta)\partial_z f-(\bar{z}-\bar{\zeta})\bar{\partial}_z f = -[(\zeta-z)\partial_\zeta f-(\bar{\zeta}-\bar{z})\partial_{\bar{\zeta}}f]$$

or, if

$$(z-\zeta)[\partial_z f-\partial_\zeta f] = (\bar{z}-\bar{\zeta})[\bar{\partial}_z f-\bar{\partial}_\zeta f]$$

for any z and ζ. Here $\partial_z f=\partial_z f(z)$ and $\partial_\zeta f=\partial_\zeta f(\zeta)$. However, the general smooth solution of this equation is given by

$$f(z) = \alpha|z|^2 + \beta z + \gamma\bar{z} + \delta$$

where α,β,γ and δ are arbitrary complex constants. This suggests defining the following quantities: The "centre of vorticity"

$$\hat{z}(t) = K_z(t)/K_1(t) = (2\pi\kappa)^{-1}\int z\omega(z)d\sigma(z) \tag{3.2}$$

and the "angular-momentum"

$$M(t) = K_{|z|^2}(t)/K_1(t) = (2\pi\kappa)^{-1}\int|z|^2\omega(z)d\sigma(z). \tag{3.3}$$

When the circulation κ is zero these quantities are either infinite or indeterminate. From the foregoing analysis we conclude (compare [2, pp. 527-530]):

Proposition 1. The following hold:

(i) the circulation κ is always invariant;

(ii) the centre of vorticity $\hat{z}(t)$ is invariant if and only if

$$\int \bar{\partial} R\omega(z)d\sigma(z) = 0,$$

i.e., when the velocity of the frame is orthogonal to the vorticity of the flow;

(iii) the angular-momentum $M(t)$ is invariant if and only if

$$\int \mathrm{Im}(z\bar{\partial}R)\omega(z)d\sigma(z) = 0,$$

i.e., when the "angular momentum of the frame" is orthogonal to the vorticity of the flow .

Another important quantity is the "effective-energy"

$$E(t) = -\frac{1}{2}\int\omega(z)[\psi_o(z)+2R(z)]d\sigma(z), \tag{3.4}$$

where, again, the dependence on time t in the functions of the integrand is suppressed. This quantity is also known as the "Kirchoff-invariant" (see also [2, p. 530]) in the special case that $R(z,t)\equiv 0$. It is interesting to note, however, that while the correction term for the natural stream function ψ_o is R, the corresponding correction in E(t) is 2R. This factor of 2 is essential and its appearance may be explained by the double integration in (3.4) due to ψ_o in (2.4).

Proposition 2. We have $\dot{E}(t) = -\int\omega(z)\partial_t R(z)d\sigma(z)$ and, therefore, the effective-energy is invariant when $\partial_t R$ is orthogonal to the voricity of the core.

Proof. By (2.1), (2.2) and (3.4),

$$\dot{E}(t) = -\frac{1}{2}\int\omega(z)[\dot{\psi}_o(z)+2\dot{R}(z)]d\sigma(z).$$

On the other hand, by (2.1)-(2.4), we deduce that

$$\dot{\psi}_o(z) = \frac{1}{2\pi}\mathrm{Re}\int\omega(\zeta)\frac{\dot{z}-\dot{\zeta}}{z-\zeta}d\sigma(\zeta) = -\mathrm{Im}\int\omega(\zeta)\frac{\bar{\partial}_z\psi-\bar{\partial}_\zeta\psi}{z-\zeta}d\sigma(\zeta).$$

Also, from (2.1) we have

$$\dot{R}(z) = \partial_t R + 2\mathrm{Re}(\dot{\mathbf{z}}\partial R) = \partial_t R - 4\mathrm{Im}(\bar{\partial}\psi\partial R).$$

Therefore

$$\dot{E}(t) = \frac{1}{2}\mathrm{Im}\{\frac{1}{\pi}\iint\omega(z)\omega(\zeta)\frac{\bar{\partial}_z\psi-\bar{\partial}_\zeta\psi}{z-\zeta}d\sigma(z)d\sigma(\zeta) + 8\int\omega(z)\bar{\partial}Rd\sigma(z)\}-\int\omega(z)\partial_t R(z)d\sigma(z)$$

and so, by symmetrization and (2.4),

$$\dot{E}(t) = 4\mathrm{Im}\int\omega(z)\bar{\partial}\psi[\partial\psi_o + \partial_z R]d\sigma(z) - \int\omega(z)\partial_t R(z)d\sigma(z).$$

Finally, from (2.5),

$$\dot{E}(t) = 4\mathrm{Im}\int\omega(z)|\partial\psi|^2 d\sigma(z) - \int\omega(z)\partial_t R(z)d\sigma(z) = -\int\omega(z)\partial_t R(z)d\sigma(z)$$

and the proposition follows.

The relationship of the effective-energy E with the energy T of the system is as follows: Let $S = \{z\varepsilon\mathbb{C}: |z| < r\}$ be a large disk containing the support D(t) of ω and let T_S be that kinetic energy of the fluid enclosed in S. Then, by (2.1),

$$T_S = \frac{1}{2}\int_S|\dot{z}|^2 d\sigma(z) = 2\int_S|\partial\psi|^2 d\sigma(z)$$

and thus, using Green's formula,

$$T_S = -i\int_{\partial S}\psi\partial\psi dz - 2\int_S\psi\partial\bar{\partial}\psi d\sigma(z).$$

A use of (2.5), (3.4) and another application of Green's formula shows that

$$T_S = E(t) + 2\int_S|\partial R|^2 d\sigma(z) + \mathrm{Im}\int_{\partial S}(\psi+R)\partial\psi_o dz.$$

However,

$$\mathrm{Im}\int_{\partial S}\psi_o\partial\psi_o dz = [\alpha_o^2\log r - \frac{1}{2}\sum_{n=1}^{\infty}\frac{|\alpha_n|^2}{nr^{2n}}],$$

where

$$\alpha_n = (2\pi)^{-1}\int\omega(z)z^n d\sigma(z) \qquad (n=0,1,\ldots),$$

and thus, by (3.1), $\kappa = \alpha_o$. Consequently,

$$\lim_{r\to\infty}\{T_S - 2\int_S|\partial R|^2 d\sigma(z) - 4\mathrm{Re}\int_S\bar{\partial}(R\partial\psi_o)d\sigma(z) - \pi\kappa^2\log r\} = E(t).$$

In particular, when R satisfies

$$\lim_{r\to\infty}\int_S\{|\partial R|^2 + 2\mathrm{Re}(\partial R\bar{\partial}\psi_o)\}d\sigma(z) = 0, \tag{3.5}$$

we have (compare [2, p. 529])

$$\lim_{r\to\infty}[T_S - \pi\kappa^2\log r] = -\frac{1}{2}\int\omega(z)\psi_o(z)d\sigma(z). \tag{3.6}$$

In the specialized case of (3.5), we see from (3.6) that when $\kappa \neq 0$, $T_\infty = +\infty$ while when $\kappa = 0$ we have $T_\infty = 2^{-1}\int\omega(z)\psi_o(z)d\sigma(z)$ which is finite.

4. REGIONS OF CONSTANT VORTICITY

In many applications (see, for example, [3]), the function R, being smooth in the entire plane, is chosen as

$$R(z,t) = -\frac{1}{2}\Omega|z-\hat{z}|^2 + \mathrm{Re}L(z,t) \tag{4.1}$$

where Ω is a constant, $\hat{z} = \hat{z}(t)$, the centre of vorticity and L(z,t) is an entire function for every fixed t. The velocity of the frame is therefore

$$-2i\bar{\partial}R = i[\Omega(z-\hat{z}) - \overline{\partial L(z,t)}]. \tag{4.2}$$

In this case $\Delta R = -2\Omega$ and, therefore, Ω represents the angular velocity about $\hat{z}$ of the frame of reference for which the flow is at rest at infinity. In consistency with (2.2)-(2.3) and (2.6)-(2.7) we choose ω to be a piecewise constant. This suggests the introduction of a region of constant vorticity $D \equiv D(t)$ described below. On each component D_j of D, the vorticity $\nu = \Delta\psi$ is given by $\chi_{D_j}\omega_j - 2\Omega$ where χ_{D_j} is the characteristic function of the closure of D_j.

Let $A_j = A(D_j)$ be the area of the component D_j and let A be the total area of D. Thus $A = A(D) = \Sigma A_j$. We also denote by $C_j = \partial D_j$ the boundary of D_j. The circulation κ_j of D_j is $\kappa_j = \omega_j A_j$ while the total circulation is $\kappa = \Sigma\kappa_j$. Moreover, the centre of vorticity and the angular momentum are now $\hat{z}(t) = \Sigma\hat{z}_j(t)$ and $M(t) = \Sigma M_j(t)$, respectively. Now, according to Proposition 1, κ is always invariant while, using (4.1), $\hat{z}(t)$ and $M(t)$ are invariant if and only if $\int\omega(z)\partial L(z,t)d\sigma(z) = 0$ and $\text{Im}\int\omega(z)z\partial L(z,t)d\sigma(z) = 0$, respectively.

The natural stream function of (2.4) is expressible here are

$$\psi_o(z) = \frac{1}{2\pi}\sum\omega_j\int_{D_j}\log|z-\zeta|d\sigma(\zeta)$$

and, therefore,

$$\partial\psi_o(z) = \sum\omega_j\phi_j(z) \quad , \quad \phi_j(z) = \frac{1}{4\pi}\int_{D_j}(z-\zeta)^{-1}d\sigma(\zeta). \tag{4.3}$$

By a classical inequality due to Ahlfors and Beurling [1] we conclude that

$$|\phi_j(z)| \leq \frac{1}{4\pi}\sqrt{\pi A_j} \tag{4.4}$$

for every $z\in\mathbb{C}$ with equality at one point z_o, if and only if D_j is a disk whose radius is $\sqrt{A_j/\pi}$, and, in that case z_o is necessarily on C_j, the boundary of D_j.

Using the norm $||\omega||^2 = \Sigma\omega_j^2$, (4.3)-(4.4) and Cauchy-Schwarz inequality we obtain:

<u>Proposition 3</u>. Let D be a region of constant vorticity as described before. Then $|\partial\psi_o|^2 \leq (16\pi)^{-1}A||\omega||^2$ for every $z\in\mathbb{C}$. Equality at one point z_o occurs if and only if D is a disk whose radius is $\sqrt{A/\pi}$, and, in that case z_o is necessarily on the boundary of D and $D=D_j$, $\omega=\omega_j$ for all j.

This result may be used to estimate the kinetic energy T_S of the fluid enclosed in a measurable subset S of the plane. For this purpose we introduce the norm of any square-integrable function f on S:

$$||f||_S^2 = \int_S |f(z)|^2 d\sigma(z),$$

and thus $T_S = 2||\partial\psi||_S^2 = 2||\partial\psi_o + \partial R||_S^2$. Then, a use of the triangle inequality and Proposition 3 leads to:

Proposition 4. Let D be a region of constant vorticity $\omega_j - 2\Omega$ on each component D_j, let A_j be the area of D_j and assume that the velocity of the frame $-2i\bar{\partial}R$ is as in (4.2). Let S be a measurable subset of the plane whose are A(S) is finite. Then the energy T_S enclosed in S satisfies

$$T_S \leq \frac{1}{8\pi} ||\omega||^2 AA(S)\{1 + \frac{2}{||\omega||} (\frac{\pi}{A})^{1/2} \mu(S)\}^2$$

where $||\omega||^2 = \Sigma\omega_j^2$, $A \equiv A(D) = \Sigma A_j$ and $\mu(S) = \sup_{z \in S} |\Omega(z-\hat{z}) - \overline{\partial L(z,t)}|$.

REFERENCES

[1] L. Ahlfors, and A. Beurling, Conformal invariants and function theoretic null sets, Acta Math. 83 (1950), 101-129.

[2] G. K. Batchelor, Fluid Dynamics, Cambridge Univ. Press, Cambridge, 1979.

[3] J. Burbea, Vortex motions and their stability, Proc. Nonlinear Phenomena in Math. Sci., Arlington 1980, Academic Press, to appear.

INSTABILITY AND STABILITY IN VOLTERRA EQUATIONS

T. A. Burton

Department of Mathematics
Southern Illinois University
Carbondale, Illinois

W. E. Mahfoud

Department of Mathematics
Murray State University
Murray, Kentucky

ABSTRACT

We consider a system of Volterra integrodifferential equations of the form

$$x' = A(t)x + \int_0^t C(t,s)x(s)ds$$

in which A and C are continuous $n\times n$ matrices. Necessary and sufficient conditions for stability of the zero solution are given.

INTRODUCTION

We consider a system of Volterra equations

$$x' = A(t)x + \int_0^t C(t,s)x(s)ds \tag{1}$$

in which A is an $n\times n$ matrix of functions continuous for $0 \le t < \infty$ and C is an $n\times n$ matrix of functions continuous for $0 \le s \le t < \infty$. The purpose of the paper is to give conditions on A and C which imply stability or instability of the zero solution of (1).

Basic existence, uniqueness, and stability theory for (1) may be found in Driver [5]. Briefly, for each $t_0 \ge 0$ and each continuous function $\varphi[0,t_0] \to R^n$ there is a unique solution $x(t,t_0,\varphi)$ of (1) on an interval $[t_0,T)$ with $x(t,t_0,\varphi) = \varphi(t)$ for $0 \le t \le t_0$. If $x(t,t_0,\varphi)$ remains bounded, then $T = \infty$.

SCALAR EQUATIONS

It is convenient to separate the results into scalar and vector classes. Thus, we first consider

$$x' = A(t)x + \int_0^t C(t,s)x(s)\,ds \tag{2}$$

in which A, C, and x are all scalars, while A and C are continuous as in (1).

THEOREM 1. Suppose that

$$\int_t^\infty |C(u,t)|\,du \quad \text{is defined for } t \geq 0$$

and that there is a constant $\alpha > 0$ with

$$\int_0^t |C(t,s)|\,ds + \int_t^\infty |C(u,t)|\,du - 2|A(t)| \leq -\alpha. \tag{3}$$

Then the zero solution of (2) is stable if and only if $A(t) < 0$.

Proof. Suppose $A(t) < 0$ and consider the functional

$$V(t,x(\cdot)) = x^2 + \int_0^t \int_t^\infty |C(u,s)|\,du\, x^2(s)\,ds.$$

A calculation will show that the derivative of V along a solution of (2) satisfies $V'_{(2)}(t,x(\cdot)) \leq -\gamma x^2$, $\gamma > 0$. The stability now follows by standard arguments [3].

Now, suppose that $A(t) > 0$ and consider the functional

$$W(t,x(\cdot)) = x^2 - \int_0^t \int_t^\infty |C(u,s)|\,du\, x^2(s)\,ds.$$

The derivative of W along a solution x(t) of (2) satisfies $W'(t,x(\cdot)) \geq \mu x^2$ for some $\mu > 0$. Thus, given any $t_0 \geq 0$ and any $\delta > 0$ we can find a continuous function $\varphi : [0,t_0] \to R$ with $|\varphi(t)| < \delta$ and $W(t_0,\varphi(\cdot)) > 0$. If $x(t) = x(t,t_0,\varphi)$, then

$$\begin{aligned} x^2(t) &\geq W(t,x(\cdot)) \geq W(t_0,\varphi(\cdot)) + \mu\int_{t_0}^t x^2(s)\,ds \\ &\geq W(t_0,\varphi(\cdot)) + \mu\int_{t_0}^t W(t_0,\varphi(\cdot))\,ds \\ &\geq W(t_0,\varphi(\cdot)) + \mu W(t_0,\varphi(\cdot))(t - t_0) \end{aligned}$$

which implies the instability. This completes the proof.

Results of this type for convolution equations were obtained by Brauer [2] using Laplace transform techniques.

We select a continuous function G(t,s) with

$$\partial G(t,s)/\partial t = C(t,s) \tag{4}$$

and let

$$Q(t) = A(t) - G(t,t) \tag{5}$$

so that (2) may be written as

$$x' = Q(t)x + (d/dt)\int_0^t G(t,s)x(s)\,ds. \tag{6}$$

THEOREM 2. Suppose that (4) and (5) hold and there are constants Q_1, Q_2, J, and R with $R < 2$ such that

(i) $0 < Q_1 \leq |Q(t)| \leq Q_2$,

(ii) $\int_0^t |C(t,s)|\,ds \leq J < 1$,

and

$$\text{(iii)}\quad \int_0^t |G(t,s)|\,ds + \int_t^\infty |G(u,t)|\,du \le RQ_1/Q_2$$

for $0 \le t < \infty$.

Suppose also that there is a continuous function $h:[0,\infty) \to [0,\infty)$ with $|G(t,s)| \le h(t-s)$ and $h(u) \to 0$ as $u \to \infty$. Then the zero solution of (6) is stable if and only if $Q(t) < 0$.

Proof. If $Q(t) < 0$, define

$$V(t,x(\cdot)) = \left(x - \int_0^t G(t,s)x(s)\,ds\right)^2 + Q_2\int_0^t \int_t^\infty |G(u,s)|\,du\; x^2(s)\,ds$$

and obtain $V'_{(6)}(t,x(\cdot)) \le -\beta x^2$, some $\beta > 0$. A substantial argument [4] is required to show stability from this V as it is non-negative, but not positive definite.

If $Q(t) > 0$, define

$$W(t,x(\cdot)) = \left(x - \int_0^t G(t,s)x(s)\,ds\right)^2 - Q_2\int_0^t \int_t^\infty |G(u,s)|\,du\; x^2(s)\,ds$$

and obtain $W'_{(6)}(t,x(\cdot)) \ge \gamma x^2$, some $\gamma > 0$. Again, it requires considerable argument [4] to show instability.

REMARK 1. We call (3) the covering assumption for Theorem 1, while (i), (ii), and (iii) are called the covering assumptions for Theorem 2. Frequently the covering assumptions in a necessary and sufficient theorem are so restrictive that the final result says very little. That is not true in Theorem 1. For suppose (ii) holds and C is of convolution type. Under mild conditions, when $A(t) < 0$ and when (3) is replaced by

$$\int_0^t |C(t,s)|\,ds + \int_t^\infty |C(u,t)|\,du - 2|A(t)| \ge \alpha > 0,$$

then the conditions of Theorem 2 hold with $Q(t) > 0$, implying instability.

REMARK 2. Theorem 1 shows that (2) inherits the stability or instability of the O.D.E. $x' = A(t)x$, while Theorem 2 shows that (6) inherits the stability or instability of $x' = Q(t)x$. These two results may be thought of as the boundary results. An infinite number of results lie between them. Let

$$x' = A(t)x + \int_0^t [C_1(t,s) + C_2(t,s)]x(s)\,ds \tag{7}$$

and select a matrix $H(t,s)$ with

$$\partial H(t,s)/\partial t = C_2(t,s) \tag{8}$$

and let

$$L(t) = A(t) - H(t,t) \tag{9}$$

so that (7) may be written as

$$x' = L(t)x + \int_0^t C_1(t,s)x(s)\,ds + (d/dt)\int_0^t H(t,s)x(s)\,ds. \tag{10}$$

Under appropriate smallness conditions on C_1 and H the zero solution of (10) is stable if and only if $L(t) < 0$. The proof is accomplished by using the Liapunov functional

$$V(t,x(\cdot)) = \left(x - \int_0^t H(t,s)x(s)ds\right)^2$$
$$+ M\int_0^t\int_t^\infty |C_1(u,s)|du\, x^2(s)ds + K\int_0^t\int_t^\infty |H(u,s)|du\, x^2(s)ds,$$

together with liberal application of the Schwarz inequality.

When C_2 is chosen to be zero we obtain Theorem 1, while $C_1 = 0$ yields Theorem 2.

VECTOR EQUATIONS

We now consider the system

$$x' = Ax + \int_0^t C(t,s)x(s)ds \tag{11}$$

with A being n×n and constant, while C is n×n and continuous for $0 \le s \le t < \infty$. The first objective is to show that (11) inherits the stability or instability of $x' = Ax$ when C is small. To this end we seek a symmetric positive definite matrix D such that

$$A^TB + BA = -D \tag{12}$$

has a solution $B = B^T$. Barbashin [1] claims that for any positive definite matrix D, Equation (12) can be uniquely solved for B provided that the characteristic roots $\lambda_1,\ldots,\lambda_n$ of A satisfy $\lambda_i + \lambda_j \neq 0$ for any i and j.

It is readily verified that (12) has no solution when A has a characteristic root with zero real part, regardless of the choice of positive definite matrix D. Moreover, one may construct examples in which $\lambda_i + \lambda_j = 0$, while (12) has a solution B for certain choices of D; the matrix B may not be unique.

Thus, although it is more general to consider (12) with a matrix D to be selected, to be definite we ask that

$$A^TB + BA = -I \tag{13}$$

have at least one solution $B = B^T$.

THEOREM 3. Let B satisfy (13) and let

$$|B|\left(\int_0^t |C(t,s)|ds + \int_t^\infty |C(u,t)|du\right) \le M < 1.$$

Then the zero solution of (11) is stable if and only if B is positive definite.

Proof. We consider the functional

$$V(t,x(\cdot)) = x^TBx + |B|\int_0^t\int_t^\infty |C(u,s)|du|x(s)|^2ds$$

and find that $V'_{(11)}(t,s(\cdot)) \le -\alpha|x|^2$, some $\alpha > 0$.

If B is positive definite, then it is easy to argue that $x = 0$ is stable.

If B is not positive definite, then one may argue that there is an $x_0 \neq 0$ with $x_0^T B x_0 \leq 0$ and then argue that, as V' is negative, there is an x_0 with $x_0^T B x_0 < 0$. Moreover, for this last x_0, if $k \neq 0$, then $(k x_0)^T B(k x_0) < 0$. Thus, there are x_0 arbitrarily near zero with $V(0,x_0) < 0$. It can then be shown that $x(t,0,x_0)$ is bounded strictly away from zero and, as $V' \leq \alpha|x|^2$, the theorem follows.

Results of this type in the convolution case may be found in Jordan [6].

The counterpart of Theorem 2 for systems is more complicated and we ask that C be of convolution type so that we are considering

$$x' = Ax + \int_0^t C(t-s)x(s)\,ds \tag{14}$$

with A being $n \times n$ and constant, while C is $n \times n$ and continuous for $0 \leq t < \infty$.

Select an $n \times n$ matrix $G(t)$ with

$$G'(t) = C(t) \tag{15}$$

and set

$$Q = A - G(0) \tag{16}$$

so that (14) takes the form

$$x' = Qx + (d/dt)\int_0^t G(t - s)x(s)\,ds. \tag{17}$$

We ask that there exists a symmetric matrix D satisfying

$$Q^T D + DQ = -I. \tag{18}$$

THEOREM 4. Let (15), (16), and (18) hold. Suppose

(i) $2|DQ|\int_0^\infty |G(v)|\,dv < 1$

and

(ii) $G(t) \to 0$ as $t \to \infty$.

Then the zero solution of (17) is stable if and only if D is positive definite.

Proof. Define

$$V(t,x(\cdot)) = \left(x - \int_0^t G(t-s)x(s)\,ds\right)^T D\left(x - \int_0^t G(t-s)x(s)\,ds\right) + |DQ|\int_0^t\int_t^\infty |G(u - s)|\,du\,|x(s)|^2\,ds$$

so that along a solution of (17) we have $V'_{(17)}(t,x(\cdot)) \leq -\mu|x|^2$, some $\mu > 0$. The result then follows from considerable hard analysis [4] concerning the properties of V.

REMARK 3. Theorems 3 and 4 are the boundary results for the vector equations. We may obtain the analog of Equation (10) for vector equations and write

$$x' = Lx + \int_0^t C_1(t - s)x(s)ds + (d/dt)\int_0^t H(t - s)x(s)ds \tag{19}$$

and define a Liapunov functional

$$V(t,x(\cdot)) = (x - \int_0^t H(t - s)x(s)ds)^T D(x - \int_0^t H(t - s)x(s)ds)$$
$$+ M\int_0^t \int_t^\infty |C_1(u - s)|du|x(s)|^2 ds$$
$$+ K\int_0^t \int_t^\infty |H(u - s)|du|x(s)|^2 ds.$$

With smallness conditions on C_1 and H, then the zero solution of (19) is stable if and only if D is positive definite.

In addition to providing details required in this work, the paper [4] gives a number of qualitative results concerning the behavior of derivatives of solutions. It also shows the equivalence of numerous types of stabilities of Volterra equations.

REFERENCES

[1] E. A. Barbashin, The construction of Lyapunov functions, Differential Equations 4(1968), 1097-1112 (This is the translation of Differential'nye Uravneniya 4(1968), 2127-2158.)

[2] F. Brauer, Asymptotic stability of a class of integrodifferential equations, J. Differential Equations 28(1978), 180-188.

[3] T. A. Burton, Stability theory for Volterra equations, J. Differential Equations 32(1979), 101-118.

[4] T. A. Burton and W. E. Mahfoud, Stability criteria for Volterra equations, to appear.

[5] R. D. Driver, Existence and stability of solutions of a delay-differential system, Arch. Rational Mech. Anal. 10(1962), 401-426.

[6] G. S. Jordan, Asymptotic stability of a class of integrodifferential systems, J. Differential Equations 31(1979), 350-365.

ALTERNATING BOUNDS CONVERGING TO THE SOLUTION OF A SEMILINEAR EULER-POISSON-DARBOUX EQUATION

C. Y. Chan

Department of Mathematics and Computer Science
Florida State University
Tallahassee, Florida

ABSTRACT

An alternating sequence consisting of two monotone subsequences of upper and lower bounds for solutions of a singular Cauchy problem involving a semilinear Euler-Poisson-Darboux equation is constructed by using an iteration of the Picard type. This sequence is shown to converge uniformly to establish an existence and uniqueness theorem.

1. INTRODUCTION

Recently, Chan and Young [2] established existence of a maximal and a minimal nonnegative solution of the following semilinear singular Cauchy problem

$$Lu = g(x,y,t;u),\ 0 < t \leq T < \infty, \tag{1}$$

$$u(x,y,0) = 0,\ u_t(x,y,0) = 0, \tag{2}$$

where L is the two-dimensional Euler-Poisson-Darboux operator given by

$$Lu \equiv u_{tt} + (k/t)u_t - u_{xx} - u_{yy}$$

with a real parameter $k > 0$; conditions under which the problem had at most one solution were also given. The restriction $k > 0$ was made so that the corresponding linear problem had a unique solution (cf. Diaz and Ludford [4]). The case $k = 0$ was omitted since for $k = 0$, the problem was no longer singular.

In section 2 here, we show that an iteration scheme of the Picard type as in the above-mentioned paper under different assumptions on g gives rise to an alternating sequence (cf. Chan [1] for parabolic equations, and Keller [5] for elliptic equations). This sequence consists of two monotone subsequences bounding any solution of the problem (1) and

(2) from above and below. Thus, each successive iteration gives a more accurate pointwise upper or lower bound. This sequence is shown to converge uniformly to obtain an existence and uniqueness theorem. The solution thus obtained need not be nonnegative, in contrast to that obtained by Chan and Young [2].

2. ALTERNATING BOUNDS

In what follows, we shall use the following conditions:

(i) $g(x,y,t;u) \leq g(x,y,t;v)$ whenever $u > v$ at (x,y,t);

(ii) g is twice continuously differentiable, and there exists a positive constant b such that

$$|g(x,y,t;u)| \leq b;$$

(iii) there exists a positive constant c such that

$$g(x,y,t;v) - g(x,y,t;u) \leq c[u(x,y,t) - v(x,y,t)]$$

whenever $u > v$ at (x,y,t).

Since g is twice continuously differentiable, we can use the solution formula given by Diaz and Ludford [4] for a linear problem to convert the problem (1) and (2) into the integral equation

$$u(x,y,t) = \int_D K(x,y,t;\xi,\eta,\tau)g(\xi,\eta,\tau;u(\xi,\eta,\tau))d\xi d\eta d\tau, \tag{3}$$

where $K(x,y,t;\xi,\eta,\tau)$ is the kernel (cf. Davis [3] and Young [7]) of the operator L given by

$$K(x,y,t;\xi,\eta,\tau) = \frac{2^{k-1}\tau^k R_-^{-1/2}}{\pi R_+^{k/2}} F(\frac{k}{2},\frac{k-1}{2};\frac{1}{2};z),$$

$$R_- = (t - \tau)^2 - (\xi - x)^2 - (\eta - y)^2,$$

$$R_+ = (t + \tau)^2 - (\xi - x)^2 - (\eta - y)^2,$$

$$z = R_-/R_+,$$

F is a hypergeometric function, and D is the domain bounded by the retrograde cone

$$(\xi - x)^2 + (\eta - y)^2 = (t - \tau)^2, \quad 0 < \tau < t,$$

and the plane $t = 0$. We note that since $0 < z < 1$ in D, the hypergeometric function is positive (cf. Rainville [6, p. 45 for $k \geq 1$, and p. 60 for $0 < k < 1$]), and hence the kernel $K(x,y,t;\xi,\eta,\tau)$ is also positive.

Let us construct a sequence $\{u_i(x,y,t)\}$ by

$$Lu_0(x,y,t) = b, \quad 0 < t \leq T,$$

$$u_0(x,y,0) = 0, \quad \partial u_0(x,y,0)/\partial t = 0,$$

and for $i = 0, 1, 2, \ldots,$

$$Lu_{i+1}(x,y,t) = g(x,y,t;u_i), \quad 0 < t \leq T, \tag{4}$$

$$u_{i+1}(x,y,0) = 0, \quad \partial u_{i+1}(x,y,0)/\partial t = 0.$$

Since the linear problem has a unique solution, we can easily verify that

$$u_0(x,y,t) = bt^2/[2(k + 1)] > 0, \quad 0 < t \leq T. \tag{5}$$

It follows from (3) with $g(\xi,\eta,\tau;u)$ being replaced by $g(\xi,\eta,\tau;u_i)$ that the sequence $\{u_i\}$ is well defined.

The following theorem gives the alternating bounds.

THEOREM 1. Under conditions (i) and (ii), any solution u of the problem (1) and (2) satisfies

$$-u_0 \leq u_1 \leq \cdots \leq u_{2i+1} \leq \cdots \leq u \leq \cdots \leq u_{2i} \leq \cdots \leq u_0, \quad 0 < t \leq T. \tag{6}$$

Proof. Since

$$L(u_0 - u) = b - g(x,y,t;u) \geq 0, \quad 0 < t \leq T,$$

$$u_0(x,y,0) - u(x,y,0) = 0, \quad \partial(u_0 - u)(x,y,0)/\partial t = 0,$$

it follows from (3) that $u_0 \geq u$ for $0 < t \leq T$. This in turn implies that

$$L(u - u_1) = g(x,y,t;u) - g(x,y,t;u_0) \geq 0$$

by using condition (i). From (3), $u \geq u_1$ for $0 < t \leq T$. Also,

$$L(u_1 + u_0) = g(x,y,t;u_0) + b \geq 0$$

implies $u_1 \geq -u_0$. Thus, we have

$$-u_0 \leq u_1 \leq u \leq u_0, \quad 0 < t \leq T.$$

To complete the proof of the theorem, we use the principle of mathematical induction. Let us assume that for a particular value of i, say j, we have for $0 < t \leq T$,

$$u_1 \leq \cdots \leq u_{2j+1} \leq u \leq u_{2j} \leq \cdots \leq u_0.$$

Then for $i = j + 1$, we have from condition (i) that

$$L(u_{2j+2} - u) = g(x,y,t;u_{2j+1}) - g(x,y,t;u) \geq 0,$$

which implies $u_{2j+2} \geq u$. By using the above procedure for $u_{2j} - u_{2j+2}$, $u - u_{2j+3}$, and $u_{2j+3} - u_{2j+1}$ respectively, we have

$$u_{2j} \geq u_{2j+2}, \quad u \geq u_{2j+3}, \quad u_{2j+3} \geq u_{2j+1}.$$

Hence (6) follows.

The subsequence $\{u_{2i+1}\}$ is monotone nondecreasing and bounded above by u_0 while the other subsequence $\{u_{2i}\}$ is monotone nonincreasing and bounded below by u_1. Even if the odd and the even subsequences converge respectively to a lower bound and an upper bound, neither of them may be a solution. The next result shows that under an additional assumption, the alternating sequence does converge to a unique solution, which need not be nonnegative.

THEOREM 2. Under the hypotheses of Theorem 1, if condition (iii) holds, then the problem (1) and (2) has a unique solution, to which the sequence $\{u_i\}$ converges uniformly.

Proof. From (5) and (6),

$$u_0 - u_1 \leq bt^2/(k+1).$$

Using (4) and condition (iii), we have

$$\begin{aligned} u_2(x,y,t) - u_1(x,y,t) &= \int_D K(x,y,t;\xi,\eta,\tau)[g(\xi,\eta,\tau;u_1) - g(\xi,\eta,\tau;u_0)]d\xi d\eta d\tau \\ &\leq c\int_D K(x,y,t;\xi,\eta,\tau)[u_0(\xi,\eta,\tau) - u_1(\xi,\eta,\tau)]d\xi d\eta d\tau \\ &\leq \frac{bc}{k+1}\int_D K(x,y,t;\xi,\eta,\tau)\tau^2 d\xi d\eta d\tau. \end{aligned}$$

Let

$$\begin{aligned} \xi - x &= \alpha(t-\tau), \\ \eta - y &= \beta(t-\tau), \\ \gamma^2 &= \alpha^2 + \beta^2. \end{aligned}$$

Then,

$$u_2(x,y,t) - u_1(x,y,t) \leq \frac{bc}{k+1}\int_0^t \iint_{\alpha^2+\beta^2 \leq 1} (t-\tau)^2 K(x,y,t;\xi,\eta,\tau)\tau^2 d\alpha d\beta d\tau.$$

From Diaz and Ludford [4],

$$(t-\tau)^2 K(x,y,t;\xi,\eta,\tau) \leq At/[2\pi(1-\gamma^2)^{1/2}],$$

where $0 < \gamma < 1$, $0 < \tau < t$, and A is a constant depending on k. Thus,

$$\begin{aligned} u_2(x,y,t) - u_1(x,y,t) &\leq \frac{Abct}{2\pi(k+1)}\int_0^t \tau^2 d\tau \iint_{\alpha^2+\beta^2 \leq 1} (1-\gamma^2)^{-1/2} d\alpha d\beta \\ &= \frac{Abct}{2\pi(k+1)}\int_0^t \tau^2 d\tau \int_0^{2\pi} d\theta \int_0^1 \gamma/(1-\gamma^2)^{1/2} d\gamma \\ &= \frac{Abc}{3(k+1)}t^4. \end{aligned}$$

It follows from the principle of mathematical induction that

$$|u_{n+1} - u_n| \le \frac{bA^n c^n t^{2(n+1)}}{(k+1)(1)(3)(5)\cdots(2n+1)}. \tag{7}$$

By the ratio test, the series

$$\sum_{n=0}^{\infty} (u_n - u_{n+1})$$

converges uniformly and absolutely. Since

$$\sum_{n=0}^{\infty} (u_n - u_{n+1}) = \lim_{n \to \infty} (u_0 - u_{n+1}),$$

it follows that

$$\lim_{n \to \infty} u_n$$

exists. From the Lebesgue Convergence Theorem, this limit is a solution of the integral equation (3), and hence is a solution of the problem (1) and (2).

To prove uniqueness of the solution, we observe from (6) and (7) that

$$0 \le u_{2n} - u_{2n+1} \le \frac{bA^{2n} c^{2n} t^{2(2n+1)}}{(k+1)(1)(3)(5)\cdots(4n+1)}$$

$$\le \frac{bT^2}{k+1} \times \frac{AcT^2}{3} \times \frac{AcT^2}{5} \times \frac{AcT^2}{7} \times \cdots \times \frac{AcT^2}{4n+1},$$

which tends to zero as n tends to infinity. Thus,

$$\lim_{n \to \infty} u_{2n} = \lim_{n \to \infty} u_{2n+1},$$

from which uniqueness follows.

REFERENCES

[1] C. Y. Chan, Temperatures of linearly and nonlinearly radiating semi-infinite rods, Quart. Appl. Math. 34 (1976), 173-181.

[2] C. Y. Chan and E. C. Young, Maximal and minimal nonnegative solutions of a semilinear Euler-Poisson-Darboux equation, J. Math. Anal. Appl., in press.

[3] R. M. Davis, On a regular Cauchy problem for the Euler-Poisson-Darboux equation, Ann. Mat. Pura Appl., Ser. IV, 42 (1956), 205-226.

[4] J. B. Diaz and G. S. S. Ludford, On the singular Cauchy problem for a generalization of the Euler-Poisson-Darboux equation in two space variables, Ann. Mat. Pura Appl., Ser. IV, 38 (1955), 33-50.

[5] J. B. Keller, Elliptic boundary value problems suggested by nonlinear diffusion processes, Arch. Rational Mech. Anal. 35 (1969), 363-381.

[6] E. D. Rainville, "Special Functions," Macmillan Co., New York, 1960.

[7] E. C. Young, On a method of determining the Riesz kernel for the Euler-Poisson-Darboux operator, J. Math. Anal. Appl. 16 (1966), 355-362.

COMPARISON THEOREMS AND STOCHASTIC BOUNDARY VALUE PROBLEMS

Jagdish Chandra

U. S. Army Research Office
Research Triangle Park, North Carolina

G. S. Ladde*

Department of Mathematics
The University of Texas
Arlington, Texas

1. INTRODUCTION

It is well known that the comparison principle provides an important tool in the study of initial-value problems for ordinary differential equations. A comprehensive review of these results can be found in [6] and [5] respectively for the deterministic and the stochastic cases. Attempts have been made recently for obtaining similar results for boundary value problems. In particular, in the case of stochastic boundary value problems (SBVP), this was motivated by a desire to analyze the effects of roughness in a compressible gas lubricated slider bearing problem [1,2]. The pressure in such a bearing is described by a nonlinear Reynolds equation with random coefficients modeling the roughness of the sliding surfaces. A more realistic bearing geometry (namely, finite width) leads to a nonlinear partial differential equation of elliptic type.

In this paper we present a general comparision principle for the stochastic boundary value problem of the elliptic type. As a corollary to this result we extend the classical maximal principle for elliptic equations [7] to the corresponding stochastic case. Our results are obtained in the context of sample calculus approach [5] which has proved to be quite useful in modeling systems whose dynamics is governed by random forces.

2. NOTATIONS AND DEFINITIONS

We shall denote by R^m, m-dimensional Euclidean space; points $x \in R^m$

*Research partially supported by U. S. Army Research Grant #DAAG29-81-G-0008.

are written as $(x_1,x_2,\ldots,x_m)^T = x$, and $\|x\|$ is defined by $\|x\| = \left(\sum_{i=1}^{m} x_i^2\right)^{1/2}$, where T stands for the transpose of a vector or matrix. (Ω,F,P) stands for a complete probability space. We shall denote by D, an open connected and bounded subset of R^m, $\bar{D}$ and ∂D denote the closure and the boundary of D, respectively. Without loss in generality, all random functions that are considered in this paper are separable processes.

Definition 2.1. For $A \subseteq D$ a random function $u(x,\omega)$ defined on $Ax(\Omega,F,P)$ into R is said to be almost-surely sample Hölder-continuous of exponent α, $0 < \alpha < 1$, if there exists a random variable

$$(2.1) \qquad |u(x,\omega)-u(y,\omega)| \leq H(\omega)\|x-y\|^{\alpha} \quad \text{for all } x,y \in A.$$

Note that the quantity $H_A^{\alpha}(u,\omega)$ defined by

$$H_A^{\alpha}(u,\omega) = \sup_{\substack{x,y\ A \\ x\neq y}} \left\{\frac{|u(x,\omega)-u(y,\omega)|}{\|x-y\|^{\alpha}}\right\}$$

is a random variable, and $H_A^{\alpha}(u,\omega)$ is the smallest of $H(\omega)$.

We say that $u \in C^{k,\alpha}[A,R[(\Omega,F,P),R]]$, if the sample partial derivatives of u up to order k are almost-surely sample continuous on A and the k-th sample partial derivatives are almost-surely sample Hölder continuous with an exponent α. $R[(\Omega,F,P),R]$ stands for the collection of all random variables defined on (Ω,F,P) into R. For $u \in C^1[A,R[\Omega,F,P),R]]$, we shall use the notation $\Delta u = (\frac{\partial u}{\partial x_1}, \frac{\partial u}{\partial x_2},\ldots,\frac{\partial u}{\partial x_m})^T$.

Let a_{ij}, b_i and $c \in C^{\alpha}[\bar{D},R[(\Omega,F,P),R]]$, $c(x,\omega) \leq 0$ in $\bar{D}$ with probability one (w.p. 1). Let $L(\Omega)$ be a second order random elliptic operator defined by

$$(2.2) \qquad L(\omega) = \sum_{i,j=1}^{m} a_{ij}(x,\omega)\frac{\partial^2}{\partial x_i \partial x_j} + \sum_{i=1}^{m} b_i(x,\omega)\frac{\partial}{\partial x_i} + c(x,\omega).$$

Definition 2.2. The stochastic operator $L(\omega)$ defined by (2.2) is said to be elliptic with probability one at a point $x \in \bar{D}$, if the coefficient random matrix process $(a_{ij}(x,\omega))$ is almost-surely positive-definite, that is, there exist two random functions such that

$$(2.3) \qquad 0 < \lambda(x,\omega)\|y\|^2 \leq \sum_{i,j=1}^{m} a_{ij}(x,\omega)y_i y_j \leq \Lambda(x,\omega)\|y\|^2, \quad \text{w.p. } 1,$$

for all $y \in R^m - \{0\}$. If $\lambda(x,\omega) > 0$ on D, w. p. 1, then L is almost-surely elliptic in D, and almost-surely sample elliptic in D, if $\lambda(x,\omega) \geq \lambda_0(\omega) > 0$ for some random number $\lambda_0(\omega)$. If $\frac{\Lambda(x,\omega)}{\lambda(x,\omega)} \leq M(\omega)$ in D w.p. 1, we shall say that L is almost-surely uniformly sample elliptic in D, that is, (2.3) can be written as

$$(2.4) \qquad M^{-1}(\omega)\|y\|^2 \leq \sum_{i,j=1}^{m} a_{ij}(x,\omega)y_iy_j \leq M(\omega)\|y\|^2, \quad \text{w.p. } 1$$

for all $y \in R^m$.

Let $\nu(x) = (\nu_1(x),\nu_2(x),\ldots,\nu_m(x))^T$ denote the unit normal vector field on ∂D. Let $p,q \in C^{1,\alpha}[\partial D,R[(\Omega,F,P),R]]$ be non-negative real-valued processes which do not vanish simultaneously with probability one. For $u \in C^1[\bar{D},R[(\Omega,F,P),R]]$, we define

$$(2.5) \qquad B(\omega)u = p(x,\omega)u + q(x,\omega)\frac{du}{d\nu},$$

where $\frac{du}{d\nu}$ denotes the sample normal derivative of u.

Consider the stochastic boundary value problem (SBVP)

$$(2.6) \qquad L(\omega)u = f(x,u,\Delta u,\omega) \quad \text{in } D, \quad \text{w.p. } 1,$$

and

$$(2.7) \qquad B(\omega)u = \phi(x,\omega), \; x \in \partial D, \quad \text{w.p. } 1,$$

where $L(\omega)$ and $B(\omega)$ are as defined in (2.1) and (2.5), respectively; $\phi \in C^{1,\alpha}[\bar{D},R[(\Omega,F,P),R]]$, $f \in C^{\alpha}[\bar{D}\times R\times R^m,R[(\Omega,F,P),R]]$.

3. COMPARISON THEOREMS

Let us introduce the comcepts of upper and lower solution processes of (2.6) - (2.7).

Definition 3.1. Let $\beta \in C^2[\bar{D},R[(\Omega,F,P),R]]$. $\beta(x,\omega)$ is said to an upper solution process of (2.6) - (2.7), if $\beta(x,\omega)$ satisfies

$$(3.1) \qquad L(\omega)\beta(x,\omega) \leq f(x,\beta(x,\omega),\Delta\beta(x,\omega),\omega) \quad \text{in } D,$$

and

$$(3.2) \qquad B(\omega)\beta(x,\omega) \geq \phi(x,\omega) \quad \text{on } \partial D, \quad \text{w.p. } 1;$$

similarly, for $\alpha \in C^2[\bar{D},R[(\Omega,F,P),R]]$, $\alpha(x,\omega)$ is called a lower solution process of (2.6) - (2.7), if $\alpha(x,\omega)$ satisfies

$$(3.3) \qquad L(\omega)\alpha(x,\omega) \geq f(x,\alpha(x,\omega),\Delta\alpha(x,\omega),\omega) \quad \text{in } D,$$

and

$$(3.4) \qquad B(\omega)\alpha(x,\omega) \leq \phi(x,\omega) \quad \text{on } \partial D, \quad \text{w.p. } 1.$$

Now, we are in a position to state the following comparison theorem:

<u>Theorem 3.1.</u> Assume that

(i) $\alpha,\beta \in C^2[\bar{D},R[(\Omega,F,P),R]]$, α and β respectively are the lower and the upper solutions processes of (2.6) - (2.7);

(ii) $f(x,u,v,\omega)$ is as defined in (2.6) and is almost-surely sample increasing in u for fixed $(x,v) \in \bar{D} \times R^m$;

(iii) $L(\omega)$ is almost-surely sample elliptic in D and $c(x,\omega)$ in $L(\omega)$ is non-positive process with probability one.

Then, for

(3.6) $$B(\omega)\alpha(x,\omega) \leq B(\omega)\beta(x,\omega) \quad \text{on } \partial D, \text{ w.p. } 1$$

implies

(3.7) $$\alpha(x,\omega) \leq \beta(x,\omega) \quad \text{on } \bar{D}, \text{ w.p. } 1.$$

<u>Proof:</u> Assume that the conclusion of the theorem is false. Then there exist $x_0 \in \bar{D}$, a positive random variable $\varepsilon(\omega)$ and an event $\Omega_0 \in F$ such that

(a) $P(\Omega_0) > 0$, (b) $\omega \in \Omega_0$, $\varepsilon(\omega) > 0$ w.p. 1, (c) $\alpha(x_0,\omega) - \beta(x_0,\omega) = \varepsilon(\omega)$ and $\alpha(x,\omega) - \beta(x,\omega) \leq \varepsilon(\omega)$ for $x \in \bar{D}$, $\omega \in \Omega_0$. From (c), for $x_0 \in D$, we have

(3.8) $$\alpha(x_0,\omega) > \beta(x_0,\omega),\ \Delta\alpha(x_0,\omega) = \Delta\beta(x_0,\omega)$$

and

$$\sum_{i,j=1}^{m} a_{ij}(x_0,\omega)\left(\frac{\partial^2\alpha(x_0,\omega)}{\partial x_i \partial x_j} - \frac{\partial^2\beta(x_0,\omega)}{\partial x_i \partial x_j}\right) \leq 0.$$

From the above discussion and assumption (iii)

(3.9) $$L(\omega)\alpha(x_0,\omega) \leq L(\omega)\beta(x_0,\omega) \quad \text{for } \omega \in (\Omega_0 - N)$$

and $N \subset \Omega_0$ with $P(N) = 0$.

On the other hand, from (3.8), assumptions (i) and (ii), we get

$$\begin{aligned} L(\omega)\alpha(x_0,\omega)-L(\omega)\beta(x_0,\omega) &\geq f(x_0,\alpha(x_0,\omega),\Delta\alpha(x_0,\omega),\omega)-f(x_0,\beta(x_0,\omega),\Delta\beta(x_0,\omega),\omega) \\ &\geq f(x_0,\beta(x_0,\omega),\Delta\beta(x_0,\omega),\omega)-f(x_0,\beta(x_0,\omega),\Delta\beta(x_0,\omega),\omega) \\ &> 0 \quad \omega \in \Omega_0 \text{ w.p. } 1. \end{aligned}$$

This contradicts (3.9). Thus the maximum of $\alpha(x,\omega) - \beta(x,\omega)$ can not occur in the interior of $\bar{D}$ with positive probability. This implies that $\alpha(x,\omega) - \beta(x,\omega) \leq 0$ on $\bar{D}$, w.p. 1. To complete the proof of theorem, we need to verify that the maximum of the process $\alpha(x,\omega) - \beta(x,\omega)$ with positive probability can not be attained at the boundary of D. We assume that $x_0 \in \partial D$. Because of the sample continuity of the stochastic process $\alpha(x,\omega) - \beta(x,\omega)$ on $\bar{D}$, one can conclude that

$$\alpha(x,\omega) - \beta(x,\omega) < \alpha(x_0,\omega) - \beta(x_0,\omega) \quad \text{on} \quad \bar{D} \cap S_\rho$$

where $S_\rho = \{x \in R^m \colon \|x-x_0\| < \rho\}$. This implies that $\frac{d}{d\nu}(\alpha(x_0,\omega)-\beta(x_0,\omega)) > 0$. Therefore,

$$\frac{d\alpha}{d\nu}(x_0,\omega) > \frac{d\beta}{d\nu}(x_0,\omega) \quad \text{on} \quad (\Omega_0-N) \quad \text{with} \quad P(N) = 0.$$

Thus $B(\omega)\alpha(x_0,\omega) > B(\omega)\beta(x_0,\omega)$ on (Ω_0-N) with $P(N) = 0$. This contradicts the fact that $B(\omega)\alpha(x_0,\omega) \leq B(\omega)\beta(x_0,\omega)$ with probability one. This completes the proof of the theorem.

A special case of Theorem 3.1, we present a stochastic version of the deterministic maximum principle [4,7] as follows:

Corollary 3.1. Let $L(\omega)$ be as defined in Theorem 3.1. Assume that $\phi \in C^2[\bar{D}, R[(\Omega,F,P),R]]$,

$$L(\omega)\phi \geq 0 \quad \text{in} \quad D, \tag{3.10}$$

and

$$B(\omega)\phi \geq 0 \quad \text{on} \quad \partial D, \quad \text{w.p. 1.} \tag{3.11}$$

Then

$$\phi(x,\omega) \geq 0 \quad \text{on} \quad \bar{D} \quad \text{w.p. 1.} \tag{3.12}$$

We present another special case of Theorem 3.1. This result provides a sample estimate on a process by means of a solution process of an auxiliary SBVP. We simply state such a result whose proof can be readily formulated based on the proof of Theorem 3.1.

Theorem 3.2. Assume that

(i) $v \in C^2[\bar{D}, R[(\Omega,F,P),R]]$ and satisfies

$$L(\omega)v(x,\omega) \geq g(x,v(x,\omega),\Delta v(x,\omega),\omega) \quad \text{on} \quad D,$$

and

$$B(\omega)v(x,\omega) \leq B(\omega)r(x,\omega) \quad \text{on} \quad \partial D, \quad \text{w.p. 1,}$$

where $L(\omega)$ and $B(\omega)$ are as defined before;

(ii) $r(x,\omega)$ is the maximal sample solution process of

$$L(\omega)u(x,\omega) = g(x,u(x,\omega),\Delta u(x,\omega),\omega) \quad \text{in} \quad D \tag{3.13}$$

with $F(\omega)u(x,\omega) = r(x,\omega)$ on ∂D, w.p. 1, where $F(\omega)$ is defined similar to $B(\omega)$ in (2.7);

(iii) $g \in C^{\alpha}[\bar{D}\times R\times R^m, R[(\Omega,F,P),R]]$ and $g(x,u,y,\omega)$ is almost-surely sample increasing in u for fixed $(x,y) \in \bar{D} \times R^m$;

(iv) $L(\omega)$ satisfies hypothesis (iii) of Theorem 3.1. Then

$$v(x,\omega) \le r(x,\omega) \quad \text{on} \quad \bar{D}, \quad \text{w.p. } 1. \tag{3.14}$$

Remark 3.1. A comparison result dual to Theorem 3.2 holds with appropriate modifications yielding $\rho(x,\omega) \le v(x,\omega)$ on $\bar{D}$ w.p. 1, where $\rho(x,\omega)$ is the minimal solution process and $v(x,\omega)$ is an upper solution process of (3.13).

Finally, we present a comparison result which is useful in obtaining error estimates and in the study of stability and uniqueness of solution processes of (2.6) - (2.7).

Theorem 3.3. Assume that

(i) $V \in C^2[\bar{D}\times R, R_+]$ and

$$L(\omega)V(x,u-m) \ge g(x,V(x,u-m),\Delta V(x,u-m),\omega) \quad \text{in } D \text{ w.p. } 1,$$

where $\Delta v = \left(\frac{\partial V}{\partial x_1} + \frac{\partial V}{\partial z}\cdot\frac{\partial z}{\partial x_1},\ldots,\frac{\partial V}{\partial x_m} + \frac{\partial V}{\partial z}\cdot\frac{\partial z}{\partial x_m}\right)^T = \left(\frac{\partial v}{\partial x_1},\ldots,\frac{\partial v}{\partial x_m}\right)^T,$

$$\frac{\partial^2 v}{\partial x_i \partial x_j} = U_{ij}(x,u-m,\Delta(u-m)) + \frac{\partial V}{\partial z}\frac{\partial^2 z}{\partial x_i \partial x_j},\quad U_{ij} = \frac{\partial^2 V}{\partial x_i \partial x_j} + \frac{\partial^2 V}{\partial z \partial x_i}\cdot\frac{\partial z}{\partial x_i} +$$

$$\frac{\partial^2 V}{\partial x_i \partial z}\cdot\frac{\partial z}{\partial x_j} + \frac{\partial^2 V}{\partial z^2}\cdot\frac{\partial z}{\partial x_i}\frac{\partial z}{\partial x_j},$$

$$\underset{(f-\hat{f})}{L(\omega)} V(x,\omega) = \sum_{i,j=1}^{m} a_{ij}(x,\omega)U_{ij} + \sum_{i=1}^{m} b_i(x,\omega)\frac{\partial V}{\partial x_i} + \frac{\partial V}{\partial z}[(f-\hat{f})+DL(\omega)m(x)],$$

and $DL(\omega) = \sum_{i,j=1}^{m} (\hat{a}_{ij}(x)-a_{ij}(x,\omega))\frac{\partial^2}{\partial x_i \partial x_j} + \sum_{i=1}^{m} (\hat{b}_i(x)-b_i(x,\omega))\frac{\partial}{\partial x_i} +$

$(\hat{c}(x)-c(x,\omega))$, $\hat{a}_{ij}$, $\hat{b}_i$, $\hat{c}$ and $\hat{f}$ are as defined in (iii);

g is as defined in Theorem 3.2;

(ii) the boundary conditions (2.7) imply

$$F(\omega)V(x,u(x,\omega)-m(x)) \le r(x,\omega) \quad \text{on } \partial D, \quad \text{w.p. } 1;$$

(iii) $u(x,\omega)$, $r(x,\omega)$ and $m(x)$ are sample solution process of (2.6) - (2.7), the maximal solution process of (3.13), and a solution of the corresponding deterministic elliptic boundary value problem

$$Lm = \hat{f}(x,m,\Delta m) \quad \text{on} \quad D \tag{3.15}$$

with

$$Bm(x) = \hat{\phi}(x) \quad \text{on} \quad \partial D,$$

respectively, where L is defined by

$$L = \sum_{i,j=1}^{m} \hat{a}_{ij}(x)\frac{\partial^2}{\partial x_i \partial x_j} + \sum_{i=1}^{m} \hat{b}_i(x)\frac{\partial}{\partial x_i} + \hat{c}(x)$$

and B is defined by $Bm(x) = \hat{p}(x)m(x) + \hat{q}(x)\frac{dm}{d\nu}(x)$.

Then,

$$V(x,u(x,\omega)-m(x)) \le r(x,\omega) \quad \text{on} \quad \overline{D}, \quad \text{w.p. 1.}$$

Proof. Let $u(x,\omega)$ and $m(x)$ be solutions of (2.6) - (2.7) and (3.15), respectively. Now, by setting $v(x,\omega) = V(x,u(x,\omega)-m(x))$, one can verify that $v(x,\omega)$ satisfies the hypotheses of Theorem 3.2. Therefore, by the application of Theorem 3.2, one can conclude the validity of the theorem.

4. APPLICATIONS

Consider a gas lubricated (finite) slider-bearing problem governed by the nonlinear stochastic Reynolds equation in two-dimension

$$\frac{\partial}{\partial x_1}\left(H_1^3(x_1,x_2,\omega)p\frac{\partial p}{\partial x_1}\right) + \frac{\partial}{\partial x_2}\left(H_1^3(x_1,x_2,\omega)p\frac{\partial p}{\partial x_2}\right) = \Lambda\frac{\partial}{\partial x_1}(H_2(x_1,x_2,\omega)p) \tag{4.1}$$

subject to the boundary conditions

$$\begin{cases} p(0,x_2,\omega) = p(1,x_2,\omega) = 1, \\ p(x_1,-\ell,\omega) = p(x_1,\ell,\omega) = 1. \end{cases} \tag{4.2}$$

Here $p(x_1,x_2,\omega)$ is the unknown pressure; $H_1(x_1,x_2,\omega)$ is the known profile of the film thickness process. We shall assume that

$$H_1(x_1,x_2,\omega) = h(x_1,x_2) + \delta_1(x_1,x_2,\omega) + \delta_2(x_1,x_2,\omega),$$

and

$$H_2(x_1,x_2,\omega) = h(x_1,x_2) + v(\delta_1(x_1,x_2,\omega)-\delta_2(x_1,x_2,\omega));$$

$E[H_i(x_1,x_2,\omega)] = h(x_1,x_2)$; $v = (v_2-v_1)/(v_2+v_1)$, v_1 and v_2 are the rolling speeds of the lower and upper surfaces; $|\delta_1(x_1,x_2,\omega)| + |\delta_2(x_1,x_2,\omega)| \le h((x_1,x_2)$; $\Lambda = \sigma\mu(v_1+v_2)$ and μ is the viscosity of the lubricant.

By setting $p^2 = 2u$, the SBVP (4.1) - (4.2) is transformed into the quasilinear problem given by

$$\frac{\partial}{\partial x_1}\left(H_1^3\frac{\partial u}{\partial x_1}\right) + \frac{\partial}{\partial x_2}\left(H_1^3\frac{\partial u}{\partial x_2}\right) = \Lambda\frac{\partial}{\partial x_1}(H_2\sqrt{2u}\,) \tag{4.3}$$

and

$$\begin{cases} u(0,x_2,\omega) = u(1,x_2,\omega) = \frac{1}{2}, \\ u(x_1,-\ell,\omega) = u(x_1,\ell,\omega) = \frac{1}{2}. \end{cases} \tag{4.4}$$

The obvious choice of a lower solution process of (4.3) - (4.4) is $\alpha(x_1,x_2,\omega) = \frac{1}{2}$ if $\frac{\partial H_2}{\partial x_1}(x_1,x_2,\omega) \le 0$, w.p. 1. A choice of an upper solution process $\beta(x_1,x_2,\omega)$ is given by

$$\beta(x_1,x_2,\omega) = \frac{1}{2}\Big[1 + \Lambda\int_0^{x_1} H_1^{-3}(x_1,s,\omega)H_2(x_1,s,\omega)ds + \Lambda\int_{-\ell}^{x_2} H_1^{-3}(x,s,\omega)H_2(x_1,s,\omega)ds\Big]^2,$$

if

$$\frac{\partial}{\partial x_1}\left\{H_1^3\left[1 + \Lambda\int_0^{x_1} H_1^{-3}H_2 ds + \Lambda\int_{-\ell}^{x_2} H_2^{-3}ds\right]\left[\int_{-\ell}^{x_2}\frac{\partial}{\partial x_1}(H_1^{-3}H_2)ds\right]\right\} +$$

$$\frac{\partial}{\partial x_2}\left\{\left[1 + \Lambda\int_0^{x_1} H_1^{-3}H_2 ds + \int_{-\ell}^{x_2} H_1^{-3}H_2 ds\right]\left[H_2 + H_1^3\int_0^{x_1}\frac{\partial}{\partial x_2}(H_1^{-3}H_2)ds\right]\right\} \le 0.$$

We note that $\alpha(x_1,x_2,\omega) \le \beta(x_1,x_2,\omega)$ and $\alpha(x_1,x_2,\omega) \le u(x_1,x_2,\omega) \le \beta(x_1,x_2\omega)$ on ∂D w.p. 1, where $D = \{(x_1,x_2) \in R^2 \colon 0 < x_1 < 1, -\ell < x_2 < \ell\}$. By considering $V(x,u-m) = (u-m)^2$, one can complete $L(\omega)_{(f-\hat{f})}V$ and obtain the error estimates. The details are under investigation and will be reported elsewhere.

REFERENCES

[1] J. Chandra, G. S. Ladde and V. Lakshmikantham, "On the Fundamental Theory of Nonlinear Second Order Stochastic Boundary Value Problems," Stochastic Anal. and Appl, 1, (1983), 1-19.

[2] J. Chandra, G. S. Ladde, V. Lakshmikantham, "Stochastic Analysis of a Compressible Gas Lubricated Slider Bearing Problem," SIAM J. Appl. Math. (in press).

[3] J. Chandra, V. Lakshmikantham and S. Leela, "Comparison Principle and Theory of Nonlinear Boundary Value Problems," Nonlinear Phenomena in Mathematical Sciences (Edited by V. Lakshmikantham), Academic

Press, New York, (1982), 241-248.

[4] D. Gilberg and N. S. Trudinger , Elliptic Partial Differential Equations of Second Order, Springer-Verlag, Berlin Heidelbert, New York, 1977.

[5] G. S. Ladde and V. Lakshmikantham, Random Differential Inequalities, Academic Press, New York, 1980.

[6] V. Lakshmikantham and S. Leela, Differential and Integral Inequalities I and II, Academic Press, New York, 1969.

[7] M. H. Protter and H. F. Weinberger, Maximum Principles in Differential Equations, Prentice-Hall, Inc., New Jersey, 1967.

[8] G. Vidossich, "Comparison, existence, uniqueness and successive approximations for the Dirichlet problem of elliptic equations," The University of Texas at Arlington, Technical Report #119, 1979.

CONSTRAINED GLOBAL NULL CONTROLABILITY OF NONLINEAR DELAY SYSTEMS

E.N. Chukwu

Federal University of Technology
Yola, Nigeria

ABSTRACT

We state sufficient conditions for the Euclidean null-controllability of differential delay systems when it is assumed that the controls are "small" with values in a unit m-dimensional Euclidean cube. The systems considered have delays in the state as well as in the controls. Both linear and nonlinear systems are considered. The results extend classical theorems for ordinary differential equations.

1. INTRODUCTION

In this paper we consider the global null controllability, with controls whose values are contained in an m-dimensional Euclidean cube, of the linear delay system,

$$\dot{x}(t) = L(x,u), \tag{1}$$

and its nonlinear perturbation,

$$\dot{x}(t) = L(x,u)+f(t,x(t),x(t-1),u(t),u(t-h)), \tag{2}$$

where the operator L is defined by

$$L(x,u) = A(t)x(t)+B(t)u(t-1) + \int_{-1}^{o} K(t,s)x(t+s)ds + C(t)u(t) + D(t)u(t-h).$$

We will show that if the system (1) is completely Euclidean controllable, and if the delay system

$$\dot{x}(t)=L(x,o)=A(t)x(t)+B(t)x(t-1) + \int_{-1}^{o} K(t,s)x(t+s), \tag{3}$$

is globally uniformly asymptotically stable, so that every solution $x(.,\sigma,\phi)$ of (3) with initial function ϕ at time σ satisfies

$$x(t,\sigma,\phi) \leq k\, e^{-\alpha(t-\sigma)}\, ||\phi||, \text{ for all } t \geq \sigma, \tag{4}$$

where $\alpha>o$, $k>1$ then (1) is globally Euclidean null controllable with controls whose values lie in

$$C^m = \{u \in E^m : \ |u_j| \leq 1,\ j = 1,\dots,m\} \tag{5}$$

where u_j is the jth component of the vector u. We shall show further that if (1) is completely Euclidean controllable, and if (3) is globally uniformly asymptotically stable so that solutions of (3) satisfy (4) then the system (2) is globally null controllable with controls in (5) provided f satisfies some growth conditions. For systems without delay the problem studied here is the predecessor of the classical result of Lasalle [1], and Markus [2]. For systems without delay in the control, an investigation was made in Chukwu [3] and [4]. Problems of controllability of the system (1) and (2) were studied in [5]. In [5] it was assumed that the controls are "big" and without constraints. Our method of solution here differs from those in [5], and exploits the properties of the reachable sets of the given control systems.

2. PRELIMINARIES

Throughout the sequel E^r represents the Euclidean r-dimensional space. The state x(t) is contained in E^n, and the control u(t) is contained in E^m. The matrix functions A,B,C and D are continuous on $[o,\infty)\equiv J$ and are of appropriate dimensions. The nxn matrix function K is continuous on J x [-1,o]. The function

$$f : J \times E^n \times E^n \times E^m \times E^m \to E^n$$

is continuous, and Lipschitzian in the last four arguments and is such that the solution $x(t,\sigma,\phi,u)$ if (2) is continuous in t,σ,ϕ,u. Functions $u:[-h,\infty)\to E^m$, h>o will be called admissible if u(t) is square integrable on $[-h,\infty)$ with u(t) = o for $t \in [-h,o)$.

We know from Banks [6] that if ϕ is an n-vector initial function which is continuous on [-1,o] then there exists a unique solution of the system (1) on J which satisfies $x(t) = \phi(t)$ $t \in [-1,o]$. This solution denoted by x_L is given for $t\in J$ $t\geq\sigma$ by

$$x_L(t) = x_L(t,\sigma,\phi,u)$$

$$x_L(t) = X(t,o)\phi(o) + \int_{-1}^{o} X(t,s+1)\ B(s+1)\phi(s)ds$$

$$+\int_{-1}^{o}\int_{o}^{w+1} X(t,s)\,K(s,w-s)ds\,dw +$$

$$+\int_{o}^{t} X(t,s)\,[C(s)u(s) + D(s)u(s-h)]ds \tag{8}$$

where $X(t,s)$ is an nxn matrix function which satisfies

$$\partial X(t,s)/\partial t = A(t)X(t,s)+B(t)X(t-1,s) + \int_{-1}^{o} K(t,w)X(t+w,s)ds$$

for $o \leq s < t < \infty$, (9)

such that $X(t,t)= I$, the identity matrix, $X(t,s)=o$ for $t<s$. We now define the matrix Z by

$$Z(t,s)=\begin{cases} X(t,s)C(s) + X(t,s+h)D(s+h), & o \leq s \leq t-h \\ X(t,s)\,C(s) \text{ for } t-h<s<t \end{cases} \tag{10}$$

It is clear that $Z(t,s)$ is piecewise continuous in s for fixed t and the only point of discontinuity is $s = t-h$. Observe that $u(t)=o$ for $t<o$ so that the solution of (1) corresponding to u can be deduced from (8) to be

$$x_L(t) = x_L(t,\sigma,o) + \int_{o}^{t} Z(t,s)u(s)ds \quad t\varepsilon J \tag{11}$$

where

$$x_L(t,\phi,o) = X(t,o)\phi(o) + \int_{-1}^{o} X(t,s+1)B(s+1)\phi(s)ds + \int_{-1}^{o}\int_{o}^{w+1} X(t,s)\,K(s,w-s)\phi(w)ds\,dw \tag{12}$$

Obviously the solution of the nonlinear perturbed system is given by

$$x(t)=x_L(t) + \int^{t} Z(t,s)f(s,x(s),x(s-1),u(s),u(s-1)) \tag{13}$$

for $t \geq o$ and $x(t) = \phi(t)$ for $t\varepsilon[-1,o]$.

Definition 2.1. The system (1) is said to be Euclidean controllable on $[\sigma,t_1]$ if for each continuous function $\phi:[-1.o]\to E^n$ and each point $\xi \in E^n$ there is an admissible control $u \in L_2(\sigma,t_1],E^m)$ such that the solution $x(.,\sigma,\phi,u)$ of (1) satisfies $x_\sigma(.,\sigma,\phi,u)=\phi$, $x(t_1,\sigma,\phi,u)=\xi$. The symbol $L_2([\sigma,t_1],E^m)$ represents the space of square integrable functions $u:[\sigma,t_1]\to E^m$. The symbol $x_\sigma(\;)$ represents the function $x_\sigma:[-1,o]\to E^n$ such that $x_\sigma(s) = x(\sigma+s)$ $s\varepsilon[-1,o]$. The system is Euclidean null-controllable if the preceding definition can be met by $\xi = o$. We drop the qualifying phrase "on the interval $[\sigma,t_1]$ in the definitions above if the definitions hold on every interval $[\sigma,t_1],t_1>\sigma$. The above definitions can be formulated with controls in $L([\sigma,t_1],C^m)$, the space of square integrable functions $u:[\sigma,t_1]\to C^m$. In that case we say that (1) is controllable on $[\sigma,t_1]$ with constraints.

Definition 2.2. The reachable set of (1) on $[\sigma,t]$ is a subset of E^n given by

$$A(t,\sigma) = \{\int_\sigma^t Z(t,s)u(s)ds \; : \; u\varepsilon L_2([\sigma,t],E^n)\} \qquad (14)$$

If the controls are in $L_2([\sigma,t_1],C^m)$ we define the constrained reachable set by

$$IR(t,\sigma) = \{\int_\sigma^t Z(t,s)u(s)ds \; : \; u\varepsilon L_2([\sigma,t],C^m) \qquad (15)$$

Note that $A(t,\sigma)$ is a subspace of E^n which is symmetric about zero. Also $IR(t,\sigma)$ is a proper subset of E^n which is symmetric about zero.

Theorem 2.1. The following are equivalent

(1) The system (1) is Euclidean controllable on $[\sigma,t]$

(2) The matrix

$$W = \int_\sigma^t Z(t,s)\, Z^*(t,s)ds, \qquad (16)$$

where Z^* denotes the transpose of Z, is nonsingular. The proof of this is contained in Proposition 3.1 of Dauer and Gahl in [5].

Definition 2.3. The system (1) is said to be proper on $[\sigma,t]$ if and only if $o \in \text{Int } IR(t,\sigma)$, where Int $IR(t,\sigma)$ denotes the interior relative to E^n of $IR(t,\sigma)$. The system is proper if it is proper on each interval $[\sigma,t], t > \sigma$.

The next observation is crucial in what follows. It asserts that if (1) is controllable then it is proper and vice versa. As a consequence the domain of null controllability with constrained controls contains zero in its interior.

Theorem 2.2. The system (1) is Euclidean controllable on $[\sigma,t], t>\sigma$, if and only if (1) is proper.

Proof. Suppose (1) is Euclidean controllable on $[\sigma,t]$, $t>\sigma$ then

$$A(t,\sigma) = E^n.$$

Hence the mapping $T:L_2([\sigma,t],E^n)\to E^n$ given by

$$Tu = \int_o^t Z(t,s)u(s)ds$$

is a surjection. Also $T(L_2([\sigma,t],C^m)) = \mathit{IR}(t,\sigma)$. Let 0 be an open ball around zero with $0 \subseteq L_2([\sigma,t],C^m)$. Then

$$T(0) \subseteq T(L_2([\sigma,t],C^m) = \mathit{IR}(t,\sigma).$$

But then T is a continuous linear map, by a simple deduction from Corollary 1.1 of Hale [7:143]. Therefore (see [7, p 99]), T is an open map, so that T(0) is open and contains zero. This proves that $o \in \text{Int } \mathit{IR}(t,\sigma)$. Conversely assume that $o \in \text{Int } \mathit{IR}(t,\sigma)$. Then clearly $o \in \text{Int } A(t,\sigma)$. But since $A(t,\sigma)$ is a subspace which is symmetric about zero we have $A(t,\sigma) = E^n$, proving complete Euclidean controllability,

Definition 2.4. The domain of null controllability of (1) is the collection of initial continuous functions $\phi:[-1,o]\to E^n$ such that the solution $x(.,\sigma,\phi,u)$ of (1) for some t and some $u \in L([\sigma,t],C^m)$ satisfies

$$x_\sigma(.,\sigma,\phi,u) = \phi, \; x(t,\sigma,\phi,u) = 0.$$

If $\phi \in D$ then from (11)

$$x(t,\sigma,\phi,o) = -\int_\sigma^t Z(t,s)u(s)ds$$

so that

$$x(t,\sigma,C,o) \subseteq \mathit{IR}(t,\sigma), \; t > \sigma,$$

where $x(t,\sigma,C,o) = \{x_L(t,\sigma,\phi): \phi\epsilon C = C[-1,o],E^n)$

$x_L(t,\sigma,\phi)$ is a solution of (1)$\}$

Proposition 2.1. Assume that (1) is proper. Then the domain D of null controllability of (1) contains zero in its interior.

Proof. Assume (1) is proper. Then $o \in \text{Int } \mathit{IR}(t,\sigma)$ $t>\sigma$. Assume now that zero is not interior to D. Then there is a sequence $\{\phi_m\}^\infty$ of continuous functions $\phi_m:[-1,0]\to E^n$ such that $\phi_m\to o$ as $m\to\infty$, $\phi_m \notin D$, with $o \neq \phi_m$. Let $y_m=x(t,\sigma,\phi_m,o)$. Then $y_m\neq o$, $y_m \notin \mathit{IR}(t_1,\sigma)$ for any $t>\sigma$. Since $x(t,\sigma,\phi_m,o)$ is continuous in its arguments [6 p.143] we have produced a sequence $\{y_m\}_1^\infty \subseteq E^n$, $y_m\neq o$, $y_m\to o$ as $m\to\infty$, and no y_m is in $\mathit{IR}(t,\sigma)$ for any $t>\sigma$. This proves that zero is not interior to $\mathit{IR}(t,\sigma)$ for any $t>\sigma$. We conclude from this contradiction that $o \in \text{Int } D$.

3. MAIN RESULTS

Theorem 3.1. Assume that (i) (1) is Euclidean controllable; (ii) (3) is uniformly asymptotically stable.

Then (1) is Euclidean null controllable with constraints.

Proof. By assumption (1) is Euclidean controllable. By Theorem 2.2, (1) is proper, so that by Proposition 2.1, the domain D of null controllability of (1) is such that $o \in \text{Int } D$. Because (3) is assumed uniformly asymptotically stable every solution of (3) which is of course a solution of (1) with $u=o \in L_2([\sigma,t),C^m)$ for each t, is such that $x(t,\sigma,\phi,o)\to o$ as $t\to\infty$ Hence there exists a finite t_o such that $x_{t_o}(.,\sigma,\phi,o)\equiv \psi \in \text{Int } D$. With this ψ as an initial point some control $u\in L_2([t_o,t_1],C^m)$, for some t_1 will steer ψ to zero. Thus the control

$$v = o \text{ in } [\sigma,t_o], \quad v = u \text{ in } [t_o,t_1]$$

will steer ϕ to zero in finite time. Hence there exist a time t_1 and a control $v\in L_2([\sigma,t_1],C^m)$ such that the solution $x(t,\sigma,\phi,v)$ satisfies $x_\sigma(.,\sigma,\phi,v) = \phi, x(t,\sigma,\phi,u) = 0$. Hence (1) is Euclidean null controllable with constraints.

We now study the nonlinear system (2). We note that the basic assumption in the introductions and preliminaries hold.

Theorem 3.2. In (2) assume that

(i) the system (1) is Euclidean controllable,

(ii) the free system (3) is uniformly asymptotically stable so that there exist some constants $k\geq 1$, $\alpha>o$ such that every solution $x(.,\sigma,\phi,o)$ of (3) satisfies

$$|x(t,\sigma,\phi,o)|\leq k||\phi|| \; e^{-\alpha(t-\sigma)}, \quad t\geq\sigma;$$

(iii) $f(t,o,o,o,o) = o$;

(iv) $f(t,x(t),x(t-1),o,o)=f_1(t,x(t),x(t-1)+f_2(t,x(t),x(t))$, where,

$$|f_1(t,x(t),x(t-1))|\leq \pi(t) \;\{|x(t)| + |x(t-1)|\}$$

$$|f_2(t,x(t),x(t-1))|\leq \varepsilon \;\{|x(t)| + |x(t-1)|\},$$

with $\varepsilon = \alpha/_{2k}, \Pi=\int_\sigma^\infty (t)dt < \infty$

Then the system (2) is null controllable with constraints.

Proof. To begin, we motivate the definition of certain sets. Suppose that the solution x of (3) with $x_\sigma(.,\sigma,\phi,u) = \phi$ satisfies $x(t_1,\sigma,\phi,u) = 0$ for some t_1 and some u then by (13),

$$x(t_1,\sigma,\phi,u) = x(t_1,\sigma,\phi,o) + \int_\sigma^{t_1} Z(t_1,s)u(s)ds$$

$$+\int_{\sigma}^{t_1} Z(t_1,s)\, f(s,x(s),x(s-1), u(s),u(s-t))ds = 0$$

so that we have

$$x(t_1,\sigma,\phi,o) = -\int_{\sigma}^{t_1} Z(t_1,s)u(s)ds$$

$$-\int_{\sigma}^{t_1} Z(t_1,s)\, f(s,x(s), x(s-1), u(s),u(s-1))$$

If we now define the set $Q(t_1,\sigma)$ as follows

$$Q(t_1,\sigma) = \{-\int_{\sigma}^{t_1} Z(t_1,s)u(s)ds - \int_{\sigma}^{t_1} Z(t_1,s)f(s,x(s),x(s-1),u(s),u(s-1))ds$$

$$: u\varepsilon L_2(\ ,t_1],C^m)\ \phi \ \varepsilon\ C: x_\sigma = \phi\}$$

and if we recall that

$$IR(t_1,\sigma) = \{\int_{\sigma}^{t_1} Z(t_1,s)u(s)ds : u\varepsilon L_2([\sigma,t_1],C^m\}$$

then obviously $Q(t_1,\sigma)$ is the sum of $-IR(t_1,\sigma) = IR(t_1,\sigma)$ and a second set which is denoted by $F(t_1,\sigma)$. Therefore $Q(t_1,\sigma) = IR(t_1,\sigma) + F(t_1,\sigma)$. If we now define the domain D of null controllability of (2) to be the collection $\phi\ \varepsilon\ C([-1,o],E^m)$ such that there exist t_1 and some $u\varepsilon L_2([\sigma,t_1],C^m)$ such that the solution $x(t,\sigma,\phi,u)$ of (2) satisfies $x(t_1,\sigma,\phi,u) = o$, $x_\sigma(.,\sigma,\phi,u) = 0$ then D is a subset of $C=C([-1,o],E^m)$ such that

$$x(t_1,\sigma,D,o) \subseteq IR(t_1,\sigma) + F(t_1,\sigma) \overset{\text{def}}{=\!=} Q(t,\sigma)$$

where

$x(t_1,\sigma,D,o) = \{x(t_1,\sigma,\phi,o) : \phi\ \varepsilon\ D$ and $x(t_1,\sigma,\phi,o)$ is a solution of (3)$\}$. Note that $f(t,o,o,o,o) = o$ so that $x(\sigma,o,o,o) = o$ is a trivial solution of (2). Hence $o\ \varepsilon\ Q(t_1,\sigma)$ for each $t > o$. We now invoke assumption (i) of Theorem 3.2 to deduce from Corollary 2.1 that there is an open ball P around the zero point in E^n such that $P \subseteq IR(t,\sigma)$ for each $t > \sigma$.

Clearly

$$P+F(t,\sigma) \subseteq IR(t,\sigma) + F(t,\sigma) \equiv Q(t,\sigma)$$

This proves that $P+F(t,\sigma)$ is a ball around $F(t,\sigma)$, so that $o\ \varepsilon\ F(t,\sigma) \subseteq \text{Int}\ Q(t,\sigma)$ for each $t>\sigma$. That $o\ \sigma$ Int D is almost immediate. We argue as in the proof of Proposition 2.1 or as in [4] Theorem 4.2. We now conclude that $o\ \varepsilon$ Int D. We claim that under assumption (ii) (iii) and (iv) every solution of $x(t,\sigma,\phi,o)$ of

$$\dot{x} = L(t,x,o) + f(t,x(t),x(t-1),o,o) \tag{18}$$

satisfies

$$x(t,\sigma,\phi,o) \to \text{ as } t \to \infty, \tag{19}$$

so that at some finite time t_o

$$x_{t_o}(.,\sigma,\phi,o) \equiv \psi \in IB \subseteq \text{Int } D.$$

With ψ as an initial point in D, some control $u\varepsilon L_2([t_o,t_1],C^m)$ can steer ψ to zero in some time t_1.

Thus with

$$v = o \text{ in } [\sigma,t_o], \quad v = u \text{ in } [t_o,t_1]$$

we steer ϕ to zero in finite time.

We now conclude the proof by noting that (19) holds for (18) because of Theorem 3.1. of Chukwu [4].

REFERENCES

[1] LASALLE, J.P., The Time Optimal Control Problem, Theory of Nonlinear Oscillation, 5 (1959), Princeton University Press, Princeton, New Jersey.

[2] MARKUS, L., Controllability of Nonlinear Processes, J. Soc. Ind. Appl. Math. See a Control 5 1965, 78-90.

[3] CHUKWU, E.N., Euclidean Controllability of Linear Delay Systems with Limited Control, IEEE Trans. Automatic Contr. AC-24, No.5 (1979), 798-800.

[4] CHUKWU, E.N., On the Null Controllability of Nonlinear Delay Systems with Restrained Controls, J.Math.Anal. Appl. 76, (1980) 283-296.

[5] DAUER, J.P. and GAHL, Controllability of Nonlinear Delay Systems, J. Optimization Theory Appl. 21 No.1 (1967) 59-70.

[6] BANKS, H.T., The Representation of Solutions of Linear Functional Differential Equations J. Differential Equations 5 (1969) 399-410.

[7] HALE, J. Theory of Functional Differential Equations Applied Mathematical Sciences 3 (1977) Springer-Verlag New York, New York.

[8] RUDIN, W., Real and Complex Analysis, McGraw Hill, New York, 1966.

SPECTRAL THEORY FOR SYMMETRIC PAIRS OF ORDINARY DIFFERENTIAL OPERATORS

Earl A. Coddington

University of California
Los Angeles, California

ABSTRACT

Selfadjoint eigenvalue problems associated with pairs L, M of ordinary differential expressions are considered. If M is positive in some sense, these problems can be considered in a Hilbert space generated by a positive selfadjoint extension of the minimal operator for M. Associated with L, M are certain selfadjoint multivalued operators in this Hilbert space, and these determine eigenfunction expansions involving the solutions of the equation $Lf = \lambda Mf$.

1. INTRODUCTION

This is a description of some work by A. Dijksma, H.S.V. de Snoo and myself.

Let L be an ordinary differential expression of order n on $\iota = (a,b)$, which is formally symmetric, and consider problems of the type $Lf = \lambda mf$, where $m(x) > 0$, $x \in \iota$, and f subject to some selfadjoint boundary conditions. A standard way to look at this is to note that $S = m^{-1}L$ on $C_0^\infty(\iota)$ (C^∞ functions with compact support on ι) is symmetric in the Hilbert space

$$L^2(m) = \left\{f \,\middle|\, \int_\iota |f|^2 m\, dx < \infty\right\}, \qquad (f,g) = \int_\iota g^* f m\, dx .$$

For $f,g \in C_0^\infty(\iota)$ we have

$$(Sf,g) = \int_\iota g^*(m^{-1}Lf)m = \int_\iota g^* Lf = \int_\iota (Lg)^* f = \int_\iota (m^{-1}Lg)^* fm = (f,Sg),$$

and then we can analyse selfadjoint extensions A of S in $L^2(m)$. The results are similar to the case $m = 1$.

Here we consider problems of the type $Lf = \lambda Mf$, where M is a formally symmetric differential expression of order $\nu > 0$. If $M > 0$ in some sense, we would expect similar success. If $H = H^*$ is a selfadjoint extension of M_{min} in $L^2(\iota)$ such that $H > 0$, and if H^{-1} exists on all of $L^2(\iota)$, then $S = H^{-1}L$ on $C_0^\infty(\iota)$ is symmetric in $\mathfrak{H}_H$, the completion of $\mathfrak{D}(H)$, the domain of H, with inner product

$$(f,g) = (Hf,g)_2 = \int_\iota g^* Hf \ .$$

Indeed, we have for $f,g \in C_0^\infty(\iota)$,

$$(Sf,g) = (HH^{-1}Lf,g)_2 = (Lf,g)_2 = (f,Lg)_2 = (f,HH^{-1}Lg)_2 = (Hf,H^{-1}Lg)_2$$
$$= (f,Sg) \ .$$

There are several new problems which arise:

(i) there are many choices for H generated by M, and hence for $\mathfrak{H}_H$,

(ii) once $\mathfrak{H}_H$ is chosen, there are many choices for $A = A^*$ which are extensions of S, and these need not be (single-valued) operators.

A hint as to how the latter problem can arise is as follows. We identify operators with their graphs, and put

$$L_1 = \{\{f,Lf\} \mid f \in C^n(\iota)\} \ , \qquad M_1 = \{\{g,Mg\} \mid g \in C^\nu(\iota)\} \ ,$$

and consider

$$\begin{aligned} A_1 = M_1^{-1} L_1 &= \{\{f,g\} \mid \{f,h\} \in L_1, \ \{h,g\} \in M_1^{-1}, \ \text{some } h\} \\ &= \{\{f,g\} \mid \{f,h\} \in L_1, \ \{g,h\} \in M_1, \ \text{some } h\} \\ &= \{\{f,g\} \mid f \in C^n(\iota), \ g \in C^\nu(\iota), \ h = Lf = Mg\} \ . \end{aligned}$$

Now A_1 need not be the (graph) of a single-valued operator, since

$$A_1(0) = \{g \mid \{0,g\} \in A_1\} = \{g \in C^\nu(\iota) \mid Mg = 0\}$$

need not reduce to $\{0\}$.

We remark that we consider both cases $n > \nu$ (right definite case), and $n < \nu$ (left definite case).

2. THE DIFFERENTIAL OPERATOR M

Let $\iota = (a,b) \subset \mathbb{R}$, and

$$M = \sum_{k=0}^{\nu} Q_k D^k , \qquad D = d/dx ,$$

where the Q_k are $m \times m$ complex matrix-valued functions of class $C^k(\iota)$, with $Q_\nu(x)$ invertible for $x \in \iota$. Let

$$L^2(\iota) = \left\{ f : \iota \to C^m(m \times 1) \;\middle|\; \int_\iota f^* f < \infty \right\} ,$$

with immer product

$$(f,g)_2 = \int_\iota g^* f .$$

We assume

$$(A_0) \qquad (Mf,f)_2 \geq 0 , \qquad f \in C_0^\infty(\iota) .$$

This implies that $\nu = 2\mu$,

$$M = M^+ = \sum_{k=0}^{\nu} (-1)^k D^k Q_k^* ,$$

and we can write M as

$$M = \sum_{j,k=0}^{\mu} (-1)^j D^j Q_{jk} D^k , \qquad Q_{jk}^* = Q_{kj} .$$

If $M_0 = \{\{f,Mf\} \mid f \in C_0^\infty(\iota)\}$, then (A_0) just says that M_0 is symmetric ($M_0 \subset M_0^*$) and $M_0 \geq 0$. We define $M_{min} = (M_0)^c$, the closure of M_0 in $L^2(\iota) \oplus L^2(\iota)$, and $M_{max} = M_0^*$. We say M is regular if $\iota = (a,b)$ is finite, $\bar{\iota} = [a,b]$, $Q_k \in C^k(\bar{\iota})$, and $Q_\nu(x)$ is invertible for $x \in \bar{\iota}$. Otherwise we say M is singular.

3. HILBERT SPACES GENERATED BY M

Selfadjoint extensions $H = H^*$ of M_0 exist in $L^2(\iota) \oplus L^2(\iota)$. One of these is the Friedrichs extension M_F given by

$$M_F = \{\{f,Mf\} \in M_{max} \mid \|f_n - f\|_2 \to 0, \ (M(f_n - f_m), f_n - f_m)_2 \to 0, \text{ some } f_n \in \mathfrak{D}(M_0)\} .$$

Let $M_0 \subseteq H = H^* \subset L^2(\iota) \oplus L^2(\iota)$. We further assume

$$(A_1) \qquad (Hf,f)_2 \geq c^2(f,f)_2 , \qquad f \in \mathfrak{D}(H), \quad \text{some} \quad c > 0 .$$

This implies H^{-1} exists as a bounded operator on all of $L^2(\iota)$, and a stronger version of (A_0), namely,

$$(A_0') \qquad (Mf,f)_2 \geq c^2(f,f)_2 , \qquad f \in C_0^\infty(\iota) ,$$

or $M_0 \geq c^2 > 0$. Let $\mathfrak{H}_H$ be the completion of $\mathfrak{D}(H)$, with $(f,g) = (Hf,g)_2$, $f,g \in \mathfrak{D}(H)$. This is a Hilbert space and (A_1) extends to

$$\|f\| \geq c\|f\|_2 , \qquad f \in \mathfrak{H}_H ,$$

which shows that $\mathfrak{H}_H \subset L^2(\iota)$. Now $C_0^\infty(\iota) \subset \mathfrak{D}(H) \subset \mathfrak{H}_H$, and if we put $\mathfrak{H}_M = (C_0^\infty(\iota))^c$, the closure of $C_0^\infty(\iota)$ in $\mathfrak{H}_H$, we have the following result.

<u>Theorem 1.</u> <u>We have</u> $\mathfrak{H}_M = \mathfrak{H}_{M_F}$,

$$\mathfrak{N}_H = \mathfrak{H}_H \ominus \mathfrak{H}_M = \{f \in C^\nu(\iota) \cap \mathfrak{H}_H \mid Mf = 0\} ,$$

$$\dim \mathfrak{N}_H \leq \nu m .$$

For $f,g \in C_0^\infty(\iota)$ we have

$$(f,g) = \int_\iota \sum_{j,k=0}^{\mu} (D^j g)^* Q_{jk} D^k f = (f,g)_D ,$$

the Dirichlet inner product.

Abstractly, $\mathfrak{H}_H = \mathfrak{D}(H^{1/2})$, where $H^{1/2}$ is the positive square root of H.

<u>Theorem 2.</u> <u>There exists a linear differential expression</u> N <u>of order</u> μ <u>such that</u> $M = N^+N$, <u>and we have</u>

$$\text{(i)} \quad M_F = N^+_{max} N_{min} , \qquad \text{(ii)} \quad \mathfrak{H}_M = \mathfrak{D}(N_{min}) ,$$

$$\text{(iii)} \quad \|f\| = \|N_{min} f\|_2 , \qquad f \in \mathfrak{H}_M .$$

This shows that the elements in $\mathfrak{H}_H$ have derivatives up to order μ $(\nu = 2\mu)$. The norm on $\mathfrak{N}_H$ can also be described, but this is a bit more complicated.

Although $(f,f) = (f,f)_D$ for $f \in C_0^\infty(\iota)$, it does not follow that $(f,f) = (f,f)_D$ for $f \in \mathfrak{H}_M = (C_0^\infty(\iota))^c$. There exists an example of an M satisfying (A_0') and $f \in \mathfrak{H}_M$ such that $(f,f)_D = -\infty$, but

$0 < (f,f) = (Nf,Nf)_2 < \infty$.

In the regular case $\mathfrak{H}_H$ is the set of all $f : \iota \to \mathbb{C}^m$ such that $f^{(\mu-1)}$ is absolutely continuous on $\bar{\iota}$, $f^{(\mu)} \in L^2(\iota)$, and f satisfies certain homogeneous boundary conditions in $f, f', \ldots, f^{(\mu-1)}$. For $f,g \in \mathfrak{H}_H$ we have $(f,g) = (f,g)_D$ + boundary terms in $f,f',\ldots,f^{(\mu-1)}$ and $g,g',\ldots,g^{(\mu-1)}$.

4. SELFADJOINT SUBSPACES A DETERMINED BY L, M

Let M be as above, and $H = H^*$ fixed with (A_1) assumed, $\mathfrak{H} = \mathfrak{H}_H$, $\mathfrak{H}^2 = \mathfrak{H} \oplus \mathfrak{H}$. We consider a second formally symmetric differential expression of order n,

$$L = \sum_{k=0}^{n} P_k D^k = L^+ ,$$

with $P_k \in C^k(\iota)$, $P_n(x)$ invertible, $x \in \iota$, if $n > \nu$. The maximal linear manifold for L, M is

$$T = \{\{f,g\} \in \mathfrak{H}^2 \mid f \in C^r(\iota),\quad g \in C^{\nu}(\iota),\quad Lf = Mg\} ,$$

where $r = \max(n,\nu)$, and the minimal linear manifold for L, M is

$$S = \{\{f, H^{-1}Lf\} \mid f \in C_0^{\infty}(\iota)\} .$$

Since for $g = H^{-1}Lf$ we have $Mg = Lf$, we see that $S \subset T$. Note that S is (the graph of) an operator but T need not be, since $T(0) \subset \{g \in C^{\nu}(\iota) \mid Mg = 0\}$. The maximal subspace is $T_1 = T^c$, the closure of T in $\mathfrak{H}^2$, and the minimal subspace is $T_0 = S^c$. This T_0 need not be an operator.

The adjoint A^* of a linear manifold $A \subset \mathfrak{H}^2$ is defined as

$$A^* = \{\{h,k\} \in \mathfrak{H}^2 \mid (g,h) - (f,k) = 0,\quad \text{all } \{f,g\} \in A\} ,$$

and $\nu(A)$ is the null space of A, $\nu(A) = \{f \mid \{f,0\} \in A\}$.

<u>Theorem 3</u>. <u>We have</u>

(i) $T_0 \subset T_1 = T_0^*$, $T_0 = T_1^*$,

(ii) $T_1(0) = T(0) = \mathfrak{N}_H$,

(iii) $T_1 \ominus T_0 = \{\{f,g\} \in \mathfrak{H}^2 \mid f,g \in C^r(\iota),\ Lf = Mg,\ Mf = -Lg\}$,

(iv) $\nu(T_1 - \ell I) = \nu(T - \ell I) = \{f \in C^r(\iota) \cap \mathfrak{H} \mid Lf = \ell Mf\}$,
<u>if</u> $\ell \in \mathbb{C} \setminus \mathbb{C}_e$, <u>where</u> $\mathbb{C}_e$ <u>is the set of all</u> $\ell \in \mathbb{C}$ <u>such that</u>

the leading coefficient of $L - \ell M$ *is not invertible for some* $x \in \iota$ $(\mathbb{C}_e = \emptyset,\ n > \nu,$ *and* $\mathbb{C}_e = \{0\},\ n < \nu)$.

We are interested in selfadjoint subspaces A which are extensions of T_0, $T_0 \subset A = A^*$.

Theorem 4. We have $T_0 \subset A = A^* \subset \mathfrak{H}^2$ *if and only if*

(i) $\dim M^+ = \dim M^-$,

where

$$M^{\pm} = \{\{f,g\} \in T_1 \mid g = \pm if\} .$$

In case (i) *is satisfied,*

(ii) $A = T_0 \oplus M_1 = T_1 \cap M_1^*$,

where M_1 *is a subspace such that*

(iii) $M_1 \subset T_1 \ominus T_0$, $2 \dim M_1 = \dim(T_1 \ominus T_0)$, $M_1 \subset M_1^*$.

The condition (i) need not be satisfied, even in case $M = I$. But there always exist larger Hilbert spaces $\mathfrak{K} \supset \mathfrak{H}$, and $A = A^* \subset \mathfrak{K}^2$ such that $T_0 \subset A$. Condition (i) is satisfied if L, M are regular.

In case (i) is satisfied the representation $A = T_1 \cap M_1^*$ in (ii) shows how A is obtained by restricting T_1 by $d = \dim M_1$ generalized boundary conditions:

$$A = \{\{f,g\} \in T_1 \mid (g,\alpha_j) - (f,\beta_j) = 0,\quad j = 1,\ldots,d\} ,$$

where $\{\alpha_j,\beta_j\}$, $j = 1,\ldots,d$, is a basis for M_1. These are not boundary conditions of the usual type, since T_1 in general contains elements with only μ derivatives. If we put $A_T = A \cap T$, then $A_T \subset C^r(\iota) \times C^r(\iota)$, $(A_T)^c = A$, and this smooth version of A can be characterized, in many cases, by boundary conditions of the usual type. In particular, this is true in case L, M are regular.

5. EIGENFUNCTION EXPANSIONS

Consider first the case $n > \nu$. Let $T_0 \subset A = A^* \subset \mathfrak{K}^2$, where $\mathfrak{K} \supset \mathfrak{H}$. Let

$$A(0) = \{g \in \mathfrak{K} \mid \{0,g\} \in A\},\quad A_\infty = \{\{0,g\} \in A\},\quad A_s = A \ominus A_\infty .$$

Then A_s is a densely defined selfadjoint operator in $\mathfrak{K} \ominus A(0)$, with a spectral resolution

$$A_s = \int_{-\infty}^{\infty} \lambda\, dE_s(\lambda) ,$$

where E_s is a right-continuous spectral family for A_s. We define the spectral family E for the subspace A via

$$E(\lambda)f = E_s(\lambda)f , \qquad f \in \mathfrak{K} \ominus A(0) ,$$
$$= 0 , \qquad f \in A(0) ,$$

so that $E(\infty)$ is the orthogonal projection of $\mathfrak{K}$ onto $\mathfrak{K} \ominus A(0)$. Associated with A is the generalized spectral family F given by

$$F(\lambda)f = PE(\lambda)f , \qquad f \in \mathfrak{H} ,$$

where P is the orthogonal projection of $\mathfrak{K}$ onto $\mathfrak{H}$. A central problem is to compute F, or equivalently all $F(\Delta) = F(\beta) - F(\alpha)$, for $\Delta = (\alpha,\beta] \subset \mathbb{R}$. This was carried out in detail for $A \subset \mathfrak{H}^2$ in the regular case by de Snoo and myself [2]. Recently Dijksma and de Snoo showed how to do this for any $A \subset \mathfrak{K}^2$ in the general (not necessarily regular) case. We briefly summarize these results.

We denote by $L_0^2(\iota)$ the set of all $f \in L^2(\iota)$ having compact support, and let $s(\ell) = (s_1(\ell),\ldots,s_{mn}(\ell))$ satisfy

$$(L - \ell M)s(\ell) = 0 , \qquad \tilde{s}(c,\ell) = I_{mn} ,$$

where $c \in \iota$ is fixed, I_{mn} is the $mn \times mn$ identity matrix, and

$$\tilde{s} = \begin{pmatrix} s \\ s' \\ \vdots \\ s^{(n-1)} \end{pmatrix} .$$

Theorem 5. Let $n > \nu$. There exists an $mn \times mn$ hermitian matrix-valued function ρ on $\mathbb{R}$, which is right continuous and non-decreasing, such that

$$F(\Delta)H^{-1}\varphi = \int_\Delta s(\lambda)d\rho(\lambda)\, \widehat{H^{-1}\varphi}(\lambda) , \qquad \varphi \in L_0^2(\iota) ,$$

where $(\widehat{H^{-1}\varphi})(\lambda) = (\varphi, s(\lambda))_2$.

This describes the $F(\Delta)$ on a dense set $H^{-1}L_0^2(\iota) \subset \mathfrak{H}$ in terms of a basis for the solutions of the equation $(L - \ell M)u = 0$ on ι.

For $\zeta,\eta : \mathbb{R} \to \mathbb{C}^{mn}$, considered as $mn \times 1$ matrix-valued functions, let

$$(\zeta,\eta) = \int_{-\infty}^{\infty} \eta^*(\lambda)d\rho(\lambda)\zeta(\lambda) .$$

Since $(\zeta,\zeta) \geq 0$ we have $\|\zeta\| = (\zeta,\zeta)^{1/2}$ gives a norm, and we let

$$\hat{\mathfrak{H}} = L^2(\rho) = \{\zeta \mid \|\zeta\| < \infty\} .$$

For $f \in \mathfrak{H}$ there are $\varphi_n \in L_0^2(\iota)$ such that $H^{-1}\varphi_n \to f$ in $\mathfrak{H}$, and this implies $\widehat{H^{-1}\varphi_n} \to \hat{f}$ in $\hat{\mathfrak{H}}$, since $(\|\widehat{H^{-1}\varphi}\| \leq \|H^{-1}\varphi\|$, $\varphi \in L_0^2(\iota)$. The following result gives the eigenfunction expansion associated with the extension A of T_0.

<u>Theorem 6</u>. <u>The map</u> $V : f \in \mathfrak{H} \mapsto Vf = \hat{f} \in \hat{\mathfrak{H}}$ <u>is such that</u>

$$F(\infty)f = \int_{-\infty}^{\infty} s(\lambda)d\rho(\lambda)\hat{f}(\lambda) ,$$

<u>where the integral converges in</u> $\mathfrak{H}$, <u>and</u>

$$(F(\infty)f,g) = (\hat{f},\hat{g}) , \qquad f,g \in \mathfrak{H} .$$

<u>In particular,</u> V <u>is a contraction on</u> $\mathfrak{H}$ $(\|Vf\| \leq \|f\|)$, <u>an isometry on</u> $\mathfrak{H}_1 = \nu(F(\infty) - I)$ $(\|Vf\| = \|f\|)$, <u>and</u>

$$f = \int_{-\infty}^{\infty} s(\lambda)d\rho(\lambda)\hat{f}(\lambda) , \qquad f \in \mathfrak{H}_1 .$$

The generalized spectral family F induces a splitting of $\mathfrak{H}$, $\mathfrak{H} = \mathfrak{H}_0 \oplus \mathfrak{H}_r \oplus \mathfrak{H}_1$, where

$$\mathfrak{H}_0 = \nu(F(\infty)) = \mathfrak{H} \cap A(0) = \mathfrak{H} \ominus P(A(0))^{\perp} ,$$

$$\mathfrak{H}_1 = \nu(F(\infty) - I) = \mathfrak{H} \cap (A(0))^{\perp} = \mathfrak{H} \ominus PA(0) ,$$

$$\mathfrak{H}_r = \mathfrak{H} \ominus (\mathfrak{H}_0 \oplus \mathfrak{H}_1) = PA(0) \cap P(A(0))^{\perp} .$$

We have $T_0(0) \subset \mathfrak{H}_0$ and $\mathfrak{H}_r \subset T_1(0) \ominus T_0(0)$.

An $A = A^* \subset \mathfrak{K}^2$, such that $T_0 \subset A$, is a minimal selfadjoint extension of T_0 if no nontrivial subspace $\mathfrak{M} \subset \mathfrak{K} \ominus \mathfrak{H}$ is such that $A \cap \mathfrak{M}^2$ is selfadjoint.

<u>Theorem 7</u>. <u>Let</u> $A \subset \mathfrak{K}^2$ <u>be a minimal selfadjoint extension of</u> $T_0 \subset \mathfrak{H}^2$. <u>Then</u> V <u>maps</u> $\mathfrak{H}$ <u>onto</u> $L^2(\rho)$ <u>if and only if</u> $\dim(\mathfrak{K} \ominus \mathfrak{H}) = \dim \mathfrak{H}_r$, <u>and</u> V <u>maps</u> $\mathfrak{H}_1$ <u>onto</u> $L^2(\rho)$ <u>if and only if</u> $A = A^* \subset \mathfrak{H}^2$.

In case $n < \nu$, we consider A^{-1} instead of A, and the eigenfunction expansion is in terms of a basis for the solutions of the equation $(M - \ell L)u = 0$.

6. CONCLUDING REMARKS

Almost all the results go through under a milder assumption than (A_1) on H, namely,

(A_1') for each compact $J \subset \iota$, $c(J) > 0$ exists such that

$$(Hf,f)_2 \geq (c(J))^2(f,f)_{2,J}, \qquad f \in \mathfrak{D}(H),$$

where $(f,f)_{2,J} = \int_J f^* f.$

In this case, if the identity map from $\mathfrak{D}(H) \to \mathfrak{D}(H)$ extends to an injection of $\mathfrak{H}_H$ into $L^2_{loc}(\iota)$, the set of all f such that $(f,f)_{2,J} < \infty$ for each compact $J \subset \iota$, we can consider $\mathfrak{H}_H \subset L^2_{loc}(\iota)$.

REFERENCES

[1] E.A. Coddington and H.S.V. de Snoo, Differential subspaces associated with pairs of ordinary differential expressions, J. Differential Equations 35 (1980), 129-182.

[2] E.A. Coddington and H.S.V. de Snoo, Regular boundary value problems associated with pairs of ordinary differential expressions, Lecture Notes in Mathematics 858, Springer-Verlag, Berlin, 1981.

[3] A. Dijksma and H.S.V. de Snoo, Eigenfunction expansions associated with pairs of ordinary differential expressions (1982), mss.

ALMOST PERIODIC SOLUTIONS TO SOME NONLINEAR PARABOLIC EQUATIONS

C. Corduneanu

Department of Mathematics
The University of Texas at Arlington
Arlington, Texas

ABSTRACT

Using a lemma on qualitative differential inequalities (see [1]), one can prove the following results concerning the almost periodicity of bounded solutions to some nonlinear parabolic equations [1], [2]:

Theorem 1. Consider the nonlinear parabolic equation

$$u_t = u_{xx} + f(t,x,u), \quad (t,x) \in R\times(0,1), \tag{1}$$

under boundary value conditions

$$u(t,0) = 0, \quad u(t,1) = 0, \quad t \in R. \tag{2}$$

Let $f(t,x,u)$ be an almost periodic function in t, uniformly with respect to $(x,u) \in [0,1]\times[-A,A]$, for every $A > 0$, such that

$$f_u \leq \mu < \pi^2, \tag{3}$$

with fixed μ.

If $u(t,x)$ is a bounded solution of (1), (2), such that

$$|u_{xx}(t,y)-u_{xx}(t,z)| \leq \omega(|x-z|), \quad \sqrt{n}\,\omega(\tfrac{1}{n+1}) \longrightarrow 0 \quad \text{as} \quad n \longrightarrow \infty, \tag{4}$$

then $u(t,x)$ is (Bohr) almost periodic in t, uniformly with respect to $x \in [0,1]$.

Theorem 2. Consider the nonlinear parabolic equation

$$u_t = \Delta u + f(t,x,u), \quad (t,x) \in R \times \Omega, \tag{5}$$

where $\Omega \subset R^n$ is a bounded domain with smooth boundary S, under the boundary value condition

(6) $$u(t,x) = 0,\ t \in R,\ x \in S.$$

Let $f(t,x,u)$ be a map from $R \times \Omega \times R$ into R, continuous and $L^2(\Omega)$- almost periodic in t, uniformly with respect to $u \in R$, and such that

(7) $$f_u \le \mu < \lambda_1,$$

where λ_1 is the first eigenvalue of the Dirichlet problem

(8) $$\Delta v + \lambda v = 0 \quad \text{in} \quad \Omega,\ v|_S = 0.$$

If $u(t,x)$ is a $C^{(2)}$ - solution for (5), (6), which is $L^2(\Omega)$ - bounded on R, i.e., there exists a positive M such that

(9) $$\int_\Omega u^2(t,x)dx \le M,\ t \in R,$$

then $u(t,x)$ is $L^2(\Omega)$ - almost periodic.

Remark 1. The proof of Theorem 1 is based also on an approximation procedure, using the method of lines. The proof of Theorem 2 requires the use of Poicaré's inequality. The conditions of both theorems assure the uniqueness of the bounded solution on R. For details, see [1], [2].

Remark 2. The boundary value conditions (2) or (6) can be replaced by different ones. For instance, instead of (6) one can use the condition

(6') $$\frac{\partial u}{\partial n}\Big|_S = hu\Big|_S,$$

where h is a positive function on S, and n is the inner normal. Of course, in such case the eigenvalue λ_1 should be replaced by a quantity which is defined accordingly.

Remark 3. In case $f(t,x,u)$ is periodic in t, with periodic T, the bounded solution is also periodic of period T. The proofs of both theorems cover the periodic case.

A problem arising in connection to the above results is whether a milder type of almost periodicity on $f(t,x,u)$ would suffice to obtain the Bohr or $L^2(\Omega)$ - almost periodicity for the solution. We have in mind Stepanov's type of almost periodicity, which is considerably more general than the classical type due to Bohr.

The answer to the above question can be obtained by affirmative, if instead of the Lemma in [1], we apply the following result concerning differential inequalities on the real axis.

Lemma. Consider the differential inequality

$$y'(t) \leq -\alpha y(t) + f(t)\sqrt{y(t)},\ t \in R,\ \alpha > 0, \tag{10}$$

and assume $f(t) \geq 0$ belongs to the space $M(R)$ (see [3], for instance).

If $y(t)$ is a bounded solution on R, then the following estimate holds true:

$$\sup|y(t)| \leq K\left\{\sup\int_t^{t+1} f(s)ds\right\}^2, t \in R, \tag{11}$$

with K a positive constant (depending on α). The proof of the Lemma goes on the same lines as the proof given in [2] for the linear case.

REFERENCES

[1] C. Corduneanu, "Bounded and almost periodic solutions of certain nonlinear parabolic equations," Libertas Mathematica, Vol. II, (1982), 131-139.

[2] C. Corduneanu, "Almost periodic solutions to nonlinear elliptic and parabolic equations," J. Nonlinear Analysis: TMA, (in print).

[3] J. L. Massera and J. J. Schaffer, Linear Differential Equations and Function Spaces, Academic Press, New York, 1966.

CONJUGATE POINTS AND DISCONJUGACY OF SECOND ORDER SYSTEMS

Fozi M. Dannan

Department of Mathematics
Kuwait University
Kuwait

ABSTRACT

In this paper we discuss the relation between the conjugate points of the linear vector differential equations $y'' + A(t)y = 0(*)$ and $x'' + \tilde{A}(t)x = 0$, where $A(t)$ is a continuous $n \times n$ matrix and $\tilde{A}(t) = \frac{1}{2}[A(t) + A^T(t)]$. Also we give a partial answer concerning the isolation of conjugate points of a point a relative to (*). A disconjugacy criterion for the system $(\Phi(t)P(t)x')' + \Psi(t)Q(t)x = 0$ is given.

1. INTRODUCTION

Consider the linear vector differential equations

$$y'' + A(t)y = 0 \tag{1}$$

$$x'' + \tilde{A}(t)x = 0 \tag{2}$$

where $A(t)$ is a continuous $n \times n$ matrix and

$$\tilde{A}(t) = \tfrac{1}{2}[\, A(t) + A^T(t)] \tag{3}$$

Our main purpose is to establish the relation between the conjugate points of (1) and (2). We also give a partial answer for the open problem posed by Ahmad [1] concerning the isolation of conjugate points of a point a relative to (1). Finally we find the conditions that guarantee the diconjugacy of the system

$$(\Phi(t)P(t)x')' + \Psi(t)Q(t)x(t) = 0 \tag{4}$$

when the system

$$(P(t)y')' + Q(t)y = 0 \tag{5}$$

is disconjugate on some interval J. Here P and Q are $n \times n$ continuous symmetric matrices, $P(t)$ is a positive definite matrix, Φ and Ψ are $n \times n$ continuous matrices satisfying some conditions.

Recall that a number b is said to be a conjugate point of a number a relative to (1) if $b > a$ and if there exists a nontrivial solution of (1) which vanishes at a and b. A number b is called the first conjugate point of a if it is a conjugate point of a and if there is no number c, $a < c < b$, such that c is also a conjugate point of a.

The system (5) is called disconjugate on an interval J if there exists no nontrivial solution $x(t)$ such that $x(t_1) = x(t_2) = 0$, $t_1, t_2 \in J$.

For pertinent bibliography, one might consult [2], [3] and [4].

2. CONJUGATE POINTS

Theorem 1. Let b be the first conjugate point of a relative to (2). If $y(t)$ is a nontrivial solution of (1) with $y(a) = y(b) = 0$, then $y(t)$ is a solution of (2).

Proof. Since $y(t)$ is a solution of (1) we have

$$y''(t) + A(t)y(t) \equiv 0 .$$

Multiplying both sides by $-y$ and intergrating from a to b we obtain

$$\int_a^b (< y', y' > - < \tilde{A}y, y >)dt = 0 .$$

But since b is the first conjugate point of a relative to (2), then by Lemma 2 of [2], we also have

$$\int_a^b (< h', h' > - < Ah, h >)dt \geq 0 ,$$

for all functions $h(t) \in \mathcal{A}[a,b]$, where $\mathcal{A}[a,b]$ is the set of absolutely continuous R^n-valued functions on $[a,b]$ such that $|h'| \in L^2[a,b]$ and $h(a) = h(b) = 0$.

Hence by a standard result in the calculus of variations (see e.g., [5]), $y(t)$ is a solution of (2).

Theorem 2. Let b be the first conjugate point of a relative to (2). If there exists a nontrivial solution $y(t)$ of (1) with $y(a) = y(b) = 0$, then b is the first conjugate point of a relative to (1).

Proof. Assume that $c \in (a,b)$ is a conjugate point of a relative to (1). Then there exists a nontrivial solution $z(t)$ of (1) such that $z(a) = z(c) = 0$. For this solution we have

$$\int_a^c (< z', z' > - < \tilde{A}z, z >)dt = 0 . \qquad (6)$$

Now define the function $u(t)$ as follows:

$$u(t) = \begin{cases} z(t) & \text{if } a \le t \le c \\ 0 & \text{if } c \le t \le b . \end{cases}$$

Hence

$$J [u; a,b] = \int_a^b (< u', u' > - < \tilde{A}u, u >)dt =$$

$$\int_a^c (< z', z' > - < \tilde{A}z, z >)dt = 0 .$$

Since b is the first conjugate point of a relative to (2), we have $J[h; a,b] \ge 0$ for all functions $h(t) \in \mathcal{A}[a,b]$. Therefore $u(t)$ is a solution of (2). This contradicts our assumption that b is the first conjugate point of a relative to (2).

In the selfadjoint case it is a well known fact that conjugate points are isolated [4]. Ahmad [1] raised the following question: If $A(t)$ is nonsymmetric, are the conjugate points of a point a relative to (1) isolated ?

In what follows we give a partial answer for this problem.

Theorem 3. If there exists a constant positive definite matrix K such that

$$A^T(t) = K^{-1}AK \qquad (7)$$

then the conjugate points of a point a relative to (1) are isolated.

Proof. Since K is a positive definite matrix, there exists a nonsingular matrix C such that

$$K = CC^T , \qquad (8)$$

(see e.g. [6] p. 253). Using the transformation $x(t) = Cz(t)$, equation (1) becomes

$$z'' + C^{-1}A(t)Cz = 0 \;. \tag{9}$$

From (7) and (8) we conclude that $C^{-1}A(t)C$ is a symmetric matrix, hence the conjugate points of a point a relative to (9) are isolated and consequently the conjugate points of a relative to (1) are isolated.

Remark. It seems that the class of matrices satisfying (7) is not small. As an example we take

$$A(t) = \begin{pmatrix} a(t) & b(t) \\ c(t) & a(t) + k_1 b(t) + k_2 c(t) \end{pmatrix},$$

where $a(t)$, $b(t)$ and $c(t)$ are continuous functions such that $b(t)c(t) \not\equiv 0$, k_1 and k_2 are constants such that $k_2 < 0$ and $k_1 k_2 + 1 < 0$. In this case the matrix K has the form

$$K = \begin{pmatrix} -k_2 & 1 \\ 1 & k_1 \end{pmatrix}$$

3. DISCONJUGACY OF SECOND ORDER SYSTEMS

Theorem 4. Assume the following hypotheses:

(i) The system (5) is disconjugate on an interval J.

(ii) $\Phi(t)$ is a nonsingular twice differentiable matrix such that $\Phi(t)P(t)$ is a positive definite matrix.

(iii) $L(t) = \Phi^{-1}(t)\Phi'(t)P(t)$ is a symmetric matrix.

(iv) $L'(t) + 2M(t)$ is a nonnegative definite matrix, where

$$M(t) = [I - \Phi^{-1}(t)\Psi(t)]\; Q(t)$$

and I is the identity matrix.

Then (4) is disconjugate on J.

Proof. Assume the contrary that (4) has a nontrivial solution $z(t)$ on $[a,b] \subset J$ such that $z(a_1) = z(b_1) = 0$, $a \le a_1 \le b_1 \le b$. It is clear that

$$0 = \int_{a_1}^{b_1} \langle P(t)z',\; z\rangle' \; dt \;. \tag{10}$$

But

$$< Pz', z >' = < Pz', z' > - < Qz, z > + < (\tfrac{1}{2}L' + M)z, z > - $$
$$ - \tfrac{1}{2} < Lz, z >' ,$$

hence (10) becomes

$$0 = \int_{a_1}^{b_1} (< Pz', z' > - < Qz, z >)dt + \int_{a_1}^{b_1} (< (\tfrac{1}{2}L' + M)z, z >)dt.$$

From (iv) we conclude that

$$\int_{a_1}^{b_1} (< Pz', z' > - < Qz, z >)dt \le 0 \tag{11}$$

Now define the function $u(t)$ such that $u(t) = z(t)$, $a_1 \le t \le b_1$ and $u(t) = 0$ when $t \in [a,a_1] \cup [b_1,b]$. Thus by (11) we have

$$\int_{a}^{b} (< Pu', u' > - < Qu, u >)dt \le 0 .$$

This contradicts our assumption that (5) is disconjugate on J.

ACKNOWLEDGEMENT

The author wishes to thank Professor Shair Ahmad for his valuable discussions and suggestions.

REFERENCES

[1] S. Ahmad, On positivity of solutions and conjugate points of Nonse lfadjoint systems, Bulletin de L'Academie Polonaise des Sciences, 27, No.1, 1979, 71-75.

[2] S. Ahmad and A.C. Lazer, On the components of extremal solutions of second order systems, SIAM J.Math.Anal., vol. 8, No. 1, 1977, 16-23.

[3] S. Ahmad and A.C. Lazer, An N-dimensional extension of the Sturm separation and comparison theory to a class of nonselfadjoint systems, SIAM J.Math.Anal. vol. 9, No. 6, 1978, 1137-1150.

[4] W.A. Coppel, Disconjugacy, Lecture Notes in Mathematics, vol. 220, Springer-Verlag, Berlin, 1971.

[5] M.R. Hestenes, Calculus of variations and optimal control theory, John Wiley, New York, 1966.

[6] G. Strang, Linear algebra and its applications, John Wiley, New York, 1979.

SOME RESULTS ON THE EXISTENCE OF FREE BOUNDARIES FOR PARABOLIC REACTION-DIFFUSION SYSTEMS

Ildefonso Díaz
Departamento de Matematicas
Universidad de Santander
Santander, Spain

Jesús Hernández
Departamento de Matemáticas
Universidad Autónoma
Madrid, Spain

ABSTRACT

We give a survey of some recent results concerning the existence, location and time evolution of free boundaries for a class of nonlinear parabolic reaction-diffusion systems arising in applications.

The case of degenerate parabolic problems is also considered. The main tool in the proofs is the use of local comparison techniques.

1. INTRODUCTION

We give a brief survey of some recent results obtained by the authors concerning the existence of free boundaries in some parabolic systems arising in applications, especially in combustion theory. The corresponding elliptic problem was treated by the authors in |5| and applications to combustion theory were also included. An extended version of these results, including full proofs and applications will appear in |6|.

Now we consider the model system

$$\begin{aligned}
&u_t - \Delta u + \mu^2 F(u) e^{\frac{\gamma(v-1)}{v}} = 0 \quad \text{in} \quad Q = \Omega \times (0,\infty) \\
&\lambda v_t - \Delta v - \nu\mu^2 F(u) e^{\frac{\gamma(v-1)}{v}} = 0 \quad \text{in} \quad Q \\
&u = v = 1 \quad \text{on } \Sigma = \partial\Omega \times (0,\infty) \\
&u(x,0) = u_\circ(x), v(x,0) = v_\circ(x) \text{ on } \Omega
\end{aligned} \tag{1.1}$$

where Ω is a smooth bounded domain in $\mathbb{R}^N$, μ is the Thiele number, ν the Prater temperature, γ the Arrhenius number, λ^{-1} the Lewis number, and u and v are, respectively, the concentration and temperature of the reactant. The system (1.1) is a model for a single irreversible nonisothermic reaction (cf. the book by Aris |1|). F is a non-decreasing function such that $F(0)=0$, $F(1)=1$ and $F(s)>0$ for $s>0$. The most common instance of F is $F(s)=s^p$, where $p \geq 0$ is the reaction order. If $p=0$, then F is given by $F(0)=0$ and $F(s)$ if $s>0$, and this means that in this case F is discontinuous.

A general formulation which contains (1.1) is the following

$$\begin{aligned} &u_t - \Delta u + \alpha(u)f(v) = 0 && \text{in } Q \\ &v_t - \Delta v - \beta(u)g(v) = 0 && \text{in } Q \\ &u(x,t) = \Psi_1, \quad v(x,t) = \Psi_2, && \text{on } \Sigma \\ &u(x,0) = u_o(x), v(x,0) = v_o(x) && \text{in } \Omega \end{aligned} \tag{1.2}$$

where α and β are non-decreasing continuous functions, f and g are C^1 and $f(s) \geq 0$, $g(s) \geq 0$ for $s \geq 0$ and the boundary and initial data are smooth. The case p=0 can also be handled by using maximal monotone graphs.

The main interest in this paper is to study the existence of a dead core $\Omega_o(t) = \{x \in \Omega | u(x,t)=0\}$ and to get additional information about its size and location. The boundary of the subset $\bigcup_{t>0} \Omega_o(t)$ is a <u>free boundary</u> which is <u>a priori</u> unknown and plays an important role in applications.

We obtain some different kinds of results about $\Omega_o(t)$. Some of them seem to be new in the literature and there are also extensions to the case $\tilde{f}(x,t)$ nonconstant and, more important, $N>1$, of known results. (A systematic treatment of these qualitative properties for nonlinear elliptic and parabolic problems will appear in the book |3|). Our main tool are local comparison techniques introduced in Díaz |4| and Díaz-Hernández |**5**|.

Some of our results can be collected in the following

<u>Theorem 1.1</u>: Suppose that $\alpha(s) = \mu^2|s|^{p-1}s$, $p \geq 0$, ψ_1, $\psi_2 \geq 0$, $u_o, v_o \geq 0$.

i) If (u,v) is a solution of (1.2), then $v>0$ on Q. If $p \geq 1$, then $u>0$ on Q.

ii) For $0 \leq p < 1$ and (u,v) solution of (1.2) suppose that for $\lambda > 0$ there exists $\Omega_\lambda \subset \Omega$ such that

$$\Omega_\lambda \times (0,\infty) \subset \{(x,t) \in Q \mid f(v(x,t)) > \lambda\} .$$

Then we have the estimate

$$0 \leq u(x,t) \leq |(\|u_\circ\|^{1-p}_{L^\infty(\Omega)} - \frac{\lambda\mu^2}{2} (1-p)t)^+|^{\frac{1-p}{2}} \equiv U_{\frac{\lambda\mu^2}{2}}(t)$$

for any $t > 0$ and a.a. $x \in \Omega_\lambda$ such that $d(x,\partial\Omega_\lambda) \geq \rho_\lambda(t)$ where

$$\rho_\lambda(t) = \max \{\sup_{\substack{s\in|0,t| \\ x\in\partial\Omega_\lambda \cap \partial\Omega}} \left(\frac{\psi_1(x,s) - U_{\frac{\lambda\mu^2}{2}}(t)}{K_{\frac{\lambda\mu^2}{2}}} \right)^{\frac{1-p}{2}}, 0 \text{ (if } \partial\Omega_\lambda \subset \partial\Omega),$$

$$\sup_{s\in|0,t|} \left(\frac{M - U_{\frac{\lambda\mu^2}{2}}(t)}{K_{\frac{\lambda\mu^2}{2}}} \right)^{\frac{1-p}{2}} \text{ (if } \partial\Omega_\lambda \cap (\mathbb{R}^N - \partial\Omega) \neq \emptyset)\}$$

with $M = \max \{\|u_\circ\|_{L^\infty(\Omega)} , \|\psi_1\|_{L^\infty(\Sigma)}\}$

$$K_a = \left| \frac{2 \,|1 + p + (N-1)(p-1)|}{a(1-p)^2} \right|^{\frac{1}{p-1}}$$

In particular, if $\psi_1 \equiv 0$, $u(x,t) = 0$ for $x \in \Omega_\lambda$ and $t \geq T_\circ$, $T_\circ$ given by

$$T_\circ = \frac{2 \|u_\circ\|^{1-p}_{L^\infty(\Omega)}}{\lambda\mu^2(1-p)} .$$

iii) If $0 \leq p < 1$ and if we define for $t \geq 0$

$$P(t) = \{x \in \Omega \,|\, u(x,t) > 0 \}$$

we obtain the uniform estimate

$$\Omega_\circ(t) \supset \{x \in \Omega_\lambda - P(0) \,|\, d(x, \partial(\Omega_\lambda - P(0)) - (\partial\Omega - \text{supp} \|\psi_1(x,.)\|_{L^\infty(0,\infty)}) \geq (\frac{M}{K_{\lambda\mu^2}})^{\frac{1-p}{2}} \}$$

for $t \geq 0$ and the growth of $P(t)$ is given by

$$(P(t) \cap \Omega_\lambda) \subset (P(0) \cap \Omega_\lambda) + B(0,C(\lambda)t^{\frac{1-p}{2}})$$

where $C(\lambda)$ is a constant.

iv) If $0 < p < 1$ and $\psi_1 \equiv 0$, and if for $x_\circ \in \Omega_\lambda$ we have

$$0 \leq u_\circ(x) \leq K_{\lambda\mu^2} |x - x_\circ|^{\frac{2}{1-p}} \quad \text{a.e. on } \Omega_\lambda$$

then $u(x_\circ,t) = 0$ for $t \geq 0$.

This theorem is especially meaningful if $f(s) \geq m_1 > 0$ for $s \geq 0$. In this case $\Omega_\lambda = \Omega$ for $\lambda \in (0, m_1|$ and the above estimates become completely explicit. We also point out that $K_a \longrightarrow +\infty$ if $a \longrightarrow +\infty$ and this implies that the dependence of T_o and the estimate for $\Omega_o(t)$ in iii) on the "geometry" and size of Ω can be analyzed in terms of the parameter μ.

2. MAIN RESULTS

In this section we state some theorems related with Theorem 1.1. Detailed proofs, complementary results, and applications can be found in |6|.

Now we want to study the existence of a "dead core". i.e., we shall show that under suitable assumptions the set $\{(x,t) \in Q | \ u(x,t) = 0\}$ with u solution of (1.2) has a positive measure. Our results follow easily from the case of a single equation. As we want to show the full power of our comparison techniques, we consider the (possibly degenerate) parabolic problem

$$\begin{aligned} &u_t - \Delta\Phi(u) + \tilde{f}(x,t)\alpha(u) = 0 && \text{in } Q \\ &u(x,t) = \psi(x,t) && \text{on } \Sigma \qquad (2.1) \\ &u(x,0) = u_o(x) && \text{on } \Omega \text{, where} \end{aligned}$$

$\Phi: \mathbb{R} \longrightarrow \mathbb{R}$ is a continuous and non-decreasing with $\Phi(0) = 0$ and α is a maximal monotone graph such that $0 \in \alpha(0)$.

It is clear that if (u,v) is a solution of (1.2), then u satisfies (2.1) for $\Phi(s) = s$ and $\tilde{f}(x,t) = f(v(x,t))$. Here we do not insist on existence and we assume that (2.1) has a solution which is at least continuous on $\overline{Q}$.

Theorem 2.1: Let $m \geq 1$, $0 \leq p < 1$, and let u be a solution of (2.1) for $\Phi(s) = |s|^m \text{sg } s$ and $\alpha(s) = |s|^p \text{ sg } s$. Suppose that for $\lambda > 0$ and $T > 0$ there exists $\Omega_{\lambda,T}$ Ω such that

$$\Omega_{\lambda,T} \times (0,T) \subset \{(x,t) \in \Omega \times (0,T) | \ \tilde{f}(x,t) \geq \lambda\}$$

We define for $t > 0$

$$U^{\pm}_{\frac{\lambda}{2}}(t) = |(\pm\|u_0\|^{1-p}_{L^\infty(\Omega)} + \frac{\lambda}{2}(p-1)t)^{\pm}|^{\frac{1}{1-p}}. \qquad (2.2)$$

Then we get the estimate

$$U^{-}_{\frac{\lambda}{2}}(t) \leq u(x,t) \leq U^{+}_{\frac{\lambda}{2}}(t) \qquad (2.3)$$

for any $x \in \Omega_{\lambda,T}$ such that $d(x, \partial\Omega_{\lambda,T}) \geq \rho_{\lambda,T}$ with

$\rho_{\lambda,T} = \max\{\rho^{+}_{\lambda,T}, \rho^{-}_{\lambda,T}\}$ and

$$\rho^{+}_{\lambda,T} = \max\{\sup_{\substack{t\in|0,T| \\ x\in\partial\Omega_\lambda \cap \partial\Omega}} \left(\frac{\psi^{+}(x,t)^m - U^{+}_{\lambda/2}(t)^m}{K_{\lambda/2}}\right)^{\frac{1-p/m}{2}},$$

$$\sup_{t\in|0,T|} \left(\frac{M^m_{+} - U^{+}_{\lambda/2}(t)^m}{K_{\lambda/2}}\right)^{\frac{1-p/m}{2}}$$

where $M_{+} = \max\{\|u^{+}_0\|_\infty, \|\psi^{+}\|_\infty$. ($M_{-}$ and $\rho^{-}_{\lambda,T}$ are defined analogously).

Remark 2.1: If there exists $m_1 > 0$ such that $\tilde{f}(x,t) \geq m_1$, then $\Omega_\lambda = \Omega$ for $0 < \lambda \leq m_1$ and the above estimates take a simpler form. If $0 \leq p < 1$ we extend some very particular results of Kersner |8| concerning extinction in finite time. If $\psi \equiv 0$ we obtain results due to Kalashnikov |7| and Veron |9| . If $1 < p < m$ and $\psi \equiv 0$ we get an estimate included in |2|.

Theorem 2.2: Suppose $0 < p < 1 \leq m$, For $\lambda > 0$ and

$$\varepsilon = \left(\frac{\|u_0\|^m_{L^\infty(\Omega)}}{K_{\frac{\lambda}{2}}}\right)^{\frac{1-p/m}{2}}$$

let $x_0 \in \Omega_{\lambda,\infty}$ be such that $d(x_0, \partial\Omega_{\lambda,\infty}) \geq \rho_\lambda(\infty) + \varepsilon$ with

$$\rho_\lambda(\infty) = \left(\frac{\max\{\|u_0\|_\infty, \|\psi\|_\infty\}^m - \|u_0\|^m}{K_{\frac{\lambda}{2}}}\right)^{\frac{1-p/m}{2}}$$

Then $u(x_o,t) = 0$ for $t \geq T_o(x_o,\varepsilon)$, where

$$T_o(x_o,\varepsilon) = \frac{\|u_o\|_{L^\infty(B(x_o,\varepsilon))}}{(1-p)\frac{\lambda}{2}}$$

Theorem 2.3: Suppose $0 \leq p < m$. If $\lambda > 0$, $T > 0$. we have for $t \in |0,T|$

$$\Omega_o(t) \supset \{x \in \Omega_{\lambda,T} - P(0) |\ d(x, \partial(\Omega_{\lambda,T} - P(0)) - - (\partial\Omega - \operatorname{supp}\|\psi(x,.)\|_{L^\infty(0,T)})\} \geq (\frac{M}{K_\lambda})^{\frac{m-p}{2}}$$

with $M = \max\ \{\|u_o\|_\infty, \|\psi\|_\infty\}$.

Theorem 2.4: Suppose $0 \leq p < 1 = m$, $\lambda > 0$, $T > 0$.

If u is a solution such that $u_t \in L^\infty(Q)$, then for $0 < t < T$

$$P(t) \cap \Omega_{\lambda,T} \subset (P(0) \cap \Omega_{\lambda,T}) + B(0, C(\lambda)t^{\frac{1-p}{2}}),\ C(\lambda) \text{ constant.}$$

Remark 2.2: This last result seems to be new for $0 < p < 1$.

The following theorem is related with work by Kersner |8| concerning the time evolution of the subsets $P(t)$.

Theorem 2.5: Let $p < m$ and let $x_o \in \Omega_{\lambda,T}$ such that $u_o(x_o) = 0$. We assume that

$$d(x_o, \partial\Omega_{\lambda,T}) \geq \varepsilon = \left(\frac{\max\ \{\|u_o\|_{L^\infty(\Omega)}, \|\psi\|_{L^\infty(\Sigma)}\}}{K_\lambda}\right)^{\frac{1-p/m}{2}}$$

and that for any $x \in B(x_o,\varepsilon)$

$$0 \leq u_o(x) \leq K_\lambda |x - x_o|^{\frac{2}{1-p/m}}$$

Then $u(x_o,t) = 0$ for $t \in |0,T|$.

Remark 2.3: If $\Omega_{\lambda,T} = \Omega$ and $\psi \equiv 0$ the assumption can be written

$$0 \leq u_o(x) \leq K_\lambda |x - x_o|^{\frac{2}{1-p/m}}$$

for any $x \in \Omega$. This extends Theorem 11 in |2| which is only valid for $1 \leq p < m$, $\psi \equiv 0$ and $\tilde{f} \equiv 1$.

To give an idea of our methods we sketch the proof of Theorem 2.1.

Sketch of the proof: We define the function

$$\bar{u}(x,t) = \left(h_{\frac{\lambda}{2}}(x) + (U^+_{\frac{\lambda}{2}}(t))^m\right)^{\frac{1}{m}}$$

where $h_\lambda(x) = K_\lambda |x-x_\circ|^{\frac{2}{1-p/m}}$, $U^+_{\lambda/2}$ is given by (2.2), and K_λ is obtained by replacing p by p/m in the K_a of Theorem 1.1. By using lemma 2.1 in |5| , the explicit expression of $U^+_{\lambda/2}$, and the monotonicity and concavity properties of the function $\Phi(s) = s^m$ and $\alpha(\Phi^{-1}(s)) = s^{p/m}$ we get

$$\bar{u}_t - \Delta\Phi(\bar{u}) + \lambda\alpha(\bar{u}) \geq 0.$$

On the other hand it is clear that

$$\bar{u}(x,0) \geq U^+_{\lambda/2}(0) = \|u^+_\circ\|_{L^\infty} \geq u_\circ(x)$$

Now, for $t_\circ > 0$ fixed, let $x_\circ \epsilon \Omega_\lambda$ be such that $|x-x_\circ| \geq M^+_\lambda(t_\circ)$, where $M^+_\lambda(t_\circ) = \rho^+_{\lambda,t_\circ}$. It is clear that $u \leq M$ on Q and

$$\begin{aligned}
&\bar{u}_t - \Delta\Phi(\bar{u}) + \lambda\alpha(\bar{u}) \geq u_t - \Delta\Phi(u) + \lambda\alpha(u) && \text{in } \Omega_\lambda \times (0,t_\circ) \\
&\bar{u} \geq u && \text{on } \partial\Omega_\lambda \times (0,t_\circ) \\
&\bar{u}(x,0) \geq u_\circ(x) && \text{on } \Omega_\lambda
\end{aligned}$$

This implies, by the Maximun Principle (more precisely, by the T-accretivity in $L^1(\Omega)$ of the operator formally defined by $-\Delta\Phi(u) + \alpha(u)$), that $u(x,t) \leq \bar{u}(x,t)$ on $\Omega_\lambda \times |0,t_\circ|$ and for $x = x_\circ$ this yields $u(x_\circ,t) \leq U^+_{\lambda/2}(t)$. The other inequality is proved in the same way.

REFERENCES

|1| Aris, R. The Mathematical Theory of Diffusion and Reaction in Permeable Catalysts. Oxford, Clarendon Press, 1975.

|2| Bertsch, M., Nambu, T., Peletier, L.A. Decay of Solutions of a Degenerate Nonlinear Diffusion Equation. Preprint.

|3| Díaz, J.I. Qualitative Properties of Nonlinear Parabolic and Elliptic Equations. Book in preparation.

|4| Díaz, J.I. Técnica de supersoluciones locales para problemas estacionarios no lineales: aplicación al estudio de flujos subsónicos. Memory of the Real Academia de Ciencias, Madrid, 1980.

|5| Díaz, J.I., Hernández, J. On the Existence of a Free Boundary for a Class of Reaction-Diffusion Systems. University of Wisconsin, Madison, Technical Report, and paper to appear, 1982.

|6| Díaz, J.I., Hernández, J. On the Existence and Evolution of Free Boundaries for Parabolic Reaction-Diffusion Systems. To appear.

|7| Kalashnikov, A.S. The Propagation of Disturbances in Problems of Nonlinear Heat Conduction with Absorption. Zh. Vychisl. Mat. Fiz. 14(4), 1974, 891-905.

|8| Kersner, R, The Behaviour of Temperature Fronts in Media with Nonlinear Thermal Conductivity under Absorption. Vestnik, Moskov. Univ. Matem. 33, 1978, 44-51.

|9| Véron, L. Coercivité et proprietés régularisantes des semigroupes non-linéaires dans les espaces de Banach. Publications Mathématiques de Besançon, 1977.

A NEUTRAL SYSTEM WITH STATE-DEPENDENT DELAYS

R. D. Driver

Department of Mathematics
University of Rhode Island
Kingston, Rhode Island

ABSTRACT

The two-body problem of classical electrodynamics can be modeled with a system of functional differential equations of neutral type. The delays arise from the finite speed of propagation of electromagnetic effects, and hence depend on the (unknown) trajectories. A satisfactory existence and uniqueness theorem is obtained. The proof will appear elsewhere.

NATURE OF THE MODEL

Imagine two charged particles, each moving under the sole influence of the retarded fields of the other. Delays occur in the equations of motion because of the finite speed of propagation, c, of electromagnetic effects.

Specifically, let the two particles be located, with respect to some inertial reference frame, at positions $x_1(t)$ and $x_2(t)$ in R^3 at time t. Then the electromagnetic effects reaching particle i at instant t must have been produced by particle j at an earlier instant $t - r_i(t)$, where

$$cr_i(t) = |x_i(t) - x_j(t-r_i(t))| \qquad (j \neq i).$$

Here, $|\cdot|$ is the Euclidean norm in R^3.

Introduce $v_i(t) = x_i'(t)/c$ and, for economy of notation, write x_i, v_i, and r_i instead of $x_i(t)$, $v_i(t)$, and $r_i(t)$ whenever the argument is t. Then the equations of motion for particle i under the sole influence of particle j have the form

$$x_i' = cv_i, \tag{1}$$

$$\begin{aligned} v_i' = {} & (e_ie_j/m_i)g(x_i-x_j(t-r_i),\ v_i,\ v_j(t-r_i)) \\ & + (e_ie_j/m_i)G(x_i-x_j(t-r_i),\ v_i,\ v_j(t-r_i))v_j'(t-r_i), \end{aligned} \tag{2}$$

$$cr_i = |x_i - x_j(t-r_i)|. \tag{3}$$

Here e_i and e_j are the charges on the particles, m_i is the rest mass of particle i, g is a known vector-valued function, and G is a known matrix-valued function. Both $g(\xi,\eta,\zeta)$ and $G(\xi,\eta,\zeta)$ are analytic for $\xi \neq 0$, $|\eta| < 1$, and $|\zeta| < 1$.

The specific forms of g and G are obtained by substituting the electric and magnetic fields produced by particle j, as computed from the Liénard-Wiechert potentials, into the Lorentz force law for particle i. (This model does not include "radiation reaction" of a particle on itself.)

Assuming the speeds of the particles never reach c it is possible to replace the functional equation (3) with an essentially equivalent delay differential equation

$$r_i' = \frac{u_i \cdot [v_i - v_j(t-r_i)]}{1 - u_i \cdot v_j(t-r_i)} , \tag{3'}$$

where $u_i = [x_i - x_j(t-r_i)]/cr_i$, a unit vector, and $\cdot$ means dot product.

Then one can replace the system (1), (2), (3) with a somewhat more general and tidier form as follows. Consider

$$\begin{aligned} r_i' &= f_i(r_i, y_i, y(t-r_i)), \\ y_i' &= g_i(r_i, y_i, y(t-r_i)) + G(r_i, y_i, y(t-r_i))y'(t-r_i), \end{aligned} \tag{4}$$

where $i = 1, \ldots, m$, each r_i is scalar valued, each y_i is n-vector valued, and $y = (y_1,\ldots,y_m)$. The given functions in system (4) are scalar-valued f_i, n-vector-valued g_i, and $n \times mn$-matrix-valued G_i.

In the two-body problem, $m = 2$, and $n = 6$ with $y_i = (x_i,v_i)$.

EXISTENCE AND UNIQUENESS

System (4) is considered for $t \geq 0$ together with initial conditions

$$r_i(0) = r_{0i} \quad \text{for each } i \quad \text{and } y(t) = \phi(t) \quad \text{for } \alpha \leq t \leq 0. \tag{5}$$

As usual for neutral systems, y' cannot be assumed continuous. One should only demand that y and ϕ be locally absolutely continuous and that (4) be satisfied almost everywhere for $t \geq 0$. See [1].

An important restriction, explicit in the formulation of the system, is the fact that for each i the group of n+1 equations in (4) is coupled with other such groups only through the retarded terms; and each such group involves only one delay. Further restrictions are imposed below. Some of these may look artificial, but they are actually satisfied by the equations of the two-body problem.

For each $i = 1, \ldots, m$, let D_i be an open set in $R^n \times R^{mn}$, and assume that on the set $(0,\infty) \times D_i$,

(i) f_i, g_i, and G_i are continuous,

(ii) each is locally Lipschitzian with respect to y_i, and

(iii) $f_i < 1$.

For the initial conditions (5), assume that

(iv) $0 < r_{0i} < -\alpha$ for each i,

(v) $\phi = (\phi_1,\ldots,\phi_m) : [\alpha,0] \to R^{mn}$ is absolutely continuous, and

(vi) $(\phi_i(0), \phi(-r_{0i})) \in D_i$ for each i.

A solution of (4) and (5) consists of locally absolutely continuous functions $r_1, \ldots, r_m$ on $[0,\beta)$ and y on $[\alpha,\beta)$ for some $\beta > 0$ such that (5) is satisfied, while on $[0,\beta)$, $(r_i, y_i, y(t-r_i)) \in (0,\infty) \times D_i$ and (4) is satisfied almost everywhere.

<u>Theorem 1</u>. Under hypotheses (i) - (vi), Eqs. (4) and (5) have a unique solution on $[\alpha,\beta)$ where either $\beta = \infty$, or else

$\lim_{t\to\beta} r_i(t) = 0$ for some i, or

for every choice of compact sets $A_j \subset D_j$ $(j = 1,\ldots,m)$ there is an i such that $(y_i(t), y(t-r_i(t))) \notin A_i$ for some $t \in (0,\beta)$.

The proof of this theorem begins with a <u>local</u> existence and uniqueness argument. For sufficiently small $t > 0$, $y(t-r_i)$ and $y'(t-r_i)$ in (4) are replaced by $\phi(t-r_i)$ and $\phi'(t-r_i)$, respectively. This gives a system of ordinary differential equations with discontinuous right hand side. The system does not satisfy the Carathéodory conditions nor does it satisfy Filippov's conditions [3]. However, one can define

$$\sigma(t) = \sigma_{r_i}(t) \equiv t - r_i(t)$$

and introduce new unknowns

$$q(s) \equiv r_i(\sigma^{-1}(s)) \quad \text{and} \quad w(s) \equiv y_i(\sigma^{-1}(s)).$$

This produces a Carathéodory system of ordinary differential equations. After confirming the existence of a unique solution of the latter, it is possible to reverse the change of variables and uniquely recover r_i and y_i as the solution of (4) in some interval [2].

It should be noted that this argument for local existence and uniqueness depends heavily on the fact that, when the initial function ϕ is used in (4), the system decouples into m systems each with just a single delay.

The proof that the solution can actually be extended to the value of β described will appear elsewhere, as will the details of the proof of Theorem 2 below.

Noting the location of potential singularities of the functions g and G, indicated after system (1), (2), (3), one might conjecture that Theorem 1 would give the following result: Given appropriate initial trajectories, system (1), (2), (3) will have a unique solution on $[\alpha,\beta)$ where either $\beta = \infty$ or else $\limsup_{t\to\beta} |v_i(t)| = 1$ (for $i = 1$ or 2), or $\liminf_{t\to\beta} |x_1(t) - x_2(t)| = 0$.

This is indeed true. But a much better result can be proved.

Theorem 2. Let initial trajectories $x_i(t) = \theta_i(t)$ $(i = 1, 2)$ be given on $[\alpha,0]$, having absolutely continuous derivatives with $|\theta_i'| < c$, and such that $\theta_1(0) \neq \theta_2(0)$ and Eqs. (3) have solutions at $t = 0$. Then Eqs. (1), (2), and (3) determine unique extensions of these trajectories to $\beta > 0$ where either $\beta = \infty$ or $\lim_{t\to\beta} x_1(t) = \lim_{t\to\beta} x_2(t)$ -- a collision.

The proof that neither particle can approach the speed of light unless there is a collision depends on the specific form of the functions g and G in Eq. (2). When these are used, Eq. (2) yields

$$v_i \cdot v_i' = \frac{ke_ie_j}{m_i}(1 - v_i^2)^{3/2} \left\{ \frac{[u_i \cdot v_i - v_i \cdot v_j(t-r_i)][1 - v_j^2(t-r_i)]}{r_i^2[1 - u_i \cdot v_j(t-r_i)]^3} + \frac{v_i \cdot [u_i \times ((u_i - v_j(t-r_i)) \times v_j'(t-r_i))]}{r_i[1 - u_i \cdot v_j(t-r_i)]^3} \right\}, \tag{6}$$

where k is a positive universal constant, u_i is the unit vector which occurred in Eq. (3'), and $\times$ denotes cross product.

Under the assumption that the particles do not collide, one can conclude that $r_i(t) \geq \delta > 0$ on $[0,\beta)$. This observation together with some manipulations of Eqs. (3') and (6) lead to the conclusion that $|v_i(t)| \leq A < 1$ on $[0,\beta)$.

It is interesting to note that the success of this argument depended on the factor $(1 - v_i^2)^{3/2}$ in Eq. (6). Had the exponent been anything less than $3/2$ the argument would have failed.

OPEN QUESTIONS

A number of unsolved problems remain for the model of electrodynamics used in this paper. For example:

1. What can be said about continuous dependence on initial conditions? For neutral equations with known delays, it was this question which motivated the use of absolutely continuous functions [1].

2. For particles of like sign ($e_i e_j > 0$), are collisions impossible?

3. What can be said about existence and uniqueness, _even locally_, for the three-body problem? In this case the model does not exhibit the desired "decoupling" of system (4). Now within each such subsystem there is more than one delay, and this eliminates the method of proof indicated above. (_Local_ existence and uniqueness has been proved for the special case of one-dimensional motion [4].)

REFERENCES

[1] R. D. Driver, Existence and continuous dependence of solutions of a neutral functional-differential equation, Arch. Rational Mech. Anal. _19_ (1965) 149-166.

[2] _____, Topologies for equations of neutral type and classical electrodynamics, in Differencial'nye Uravnenija s Otklonjajuščimsja Argumentom, "Naukova Dumka", Kiev, 1977, pp. 113-127.

[3] A. F. Filippov, Differential equations with discontinuous right-hand side, Mat. Sbornik _51_ (93) (1960) 99-128. Translated in Amer. Math. Soc. Trans. Ser. 2, No. 43, 199-232.

[4] M. J. Norris and R. D. Driver, A uniqueness theorem for ordinary differential equations, SIAM J. Math. Anal. _12_ (1981) 141-144.

ASYMPTOTIC PROPERTIES OF SOLUTIONS TO A CLASS OF NONLINEAR SYSTEMS OF DIFFERENTIAL EQUATIONS ARISING IN PHARMACOKINETICS AND TRACER KINETICS

J. Eisenfeld

University of Texas
Arlington, Texas

INTRODUCTION

Let $x(t)$ be a solution of the system

$$x' = F(t,x)x, \ x(0) = x_0 \ ('=d/dt), \tag{1.1}$$

which is defined for $t \geq 0$ and $x(t) \in \Omega$, where Ω is a compact invariant domain which is contained in R^n_+.

$$F: R_+ \times \Omega \longrightarrow M \tag{1.2}$$

is a continuous map, where M is the class of matrices having nonnegative off-diagonal entries and

$$F(t,x) \longrightarrow A, \ (t,x) \longrightarrow (\infty,0). \tag{1.3}$$

Moreover,

$$x(t) \longrightarrow 0, \ t \longrightarrow \infty \tag{1.4}$$

Systems having the above properties arise in pharmacokinetics and tracer kinetics [1-8]. In these applications, $F(t,x) \in ¢$, where $¢$ is the class of compartmental matrices, i.e. the subclass of M of matrices which have the property that the sum of the entries in each column is nonpositive. If $F(t,x) \in ¢$ then the system is said to be a compartmental system. It is of interest to determine the asymptotic properties of solutions to (1.1 - 1.4) with an eye toward applications to the structural identification problem in compartmental systems, which is the problem of unique estimability of parameters from observed data [7,8].

2. STATEMENTS OF RESULTS, REMARKS AND EXAMPLES

Proofs will be given in the next section. Conditions (1.1) - (1.4) are assumed throughout. Sufficient conditions for (1.4) are available [3-6]. Moreover, vacuous states are ruled out, i.e.

$$x_i(t) \not\equiv 0,\ 1 \le i \le n, \tag{2.1}$$

for if such states exist then the system can be reduced to an equivalent system in which there are no vacuous states. Condition (2.1) implies $x(t)$ 0 for t sufficiently large [6]. Thus it may be assumed, without loss of generality, that

$$x(t) > 0,\ t \ge 0, \tag{2.2}$$

i.e., $x_i(t) > 0,\ 1 \le i \le n$.

Remark 1. Suppose that instead of conditions (1.3) - (1.4) one has the seemingly more general conditions: $x(t) \longrightarrow a$ as $t \longrightarrow \infty$ and $F(t,x) \longrightarrow A$ as $(t,x) \longrightarrow (\infty,a)$, where a is a constant vector (not necessarily 0). Then setting $w = e^{-\lambda t}x$, $\lambda > 0$, transforms the system into $w' = \tilde{F}(t,w)w$ where $\tilde{F}(t,w) = F(t,we^{\lambda t})w - \lambda I$. Observe that the transformed system satisfies the conditions (1.1) - (1.4).

For any function $y(t)$, which is defined and positive for sufficiently large t, set

$$LN^-(y) \varliminf_{t\to\infty} t^{-1} \log y(t),\ LN^+ = \varlimsup_{t\to\infty} t^{-1} \log y(t).$$

Remark 2. Observe that $b^+ = LN^+(y) < \infty$ implies that for any $\varepsilon > 0$ there exists $C^+ > 0$ such that $y(t) \le C^+e^{(b^+ + \varepsilon)t}$ for sufficiently large t. Similarly, $b^- = LN^-(y) > -\infty$ implies that for any $\varepsilon > 0$ there exists $C^- > 0$ such that $y(t) \ge C^-e^{(b^- - \varepsilon)t}$ for t sufficiently large. So $y(t) \longrightarrow 0$, $t \longrightarrow \infty$, implies further that $b^- \le 0$.

If $LN^-(y) = LN^+(y)$ then set

$$LN(y) = LN^{\pm}(y) = \lim_{t\to\infty} t^{-1} \log y(t)$$

Let $|x| = |x_1| + |x_2| + \ldots + |x_n|$ denote the ℓ, norm of $x \in R^n$ (actually, any Euclidean norm suffices). If $x(t)$ satisfies a set of conditions similar to (1.1), (1.3), and (1.4) (but without requiring that $F(t,x) \in \mathcal{M}$), then $LN^+(|x|) \le b$, where b is the real part of an eigenvalue of A, and its value is called the Lyapunov number of x [9]. With condition (1.2) there is a stronger result: Proposition 1 (below).

If $M \in \mathcal{M}$ then $M + pI$ (where I denotes the $n \times n$ identity matrix) is nonnegative for p sufficiently large. It is not difficult to see (from the theory of nonnegative matrices [10]) that a matrix $M \in \mathcal{M}$

has an eigenvalue $\rho(M)$ which is real and which is greater than the real part of any other eigenvalue of M. $\rho(M)$ is called the leading eigenvalue.

Proposition 1. Under conditions (1.1) - (1.4), $LN(|x|) = \rho(A)$.

Corollary 1. Let $x(t)$ be solution of the autonomous compartmental system $x' = F(x)x$, $x(0) \geq 0$, where $F: \Omega \longrightarrow ¢$ is continuous. Suppose further that $x(t) \longrightarrow 0$ as $t \longrightarrow \infty$. Then $LN(|x|) = \rho(F(0))$.

Example 1. Consider the system $x_1' = - f(x_1)x_1 - cx_1$, $x_2' = f(x_1)x_1 - dx_2$, $x_1(0) = 1$, $x_2(0) = 0$. The system simulates the biotransformation of methyl mercury, where x_1 and x_2 represent the mass of organic and nonorganic mercury, respectively. The function f, which represents the conversion of organic mercury to nonorganic mercury, is continuous and positive on [0,1] the positive constants c and d are the decay rates for the two species.

The system has the form in (1.1) where

$$F(x) = \begin{bmatrix} - f(x_1) - c & 0 \\ f(x_1) & -d \end{bmatrix} \in ¢$$

since the system is compartmental its domain is

$$\Omega = \{x \geq 0: |x| \leq |x_0|\} \tag{2.3}$$

(in this example $|x_0| = 1$) (see [1]-[8]). Moreover, it is easy to see that $x(t) \longrightarrow 0$ as $t \longrightarrow \infty$ [3]. According to Corollary 1,

$$LN(|x|) = \rho(F(0)) = \max \{- f(0) - c, - d\} \tag{2.4}$$

Normally, the total mercury $|x(t)|$ is an observed function and so the parameter $\rho(F(0))$ is uniquely identifiable from the relation (2.4).

Criteria may also be established for determining the "Liapunov number" (or asymptotic decay rate) for individual components. The concept of reachability is needed for this purpose. Let $M \in \mathcal{M}$. (Index) j reaches i in M if either $i = j$ or there is a sequence of indices $k_1, k_2, \ldots, k_q$, with $k_1 = j$ and $k_{q+1} = i$ such that the (k_{r+1}, k_r) entries of M are nonzero, $r = 1,2,\ldots,q - 1$. Reachability is best illustrated by means of the digraph (directed graph) [11,1,3]. Let $M \in \mathcal{M}$, then the diagraph of M is constructed by a drawing an arc from index j to index $i(\neq j)$

whenever the (i,j) entry of M is nonzero thereby obtaining a graph as in Fig. 1 (below).

Proposition 2. Under conditions (1.1) - (1.4), if j reaches i in A then $LN^-(x_i) \geq LN^-(x_j)$.

A matrix $A \in M$ is said to be irreducible [10] (or equivalently, its digraph is strongly connected [11]) if every pair of indices are mutually reachable.

Proposition 3. Under condition (1.1) - (1.4), if A is irreducible then $LN(x_i) = \rho(A)$, $1 \leq i \leq n$.

The indices of a (not necessarily irreducible) matrix may be partitioned into mutually disjoint nonempty subsets $J_1, J_2, \ldots, J_q$ according to the reachability relation, i.e., each J_k is a maximal subset of mutually reachable (in A) indices. Each submatrix A_k formed from entries (i,j), for which $i,j \in J_k$, is then an irreducible component of A.

Proposition 4. Let conditions (1.1) - (1.4) be satisfied. Let $\{J_k\}$ and $\{A_k\}$ be defined as above (where A is the matrix in condition (1.3)). Suppose j reaches i where $j \in J_j$ and $\rho(A_j) = \rho(A)$, then $LN(x_i) = \rho(A)$.

Example 2. 5-FU is a cancer drug which is described by a system of equations [12], the particulars of which are not essential to this discussion; it is only necessary to note that conditions (1.1) - (1.4) are satisfied and the digraph of A is given in Fig. 1 (two arcs, eminating from indices 10 and 11, represent transfers to absorbing states which are omitted from this model).

In this model the partitioning referred to in Proposition 3 is $J_1 = \{1,2,4,5,7,8,11\}$, $J_2 = \{3,6,9,10\}$. Observe that $\rho(A) = \max\{\rho(A_1), \rho(A_2)\}$. If $\rho(A_2) \leq \rho(A_1)$ then, according to Proposition 4, $LN(x_i) = \rho(A)$, $1 \leq i \leq 11$ but if $\rho(A_2) > \rho(A_1)$ then Proposition 4 yields only that $LN(x_i) = \rho(A)$, $i \in J_2$.

Remark 3. Consider the simple epidemic model $x_1' = -\alpha x_2 x_1$, $x_2' = \alpha x_2 x_1 - \gamma x_2$, $x(0) = x_0$, where x_1 and x_2 represent the susceptible and the infectious populations, respectively, α is the infectious rate and γ is the removal rate. It is known [13] that x(t)

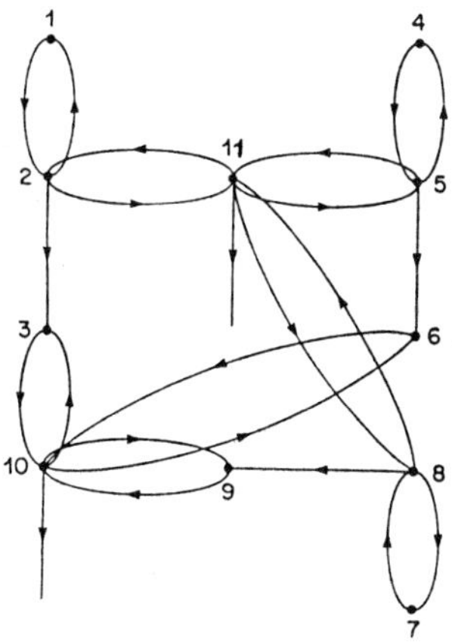

Figure 1. Digraph representation of 5-FU kinetics.

tends to a constant vector $(s,0)$, where $s > 0$ whenever, $x_1(0) > 0$. This example may be transformed into a system satisfying conditions (1.1)-(1.4), as explained in Remark 1, and so Proposition 4 applies. It is not difficult to show that $LN(x_1) = 0$ and $LN(x_2) = \alpha s - \gamma < 0$. It is interesting that $LN(x_2)$ is not determined by the limiting matrix A, which indicates the dificulties that would be encountered in attempting to strengthen Proposition 4.

Let $p(\lambda) = (\lambda-\lambda_1)^{m_1}(\lambda-\lambda_2)^{m_2}\ldots(\lambda-\lambda_k)^{m_k}$ be the minimal polynomial of a matrix A. An eigenvalue λ_j is said to have a <u>simple elementary divisor</u> if $m_j = 1$.

<u>Proposition 5</u>. Let conditions (1.1) - (1.4) be satisfied and suppose further that (i) there is a positive constant L such that $|F(t,x)-A| \leq L|x|$, $(t,x) \in R_+ \times \Omega$, (ii) A has a simple elementary divisor, and (iii) $b = \rho(A) < 0$. Then $x(t) \sim ze^{bt}$ as $t \to \infty$ (i.e., $\lim_{t\to\infty} x(t)e^{-bt} = z$), where z is a nonnegative vector such that $Az = bz$.

<u>Remark 4</u>. Condition (i) is satisfied if the Jacobian $\partial F(t,x)x/\partial x$ is bounded in $R_+ \times \Omega$. In particular, condition (i) is satisfied in an autonomous system with a continuous Jacobian defined over a compact domain, as is the case in a compartmental system where the domain is given in (2.3).

<u>Remark 5</u>. Conditions $x' = F(t,x)x$, $F(t,x) \sim A$, $x \sim ze^{bt}$ imply that $x' \sim Aze^{bt} = bze^{bt}$.

Remark 6. In general, $x(t)$ is not asymptotic to the solution of the associated linear equation $z' = Az$, $z(0) = x_0$. For instance, the solution of $x' = -(a+x)x$, $x(0) = 1$, is $x(t) = ae^{-at}/(1+a-e^{-at}) \sim ae^{-at}/(1+a)$, whereas the solution of $z' = -az$, $z(0) = 1$, is $z(t) = e^{-at}$.

3. PROOFS OF MAJOR RESULTS

Proposition 1. For each $p > 0$ let

$$G(p) = \inf_{t \geq p} F(t,x(t)), \quad H(p) = \sup_{t \geq p} F(t,x(x)).$$

By condition (1.3),

$$G(p), H(p) \longrightarrow A, \; p \longrightarrow \infty \; . \tag{3.1}$$

By the Comparison Theorem, for $t \geq p$,

$$z_p(t) \equiv x(p) \exp(tG(p)) \leq x(t) \leq x(p) \exp(tH(p)) \equiv w_p(t).$$

Thus

$$LN^-(|z_p|) \leq LN^-(|x|) \leq LN^+(|x|) \leq LN^+(|w_p|).$$

Now $w_p(t) = O(t^{m-1}e^{\rho(H(p))t})$, where m is the order of the largest Jordan block in $H(p)$ associated with the leading eigenvalue, $\rho(H(p))$. Consequently $LN^+(|w_p|) \leq \rho(H(p)) \longrightarrow \rho(A)$ as $p \longrightarrow \infty$, by (3.1), and so

$$LN^+(|x|) \leq \rho(A).$$

Since $G(p) \in M$, $\exp(tG(p)) \geq 0$ and because $x(p) > 0$ (see (2.2)) $|z_p(t)| \sim bt^{q-1}e^{(G(p))t)}$, where $b > 0$ and q is a positive integer. Consequently, $LN^-(|x|) \geq LN^-(|z_p|) \geq \rho(G(p)) \longrightarrow \rho(A)$ as $p \longrightarrow \infty$, and so

$$LN^-(|x|) \geq \rho(A).$$

Proposition 2. By induction on the length of the path from j to i it suffices to verify the result for the case in which the (i,j) entry of A, $a_{ij} > 0$. Let $b = LN^-(x_j)$. Then $\forall\, \varepsilon > 0 \; \exists\, t_0 > 0 \ni x_j(t) \geq e^{(b-\varepsilon)t}, t \geq t_0$. In view of (3.2), $\exists\, p \geq t_0 \ni G(p) \geq A/2$. Thus $x_i' \geq qx_i + de^{(b-\varepsilon)t}, t \geq p$, where $q \leq G_{ii}(p)$, and $d = a_{ij}/2 > 0$. Consequently, $x_j \geq de^{qt} \int_p^t e^{(-q+b-\varepsilon)s} ds$, from which the result readily follows.

Proposition 3. By Proposition 1, $LN^+(x_i) \leq \rho(A)$, $1 \leq i \leq n$, and $\exists$ at least one $j \ni LN^-(x_j) = \rho(A)$. Since every i is reached by j, the result is a consequence of Proposition 2.

Proposition 4. In view of Proposition 2, it suffices to show that $LN(x_j) = \rho(A)$. Now x_j belongs to a system for which $w' \geq F_j(t,x)w$ where $F_j(t,x(t)) \longrightarrow A_j$, $t \longrightarrow \infty$, and w is the projection of x onto the J_j subspace. It may be seen from the proof of Proposition 1 that $LN^-(|w|) \geq \rho(A_j)$. Since $\rho(A_j) = \rho(A)$ and, (from Proposition 1), $LN^+(|w|) \leq \rho(A)$, it follows that $LN(|w|) = \rho(A)$. Since A_j is irreducible it is not difficult to show, by the arguments in Proposition 3, that $LN(x_j) = \rho(A)$, $j \in J_j$.

Proposition 5. Let $w = xe^{-bt}$, where $b = \rho(A)$. Then

$$w' = \tilde{F}(t,w)w,$$

$$\tilde{F}(t,w(t)) = F(t,x(t)) - bI \longrightarrow A - bI \equiv \tilde{A},\ t \longrightarrow \infty. \tag{3.2}$$

Observe first that if $w(t) \longrightarrow z$ (a constant vector), then $w'(t) \longrightarrow \tilde{A}z$. It should be clear that if both w and w' both tend to constant vectors as $t \longrightarrow \infty$ then $w'(t) \longrightarrow 0$. Thus if $w(t) \longrightarrow z$, $\tilde{A}z = 0$, i.e., $Az = bz$. It remains then to show that $w(t)$ tends to a limit.

Now $w' = \tilde{A}w + g(t)$, where $g(t) = e^{-bt}[F(t,x(t)-A]x(t)$ and $w(t) = e^{t\tilde{A}}x_0 + \int_0^t e^{(t-s)\tilde{A}}g(s)ds$. Since $\rho(\tilde{A}) = 0$ has a simple elementary divisor (condition (ii)), $e^{t\tilde{A}}x_0$ tends to the projection of x_0 onto the subspace spanned by the eigenvectors of A associated with b. It suffices then to show that $\int_0^t e^{(t-s)\tilde{A}}q(s)ds$ tends to a limit. Employing condition (ii) again, $e^{t\tilde{A}}$ is of order one as $t \longrightarrow \infty$ so it suffices to show that g is absolutely integrable over R_+. By condition (i), $|F(t,x(t))-A| = 0(|x(t)|)$ and so $g(t) = 0(|x(t)|^2e^{-bt})$. By Proposition 1 and Remark 2, $|x(t)| = 0(e^{(b+\varepsilon)t})$ for any $\varepsilon > 0$ and so $g(t) = 0(e^{(b+2\varepsilon)t})$, $\varepsilon > 0$. Since $b < 0$, $g(t)$ is absolutely integrable and the proof is complete.

REFERENCES

[1] J. A. Jacquez, Compartmental Analysis in Biology and Medicine, Elsevier, New York, 1972.

[2] R. Bellman, "Topics in pharmacokinetics I: concentration-dependent rates," Math. Biosci., 6 (1970), pp. 13-17.

[3] J. Eisenfeld, "On washout in nonlinear compartmental systems," Math. Biosci., 58 (1982), pp. 259-275.

[4] J. Eisenfeld, "On approach to equilibrium in nonlinear compartmental systems," in Differential Equations and Applications in Ecology, Epidemics and Population Problems, S. N. Bussenberg, K. L. Cooke (eds.), Academic Press, New York, 1981, pp. 229-247.

[5] J. Eisenfeld, "On non-approach to equilibrium in nonlinear compartmental systems," IEEE Trans. Circuits Syst., CAS-29 (1982), pp. 196-198.

[6] J. Eisenfeld and S. W. Du, "On the qualitative properties of nonlinear compartmental systems," Proc. Int'l Conf. Nonlinear Phenomena in Mathematical Sciences, V. Lakshmikantham (ed.), Academic Press, New York, 1982, pp. 335-342.

[7] J. Eisenfeld, "Structural identifiability criteria for large linear and nonlinear compartmental systems," Proc. Tenth IMACS World Congress on System Simulation and Scientific Computation, 2 , 1982, pp. 225-228.

[8] J. Eisenfeld, "New techniques for structural identifiability for large linear and nonlinear compartmental systems," MSC, to appear.

[9] E. A. Coddington and N. Levinson, Theory of Ordinary Differential Equations, McGraw Hill, New York, 1955, pp. 314-347.

[10] A. Berman and R. J. Plemmons, Nonnegative Matrices in the Mathematical Sciences, Academic Press, New York, 1979.

[11] F. Harary, R. Z. Norman, and D. Cartwright, Structural Models: An Introduction to the Theory of Directed Graphs, Wiley, New York, 1965.

[12] D. Young and W. Wolf, "QLAR of 5-Fluorauracil," Radiopharmacueticals: Structure Activity Relationships, R. P. Spencer (ed.), Grune and Stratton, Inc., 1981, pp. 665-673.

[13] P. Waltman, Deterministic Threshold Models in the Theory of Epidemics, Springer-Verlag, New York, 1974.

SOME REMARKS ON SEMIFLOWS*

Saber Elaydi

Kuwait University
Kuwait

ABSTRACT

Semiflows with global backward extensions are studied. It is shown that such semiflows can be extended to flows which are the inverse limits of the semiflows. Topological as well as dynamical properties of the inverse limit flow are investigated. Then attainability sets and functions are defined and studied for this subclass of semiflows. An attempt is made to define and study some negative versions of stability.

By a semiflow we mean a pair (X,π), where X is a topological space and $\pi : X \times R^+ \to X$ is a continuous map with $x\pi 0 = x$ and $(x\pi t)\pi s = x\pi(t + s)$ for $x \in X$, $t, s \in R^+$. One of the principal motivations behind the study of semiflows (or semidynamical systems) is that they describe some classes of functional differential equations, partial differential equations of evolution [1] , Volterra integral equations [10], and control systems [7]. We will be concerned here with a subclass of semiflows; namely those semiflows with global backward extensions in which every negative solution [1] is principal and through each point in the phase space a principal negative solution passes. It was pointed out by Hajek [7] that such semiflows arise from certain control systems. Let I be any interval of the forms $(-\infty,0]$, $(a,0]$, $[a,0]$. Then a function $\sigma : I \to X$ is called a negative solution of π if $\sigma(-t + s) = \sigma(-t)\pi s$ whenever $s \in R^+$, $-t \in I$ and $-t + s \in I$. If domain $\sigma = (-\infty,0]$, then σ is called a principal negative solution. If $\sigma(0) = x$ for some $x \in X$, then $\sigma(-\infty,0]$ is called a principal negative trajectory through x [1] . The attainability set of $A \subset X$ at $S \subset R^-$ is defined as $F(A,S) = \{y \in X \mid y\pi s = x,$ for some $x \in A$, $s \in S\}$. A point $x \in X$ is said to be a start point if $F(x,R^-) = \phi$ [1] .

* This research was supported by Kuwait University under project SM 022.

The main purpose of this paper is to study the properties of attainable sets and negative trajectories and to clarify the relation between them. This is done by means of constructing a new flow as an inverse limit of the semiflow under consideration. Brown [2] gave a somewhat different construction of the inverse limit for a discrete semiflow. Properties of the constructed inverse limit are established. At the end we introduce several stability and attraction notions. It is assumed throughout the paper that X is locally compact and Hausdorff and the maps $\pi^t : X \to X$ defined by $\pi^t(x) = x\pi t$ are surjective for all $t \in R^+$.

§1. CONSTRUCTION OF THE INVERSE LIMIT

Let (X,π) be a semiflow and let $X = \Pi\{X_t \mid t \in R^+\}$ be the product space over the index set R^+ directed by the usual order relation on the reals and $X_t = X$ for all $t \in R^+$. Define the map $\tilde{\pi}: \tilde{X} \times R^+ \to \tilde{X}$ by $\tilde{\pi}(x,s) = (x_t\pi s)$, where $x = (x_t) \in \tilde{X}$ and $s \in R^+$. Then $(\tilde{X},\tilde{\pi})$ is a semidynamical system as it can be easily seen. Let $X_\infty = \{x = (x_t) \in \tilde{X} \mid x_t = x_s\pi(s - t)$ for $s,t \in R^+$ with $t < s\}$. Define $\pi_\infty: X \times R \to X$ as follows: for $r \in R^+$ and $x = (x_t) \in X_\infty$ we have

$$(1)\quad x\pi_\infty r = \left(y_t \;\middle|\; y_t = \begin{cases} x_0\pi(r - t) & \text{if } 0 \le t \le r \\ x_{t-r} & \text{if } r \le t \end{cases} \right),$$

$$(2)\quad x\pi_\infty(-r) = (y_t \mid y_t = x_{t+r}) .$$

It can be shown that (X_∞,π_∞) is a dynamical system. The set X_∞ is closed and invariant in $(\tilde{X},\tilde{\pi})$. For $t < s$ let $\pi_{ts} : X_s \to X_t$ be defined by $\pi_{ts}(a) = a\pi(s - t)$. Then π_{ts} is a surjective homomorphism for all $t < s$. Thus the triple $(R^+, (X_t,\pi), \pi_{ts})$ is an inverse spectrum [4] and $(X_\infty,\pi_\infty) = \operatorname{inv}\lim_{t \in R^+}\{(X_t,\pi)\}$ [4].

<u>Theorem 1.1.</u> X is locally compact iff X_∞ is locally compact.

<u>Theorem 1.2.</u> (X,π) is positively minimal [1] iff (X_∞,π_∞) is positively minimal.

<u>Theorem 1.3.</u> (X,π) is positively equicontinuous iff (X_∞,π_∞) is positively equicontinuous [5].

<u>Theorem 1.4.</u> (X,π) is of strong characteristic 0^+ iff (X_∞,π_∞) is of strong characteristic 0^+ [6].

§2 ATTAINABILITY SETS AND FUNCTIONS

Since X_∞ is locally compact and Hausdorff, then one can extend it to its one point compactification $X^* = X_\infty \cup \{\infty\}$. Then a flow $(X_\infty^*, \pi_\infty^*)$ can be defined by letting $x\pi_\infty^* t = x\pi_\infty t$ for $x \in X_\infty$ and $t \in R$ and $\infty\pi_\infty^* t = \infty$ for all $t \in R$. Such extension is not valid in general for semiflows [8]. However, it turns out that the subclass of semiflows under our consideration possesses such extension as stated in the following Theorem.

Theorem 2.1. The following statements are valid.

(1) Each $b \in X$ has at least one principal negative trajectory passing through it and all negative trajectories passing through b are principal.

(2) (X,π) has no start points and can be extended to (X^*,π^*), where X^* is the one point compactification of X and π^* is defined in a similar way to that of π_∞^* on X_∞^*.

(3) For each $x \in X$, the negative escape time $N(x) = \infty$ [8].

The preceding theorem shows, among other things, that there is a one to one correspondence between X_∞ and the set Γ of all negative solutions in X. The correspondence $g : X_\infty \to \Gamma$ is defined by letting $g(x) = \nu$, where $x = (X_t)$ and $\nu(-t) = x_t$ for each $t \in R^+$. This reveals the whole mystery about using inverse limits as a major tool in studying semiflows. For instead of dealing with negative trajectories we deal with more simpler objects; the points of X_∞.

The following scheme of topologizing a collection of closed sets is due to Michael [9]. Let 2^X be the set of all nonempty closed subsets of X. Then the sets $L(U) = \{A \in 2^X \mid A \cap U \neq \phi\}$, where U is an open subset of X, generates a topology, as a subbase, on 2^X which is called the lower semifinite topology T_ℓ and the sets $M(U) = \{A \in 2^X \mid A \subset U\}$ generates a topology, as a base, on 2^X which is called the upper semifinite topology T_u. The smallest topology on 2^X containing both T_ℓ and T_u is the finite topology T_f. A setvalued function $f : X \to Y$ is said to be {lower semicontinuous l.s.c.} {upper semicontinuous u.s.c.} {continuous} if the corresponding function $\hat{f} : X \to 2^Y$ defined by $\hat{f}(x) = f(x)$ is $\{T_\ell\text{-continuous}\}$ $\{T_u\text{-continuous}\}$ $\{T_f\text{-continuous}\}$, respectively.

Note that the attainability sets $F(x,-t)$ defined before can be regarded as images of a setvalued function $F : X \times R^- \to X$. The map F is called the attainability function.

Theorem 2.2. The attainability function F is continuous in $t \in R^-$, u.s.c. in both $x \in X$ and $(x,t) \in X \times R^-$.

Theorem 2.3. The set $F(A,P)$ is compact if both $A \subset X$ and $P \subset R^-$ are compact. Furthermore, if in addition to the above A is connected and $P = [-t,0]$, then $F(A,P)$ is connected.

§3. STRONG NEGATIVE STABILITY

Throughout this section X is assumed to be first countable. For $a \in X$ we have the following definitions :

(1) The negative limit set of a;

$L^-(a) = \{b \in X \mid$ there exist sequences $\{t_i\}$ in R^+ with $t_i \to \infty$ and $\{b_i\}$ in X with b_i $F(a,-t_i)$ and $b_i \to b\}$.

(2) The negative prolongation set of a;

$D^-(a) = \{b \in X \mid$ there exists a sequence of principal negative solutions $\{\sigma_i\}$ with $\sigma_i(0) \to a$ and a sequence $\{t_i\}$ in R^+ with $\sigma_i(-t_i) \to b\}$.

(3) The negative prolongation limit set of a;

$J^-(a) = \{b \in X \mid$ there exists a sequence of principal negative solutions $\{\sigma_i\}$ with $\sigma_i(0) \to a$ and a sequence $\{t_i\}$ in R^+ with $t_i \to \infty$ and $\sigma_i(-t_i) \to b\}$.

(4) $K^-(a) = \overline{F(a,R^-)}$

For simplicity $F(a,R^-)$ will be denoted by $F(a)$ is called the funnel through a [1] .

We observe here that $b \in D^-(a)$ iff $a \in D^+(b)$ and $b \in J^-(a)$ iff $a \in J^+(b)$. Thus our definitions of $J^-(a)$ and $D^-(a)$ coincide with those of [1] and [3] . Furthermore, it can be shown easily that $K^-(a) = F(a) \cup L^-(a)$ and $D^-(a) = F(a) \cup J^-(a)$. It is known from [3] that $J^-(a)$ is closed, positively and negatively invariant. The same properties are possesses by $L^-(a)$. The set $D^-(a)$ is closed and weakly negatively invariant. However, $D^-(a)$ is not positively invariant as it can be seen in simple examples.

Definition 3.1. Let M be a subset of X. Then

(i) M is strongly negatively Liapunov stable if for each neighborhood U of M there exists a neighborhood V of M with $F(V) \subset U$.

(ii) M is strongly $\{D^-\text{-stable}\}$ $\{J^-\text{-stable}\}$ if for each neighborhood U of M there exists a neighborhood V of M such that $\{D^-(V) \subset U\}$ $\{J^-(V) \subset U\}$.

Theorem 3.2. A compact subset M of X is strongly D^--stable iff it is negatively strongly Liapunov stable.

Theorem 3.3. A compact subset M of X is strongly D^--stable iff $D^-(M) = M$.

Theorem 3.4. A compact negatively invariant subset M of X is strongly J^--stable iff $J^-(M) = M$.

Corollary 3.5. A compact negatively invariant subset M of X is strongly D^--stable iff it is strongly J^--stable.

Theorem 3.6. A compact subset M of X is strongly negatively Liapunov stable iff every component of M is strongly negatively Liapunov stable.

Final remarks. There are a lot of problems to be investigated. One problem is to define and study weaker version of stabilities by defining $D_\sigma^-(a)$, $J_\sigma^-(a)$ and $L_\sigma^-(a)$ relative to a negative solution σ ($L_\sigma^-(a)$ has been already defined in [1] and [3]). One can also study strong and weak versions of attraction notions. Negative versions of recurrence and characteristic 0 may be defined and studied. One may say that $a \in X$ is of characteristic 0^- if $D^-(a) = K^-(a)$. It is also desirable to find more properties that can be passed back and forth from (X,π) to (X_∞,π_∞).

REFERENCES

[1] P.N. Bhatia and O. Hajek, Local semidynamical systems, Lecture Notes in Mathematics (90), Springer-Verlag, Berlin (1968).

[2] J.R. Brown, Inverse limits and weak isomorphism for discrete dynamical systems, Trans.Amer.Math.Soc. vol. 164, 1972, 55-66.

[3] K.M. Das and S.S. Lakshmi, On start sets and negative prolongational sets in semidynamical systems, Funkc. Ekvac. 17 (1974), 25-29.

[4] J. Dugundji, Topology, Allyn and Bacon Inc., Boston, 1966.

[5] S. Elaydi, Semidynamical systems with nonunique global backward extensions, Preprint.

[6] S. Elaydi and S.K. Kaul, Flows of almost strong characteristic 0 with generative phase groups, to appear in the Journal of Nonlinear Analysis.

[7] O. Hajek, Geometric theory of time optimal control, Siam J. Control. vol. 9, No. 3, 1971, 339-350.

[8] R.C. McCann, Negative escape time in semidynamical systems, Funkc. Ekvac. 20 (1970), 39-47.

[9] E. Michael, Topologies on spaces of subsets, Trans.Amer.Math.Soc. vol. 71 (1951), 461-466.

[10] R.K. Miller and G.R. Sell, Volterra integral equations and topological dynamics, Memoirs of the American Mathematical Society, No. 102, Providence, R.I. (1970).

A TWO DIMENSIONAL MODEL FOR TURBULENCE

W. E. Fitzgibbon

Department of Mathematics
University of Houston
Houston, Texas

ABSTRACT

Burgers proposed a coupled system of one ordinary and two partial differential equations to model the phenomenon of turbulence. The theory of analytic semigroups is used to provide a variation of parameters representation of the solution to the system. It is shown that as a certain parameter passes through a critical value the steady state loses stability and its stability is transferred to a family of periodic solutions.

We shall be concerned with the following nonlinear system coupling an ordinary and two partial differential equations

$$\text{(1.1 a)} \quad dU(t)/dt = P - \nu U(t) - \int_0^{\pi} [v^2(t,x) + w^2(t,x)]\, dx$$

$$\text{(1.1 b)} \quad \partial v(t,x)/\partial t = U(t)[v(t,x) - w(t,x)] + \nu\partial^2 v(t,x)/\partial x^2 - \partial/\partial x[v^2(t,x) - w^2(t,x)]$$

$$\text{(1.1 c)} \quad \partial w(t,x)/\partial t = U(t)[v(t,x) + w(t,x)] + \nu\partial^2 w(t,x)/\partial x^2 + \partial/\partial x[2v(t,x)w(t,x)]$$

$$\text{(1.1 d)} \quad U(0) = U_0 \ , \ v(0,x) = v_0(x) \ , \ w(0,x) = w_0(x)$$

subject to the Dirichlet boundary conditions

$$\text{(1.1 e)} \quad v(t,0) = v(t,\pi) = w(t,0) = w(t,\pi) = 0$$

The functions U( ), v(,) and w(,) are real valued and defined for $t \geq 0$ and $x \in [0,\pi]$.

This system was proposed by Burgers [1] to model turbulence and is intended to portray flow in a channel ($0 \le x \le \pi$). In the model $U(t)$ denotes the principle flow in the channel under the action of a constant pressure drop ($p > 0$). The first equation represents a balance between the acceleration of the principle flow, the pressure gradient, viscuous damping($\nu > 0$) and the trubulent damping of the integral term. The functions $v(\ ,\ )$ and $w(\ ,\ )$ represent components of secondary or turbulent motion. The acceleration of these components is balanced by viscuous diffusion and convective coupling.

In [6] Dlotko utilizes Galerkin methods to obtain the global existence of weak solutions to (1.1 a - e) and provides conditions sufficient to quarantee that the weak solutions are indeed classical solutions; global behavior of these solutions is then discussed. In our study we use the theory of analytic semigroups to provide a variation of parameters representation for solutions to (1.1 a - e). Exploitation of the semilinear structure of the problem allows us to provide an exponential representation of solutions and to further discuss the qualitative behavior of the solutions.

There exists an extensive literature on Burger's equations and the one component system related to it (cf [3], [4], [5], [2], [10], [13]).

We begin by outlining some energy arguments which provide apriori bounds for solutions to (1.1 a - e). Multiplying equations (1.1 b - c) by v and w respectively, adding the results and integrating on $[0,\pi]$ we obtain,

$$(1.2)\quad 1/2\ d/dt(||v||^2+||w||^2) + \nu(||v_x||^2+||w_x||^2) = U(||v||^2+||w||^2),$$

where $||\ ||$ denotes the L^2 norm. Dlotko [6] further shows that if $L(t) = 1/2\ \{U^2(t) + ||v(t,\ )||^2 + ||w(t,\ )||^2\}$ then

$$(1.3)\quad L(t) \le \{L(0) + \frac{P}{2\nu}\sup[p/\nu,U_0]\}e^{-2\nu t} + \frac{P}{2\nu}\sup[p/\nu,U_0]$$

and thereby derives the asymptotic estimate

$$(1.4)\quad \limsup_{t\to\infty} L(t) \le P/2\nu^2$$

Utilizing standard techniques together with (1.2) it is not difficult to obtain gradient bounds

Lemma 1.5 Assume that for $t \in [0,T)$, $v(t,x)$, $w(t,x)$ belong to $H_0^1(0,\pi) \cap H^2(0,\pi)$. If $[U(t), v(t,x), w(t,x)]$ satisfies (1.1 a - e) for a·e $t \in [0,T)$ then there exists a continuous nonnegative function $M(\ ,\ ,\ )$ so that for any $\tau > 0$

$$\sup_{t\in[0,T)}(||v_x(t,\ )||^2 + ||w_x(t,\ )||^2) \le M(t,||v_x(\tau,\cdot)||^2,||w_x(\tau,\cdot)||^2)$$

It is our intention to write as a Banach space differential equation of the form:

$$\phi'(t) = A\phi(t) + J\phi(t) \qquad \text{for a·e } t \in [0,T]$$
$$\phi(t) = \phi_0 \in X \tag{1.6}$$

where $\phi(\ )$ maps the interval $[0,T]$ to a Banach space X, A is the infinitesimal generator of an analytic semigroup on X and J is an unbounded nonlinear operator mapping X to itself. By a strong solution to (1.6) we mean $\phi(\ )$: $[0,T] \to X$ having the following properties: (i) $\phi(\ )$ is continuous on $[0,T]$ and $\phi(0) = \phi_0$, (ii) $\phi(\ )$ is absolutely continuous on compact subsets of $(0,T)$; and (iii) $\phi(\ )$ is differentiable for a.e.t $\in [0,T)$ and satisfies the differential equation for a:e.t $\in [0,T]$. Whenever strong solutions to (1.6) exist they have variation of parameters representation

$$\phi(t) - T(t)\phi_0 + \int_0^t T(t-s)\ J\phi(s)\ ds \tag{1.7}$$

where $\{T(t)|t \ge 0\}$ is the semigroup of linear transformation on X generated by A. If $0 \in \rho(A)$ we can define fractional powers of A. For $0 < \gamma < 1$ we let

$$A^{-\gamma} = 1/\Gamma(\gamma)\int_0^\infty t^{\gamma-1}\ T(t)\ dt$$

where $\Gamma(\ )$ is the gamma function. The $A^{-\gamma}$ are bounded one-to-one linear operators on X. Therefore we can define $A^\gamma = (A^{-\gamma})^{-1}$. The fractional powers have the desired algebraic properties and it can be shown that they are themselves infinitesimal generators of analytic semigroups. Since they are one to one we can define on interpolation norm on X. For $0 < \gamma < 1$, we let

$$X_\gamma = \{x \in X \mid x \in D/(A^\gamma)\}$$

with norm

$$||x||_\gamma = ||A^\gamma x||$$

It is clear that X_γ is dense and continuously embedded in X. The material discussed in this paragraph is well known; nice treatments of the subjects may be found in [7], [8], [9], [14], [15].

We now apply the abstract theory to (1.1 a - e). We begin by specifying a densely defined operator on $L^2[0,\pi]$. We set

$$(Au)(x) = \nu u''(x) \qquad \text{for } x \in [0,\pi]$$

where

$$D(A) = \{u \in L^2(0,\pi) \mid u \in H_0^1(0,\pi) \cap H^2(0,\pi)\}$$

It is well known that A generates an analytic semigroup $\{T(t) \mid t \geq 0\}$ on $L^2(0,\pi)$. We now let $X = R + L^2[0,\pi] + L^2[0,\pi]$ and for $\nu > 0$ specify $\hat{A}: \hat{X} \to \hat{X}$ by

$$\hat{A} = \begin{pmatrix} -\nu & 0 & 0 \\ 0 & A & 0 \\ 0 & 0 & A \end{pmatrix} \tag{1.8}$$

with

$$D(\hat{A}) = R + D(A) + D(A)$$

It is not difficult to show that $\hat{A}$ is the infinitesimal generator of an analytic semigroup $\{\hat{T}(t) \mid t \geq 0\}$ on X. The interpolation spaces X_γ will have norm

$$||[u,v,w]||_\gamma = \nu^\gamma |U| + ||v||_\gamma + ||w||_\gamma$$

Henceforth we shall use $\gamma = 1/2$.

We define a nonlinear $\hat{J}$ on a subset of $\hat{X}$ by

$$\hat{J}\phi = \begin{cases} p - (||v||^2 + ||w||^2) \\ U(v-w) - (v^2-w^2)_x \\ U(v+w) + (2vw)_x \end{cases} \tag{1.9}$$

For $\phi = [U,v,w] \in \hat{X}$.

We have the following global existence result.

Theorem 1. If $\hat{A}$ and $\hat{J}$ are defined via (1.8) and (1.9) and $\phi_0 = [U_0,v_0,w_0] \in \hat{X}_{1/2}$ the for any $T > 0$ there exists only one continuons function ϕ: $[0,T] \to \hat{X}_{1/2}$ satisfying

$$\phi(t) = \hat{T}(t)\phi_0 + \int_0^t \hat{T}(t-s)\hat{J}\phi(s)\, ds \tag{1.10}$$

If $\varepsilon > 0$ then ϕ is continuously differentiable on $[\varepsilon,T]$ and satisfies

$$\phi'(t) = \hat{A}\phi(t) + \hat{J}\phi(t) \tag{1.11}$$

Moreover if $\{\hat{F}_t(\ )\,|\,t \geq 0\}$ denotes the flow (semigroup) on $\hat{X}_{1/2}$ defined by (1.10) then for fixed $t > 0$, $\hat{F}_t(\ )$ is a C^∞ mapping of $\hat{X}_{1/2}$ to itself.

Indication of Proof We use the abstract parabolic theory of Sobolevskii [15] to obtain a local result. Global continuation will follow from the apriori bounds of Lemma 1.5. The smoothness property is a consequence of the embedding of $D(\hat{A})$ in $\hat{X}_{1/2}$, cf [12].

We define the infinitesimal generator of the flow on $\hat{X}_{1/2}$ induced by solutions of (1.10) by

$$\hat{G}\phi = \lim_{t\to 0} (\hat{F}_t(\phi) - \phi)\,/\,t \quad \text{whenever the limits exists in } \hat{X}_{1/2}$$

Proposition 2.3 of [16] shows that

$$D(\hat{G}) = \{\phi \in \hat{X}_{1/2} \mid \hat{A}\phi + \hat{J}\phi \in \hat{X}_{1/2}\}$$

Thus as an immediate consequence of the exponential representation theorem of [16] we have:

Proposition 1.2 If $\phi_0 \in \hat{X}_{1/2}$ and $t \in [0,T]$ then

$$F_t(\phi_0) = \lim_{n\to\infty} (I + t/n\hat{G})^{-n}\phi_0 \tag{1.13}$$

We now turn our attention to the qualitative behavior of solutions to our system. It is immediate that $(P/\nu,0,0)$ is a steady state solution to our system (1.1 a - e). This solution corresponds to the case of laminar flow. Burgers [1] has shown that this is the only steady state solution to the problem and has conjectured that periodic solutions exist. Dlotko [6] shows that $(P/\nu,0,0)$ is globally exponentially asymptotically stable when $P/\nu < \nu$ and that $(P/\nu,0,0)$ is an asymptotically stable when $P/\nu = \nu$. We shall be concerned with behavior above criticallity.

It is convenient to define $W(t) = U(t) - P/\nu$. In this context our system has the equivalent reformulation as

$$dW(t)/dt = -\nu W(t) - \int_0^\pi (v^2(t,x)+w^2(t,x))\, dx \tag{1.14 a}$$

(1.14 b) $\partial v(t,x)/\partial t = (P/\nu + W(t))[v(t,x) - w(t,x)]$

$$+ \nu\partial^2 v(t,x)/\partial x^2 - \partial/\partial x[v^2(t,x) - w^2(t,x)]$$

(1.14 c) $\partial w(t,x)/\partial t = (P/\nu + W(t))[v(t,x) + w(t,x)]$

$$+ \nu\partial^2 w(t,x)/\partial x^2 + \partial/\partial x[2v(t,x)w(t,x)]$$

(1.14 d) $v(t,0) = v(t,\pi) = w(t,0) = w(t,\pi) = 0$

(1.14 e) $W(0) = W_0 \ ; \ v(0,x) = v_0(x) \ ; \ w(0,x) = w_0(x)$

In this setting (0,0,0) is the steady state corresponding to laminar flow. To use the foregoing abstract framework we let

$$\hat{J}_\phi = \hat{J}[W,v,w] = \begin{cases} - \int_0^\pi (v^2+w^2)\,dx \\ P/\nu[v-w] + W[v-w] - [v^2-w^2]_x \\ P/\nu[v+w] + W[v+w] + (2vw)_x \end{cases}$$

If we linearize the system about the critical point (0,0,0) we obtain the abstract equation

$$\phi'(t) = \hat{A}\phi(t) + \hat{B}\phi(t)$$

where $\hat{B}: \hat{X} \to \hat{X}$ is the bounded linear operator given by

$$\hat{B}\rho = \begin{cases} 0 \\ P/\nu[v-w] \\ P/\nu[v+w] \end{cases} \qquad \text{for } \rho = [W,v,w] \in \hat{X}_{1/2}$$

Elementary calculations yield the spectrum of $\hat{A} + \hat{B}$

Lemma 1.15 The eigenvalues of A + B are $\{P/\nu - \nu j^2 \pm iP/\nu \mid j = 1,2\ldots\}$

Standard techniques [9] yield the following result

Proposition 1.16 If $P/\nu > \nu$, then the critical point (0,0,0) is unstable, i.e. there exists an $\varepsilon_0 > 0$ and a sequence $\xi\phi_n = [W_n,v_n,w_n]$ such that $\sup_{t\to\infty} ||\phi||_{\hat{X}_{1/2}} \geq \varepsilon_0$.

Thus the equilibrium point (0,0,0) loses stability as the parameter $= P/\nu - \nu$ increases through zero. Our final result demonstrates that a Hopf bifurcation occurs and a one parameter family of periodic solution develop as μ increases through zero. We denote the dependence of the flow or the parameter μ by F_t^u .

Theorem 2 There exists an $\varepsilon > 0$ and a one parameter family of closed orbits depending on $\mu = P/\nu - \nu \in (0,\varepsilon]$. For each μ the orbit is unique and the orbits vary continuously with μ. They are locally attracting and hence stable.

REFERENCES

[1.] C.M. Burgers, "A mathematical model illustrating the theory of turbulence, "Advances in applied Mechanics (R. von Miles and T. von Kármán, ed.) vol 1, Academic Press, New York (1948), 171-199.

[2.] J.D. Cole, "On a quasilinear parabolic equation occuring in aerodynamics" Quart, Appl. Math. 9 (1951), 225-236.

[3.] T. Doltko, "On the one dimensional Burgers equation existence, uniqueness, stability", Zeszyty Nauk. Univ. Jagiellon Prace Mat 23 (in press).

[4.] ________, "On classical solution of the one dimensional Burgers equation" Zeszyty Nauk Univ. Jagiellon Prace Mat. 23 (in press).

[5.] ________, "Some remarks concerning the unstability of the one dimensional Burgers equation" Ann. Polon. Math (to appear).

[6.] ________, "The two dimensional Burger's turbulence model" J. Math. Kyoto Univ. 21 (1981), 809-823.

[7.] A.Friedman, Partial Differential Equations Holt, Phinehart and Winston, New York, 1969.

[8.] J.Goldstein, "Semigroups of operators and abstract Cauchy problems" lecture notes, Tulene University, 1970.

[9.] D.Henry Geometric Theory of Semilinear Parabolic Equations, Lecture Notes in Mathematics, vol 840 Springer-Verlag, Berlin, 1981.

[10.] C.O.Horgan and W.E. Olmstead, "Stability and uniqueness for a turbulence model of Burgers" Quart Appl. Math. (1978), 121-127.

[11.] J.Marsden, "The Hopf bifurcation for nonlinear semigroups", Bull. Amer. Math. Soc., 79 (1973), 537-541.

[12.] J.Marsden and M. McCracken, The Hopf Bifurcation and Its Applications, Applied Mathematicsl Sciences 19, Springer-Verlag, Berlin, 1976.

[13.] W.E.Olmstead and S.H. Davis, "Stability and bifurcation in a Modulated Burgers system" Applied Mathematics Technical No 7925, Northwestern University, 1980.

[14.] A. Pazy "Semigroups of linear operators and applications to partial differential equations", Lecture Notes No 10, University of Maryland, College Part, MD, 1974.

[15.] P.E. Sobolevskii "On equations of parabolic type in a Banach space" Trudy Moscor. Mat Oršč 10 (1961), 297-350. Transl. Amer. Math Soc. 240 (1978), 129-143.

[16.] G.F. Webb, "Exponential representation of solutions to an abstract semi-linear differential equation," Pac. J. Math. 70 (1977), 269-279.

ASYMPTOTIC PROPERTIES OF INHOMOGENEOUS WAVE EQUATIONS

Jerome A. Goldstein

Department of Mathematics
Tulane University
New Orleans, Louisiana

James T. Sandefur, Jr.

Department of Mathematics
Georgetown University
Washington, D.C. &
Center for Applied Mathematics
Cornell University
Ithaca, New York

ABSTRACT

Of concern is the asymptotic behavior of equations of the form

$$d^2u(t)/dt^2 + A^2u(t) = g(t)$$

where A is a self-adjoint operator on a Hilbert space. Of especial concern are the cases when g is integrable, i.e. $\int_0^\infty \|g(t)\| \, dt < \infty$, and when g is p-periodic, i.e. $g(t + p) = g(t)$ for all t. The limiting behavior of $K_u(t)/P_u(t)$ is calculated as $t \to \infty$, where $K_u(t)$ [resp. $P_u(t)$] is the kinetic [resp. potential] energy at time t. The nonlinear case of $g = g(t,u(t),du(t)/dt)$ is briefly considered.

1. INTRODUCTION

This paper deals with the asymptotic behavior of the potential and kinetic energies associated with equations of the form

$$u''(t) + A^2u(t) = g(t) \tag{1}$$

where A is a self-adjoint operator on a Hilbert space H and prime denotes differentiation with respect to the time t. In the motivational Section 1 we review how such inhomogeneous equations arise in electromagnetism. Section 2 reviews the basic results for the homogeneous problem (i.e. when $g = 0$). Sections 3 and 4 deal, respectively, with the cases when g is integrable over $t \in \mathbb{R}^+ = [0,\infty)$ and when g is periodic. Section 5 combines these two cases, and Section 6 deals with nonlinear problems.

2. AN EXAMPLE FROM ELECTROMAGNETISM

Let $\vec{E}$, $\vec{B}$ be the electric and magnetic vectors in an electromagnetic field. These are $\mathbb{R}^3$-valued (or more generally $\mathbb{C}^3$-valued) functions of time and space $((t,x) \in \mathbb{R} \times \mathbb{R}^3)$. Maxwell's equations take the form

$$\nabla \times \vec{B} = \partial\vec{E}/\partial t + 4\pi\vec{j}\ , \quad \nabla \cdot \vec{B} = 0\ ,$$

$$\nabla \times \vec{E} = -\partial\vec{B}/\partial t\ , \quad \nabla \cdot \vec{E} = 4\pi\rho\ ,$$

where $\vec{j} : \mathbb{R} \times \mathbb{R}^3 \to \mathbb{C}^3$ is the current density, $\rho : \mathbb{R} \times \mathbb{R}^3 \to \mathbb{C}$ is the charge density, and units are chosen so that the speed of light is one. A vector potential $\vec{V} : \mathbb{R} \times \mathbb{R}^3 \to \mathbb{C}^3$ and a scalar potential $\phi : \mathbb{R} \times \mathbb{R}^3 \to \mathbb{C}$ can be chosen so that

$$\vec{E} = -\partial\vec{V}/\partial t - \nabla\phi\ , \quad \vec{B} = \nabla \times \vec{V}\ .$$

The potentials $\vec{V}$ and ϕ determine $\vec{E}$ and $\vec{B}$ and may be chosen to satisfy the Lorentz gauge condition

$$\nabla \cdot \vec{V} + \partial\phi/\partial t = 0\ .$$

It then follows that $\vec{V}$, ϕ satisfy

$$\partial^2\vec{V}/\partial t^2 - \Delta\vec{V} = 4\pi\vec{j}\ ,$$

$$\partial^2\phi/\partial t^2 - \Delta\phi = 4\pi\rho\ .$$

These are both equations of the form (1), the first in $H = [L^2(\mathbb{R}^3)]^3 = L^2(\mathbb{R}^3; \mathbb{C}^3)$ and the second in $L^2(\mathbb{R}^3)$.

The equations for $\vec{V}$, ϕ reduce to $u'' + A^2u = 0$ in the absence of electrical charges (or in a vacuum) since then $\vec{j} = 0$, $\rho = 0$.

3. BACKGROUND

Let A be a self-adjoint operator on a complex Hilbert space H. The initial value problem

$$u''(t) + A^2 u(t) = 0 \qquad (t \in \mathbb{R}), \tag{2}$$

$$u(0) = f_1, \quad u'(0) = f_2 \tag{3}$$

is well-posed (for $f_1 \in \mathrm{Dom}(A^2)$, $f_2 \in \mathrm{Dom}(A)$), and the unique solution is given by

$$u(t) = \cos(tA) f_1 + \sin(tA) A^{-1} f_2 .$$

Note that $F_t(A) = \sin(tA)A^{-1}$ is a well-defined bounded self-adjoint operator (since $\sin(tx)/x$ is bounded continuous function of $x \in \mathbb{R}$, equaling t at $x = 0$). Define the *kinetic energy*, the *potential energy*, and the *total energy* of u to be

$$K_u(t) = \|u'(t)\|^2, \quad P_u(t) = \|Au(t)\|^2, \quad E_f = K_u(t) + P_u(t);$$

here E_f is independent of t and depends only on the data $f = (f_1, f_2)$. In many concrete examples, $-A^2$ is a self-adjoint realization of the Laplacian $\Delta = \nabla^2$, and $P_u(t) = \|Au(t)\|^2 = \langle -\Delta u(t), u(t)\rangle = \|\nabla u(t)\|^2$. According to [2], [3], we have *equipartition of energy*, i.e.,

$$\lim_{t\to\pm\infty} K_u(t) = \lim_{t\to\pm\infty} P_u(t) = \frac{1}{2} E_f$$

holds for all solutions of (2) (with f defined by (3)) if and only if A is a *Riemann-Lebesgue operator*, i.e. $e^{itA} \to 0$ as $t \to \pm\infty$ in the weak operator topology. Sufficient for this is that A is spectrally absolutely continuous. Furthermore,

$$\lim_{\tau\to\pm\infty} \frac{1}{\tau}\int_0^\tau K_u(t)\,dt = \lim_{\tau\to\pm\infty} \frac{1}{\tau}\int_0^\tau P_u(t)\,dt = \frac{1}{2} E_f$$

holds for all solutions of (2) if and only if A is injective.

For more general results and references to the literature see [4]. Concerning inhomogeneous equations, in 1973 Bobisud and Calvert [1] obtained energy bounds, and Kwong [5] made remarks about partition of energy. Kwong's unpublished thesis [5], which contains interesting results about the asymptotics of symmetric hyperbolic systems, deserves to be more widely known.

4. DECAYING INHOMOGENETIES

For simplicity we restrict our attention to $t \in \mathbb{R}^+ = [0,\infty)$. Let A be self-adjoint on H. The unique solution of

$$u''(t) + A^2u(t) = g(t) \qquad (t \in \mathbb{R}^+) \tag{4}$$

and (3) is

$$u(t) = \cos(tA)f_1 + \sin(tA)A^{-1}f_2 + \int_0^t \sin(t - s)A)A^{-1}g(s)ds .$$

Here, minimally, g satisfies $g \in L^1_{loc}(\mathbb{R}^+; H)$.

We shall deal with two cases.

Case 1: g decays at infinity in the sense that $g \in L^1(\mathbb{R}^+; H)$.

Case 2: g is p-periodic, i.e. $g(t + p) = g(t)$ for all $t \geq 0$; here $p > 0$ is fixed. (This case will be dealt with in the next section.)

Consider now Case 1 and assume $g \in L^1(\mathbb{R}^+; H)$. First note that

$$v(t) = \int_0^t \sin(t - s)A)A^{-1}g(s)ds \tag{5}$$

satisfies

$$Av(t) = \frac{1}{2i} e^{-itA} \left\{e^{2itA} \int_0^t e^{-isA}g(s)ds - \int_0^t e^{isA}g(s)ds\right\} , \tag{6}$$

$$v'(t) = \frac{1}{2} e^{-itA} \left\{e^{2itA} \int_0^t e^{-isA}g(s)ds + \int_0^t e^{isA}g(s)ds\right\} . \tag{7}$$

Clearly

$$\int_0^t e^{\pm isA} g(s)ds \longrightarrow \int_0^\infty e^{\pm isA}g(s)ds \equiv z_\pm$$

as $t \to \infty$ since $g \in L^1(\mathbb{R}^+; H)$. Thus for the solution v of (4) the partial energies satisfy

$$P_v(t) = \frac{1}{4}\{\|z_-\|^2 + \|z_+\|^2 + 2\,\mathrm{Re}\,\langle e^{2itA}z_-, z_+\rangle\} + o(1)\ ,$$

$$K_v(t) = \frac{1}{4}\{\|z_-\|^2 + \|z_+\|^2 - 2\,\mathrm{Re}\,\langle e^{2itA}z_-, z_+\rangle\} + o(1)$$

as $t \to \infty$. In particular,

$$P_v(t),\ K_v(t) \longrightarrow \frac{1}{4}\{\|z_-\|^2 + \|z_+\|^2\}$$

if A is a Riemann-Lebesgue operator, and

$$\frac{1}{\tau}\int_0^\tau P_v(t)dt,\ \frac{1}{\tau}\int_0^\tau K_v(t)dt \longrightarrow \frac{1}{4}\{\|z_-\|^2 + \|z_+\|^2\}$$

as $\tau \to \infty$ if A is injective. (See [3] for the proof of the last assertion.)

The unique solution of (4), (3) is given by $u(t) = w(t) + v(t)$ where $w(t) = \cos(tA)f_1 + \sin(tA)A^{-1}f_2$. The law of cosines gives

$$\text{(8)} \quad K_u(t) = \|w'(t) + v'(t)\|^2 = K_w(t) + K_v(t) + 2\,\mathrm{Re}\langle w'(t), v'(t)\rangle\ ,$$

and (7) and the formula for w give

$$4\langle w'(t), v'(t)\rangle = \langle e^{itA}(iAf_1 + f_2) + e^{-itA}(-iAf_1 + f_2),\ e^{itA}z_- + e^{-itA}z_+\rangle$$

$$= \langle iAf_1 + f_2,\ z_-\rangle + \langle -iAf_1 + f_2,\ z_+\rangle + J$$

where J is a sum of two terms of the form $\langle e^{itA}x, y\rangle$ plus $o(1)$. Using (6) we obtain an equation similar to (8) when each K is replaced by a P. Putting everything together gives the following result.

THEOREM 1. *Let* u *satisfy* (4), (3) *with* $g \in L^1(\mathbb{R}^+; H)$. *Let*

$$K_u(t) = \|u'(t)\|^2\ , \quad P_u(t) = \|Au(t)\|^2\ ,$$

$$E_f = \|Af_1\|^2 + \|f_2\|^2\ , \quad z_\pm = \int_0^\infty e^{\pm isA}g(s)ds\ .$$

(i) $\lim_{t\to\infty} K_u(t) = \lim_{t\to\infty} P_u(t)$

$$= \frac{1}{4}\{2E_f + \|z_-\|^2 + \|z_+\|^2 + \mathrm{Re}[<z_- + z_+, f_2> + <z_- - z_+, iAf_1>]\}$$

if A *is a Riemann-Lebesgue operator.*

(ii) $\lim_{\tau\to\infty} \frac{1}{\tau}\int_0^\tau K_u(t)dt = \lim_{\tau\to\infty} \frac{1}{\tau}\int_0^\tau P_u(t)dt$

$$= \frac{1}{4}\{2E_f + \|z_-\|^2 + \|z_+\|^2 + \mathrm{Re}[<z_- + z_+, f_2> + <z_- - z_+, iAf_1>]\}$$

if A *is injective.*

5. PERIODIC INHOMOGENETIES

Let w satisfy (4) and be p-periodic where $p > 0$. Then, necessarily, g is also p-periodic. Let v satisfy the homogeneous equation (2). The law of cosines gives

$$K_u(t) = K_w(t) + K_v(t) + 2\,\mathrm{Re}\,<w'(t), v'(t)>$$

(see (8)) where $u = w + v$ is a solution of (4). Since $v'(t) = e^{itA}\phi_1 + e^{-itA}\phi_2$ for suitable ϕ_1, ϕ_2, we have $v'(t) \to 0$ weakly as $t \to \infty$ if A is a Riemann-Lebesgue operator. Assume this in what follows; averaging over t will give the corresponding results when A is merely injective.

Claim: $<v'(t), w'(t)> \to 0$ as $t \to \infty$.

Proof: Assume not. Then for some sequence $t_n \to \infty$,

$$<v'(t_n), w'(t_n)> \to c \neq 0 \;. \tag{9}$$

(Note that v' and w' are bounded.) Choose a subsequence $\{s_n\}$ of $\{t_n\}$ such that $s_n \to s$ (mod p) as $n \to \infty$. Then $w'(s_n) \to w'(s)$ strongly by the periodicity and continuity of w'. Therefore

$$|<v'(s_n), w'(s_n)>| \le |<v'(s_n), w'(s)>| + |<v'(s_n), w'(s_n) - w'(s)>| \to 0$$

as $n \to \infty$, contradicting (9). This establishes the claim.

A similar calculation gives

$$P_u(t) = P_v(t) + P_w(t) + o(1)$$

as $t \to \infty$. Consequently

$$K_u(t) - P_u(t) = K_w(t) - P_w(t) + o(1)$$

since $K_v(t) - P_v(t) = o(1)$ as $t \to \infty$. Hence we cannot deduce equipartition of energy since $K_w - P_w$ is p-periodic. But we can deduce a weak form of equipartion of energy as follows. For $J = K, P$,

$$J_u(t) = J_v(t) + J_w(t) + o(1) = \frac{1}{2} E_\psi + J_w(t) + o(1)$$

as $t \to \infty$ where $\psi = (\psi_1, \psi_2) = (v(0), v'(0))$. Let

(10) $$M_w = \mathrm{Max}\,\{||Aw(t)||^2,\ ||w'(t)||^2 : 0 \le t \le p\}\ .$$

It follows that

$$0 < c_1 \le \liminf_{t\to\infty} \frac{K_u(t)}{P_u(t)} \le \limsup_{t\to\infty} \frac{K_u(t)}{P_u(t)} \le c_2 < \infty$$

where

$$c_1 = E_\phi/(E_\phi + 2M_w)\ ,\quad c_2 = (E_\phi + 2M_w)/E_\phi$$

provided $E_\phi > 0$, i.e. $v \not\equiv 0$.

The general solution of (4) is

$$u(t) = e^{itA}\psi_1 + e^{-itA}\psi_2 + \int_0^t \sin((t-s)A)A^{-1}g(s)\,ds\ .$$

In order for there to be a p-periodic solution, g must be p-periodic and ψ_1, ψ_2 must be chosen so that

(11) $$u(0) = u(p)\ ,\quad u'(0) = u'(p)$$

both hold. We have

$$u(p) = e^{ipA}\psi_1 + e^{-ipA}\psi_2 + \int_0^p \sin((p - s)A)A^{-1}g(s)\, ds ,$$

$$u'(p) = ie^{ipA}A\psi_1 - ie^{-ipA}A\psi_2 + \int_0^p \cos((p - s)A)g(s)\, ds .$$

Defining

$$g_1 = \int_0^p \sin((p - s)A)A^{-1}\, g(s)\, ds , \tag{12}$$

$$g_2 = \int_0^p \cos((p - s)A)\, g(s)\, ds , \tag{13}$$

the requirements (11) become

$$(e^{ipA} - I)\psi_1 + (e^{-ipA} - I)\psi_2 + g_1 = 0 ,$$

$$(e^{ipA} - I)iA\psi_1 - (e^{-ipA} - I)iA\psi_2 + g_2 = 0 .$$

There are equivalent to

$$(e^{ipA} - I)iA\psi_1 = \frac{1}{2}(iAg_1 + g_2) ,$$

$$(e^{-ipA} - I)iA\psi_2 = \frac{1}{2}(iAg_1 - g_2) .$$

Since A is a Riemann-Lebesgue operator, spectral theory implies that

$$B_{\pm} = (e^{\pm ipA} - I)A \tag{14}$$

are injective operators with dense range. Consequently there is a periodic solution of (4) if and only if $g(\cdot)$ is such that

$$iAg_1 \pm g_2 \in \mathrm{Ran}(B_{\pm}) .$$

Note also that there cannot be two distinct p-periodic solutions for then the difference would be a p-periodic solution of (2) having equipartition of energy since A is a Riemann-Lebesgue operator; hence the difference would be the zero solution.

The above results can be summarized as follows.

THEOREM 2. *Let* $p > 0$ *and let* $g \in L^1_{loc}(\mathbb{R}^+; H)$ *be periodic of period* p. *Let* A *be a Riemann-Lebesgue operator on* H. *Then* (4) *has a unique* p-*periodic solution if and only if* $iAg_1 \pm g_2 \in \mathrm{Ran}(B_\pm)$ *where* g_1, g_2, $B_\pm$ *are defined by* (12)-(14). *Here* $\mathrm{Ran}(B_+) \cap \mathrm{Ran}(B_-)$ *is dense in* H. *When this condition holds and* $g \not\equiv 0$, *let* w *be the unique periodic solution of* (4) *and define* M_w *(which only depends on* g*) by* (10). *Let* u *be any solution of* (4) *other than* w. *Let*

$\phi_1 = (u - w)(0)$, $\phi_2 = (u - w)'(0)$, $E_\phi = \|A\phi_1\|^2 + \|\phi_2\|^2$. *Then*

$$0 < E_\phi/(E_\phi + M_w) \le \liminf_{t\to\infty} \frac{K_u(t)}{P_u(t)} \le \limsup_{t\to\infty} \frac{K_u(t)}{P_u(t)} \le (E_\phi + M_w)/E_\phi < \infty .$$

Minor modifications of the arguments in this section give the following result.

THEOREM 3. *In Theorem* 2 *replace the hypothesis that* A *is a Riemann-Lebesgue operator by the weaker assumption that* $A = A^*$ *does not have* $2\pi n/p$ *as an eigenvalue for any integer* n. *(Thus* $e^{\pm ipA} - I$ *are injective.)* *For* u *as in Theorem* 2,

$$0 < E_\phi(E_\phi + M_w)^{-1} \le \liminf_{T\to\infty} \int_0^T K_u(t)dt\left(\int_0^T P_u(t)dt\right)^{-1}$$

$$\le \limsup_{T\to\infty} \int_0^T K_u(t)dt\left(\int_0^T P_u(t)dt\right)^{-1} \le E_\phi^{-1}(E_\phi + M_w) < \infty .$$

6. MORE GENERAL INHOMOGENETIES

Suppose $g = g_1 + g_2$ where $g_1 \in L^1(\mathbb{R}^+; H)$ and $g_2 \in L^1_{loc}(\mathbb{R}^+; H)$ is p-periodic. Any solution u of (4) can be thought of as $u_1 + u_2$ where u_j satisfies $u_j'' + A^2u_j = g_j$ for $j = 1,2$. Thus combining the results of Sections 3 and 4, we have that if A is a Riemann-Lebesgue operator,

$$0 < (E + F)(E + F + M)^{-1} \le \liminf_{t\to\infty} K_u(t)/P_u(t)$$

$$\le \limsup_{t\to\infty} K_u(t)/P_u(t) \le (E + F)^{-1}(E + F + M) < \infty .$$

Here E, F, M are defined as follows. For $u = u_1 + u_2$ as above and w the unique periodic solution of (4), $M = M_w$, $E = E_\phi$ for $\phi_1 = (u_2 - w)(o)$, $\phi_2 = (u_2 - w)'(o)$, and $F = 4^{-1}(\|z_-\|^2 + \|z_+\|^2) + 2^{-1}\mathrm{Re}\,\{\langle z_- + z_+, \phi_2\rangle + \langle z_- - z_+, iA\phi_1\rangle\}$ where $z_\pm = \int_0^\infty e^{\pm isA} g_1(s)ds$. A similar assertion involving Césaro convergence can be made in the more general case when the point spectrum of A avoids $\{2\pi n/p\colon n$ an integer$\}$.

A more precise result can be obtained when $g = g_1 + g_2$ where $g_1 \in L^1(\mathbb{R}^+; H)$ and g_2 is a nonzero constant function, viz. $g_2(t) \equiv A^2 f_0 \neq 0$. In this case the solution of (4), (3) is given by

$$u(t) = \cos(tA)f_1 + \sin(tA)A^{-1}f_2 + \int_0^t \sin((t - s)A)^{-1} g_1(s)\, ds$$

$$+ [f_0 - \cos(tA)f_0],$$

the term is square brackets being $\int_0^t \sin((t - s)A)Af_0\, ds$. This u is of the form: a constant (f_0) plus the solution discussed in Section 3 but with f_1 replaced by $f_1 - f_0$. The reasoning used there leads to the following result.

THEOREM 4. *Let* u *satisfy* (4), (3) *with* $g = g_1 + A^2 f_0$ *where* $g_1 \in L^1(\mathbb{R}^+; H)$. *Let*

$$\tilde{E}_f = \|A(f_1 - f_0)\|^2 + \|f_2\|^2, \quad z_\pm = \int_0^\infty e^{\pm isA} g_1(s)\, ds .$$

Assume that A *is a Riemann-Lebesgue operator.*

(i) $\lim_{t\to\infty} K_u(t) = \frac{1}{4}\{2\tilde{E}_f + \|z_-\|^2 + \|z_+\|^2$

$+ \mathrm{Re}[\langle z_- + z_+, f_2\rangle + \langle z_- - z_+, iA(f_1 - f_0)\rangle]\}.$

(ii) $\lim_{t\to\infty} P_u(t) = \|Af_0\|^2 + \lim_{t\to\infty} K_u(t).$

(iii) $\lim_{t\to\infty} K_u(t)/P_u(t)$ *exists and belongs to* $[0,1]$.

Finally suppose that w *is a (hence the unique)* p-*periodic solution of* $w'' + A^2 w = g_2$ *where* g_2 *is* p-*periodic; thus* $v = u + w$ *is an arbitrary solution of* (4) *with* $g = g_1 + A^2 f_0 + g_2$.

(v) $\lim_{\tau\to\infty} p^{-1} \int_{\tau}^{\tau+p} K_v(t)\, dt = \lim_{\tau\to\infty} K_u(\tau) + p^{-1} \int_0^p \|w'(t)\|^2\, dt$,

(vi) $\lim_{\tau\to\infty} p^{-1} \int_{\tau}^{\tau+p} P_v(t)\, dt = \lim_{\tau\to\infty} P_u(\tau) + p^{-1} \int_0^p \|Aw(t)\|^2\, dt$.

Analogous assertions involving Césaro convergence hold when $A = A^*$ *is merely injective.*

7. NONLINEAR EQUATIONS

Let u be a solution of

$$u''(t) + A^2u(t) = h(t,u(t),u'(t)) \ . \tag{15}$$

Defining $g(t)$ to be $h(t,u(t),u'(t))$ we can regard (15) as an inhomogeneous equation for the known solution u and apply the results of the previous sections. We mention one specific result, assuming that A is injective. Let

$$h(t,u,v) = \alpha(t)\ k(t,u,v) + A^2f_0$$

where $\alpha \in L^1(\mathbb{R}^+)$ and $k \in L^\infty(\mathbb{R}^+ \times H \times H\ ;\ H)$. The $K_u(t)$ and $P_u(t)$ converge in the Césaro sense as $t \to \infty$ to limits which differ by $\|Af_0\|^2$ (by Theorem 4).

ACKNOWLEDGEMENT

The first author was partially supported by an NSF grant.

REFERENCES

[1] Bobisud, L. E. and Calvert, J., Energy bounds and virial theorems for abstract wave equations, *Pac. J. Math.* 47 (1973), 27-37.

[2] Goldstein, J. A., An asymptotic property of solutions of wave equations, *Proc. Amer. Math. Soc.* 23 (1969), 359-363.

[3] Goldstein, J. A., An asymptotic property of solutions of wave equations, II, *J. Math. Anal. Appl.* 32 (1970), 392-399.

[4] Goldstein, J. A. and Sandefur, Jr., J. T., Equipartition of energy for higher order abstract hyperbolic equations, *Comm. P.D.E.*, to appear.

[5] Kwong, M. K., *Asymptotic Partition of Energy Property of Solutions of Symmetric Hyperbolic Linear Differential Equations*, Ph.D. Thesis, Univ. of Chicago, 1973.

ON A PROPERTY OF ORLICZ-SOBOLEV SPACES

J.-P. GOSSEZ

Département de Mathématique,
Campus Plaine C.P.214,
Université Libre de Bruxelles,
Bruxelles, Belgium

1. The following question has been studied in the last yearsby Brézis and Browder [2], [3], [4]. Let u be a function in the Sobolev space $W_o^{m,p}(\Omega)$ and let S be a distribution in $W^{-m,p'}(\Omega) \cap L^1_{loc}(\Omega)$. Under what conditions is the function $S(x)u(x)$ integrable on Ω and if so, does $\int_\Omega S(x)u(x)dx$ equal $\langle S,u \rangle$ (where $\langle\ ,\ \rangle$ denotes the pairing in the duality between $W^{-m,p'}(\Omega)$ and $W_o^{m,p}(\Omega)$)?

It is our purpose in this talk to report on some recent work dealing with the extension of one of their results to the framework of Sobolev spaces built from Orlicz spaces. Applications of theorem 1 below are given in [7] to variational boundary value problems for second-order quasi-linear elliptic equations of the form

$$\sum_{|\alpha|\leqslant 1} (-1)^{|\alpha|} D^\alpha A_\alpha(x,u,\nabla u) + g(x,u) = f .$$

Here the first term of the left-hand side is supposed to give rise to a "good" operator within the class of Orlicz-Sobolev spaces while the second term g satisfies a sign condition $g(x,u) \geqslant 0$ but has otherwise unrestricted growth with respect to u.

2. Let Ω be an open (possibly unbounded) subset of $\mathbb{R}^N$ which satisfies the segment property. Let $W^1L_M(\Omega)$ and $W^1E_M(\Omega)$ be the Orlicz-Sobolev spaces of order 1 on Ω corresponding to a N-function M. Standard references about these spaces include [1], [9]. They will, as usual, be identified to subspaces of the product $\Pi_{|\alpha|\leq 1} L_M(\Omega) \equiv \Pi L_M$. Denoting by $\bar{M}$ the N-function conjugate to M, we define $W_0^1L_M(\Omega)$ (resp. $W_0^1E_M(\Omega)$) as the $\sigma(\Pi L_M, \Pi E_{\bar{M}})$ (resp. norm) closure of $\mathcal{D}(\Omega)$ in $W^1L_M(\Omega)$. We also consider $W^{-1}L_{\bar{M}}(\Omega)$ (resp. $W^{-1}E_{\bar{M}}(\Omega)$), the space of distributions on Ω which can be written as sums of a function in $L_{\bar{M}}(\Omega)$ (resp. $E_{\bar{M}}(\Omega)$) and of first order derivatives of functions in $L_{\bar{M}}(\Omega)$ (resp. $E_{\bar{M}}(\Omega)$). Since Ω has the segment property, $\mathcal{D}(\Omega)$ is $\sigma(\Pi L_M, \Pi L_{\bar{M}})$ dense in $W_0^1L_M(\Omega)$ (cf. [5], theorem 1.3) and consequently, the value of $S \in W^{-1}L_{\bar{M}}(\Omega)$ at an element $u \in W_0^1L_M(\Omega)$ is well-defined. It will be denoted by $\langle S,u \rangle$.

THEOREM 1. _Let_ $S \in W^{-1}L_{\bar{M}}(\Omega) \cap L^1_{loc}(\Omega)$ _and_ $u \in W_0^1L_M(\Omega)$. _Suppose that for some_ $h \in L^1(\Omega)$, $S(x)u(x) \geq h(x)$ a.e. _in_ Ω. _Then_ $Su \in L^1(\Omega)$ and

$$\langle S,u \rangle = \int_\Omega S(x)u(x)dx .$$

When $M(t) = |t|^p$ with $1 < p < \infty$, this result has been obtained in [2]. It then holds without any restriction in Ω. See also [3] for related results and [4] for the higher-order case. The introduction here of the (mild) assumption that Ω has the segment property is somehow related to the fact that, contrary to the situation in the case of the usual Sobolev spaces, the topology $\sigma(\Pi L_M, \Pi E_{\bar{M}})$ used in the definition of the space $W_0^1L_M(\Omega)$ does not allow to go the limit on u in the pairing $\langle S,u \rangle$.

The main step in the proof of theorem 1 is the following approximation lemma. Once this is obtained, the arguments can be easily adapted from [2].

LEMMA 2. _Let_ $u \in W_0^1L_M(\Omega)$. _Then there exists a sequence_ u_n _such that_ (i) $u_n \in W_0^1L_M(\Omega) \cap L^\infty(\Omega)$, (ii) supp u_n _is compact in_ Ω,

(iii) $|u_n(x)| \leq |u(x)|$ a.e. in Ω, (iv) $u_n(x)u(x) \geq 0$ a.e. in Ω,
(v) $D^\alpha u_n \to D^\alpha u$ a.e. in Ω for $|\alpha| \leq 1$, (vi) $u_n \to u$ for $\sigma(\Pi L_M, \Pi L_{\bar{M}})$.

We will sketch a proof of lemma 2 which is rather simpler than that given in [7].

PROOF OF LEMMA 2. The following two properties of Orlicz-Sobolev spaces of order one will be used repeatedly (see [7]). Let $v \in W_o^1 L_M(\Omega)$; then $|v| \in W_o^1 L_M(\Omega)$ and $\partial|v|/\partial x_i = \operatorname{sgn} v \ \partial v/\partial x_i$ a.e. in Ω, where $\operatorname{sgn} t = 1$, 0 or -1 according to the sign of t. Let $v, w \in W_o^1 L_M(\Omega)$ and write $h = \min\{v,w\}$; then $h \in W_o^1 L_M(\Omega)$ and

$$\frac{\partial h}{\partial x_i} = \begin{cases} \dfrac{\partial v}{\partial x_i} & \text{a.e. in } \{x \in \Omega \ ; \ v(x) \leq w(x)\} , \\[2ex] \dfrac{\partial w}{\partial x_i} & \text{a.e. in } \{x \in \Omega \ ; \ v(x) > w(x)\} . \end{cases}$$

Writing $u = u^+ - u^-$, we can assume without loss of generality in lemma 2 that $u \geq 0$ a.e. in Ω. By theorem 4 in [6], there exists a sequence $v_n \in \mathcal{D}(\Omega)$ such that for $|\alpha| \leq 1$ and some $\lambda > 0$, $\int_\Omega M((D^\alpha u - D^\alpha v_n)/\lambda)dx \to 0$ as $n \to \infty$. Passing to a subsequence if necessary, we can assume that for $|\alpha| \leq 1$, $D^\alpha v_n \to D^\alpha u$ a.e. in Ω. Define

$$u_n = \min\{u, |v_n|\} .$$

Then properties (i)-(iv) clearly hold. Computing the derivatives of u_n, we obtain (v). Finally, to verify (vi), we use the fact that if w_n is a sequence in $L_M(\Omega)$ with $w_n \to w$ a.e. in Ω and, for some $\Lambda > 0$, $M(w_n/\Lambda) \leq h_n$ a.e. in Ω where h_n converges in $L^1(\Omega)$, then $w \in L_M(\Omega)$ and $w_n \to w$ for $\sigma(L_M, L_{\bar{M}})$ (see [5], lemma 1.4). This applies immediately to u_n since $0 \leq u_n \leq u$ a.e. in Ω. For the derivatives, we first deduce that

$$\left|\frac{\partial u_n}{\partial x_i}\right| \leq \left|\frac{\partial u}{\partial x_i}\right| + \left|\frac{\partial v_n}{\partial x_i}\right|$$

a.e. in Ω, so that

$$M\left(\left(\frac{\partial u_n}{\partial x_i}\right)/\Lambda\right) \leqslant \frac{1}{2} M\left(4\left(\frac{\partial u}{\partial x_i}\right)/\Lambda\right) + \frac{1}{2} M\left(2\left(\frac{\partial v_n}{\partial x_i} - \frac{\partial u}{\partial x_i}\right)/\Lambda\right)$$

a.e. in Ω, and the conclusion then follows easily. Q.E.D.

It is not known whether the above results (as well as their applications to strongly nonlinear problems) are valid for higher order Orlicz-Sobolev spaces. A proof along the above lines would require the extension of the Hedberg's delicate truncation procedure [8] to the framework of Orlicz spaces.

REFERENCES

[1] R. ADAMS, Sobolev spaces, Ac. Press, New York, 1975.

[2] H. BREZIS and F. BROWDER, Sur une propriété des espaces de Sobolev, C.R.Ac.Sc. Paris, 287 (1978) 113-115.

[3] H. BREZIS and F. BROWDER, A property of Sobolev spaces, Com. Part. Diff. Eq., 9(1979) 1077-1083.

[4] H. BREZIS and F. BROWDER, Some properties of higher order Sobolev spaces, to appear.

[5] J. -P. GOSSEZ, Nonlinear elliptic boundary value problems for equations with rapidly (or slowly) increasing coefficients, Trans. Amer. Math. Soc., 190 (1974) 163-205.

[6] J. -P. GOSSEZ, Some approximation properties in Orlicz-Sobolev spaces, Studia Mat., to appear.

[7] J. -P. GOSSEZ, Strongly nonlinear elliptic problems in Orlicz-Sobolev spaces of order one, Cah. Cent. Et. Rech. Opér., to appear.

[8] L. HEDBERG, Two approximation properties in function spaces, Ark. Mat., 16 (1978) 51-81.

[9] A. KUFNER, O. JOHN and S. FUCIK, Function spaces, Academia, Praha, 1977.

AN OPTIMAL L^∞ ERROR ESTIMATE FOR GALERKIN APPROXIMATIONS TO SOLUTIONS OF TWO-POINT NONLINEAR BOUNDARY VALUE PROBLEMS

John Gregory and Marvin Zeman

Department of Mathematics
Southern Illinois University
Carbondale, Illinois

ABSTRACT

In this paper we show that the two-point nonlinear boundary value problem for $-(r(t)x'(t))' = f(t,x)$ yields an error estimate $\|x - x_h\|_{L^\infty} = O(h^2)$, where $x(t)$ and $x_h(t)$ are the respective solutions to the given problem and the numerical problem with step size h.

INTRODUCTION

Consider the two-point boundary value problem

$$\begin{cases} -(r(t)x'(t))' = f(t,x), & a < t < b, \\ x(a) = 0, \ x(b) = 0 \end{cases} \tag{1}$$

where r is in $C^3[a,b]$ f is differentiable in t and twice-differentiable in x and satisfies $\frac{\partial f}{\partial x} \geq 0$.

Let S^h be the space of functions which are linear over each interval $[a_{\ell-1}, a_\ell]$, continuous at the nodes $t = a_\ell$ and zero at $t = a$ and $t = b$ (see below). Let $x_h(t) \in S^h$ be the Galerkin approximation to $x(t)$ determined by the relation

$$\int_a^b r(t)x_h'(t)v'(t)dt = \int_a^b f(t,x_h)v(t)dt, \quad v(t) \in S^h. \tag{2}$$

Ciarlet, Schultz and Varga proved (see [1], Thm. 10), that

$$\|x - x_h\|_1 \leq Ch, \quad \text{for C independent of h,} \tag{3}$$

where $\|v\|_1^2 = \int_a^b [(v')^2 + v^2]dt$ and $x(t)$ is the unique solution to (1). By Sobolev's lemma, this implies that

$$\|x - x_h\|_{L^\infty} \doteq \max_{a \leq t \leq b} |x(t) - x_h(t)| \leq Ch. \tag{4}$$

In the linear case where $f(t,x) = p(t)x(t) + g(t)$, Douglas and Dupont [2] and M.F. Wheeler [10] improved estimate (4) to the estimate

$$\|x - x_h\|_{L^\infty} \leq Ch^2.$$

The essential tool in the proof involved the use of Green's functions. In the nonlinear case the use of this idea is no longer possible. The purpose of this note is to show that we get $O(h^2)$ convergence for the nonlinear case as well. This result is optimal in the sense that the power of h cannot be increased.

The remainder of our paper is as follows. In Section 2, we give our approximating scheme in (7) which yields the sequence $\{c_k\}$ where $x_h(a_k) = c_k\sqrt{h}$. In Section 3, we derive a second order difference equation for $e_n = e(a_n)$ where $e(t) = x(t) - x_h(t)$ is the error function. We then give a bound for e_n of the form $|e_n| \leq Ch^2$ and finally show that this yields an L^∞ bound of the form $\|e(t)\|_{L^\infty} \leq Ch^2$. Finally, we give additional error estimates. To conserve space some of our justifications are skimpy. The authors will supply additional details upon request.

THE APPROXIMATING SCHEME

We begin by defining a uniform grid or partition of an interval [a,b]. Let N be a large natural number and $h = (b - a)/N$. For each $\ell = -1,0,1,2,\ldots,N + 1$, let $a_\ell = a + \ell h$ and for each $k = 0,1,2,\ldots,N$ let

$$z_k(t) = \begin{cases} \sqrt{h} - |t - a_k|/\sqrt{h} & \text{if } t \text{ in } [a_{k-1}, a_{k+1}] \\ 0 & \text{otherwise} \end{cases} \tag{5}$$

In the continuous case, if x(t) satisfies (1) and y is any continuous piecewise C' function vanishing at t = a and t = b, then multiplying $(rx')' + f(t,x) = 0$ in (1) by $-y(t)$ and integrating by part yields

$$I(x,y) = \int_a^b r(t)x'(t)y'(t) - f(t,x(t))y(t))dt = 0 \tag{6}$$

In the discrete case, letting $y_h(t) = b_\alpha z_\alpha(t)$, $x_h(t) = c_\beta z_\beta(t)$ and noting that (6) is linear in y we have

$$0 = I(x_h,y_h) = I(x_h,b_\alpha z_\alpha) = b_\alpha I(x_h,z_\alpha).$$

(In this paper repeated indices involving α and β are summed.) Finally, noting that $z_k(t)z_\ell(t) = 0$ if $|k - \ell| > 1$ we have

$$\int_{a_{k-1}}^{a_{k+1}}\{r(t)[c_{k-1}z'_{k-1} + c_k z'_k + c_{k+1}z'_{k+1}]z'_k$$
$$-f(t,c_{k-1}z_{k-1} + c_k z_k + c_{k+1}z_{k+1})z_k\}dt = 0, \tag{7}$$
$$k = 1,\ldots,N - 1.$$

System (7) can be solved for $\{c_k\}$ by several iterative methods, all of which have been shown to be convergent [3], [7], [9].

ERROR ESTIMATES

The main purpose of this section is to show that the error function $e(t) = x(t) - x_h(t)$ satisfies $e_n = e(a_n) = O(h^2)$. There are two major steps in our proof. The first step is to derive a second order difference equation for the error expression for $\{e_n\}$. The second step is to bound this error term. Finally, we give the L^∞ error estimate.

We begin with the equality $I(x,z_k) - I(x_h,z_k) = 0$ since both quantities equal to zero. Hence, multiplying by $\sqrt{h}$ for convenience,

$$+ \int_{a_{k-1}}^{a_{k+1}} r(t)(x'(t) - x_h'(t))\sqrt{h}\, z_k'(t)dt$$
$$= \int_{a_{k-1}}^{a_{k+1}} [f(t,x(t)) - f(t,x_h(t))]\sqrt{h}\, z_k(t)dt. \tag{8}$$

Expanding r(t) in the Taylor series and letting $x(a_k) = x_k$, $r(a_k) = r_k$, $x_k' = x'(a_k)$, etc., we have

$$r(t) = r_k + r_k'(t - a_k) + r_k''(t - a_k)^2/2 + r'''(\eta(t))(t - a_k)^3/6,$$

where $a_{k-1} < \eta(t) < a_{k+1}$. Noting that $\sqrt{h}\, z_k'(t)$ is equal to 1 on (a_{k-1},a_k) and -1 on (a_k,a_{k+1}), the left hand side of (8) has summands

$$r_k \left[\int_{a_{k-1}}^{a_k} (x' - x_h')dt - \int_{a_k}^{a_{k+1}} (x' - x_h')dt\right], \tag{9}$$

$$r_k' \left[\int_{a_{k-1}}^{a_k} (t - a_k)(x' - x_h')dt - \int_{a_k}^{a_{k+1}} (t - a_k)(x' - x_h')dt\right], \tag{10}$$

$$\frac{r_k''}{2} \left[\int_{a_{k-1}}^{a_k} (t - a_k)^2(x' - x_h')dt - \int_{a_k}^{a_{k+1}} (t - a_k)^2(x' - x_h')dt\right], \tag{11}$$

and

$$\frac{1}{6} \left[\int_{a_{k-1}}^{a_k} r'''(\eta(t))(t - a_k)^3(x' - x_h')dt - \int_{a_k}^{a_{k+1}} r'''(\eta(t))(t - a_k)^3(x' - x_h')dt\right]. \tag{12}$$

We note that summand (9) is equal to $r_k(-e_{k-1} + 2e_k - e_{k+1})$. Summand (10) is equal to

$$-h(r_k'/2)(e_{k+1} - e_{k-1}) + h^4(r_k'[-\tfrac{3}{4}x''(\xi_1) + (\tfrac{1}{3})x'''(\xi_2) + D_{1,k}]$$

Where ξ_1 and ξ_2 are numbers in (a_{k-1},a_{k+1}) and $|D_{1,k}| \le \max_{a_{k-1}\le t\le a_{k+1}} |x'''(t)|$.

See Gregory and Zeman [5] for details.

To evaluate (11), we set $\bar{x}_k = x_h(a_k) = \sqrt{h}\, c_k$, expand x(t) in Taylor series about a_{k-1} and a_{k+1} and use the piecewise linearity of $x_h(t)$. Continuing in this manner the left hand side of (8) becomes

$$
\begin{aligned}
&r_k(-e_{k-1} + 2e_k - e_{k+1}) - \frac{h}{2} r_k'(e_{k+1} - e_{k-1}) \\
&+ \frac{h^2}{2} r_k''(-e_{k-1} + 2e_k - e_{k+1}) \\
&+ h^3\{-D_{2,k}e_{k-1} + (D_{2,k} + D_{3,k})e_k - D_{3,k}e_{k+1}\} + h^4 C_k,
\end{aligned} \tag{13}
$$

where C_k is bounded, independently of h, if r and $x \in C^3[a,b]$; and

$$
|D_{2,k}| \le \sup_{a_{k-1} \le t \le a_k} \frac{|r''(t)|}{24}, \quad |D_{3,k}| \le \sup_{a_k \le t \le a_{k+1}} \frac{|r''(t)|}{24}. \tag{14}
$$

To evaluate the right hand side of (8) we set

$$
g(t) = \frac{f(t,x(t)) - f(t,x_h(t))}{x(t) - x_h(t)} \tag{15}
$$

Since x(t) and $x_h(t)$ are both smooth in each of the intervals (a_{k-1}, a_k) and (a_k, a_{k+1}), we expand g(t) in Taylor series,

$$
g(t) = \begin{cases} g(a_k) + g'(\xi_3(t))(t - a_k), & a_{k-1} < t < a_k \\ g(a_k) + g'(\xi_4(t))(t - a_k), & a_k < t < a_{k+1} \end{cases}
$$

If f is twice-differentiable in x and once-differentiable in t, then $g'(\xi_3(t))$ and $g'(\xi_4(t))$ are both bounded functions in their respective intervals. Hence the right hand side of (8) becomes

$$
\begin{aligned}
&-g(a_k)\left[\int_{a_{k-1}}^{a_k} (x - x_h)(t - a_{k-1})\,dt - \int_{a_k}^{a_{k+1}} (x-x_h)(t-a_{k+1})\,dt\right] \\
&\quad - \{\int_{a_{k-1}}^{a_k} g'(\xi_3(t))(x-x_h)(t-a_k)(t-a_{k-1})\,dt \\
&\quad - \int_{a_k}^{a_{k+1}} g'(\xi_4(t))(x-x_h)(t-a_k)(t-a_{k+1})\,dt\}
\end{aligned} \tag{16}
$$

Continuing in this way where

$$
M_k = h^3\{-D_{2,k}e_{k-1} + (D_{2,k} + D_{3,k})e_k - D_{3,k}e_{k+1}\} + O(h^4), \tag{17}
$$

we have proved:

LEMMA 1. Assume r, $x \in C^3[a,b]$ and let $f(t,x) \in C^1 \times C^2$ for $t \in [a,b]$. Then the error $e_k = x(a_k) - x_h(a_k)$ satisfies

$$- r_k[-e_{k-1} + 2e_k - e_{k+1}] + \frac{h}{2} r'_k(e_{k+1} - e_{k-1})$$
$$- \frac{r''_k}{6} h^2(-e_{k-1} + 2e_k - e_{k+1}) \qquad (18)$$
$$= h^2 g(a_k)(\frac{1}{6} e_{k-1} + \frac{2}{3} e_k + \frac{1}{6} e_{k+1}) + M_k,$$

where M_k is defined by (17).

We will now prove our main result concerning the pointwise estimate $e_n = x(a_n) - x_h(a_n)$.

THEOREM 1. *If $r \in C^3[a,b]$ and $f(t,x)$ is differentiable in t and twice differentiable in x such that $\frac{\partial f}{\partial x} \geq 0$ and $|e_i| \leq C_1 h^2$ for $i = 0,1$ and some $C_1 > 0$, then there exists a constant $C_2 > 0$ independent of h such that $|e_n| \leq C_2 h^2$, $0 \leq n \leq N$, for h sufficiently small.*

Proof. We first consider M_k in (18). We will do this by appealing to the result of [1] which yields (4). Using (4), we know that $|e_k| \leq Ch$ and $|x(t) - x_h(t)| \leq Ch$ for some C independent of h. These apriori extimates imply that $M_k = C'_k h^4$, where C'_k is uniformly bounded, independent of k and h. Setting $C_k = C'_k/r(a_k)$ we have

$$e_{k+1} - 2e_k + e_{k-1} + hb_k(e_{k+1} - e_{k-1}) = h^2 c_k e_k + C_k h^4 \qquad (19)$$

where

$$b_k = \frac{3 r'(a_k)}{6r(a_k) + [r''(a_k) - g(a_k)]h^2}$$

and

$$c_k = \frac{6g(a_k)}{6r(a_k) + [r''(a_k) - g(a_k)]h^2}$$

Since $\frac{\partial f}{\partial x} \geq 0$, then $g(a_k) \geq 0$. In addition, if h is sufficiently small, $h^2(|r''(a_k)| + g(a_k)) < 6r(a_k)$, $c_k \geq 0$ and $|h\, b_k| < 1$. Equation (19) is a difference equation which is the same type encountered when studying the finite difference method for the two point boundary value problem. The same proofs can be used to show that

$|e_k| \leq C'_2 h^2$, where C'_2 is a constant depending (20)

upon $x^{(k)}$, $k = 0, 1, 2, 3$, but independent of h. For more details, see Issacson and Keller [6] and Phillips and P. J. Taylor [8].

Next, we use the error estimate at the nodes to obtain estimates for e(t) on the full interval [a,b].

THEOREM 2. *Let $e(t) = x(t) - x_h(t)$ where x(t) satisfies (1) and $x_h(t)$ is the piecewise linear approximate solution to (1) defined by (2). Let C_2*

be as defined in Theorem 1 and $C_3 = \max_{a \le t \le b} |x''(t)|$. Then

$$\|e\|_{L^\infty} \le (C_2 + \frac{1}{8} C_3)h^2. \tag{21}$$

Proof. The proof is similar to Theorem 4 of [4].

Corollary 1. Let $\|e\|_0^2 = \int_a^b (e(t))^2 dt$. Then

$$\|e\|_0 \le \sqrt{b - a}\, (C_2 + \frac{1}{8} C_3)h^2. \tag{22}$$

Proof. This follows by squaring (21) and integrating over [a,b].

REFERENCES

[1] P.G. Ciarlet, M.H. Schultz and R.S. Varga, Numerical methods of high order accuracy for nonlinear boundary value problems I. One dimensional problem. Numer. Math. 9 (1967), 394-430.

[2] J. Douglas, Jr. and T. Dupont, Galerkin approximations for the two-point boundary problem using continuous, piecewise polynomial spaces, Numer. Math. 22(1974), 99-109.

[3] P. Henrici, Discrete Variable Methods in Ordinary Differential Equations, John Wiley, 1962.

[4] J. Gregory and M. Zeman, A Galerkin approximation for the initial-value problem for linear second order differential equations, submitted for publication.

[5] J. Gregory and M. Zeman, A finite element approximation for the initial value problem for nonlinear second order differential equations, submitted for publication.

[6] E. Isaacson and H.B. Keller, Analysis of Numerical Methods, John Wiley, 1966.

[7] J.M. Ortega and M.L. Rockoff, Nonlinear difference equations and Gauss-Seidel type iterative methods, SIAM J. Numer. Anal. 3(1966) 497-513.

[8] G.M. Phillips and P.J. Taylor, Theory and Applications of Numerical Analysis, Academic Press, 1973.

[9] S. Schechter, Iterative methods for nonlinear problems, Trans. Amer. Math. Soc. 104 (1962), 179-189.

[10] M.F. Wheeler, An optimal L^∞ error estimate for Galerkin approximations to solutions of two-point boundary value problems, SIAM J. Numer. Anal. 10(1973), 914-917.

QUADRATIC SYSTEMS

R. D. S. Grisell
Texas Instruments
Dallas, Texas

ABSTRACT

Frequency domain techniques are presented for systems with kernels negligible above second order in a functional power series expansion. An analytic nonlinear system can be decomposed into a cascade of linear dynamic subsystem feeding into a no-memory (nondynamic) subsystem. Any kernel in the expansion of such a system has a rational polynomial Laplace transform of a particular type: there are no "overlapping" summation pole terms in any factorization of the denominator. This fact is used to make explicit a reduction method of D. A. George which is useful in obtaining a system response function (or frequency response function) from the transformed kernels.

Norbert Wiener showed that an analytic nonlinear system could be decomposed into a cascaded combination of a linear dynamic subsystem feeding into a no-memory (nondynamic) subsystem so that the following is not as restrictive as it may appear:

Fact 1: [1] Any kernel in the characterization of a system composed of linear dynamic subsystems and non-dynamic analytic nonlinearities, as combined by means of addition, multiplication, cascading, and feedback connections, has Laplace transform of the general form

$$\frac{N_n(s_1, s_2, \ldots, s_n)}{\prod_k (\sum_i s_i + \alpha_k)}$$

where the N_n are polynomials, and where some of the "summation-poles" may be devoid of various complex variables. Moreover, "overlapping" summation-poles do not occur, e.g. $(s_1+s_2+\alpha)$ and $(s_2+s_3+\beta)$.

George [2] illustrated his association operations

$$R_1\{G(s,s')\} \equiv \frac{1}{2\pi j} \int_{\alpha-j\infty}^{\alpha+j\infty} G(s-s',s')ds'$$

$$R_2\{F(s,s',s''\} \equiv \frac{1}{(2\pi j)^2} \int_{\alpha-j\infty}^{\alpha+j\infty} \int_{\rho-j\infty}^{\rho+j\infty} F(s-s',s'-s'',s'')ds'ds''$$

with rational polynomials and claimed the following:

<u>Fact 2:</u> A rational polynomial, as a realizable transfer function, can be reduced to a rational polynomial of one variable by means of pairwise associations of the form

$$\frac{A(s_1+s_2)}{(s_1+\alpha)^n (s_2+\beta)^m} \cdot Q(s_3,\ldots,s_n) \rightarrow \frac{A(s)Q(s_3,\ldots,s_n)}{(n-1)!(m-1)!(s+\alpha+\beta)^{n+m-1}}$$

where Q is some rational polynomial. Of course, the function A might be the identically = 1 function, or, n or m might be 1.

As part of George's "inspection" technique, as he calls it, the following partial fractions expansion is repeatedly used:

$$\frac{Q(s_1,\ldots,s_n)}{(s_i+\alpha)(s_i+\beta)} \rightarrow Q(s_1,\ldots,s_n)(\alpha-\beta)\left[\frac{1}{(s_i+\alpha)} - \frac{1}{(s_i+\beta)}\right]$$

where Q is another rational polynomial. We will call this operation "splitting." Also as part of George's inspection technique must be included an operation we will call "absorption":

$$\frac{Q(s_1,\ldots,s_n)}{(s_i+s_j+\alpha) \quad (s_i+\beta)} \rightarrow \frac{Q(s_1,\ldots,s_{i-1},s_i-\beta,\ldots,s_{j-1},\ \beta,\ldots,s_n)}{2\pi j\ (s_i+\alpha)}$$

Here, the s_i pole term appears to be swallowed by the s_i+s_j summation-pole term. This follows immediately from the definition of R_1 and Cauchy's integral formula. Thus George's claim is that there exists an appropriate sequence of the above three operations -- pairwise association, splitting, and absorption -- to reduce any realizable rational polynomial of several variables to a single variable. An example is given [2] where a particular unwary order of applications of these operations results in an indeterminate form in several variables which cannot be further reduced, but where an appropriate order of application can work. An essential reason that the inspection technique can work at all is that no overlapping summation-pole terms occur.

Since duplicated summation-poles in a term like

$$\frac{Q(s_1,\ldots,s_n)}{(s_i+s_j+\ldots+s_k+\alpha)\ (s_i+s_j+\ldots+s_k+\beta)}$$

can be split by partial fractions expansion, one will in general have only to deal with terms containing arbitrary nestings of summation-poles:

$$(s_1+s_2+\ldots+s_i+\ )^p(s_{j_1}+s_{j_2}+\ldots s_{j_J}+\ )^q\ldots(s_{k_1}+s_{k_2}+\ldots+s_{k_K}+\delta)^r,$$

where each summation pole except the left-most has the property that all of its variables are found in some one summation-pole to its left (nesting) and where some of the pole terms may be single poles (e.g. K = 1). Here, p, q, and r are any positive integers. Three things can happen at the deepest level of nesting:

1. There is at least one summation-pole term.
2. There are no summation-poles, but two or more single poles are found whose variables are all in one summation-pole term.
3. All single poles are nested in separate summation-pole terms.

Case 1 is immediately reduced to case 2 or case 3 via association. In case 2, the single poles with variables common to one summation-pole term can be pairwise-associated into a single term where the poles appear as a sum. In case 3, the single poles are simply absorbed into their corresponding higher level summation-pole terms. One notes that all such single pole terms must be absorbed before the absorber can itself be absorbed into a next higher level pole term. Therefore, the reduction of the multivariable expression must proceed along the lines of a tree-trimming algorithm where all the branches from a given node must be pruned before pruning the branch to that node. In fact, a consideration of the nesting tree allows convenient determination of which poles occur in the completely reduced expression (of one variable). This proves the useful result:

Result: Given a rational polynomial $R(s_1,s_2,\ldots,s_n)$ with nonoverlapping summation-pole terms in the denominator, it is possible (and we claim, convenient) to determine which poles occur in the completely associated form $R(s)$ by the following procedure: Split the denominator via partial fractions expansion into a sum of terms each denominator of which is a maximal nesting of pole terms. Construct the nesting lattice (tree) for each maximal nesting, and then prune the branches as follows:

Trivial association of s_j with s_i (as in case 1):

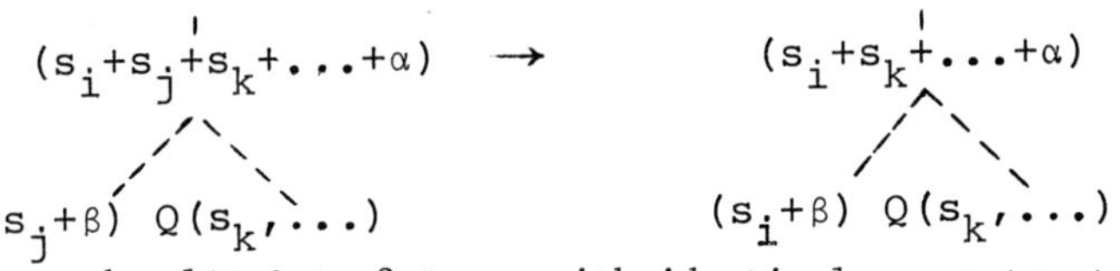

Associate and split into 2 trees with identical superstructure:

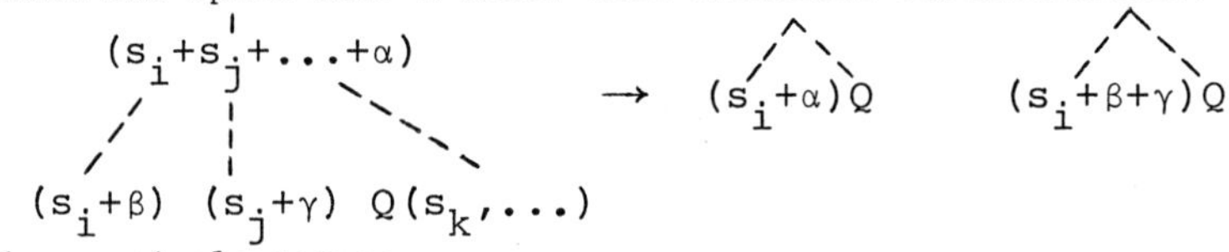

Absorb any single terms:

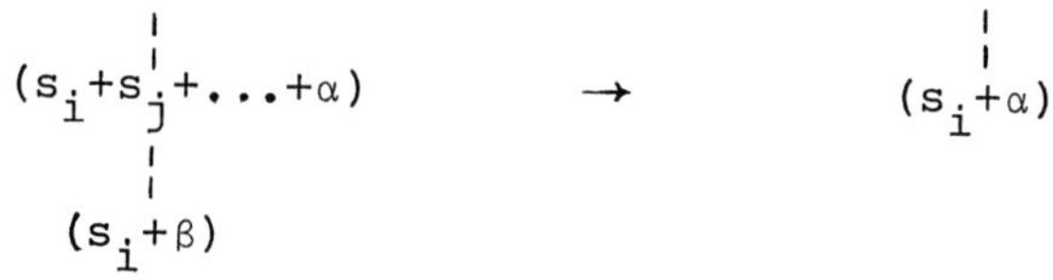

Example: Consider a single maximally nested pole structure:

$$\frac{1}{(s_1+s_2+s_3+s_4+s_5+\theta)\ (s_1+s_2+s_3+s_4+\nu)\ (s_1+s_2+\varepsilon)\ (s_3+s_4+\mu)\ (s_1+\alpha)\times}$$

$$\overline{\times(s_3+\beta)\ \ (s_4+\gamma)\ (s_5+\delta)}$$

The corresponding nested lattice appears as:

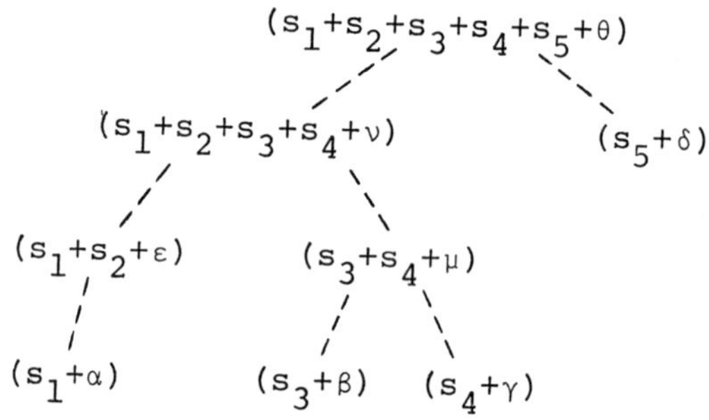

The poles of the complerely associated function would then be:

Definition: We propose the term "quadratic" to indicate those systems whose kernels vanish above second order.

A rough classification of quadratic systems with rational polynomial transfer functions is readily obtained in terms of structures of the two-dimensional Laplace transforms, which are rational. For simplicity, the linear terms will be ignored.

<u>Class 1:</u> $h_2(t_1,t_2) = f(t_1) + g(t_2)$.

This second order kernel would arise when the transform splits additively into two parts, one a function of s_1, the other of s_2.

<u>Class 2:</u> $h_2(t_1,t_2) = f(t_1)g(t_2)$.

Here, the frequency-domain function will show convolutive structure.

<u>Class 3:</u> $h_2(t_1,t_2) = \delta(t_1-t_2)f(t_2)$.

This is the main mechanism by which summation-pole terms appear. Its Laplace transform is

$$H(s_1,s_2) = F(s_1+s_2) .$$

Thus $\frac{1}{s_1 + s_2 + \alpha}$ for example, is the transform of $\delta(t_1-t_2)e^{-\alpha t_2}$.

<u>Class 4:</u> $h_2(t_1,t_2) = \int f(t_1,t_3)g(t_2-t_3)dt_3$

which has the multiplicative frequency structure

$$H_2(s_1,s_2) = \iint F(s_1,s_3)G(s_2-s_3)ds_3ds_4 .$$

<u>Class 5:</u> $h_2(t_1,t_2) = t_1^{m-1}e^{-\alpha t_1}f(t_2)$.

This gives rise to multiplicities:

$$H_2(s_1,s_2) = \frac{F(s_2)}{(s_1+ \alpha)^m} .$$

<u>Class 6:</u> $h_2(t_1,t_2) = f(t_1)g(t_2)^n$.

$$H_2(s_1,s_2) = F(s_1)\iint\cdots\int G(r_n)G(r_{n-1}-r_n)\ldots G(s_2-r_1)dr_n\ldots dr_1 .$$

To conclude this short introduction to quadratic systems, there is space only to mention some general sources of motivation. While one can only obtain linear systems by arbitrary addition, concatenation, and feedback of linear subsystems, the quadratic nonlinearity allows much richer structures. This simplest of nonlinearities can serve as a single "base" of sorts for an algebra under addition, concatenation, and feedback in combination with linear components which can approximate analytic nonlinearities:

<u>Theorem:</u> The systems obtainable via finite concatenation and addition

of quadratic systems are dense in the analytic systems whose kernels have rational polynomial transforms. This density is with respect to any integral or sup norm having compact support.

REFERENCES

[1] L. Crum and A. Heinen, "Simultaneous Reduction and Expansion of Multidimensional Laplace Transform Kernels," SIAM J. Appl. Math. 26(4) (1974), pp. 753-770.

[2] D. A. George, "Continuous Nonlinear Systems." Technical Report 355 (July 24, 1959) M.I.T. Research Laboratory of Electronics.

CONTINUATION THEOREMS FOR NONLINEAR EQUATIONS

Chaitan P. Gupta

Department of Mathematical Sciences
Northern Illinois University
DeKalb, Illinois

1. INTRODUCTION.

Let X, Y be given Banach spaces and $A(x,t)$: $X \times [0,1] \to 2^Y$ define a continuous family of mappings A_t: $X \to 2^Y$ by $A_t(x) = A(x,t)$, $t \in [0,1]$, $x \in X$. Let $t \in [0,1] \to y_t \in Y$ define a continuous curve in Y. A continuation theorem for the nonlinear equations $A(x,t) \ni y_t$ is a theorem which gives the solvability of the equation $A(x,1) \ni y_1$ from the knowledge of the solvability of the equation $A(x,0) \ni y_0$. In this paper, we study some continuation theorems when $Y = X^*$ and for each $t \in [0,1]$, A_t: $X \to 2^{X^*}$ is a mapping of monotone type, e.g., A_t is a maximal monotone mapping or a mapping satisfying condition (S). We remark that continuation theorems are of interest in the study of existence, continuous dependence and stability of solutions of nonlinear equations.

2. MAIN RESULTS.

Let X be a real reflexive Banach space and X^* denote its dual Banach space. For $x \in X$ and $w \in X^*$ we denote by (w,x) the duality pairing between X and X^*. For a given sequence $\{u_n\}$ in X and $u \in X$ we denote by "$u_n \to u$" that u_n converges to u in the norm topology of X and by "$u_n \rightharpoonup u$" that u_n converges to u in the weak-topology of X. Let A: $X \to 2^{X^*}$ be a given mapping. We denote by $D(A)$, <u>the effective domain</u> of A, the subset of X given by $D(A) = \{x \in X \mid Ax \neq \phi\}$; by $R(A)$, <u>the range of A</u>, the subset of X^* given by $R(A) = \cup\{Ax \mid x \in D(A)\}$ and by $G(A)$, <u>the graph of A</u>, the subset of $X \times X^*$ given by $G(A) = \{[x,w] \mid x \in D(A), w \in Ax\}$. A subset G of $X \times X^*$ is said to be <u>monotone</u> if for $[x_i,w_i] \in G$, $i = 1,2$, we have $(w_1-w_2,x_1-x_2) \geq 0$. A mapping A: $X \to 2^{X^*}$ is said to be monotone if its graph $G(A)$ is a monotone subset of $X \times X^*$; A is said to be maximal monotone if its graph $G(A)$ is a monotone set and is maximal among all monotone subsets of $X \times X^*$ in the sense of inclusion.

A mapping $T: X \to X^*$ is said to be bounded if it maps bounded subsets of X into bounded subsets of X^*; it is said to be demi-continuous if it is continuous from X into X^* endowed with weak-topology. $T: X \to X^*$ satisfies condition (S) if for any sequence $\{x_n\}$ in X such that $x_n \rightharpoonup x$, $x \in X$, and $\lim_{n\to\infty} (Tx_n - Tx, x_n - x) = 0$; then $x_n \to x$.

Finally, we define the duality mapping $J: X \to 2^{X^*}$ by $Jx = \{w \in X^* \mid (w,x) = \|x\|^2, \|w\| = \|x\|\}$. We note that for each $x \in X$ there exists a $w \in X^*$ satisfying the conditions $(w,x) = \|x\|^2$, $\|w\| = \|x\|$, by Hahn-Banach Theorem. Moreover, for $x_i \in X$, $w_i \in Jx_i$, $i = 1, 2$, we have $(w_1 - w_2, x_1 - x_2) \geq (\|x_1\| - \|x_2\|)^2 \geq 0$ and $J: X \to 2^{X^*}$ is single-valued if X^* is strictly-convex.

Theorem 1:- Let X be a real reflexive Banach space and X^* denote its dual Banach space. Let $A_t(x) \equiv A(x,t): X \times [0,1] \to 2^{X^*}$ be a mapping satisfying the following:

(i) For each fixed $t \in [0,1]$; $A_t: X \to 2^{X^*}$ is maximal monotone and for $t, s \in [0,1]$, $D(A_t) = D(A_s)$. Moreover, $0 \in D(A_t)$ for every $t \in [0,1]$ and $\sup \{A_t^\circ(0) : t \in [0,1]\} < +\infty$ where $A_t^\circ(0) = \inf \{\|u\| : u \in A_t(0)\}$ for $t \in [0,1]$.

(ii) For each $x \in X$, the mapping $t \in [0,1] \to A(x,t) \in 2^{X^*}$ is such that for $s \in [0,1]$ and $u_s \in A(x,s)$ there exists for each $t \in [0,1]$ a $u_t \in A(x,t)$ and $c(s,x) > 0$ such that $\|u_t - u_s\| \leq c(s,x)|t-s|$.

(iii) $0 \in A_0(0)$

(iv) there is an $r > 0$ and $\alpha > 0$ such that for $t \in [0,1]$, $x \in X$, $\|x\| = r$, $u \in A(x,t) \equiv A_t(x)$ we have $\|u\| \geq \alpha$.

Then, for every $\alpha' < \alpha$ we have $A_1(B_r) \supset \bar{B}^*_{\alpha'}$, where $B_r = \{x \in X: \|x\| < r\}$ and $\bar{B}^*_{\alpha'} = \{u \in X^*: \|u\| \leq \alpha'\}$.

Proof:- We first observe that it suffices to show that $0 \in A_1(B_r)$ to prove that $\bar{B}^*_{\alpha'} \subset A_1(B_r)$ for every $\alpha' < \alpha$. Indeed, if $u \in \bar{B}^*_{\alpha'}$ we define the mappings: $\tilde{A}(x,t): X \times [0,1] \to 2^{X^*}$ by $\tilde{A}(x,t) = A(x,t) - tu$. Then the family of mappings $\tilde{A}_t(x) = \tilde{A}(x,t)$; $X \times [0,1] \to 2^{X^*}$ satisfies conditions (i), (ii), (iii) and (iv) with α replaced by $\alpha - \alpha'$.

We shall show that for $0 < \varepsilon < \frac{\alpha}{r}$ there is an $x_\varepsilon \in X$, $\|x_\varepsilon\| \leq r$ such that $0 \in A_1(x_\varepsilon) + \varepsilon J(x_\varepsilon)$ where $J\colon X \to 2^{X^*}$ is the duality mapping. We shall then let $\varepsilon \to 0+$ to obtain an $x_0 \in X$ such that $0 \in A_1(x_0)$.

Let, now ε be fixed in the following and suppose that $0 < \varepsilon < \frac{\alpha}{r}$. For $t \in [0,1]$, set $S_{\varepsilon,t} = A_t + \varepsilon J\colon X \to 2^{X^*}$. Since $A_t\colon X \to 2^{X^*}$ is maximal monotone for each $t \in [0,1]$ we have $R(S_{\varepsilon,t}) = R(A_t+\varepsilon J) = X^*$ for every $t \in [0,1]$. (see [1]) So for each $t \in [0,1]$ there exists an $x_t \in X$ such that $0 \in A_t(x_t) + \varepsilon J(x_t) \equiv S_{\varepsilon,t}(x_t)$ or equivalently $x_t \in S^{-1}_{\varepsilon,t}(0)$. It it easy to see that $x_t \in S^{-1}_{\varepsilon,t}(0)$, $y_t \in S^{-1}_{\varepsilon,t}(0)$ imply $\|x_t\| = \|y_t\|$. We thus see that $S^{-1}_{\varepsilon,0}(0) = \{0\}$ since $0 \in A_0(0) + \varepsilon J(0) = S_{\varepsilon,0}(0)$ and that the mapping $\psi\colon [0,1] \to \mathbb{R}$ defined by $\psi(t) = \|x_t\|$ where $x_t \in S^{-1}_{\varepsilon,t}(0)$ is a well-defined mapping. Also $\psi(0) = 0$. We next note for $t \in [0,1]$, $\psi(t) \leq \frac{1}{\varepsilon} C$; where $C = \sup\{A^\circ_t(0)\colon\ t \in [0,1]\}$.

Claim 1:- $\psi\colon [0,1] \to \mathbb{R}^+$ is continuous on $[0,1]$.

Proof of Claim 1:- Let $s \in [0,1]$, $x_s \in S^{-1}_{\varepsilon,s}(0)$ be such that $\psi(s) = \|x_s\|$. Also for $t \in [0,1]$, let $x_t \in S^{-1}_{\varepsilon,t}(0)$ be such that $\psi(t) = \|x_t\|$. Again, let $u_t \in A_t(x_t)$, $\xi_t \in J(x_t)$, $u_s \in A_s(x_s)$, $\xi_s \in J(x_s)$ be such that

$$0 = u_t + \varepsilon\xi_t$$
$$0 = u_s + \varepsilon\xi_s.$$

Let, now, $u \in A_t(x_s)$ be chosen in accordance to assumption (ii). Then

$$\begin{aligned} 0 &= (u_t-u_s, x_t-x_s) + \varepsilon(\xi_t-\xi_s, x_t-x_s) \\ &= (u_t-u, x_t-x_s) + (u-u_s, x_t-x_s) + \varepsilon(\xi_t-\xi_s, x_t-x_s) \\ &\geq (u-u_s, x_t-x_s) + \varepsilon(\|x_t\|-\|x_s\|)^2 \end{aligned}$$

This gives in view of assumption (ii) and claim 2 that

$$(\|x_t\|-\|x_s\|)^2 \leq 2c(s,x_s)\,|t-s|\frac{C}{\varepsilon}.$$

This gives the continuity of ψ at $s \in [0,1]$.

Claim 2:- $r \notin [0,1]$.

Proof of Claim 2:- Indeed, let $\exists\, t \in [0,1]$ such that $\psi(t) = r$. So there is an $x_t \in S^{-1}_{\varepsilon,t}(0)$ with $\|x_t\| = r$. Let $u_t \in A_t(x_t)$, $\xi_t \in J(x_t)$ be such that $0 = u_t + \varepsilon\xi_t$. Now assumption (iii) implies that $\|u_t\| \geq \alpha$. It then follows that

$$0 = \|u_t + \varepsilon\xi_t\| \geq \|u_t\| - \varepsilon\|\xi_t\| = \|u_t\| - \varepsilon\|x_t\| \geq \alpha - \varepsilon r$$

so that $\varepsilon \geq \frac{\alpha}{r}$. This contradicts our assumption that $0 < \varepsilon < \frac{\alpha}{r}$. Hence the claim.

We thus see that the set $\psi([0,1])$ is a connected set such that $0 \in \psi([0,1])$ and $r \notin \psi([0,1])$ so $\psi([0,1]) \subset [0,r)$.

Accordingly, there exists an $x_\varepsilon \in X$, $\|x_\varepsilon\| < r$ such that $x_\varepsilon \in S^{-1}_{\varepsilon,1}(0)$ i.e. $0 \in A_1(x_\varepsilon) + \varepsilon J(x_\varepsilon)$ for every ε, $0 < \varepsilon < \frac{\alpha}{r}$. Let $u_\varepsilon \in A_1(x_\varepsilon)$, $\xi_\varepsilon \in J(x_\varepsilon)$ be such that $0 = u_\varepsilon + \xi_\varepsilon$. This gives that $\|u_\varepsilon\| = \varepsilon\|\xi_\varepsilon\| = \varepsilon\|x_\varepsilon\| < \varepsilon r \to 0$ as $\varepsilon \to 0+$. Let, now, $\{\varepsilon_n\}$ and $x_0 \in X$ be such that $\varepsilon_n \to 0+$ and $x_{\varepsilon n} \equiv x_n \to x_0$. Then $u_n \equiv u_{\varepsilon n} \to 0$. Accordingly, $0 \in A_1(x_0)$ since A_1 is maximal monotone.

Hence the theorem.//

Corollary 1:- Let X be a real reflexive Banach space and X^* its dual Banach space. Let $T: X \to X^*$ be a demi-continuous monotone mapping. Suppose that there is an $r > 0$ and an $\alpha > 0$ such that

$$\|T(x) - (1-t)T(-x)\| \geq \alpha$$

for $x \in X$, $\|x\| = r$ and $t \in [0,1]$.

Then for every $\alpha' < \alpha$: $T(B_r) \supset \overline{B}^*_{\alpha'}$. In particular, the equation $T(x) = 0$ is solvable in B_r.

Remark 1:- Corollary 1 is due to de-Figueiredo-Gupta [2].

Theorem 2:- Let X be a finite-dimensional Banach space. Let $A_t(x) \equiv A(x,t)$: $X \times [0,1] \to X^*$ be a continuous family of continuous mapping satisfying the following:

(i) there exists an $r > 0$ such that $A(x,t) \neq 0$ for $x \in X$, $\|x\| = r$ and $t \in [0,1]$.

(ii) the mapping $A_0(x) = A(x,0)$: $X \to X^*$ is odd on $\{x \in X: \|x\| = r\}$.

Then, there exists a $d > 0$ such that $A_1(B_r) \supset \overline{B}^*_d$.

Proof:- Since X is finite-dimensional we have in view of our assumptions that

$$d_0 = \min \{\|A(x,t)\| : \|x\| = r,\ t \in [0,1]\} > 0.$$

Let $d > 0$ be such that $d < d_0$. Then for $u \in \overline{B}^*_d$ we have $\|A(x,t)-u\| \geq \|A(x,t)\| - \|u\| \geq d_0 - d > 0$ for $x \in X$, $\|x\| = r$, $t \in [0,1]$. Accordingly, the Brouwer degree $\deg(A_t, B_r, u)$ is defined for every $u \in \overline{B}^*_d$ and is independent of $t \in [0,1]$. But $\deg(A_0, B_r, u)$ is odd by Borsuk's Theorem in view of assumption (ii). So for every $u \in \overline{B}^*_d$ we have $\deg(A_1, B_r, u) = \deg(A_0, B_r, u) \neq 0$. Hence $A_1(B_r) \supset \overline{B}^*_d$. This proves Theorem 2.

Theorem 3:- Let X be a separable reflexive Banach space and $A_t(x) \equiv A(x,t)$: $X \times [0,1] \to X^*$ be a family of mappings satisfying the following:

(i) for each $t \in [0,1]$, A_t: $X \to X^*$ is a bounded, demi-continuous mapping satisfying condition (S).

(ii) For any bounded subset B of X, the mapping $t \in [0,1] \to A(x,t) \in X^*$ is uniformly continuous on B.

(iii) there exists an $r > 0$ such that $A(x,t) \neq 0$ for $x \in X$, $\|x\| = r$, $t \in [0,1]$.

(iv) the mapping $A_0(x) = A(x,0)$: $X \to X^*$ is odd on $\{x \in X: \|x\| = r\}$.

Then, there exists a $d > 0$ such that $A_1(B_r) \supset \overline{B}^*_d$.

Let $\{F_n\}$ be an increasing sequence of finite dimensional subspaces of X such that $\bigcup_{n=1}^{\infty} F_n$ is dense in S. Let j_n: $F_n \to X$ denote the inclusion mapping $F_n \subset X$ and j^*_n: $X^* \to F^*_n$ be the corresponding projection mapping. For each n and $t \in [0,1]$ let $A_{tn} = j^*_n A_t j_n$: $F_n \to F^*_n$.

Lemma:- There exists a $d > 0$ and an integer $n_0 > 0$ such that for all $n \geq n_0$, $t \in [0,1]$, $x \in F_n$, $\|x\| = r$ we have $\|A_{tn}(x)\| \geq d$.

Proof of Lemma:- Suppose the lemma is not true. Then we can find an infinite sequence $\{n_j\}$ of positive integers, a sequence $\{t_j\}$ in $[0,1]$ and points $x_j \in F_{n_j}$ with $\|x_j\| = r$ such that $\|A_{t_j n_j}(x_j)\| \to 0$ as $j \to \infty$.

We may assume that there is a $t_0 \in [0,1]$ and $x_0 \in X$ such that $t_j \to t_0$ and $x_j \rightharpoonup x_0$ (weakly) in X. Now,

$$(1)\qquad |(A_{t_0}(x_j), x_j-x_0)| \le |(A_{t_0}(x_j) - A_{t_j}(x_j), x_j-x_0)| + |(A_{t_j}(x_j), x_j-x_0)|$$
$$\le |(A_{t_0}(x_j) - A_{t_j}(x_j), x_j-x_0)| + |(A_{t_j}(x_j), x_j)$$
$$+ |(A_{t_j}(x_j), x_0)|.$$

Since $t_j \to t_0$ and $||x_j|| = r$, $||x_0|| \le r$ we see from assumption (ii) that $|(A_{t_0}(x_j) - A_{t_j}(x_j), x_j - x_0)| \to 0$ as $j \to \infty$. Also since $x_j \in F_{n_m}$ we have

$$|(A_{t_j}(x_j), x_j)| = |(A_{t_j n_j}(x_j), x_j)| \to 0 \quad \text{as} \quad n \to \infty .$$

Finally, we see that for $v \in \bigcup_{n=1}^{\infty} F_n$ so that $v \in F_m$ for some m we have for $n_j \ge m$ that

$$|(A_{t_j}(x_j), v)| = |(A_{t_j n_j}(x_j), v)| \le ||A_{t_j n_j}(x_j)|| \; ||v|| \to 0$$

so that $(A_{t_j}(x_j), v) \to 0$, for every $v \in \bigcup_{n=1}^{\infty} F_n$. It then follows easily from the boundedness of A_{t_0}, $t_j \to t_0$, the assumption (ii) and the density of $\bigcup_{n=1}^{\infty} F_n$ in X that

$$|(A_{t_j}(x_j), v)| \to 0 \quad \text{as} \quad j \to \infty \text{ for every } v \in X.$$

It then follows from (1) that $|(A_{t_0}(x_j), x_j - x_0)| \to 0$ as $j \to \infty$. Since now A_{t_0} satisfies condition (S) we get that $x_j \to x_0$ and so $A_{t_0}(x_j) \rightharpoonup A_{t_0}(x_0)$ (weakly) in X* by the demi-continuity of A_{t_0}. Also $||x_0|| = \lim_{j\to\infty} ||x_j|| = r$. We next see from assumption (ii) that

$$||A_{t_j}(x_j) - A_{t_0}(x_j)|| \to 0 \quad \text{as} \quad j \to \infty$$

since $t_j \to t_0$. Also since we have shown above that $(A_{t_j}(x_j), v) \to 0$ for every $v \in X$ we get $A_{t_j}(x_j) \rightharpoonup 0$ (weakly) in X*. Thus we see that $A_{t_0}(x_j) = A_{t_0}(x_j) - A_{t_j}(x_j) + A_{t_j}(x_j) \rightharpoonup 0$ (weakly) in X^*. Hence $A_{t_0}(x_0) = 0$. But this contradicts assumption (iii) since $\|x_0\| = r$. Hence the lemma is true.

Proof of Theorem 3:- Let $d' > 0$ be such that $d' < d$ (d as in lemma above). Now we have from Theorem 2 that for every $n \geq n_0$ (n_0 as in Lemma above) that

$$A_{1n}(B_r \cap F_n) \supset j_n^*(\overline{B}_{d'}^*)$$

Let, now, $w \in \overline{B}_{d'}^*$ be fixed arbitrarily. Then, for each $n \geq n_0$, there is an $x_n \in F_n$, $\|x_n\| \leq r$ such that $A_{1n}(x_n) = j_n^*(w)$. Since A_1 is bounded and X is reflexive, we may assume that there is an $x_0 \in X$ such that $x_n \rightharpoonup x_0$ (weakly) in X and the sequence $\{A_1(x_n)\}$ is weakly convergent in X^*. Let, now, $y \in \bigcup_{n=1}^{\infty} F_n$ so that $y \in F_m$ for some m. We then see that

$$(A_1(x_n),y) = (A_{1n}(x_n),y) = (w,y)$$

for all $n \geq m$; so that

$$(A_1(x_n)-w,y) \to 0 \quad \text{for every} \quad y \in \bigcup_{n=1}^{\infty} F_n.$$

It then follows from the density of $\bigcup_{n=1}^{\infty} F_n$ in X and the boundedness of A_1 that $(A_1(x_n)-w,y) \to 0, \forall\, y \in X$. Hence $A_1(x_n) \rightharpoonup w$ in X^*. Also, for each $n \geq n_0$ we have

$$(A_1(x_n),x_n) = (A_{1n}(x_n),x_n) = (w,x_n).$$

This gives for $n \geq n_0$ that $(A_1(x_n),x_n-x_0) = (w,x_n) - (A_1(x_n),x_0) \to 0$ as $n \to \infty$. It then follows that $x_n \to x_0$ since A_1 satisfies condition (S) and $A_1(x_n) \rightharpoonup A_1(x_0)$ by the demicontinuity of A_1. Hence $A_1(x_0) = w$. Since $\|x_0\| \leq \liminf_{n\to\infty} \|x_n\| \leq r$ the proof of the theorem is complete in view of assumption (iii).

Corollary 2:- Let X be a separable reflexive Banach space and $T: X \to X^*$ be a bounded demicontinuous mapping. Suppose that for each $t \in [0,1]$ the mapping $x \in X \to T(x) - (1-t)T(-x) \in X^*$ satisfies condition (S). Suppose further that there is an $r > 0$ such that for $x \in X$, $\|x\| = r$, $t \in [0,1]$, $T(x) - (1-t)T(-x) \neq 0$.

Then there exists a $d > 0$ such that $T(B_r) \supset \overline{B}^*_d$.

Remark 3:- Corollary 2 is due to De-Figueiredo-Gupta [2].

Remark 4:- Theorem 3 is similar to a result of Browder [1], but Browder used generalized degree theory to prove his result. For results similar to theorem 3 in non-separable Banach spaces, but of the nature of a existence type continuation theorem we refer to Hess [3], Pohožaev [6], and Nečas [5].

REFERENCES

[1] Browder, F. E. Nonlinear operators and nonlinear equations of evolution in Banach spaces. Proc. Symp. Pure Math. Vol. 18, part 2, (1976) Amer. Math. Soc., Providence, R.I.

[2] deFigueiredo, D. G., and Gupta, C. P. Borsuk type theorems for nonlinear noncompact mappings in Banach spaces. Proc. Conf. Functional Analysis, I. I. T. Kanpus (1970), p. 48-55.

[3] Hess, P. On nonlinear mappings of monotone type homotopic to odd operators. Jour. of Functional Analysis 11(1972) p. 138-167.

[4] Hess, P. A homotopy argument for mappings of monotone type in Banach spaces. Math. Ann. 207(1974) 63-65.

[5] Nečas, J. A remark on the Fredholm alternative for nonlinear operators with application to nonlinear integral equations of generalized Hammerstein type. C.M.U.C. 13(1972) 109-120.

[6] Pohožaev, S. I. The solvability of nonlinear equations with odd operators. Functional Analysis & Applications, 1(1967), 66-73.

MULTIPLICATIVE PERTURBATION OF m-ACCRETIVE OPERATORS AND EVOLUTION EQUATIONS IN $L^{\infty}(\Omega)$

Ki Sik Ha

Department of Mathematics
Busan National University
Jangjeon-Dong, Busan, Korea

ABSTRACT

Let Ω be a measure space with bounded measure. To solve the evolution equation $(du/dt)+A\beta u \ni 0$, $u(0)=u_o$, we prove the m-accretiveness of $A\beta$ in $L^{\infty}(\Omega)$.

Let Ω be a measure space. $L^p(\Omega)$ denotes the usual Lebesgue space with its norm $|\cdot|_p$ $(1\leq p\leq\infty)$.

Let A be an operators of $L^p(\Omega)$ and $\beta:\Omega\times R\to 2^R$ a mapping. Define the operator $A\beta$ and βA of $L^p(\Omega)$ by

$$A\beta=\{(u,w)\in L^p(\Omega)\times L^p(\Omega)\,|\,\exists\, v\in L^p(\Omega) \text{ such that } (v,w)\in A \text{ and a.e. } x\in\Omega,\ v(x)\in\beta(x,u(x))\}$$

and

$$\beta A=\{(u,w)\in L^p(\Omega)\times L^p(\Omega)\,|\,\exists\, v\in L^p(\Omega) \text{ such that } (u,v)\in A \text{ and a.e. } x\in\Omega,\ w(x)\in\beta(x,v(x))\}.$$

To study the evolution equation

$$\begin{cases}\dfrac{du}{dt} + A\beta u \ni 0\\ u(0) = u_o\end{cases}$$

in $L^{\infty}(\Omega)$ by the semi-group theory, it is sufficient to apply the theorem of Crandall-Liggett ([10]) to $A\beta$ satisfying the condition

(#) $\begin{cases} A\beta \text{ is accretive} \\ R(I+\lambda A\beta)\supset D(A\beta) \text{ for every } \lambda>0 \end{cases}$

in $L^\infty(\Omega)$.

The condition (#) in $L^1(\Omega)$ was studied in [2], [3] and the condition

$$\begin{cases} \beta A \text{ is accretive} \\ R(I+\lambda\beta A)\supset D(\beta A) \text{ for every } \lambda>0 \end{cases}$$

in $L^\infty(\Omega)$ was studied in [5], [11].

The object of this paper is to study the condition (#) in $L^\infty(\Omega)$.

First of all, we give the main theorem.

Theorem 1: Suppose $\text{mes}\Omega<\infty$. Let A be an operator of $L^\infty(\Omega)$ satisfying:

(H1) A is m-accretive in $L^\infty(\Omega)$ and monotone in $L^2(\Omega)$.

Let $\beta:\Omega\times R\to 2^R$ be a mapping satisfying:

(H2) a.e. $x\in\Omega$, $r\in R\to\beta(x,r)\in 2^R$ is maximal monotone in R

and

(H3) for every $r\in R$, there exists $v\in L^\infty(\Omega)$ such that a.e. $x\in\Omega$, $v(x)\in\beta(x,r)$.

Suppose that for every $f\in L^\infty(\Omega)$ and $\lambda>0$, there exists at most one solution $u\in L^\infty(\Omega)$ of $u+\lambda A\beta u\ni f$. Suppose $A\beta\neq\Phi$. Then $A\beta$ is m-accretive in $L^\infty(\Omega)$.

To prove this theorem, we need some lemmas and some propositions

By the results of [4], we have:

Proposition 2:([11]) Let A satisfy (H1). Then

$$\int(\text{sgn}(v_1 - v_2))(|v_1 - v_2| - k)^+(w_1 - w_2) \geqslant 0$$

for every (v_1,w_1), $(v_2,w_2)\in A$ and $k\geqslant 0$, where $r^+=\max(r,0)$ for $r\in R$.

We also define the operator $A+\beta$ of $L^{\infty}(\Omega)$ by

$$A+\beta=\{(u,v+w)\in L^{\infty}(\Omega)\times L^{\infty}(\Omega)\,|\,(u,v)\in A \text{ and } \text{a.e.}\,x\in\Omega,\ w(x)\in\beta(x,u(x))\}.$$

Let A_2 be the closure of A in $L^2(\Omega)$. Define the operator $A_2+\beta$ of $L^2(\Omega)$ by

$$A_2+\beta=\{(u,v+w)\in L^2(\Omega)\times L^2(\Omega)\,|\,(u,v)\in A_2 \text{ and } \text{a.e.}\,x\in\Omega,\ w(x)\in\beta(x,u(x))\}.$$

Using Proposition 2, we obtain:

Lemma 3: Let A satisfy (H1) and let $\gamma:D(\gamma)\subset\Omega\times R\to R$ be a mapping that a.e. $x\in\Omega$, $r\in D_x(\gamma)\subset R\to\gamma(x,r)\in R$ is maximal monotone single-valued in R, where $D_x(\gamma)=\{r\in R\,|\,(x,r)\in D(\gamma)\}$. Suppose that for every f_1, $f_2\in L^{\infty}(\Omega)$ and $\lambda>0$, there exist v_1, $v_2\in L^2(\Omega)$ such that

$$v_1+\lambda(A_2+\gamma)v_1\ni f_1 \text{ and } v_2+\lambda(A_2+\gamma)v_2\ni f_2.$$

Then for every $\varepsilon>0$,

(1) $$|v_1(x)-v_2(x)|\leq|f_1-f_2|_{\infty}+\varepsilon \quad \text{a.e.}\,x\in\Omega,$$

and hence

(2) $$v_1-v_2\in L^{\infty}(\Omega) \text{ and } |v_1-v_2|_{\infty}\leq|f_1-f_2|_{\infty}.$$

Corollary 4: Under the hypotheses of A and γ in Lemma 3, let γ satisfy (H3) and $(0,0)\in A$, $\gamma(x,0)=0$ a.e. $x\in\Omega$. Suppose that for every $f\in L^{\infty}(\Omega)$ and $\lambda>0$, there exists $v\in L^2(\Omega)$ such that $v+\lambda(A_2+\gamma)v\ni f$. Then $v\in L^{\infty}(\Omega)$, $|v|_{\infty}\leq|f|_{\infty}$ and $v+\lambda(A+\gamma)v\ni f$.

Proposition 5: Let A satisfy (H1) and let $\gamma:D(\gamma)\subset\Omega\times R\to R$ be a mapping that a.e. $x\in\Omega$, $r\in D_x(\gamma)\to\gamma(x,r)\in R$ is maximal monotone single-valued in R with (H3). Suppose $A(I+\alpha\gamma)^{-1}\neq\Phi$ for every $\alpha>0$. Then $A(I+\alpha\gamma)^{-1}$ is m-accretive in $L^{\infty}(\Omega)$ for sufficiently small $\alpha>0$.

Proof of Proposition 5: Let $u_1+\lambda A(I+\alpha\gamma)^{-1}u_1\ni f_1$ and $u_2+\lambda A(I+\alpha\gamma)^{-1}u_2\ni f_2$ for u_1, $u_2\in D(A(I+\alpha\gamma)^{-1})$, f_1, $f_2\in L^{\infty}(\Omega)$ and α, $\lambda>0$. There exist v_1, $v_2\in L^{\infty}(\Omega)$ such that $v_1=(I+\alpha\gamma)^{-1}u_1$, $v_2=(I+\alpha\gamma)^{-1}u_2$ a.e. on Ω and $u_1+\lambda Av_1\ni f_1$, $u_2+\lambda Av_2\ni f_2$. Thus $v_1+(\lambda A+\alpha\gamma)v_1\ni f_1, v_2+(\lambda A+\alpha\gamma)v_2\ni f_2$. By (1) of Lemma 3, for $\varepsilon>0$,

$|v_1-v_2|\leq|f_1-f_2|_\infty+\varepsilon$ a.e. on Ω. By (H3), a.e. on Ω

$$\lim_{\alpha\to0+}(I+\alpha\gamma)^{-1}u_i=u_i\quad(i=1,2).$$

For sufficiently small $\alpha>0$, $|u_1-u_2|\leq|f_1-f_2|_\infty+\varepsilon$ a.e. on Ω. Thus $u_1-u_2\varepsilon L^\infty(\Omega)$ and as $\varepsilon\to0+$, $|u_1-u_2|_\infty\leq|f_1-f_2|_\infty$. Hence $A(I+\alpha\gamma)^{-1}$ is accretive in $L^\infty(\Omega)$ for sufficiently small $\alpha>0$.

To prove $R(I+A(I+\alpha\gamma)^{-1})=L^\infty(\Omega)$, we may suppose that $(0,0)\varepsilon A$ and $\gamma(x,0)=0$ a.e.$x\varepsilon\Omega$([11]). Let $f\varepsilon L^\infty(\Omega)$. Take $M>0$ such that $|f|_\infty\leq M$. By (H3), there exist w_+, $w_-\varepsilon L^\infty(\Omega)$ such that $\gamma(x,M)=w_+(x)$, $\gamma(x,-M)=w_-(x)$ a.e.$x\varepsilon\Omega$. Define $\gamma^1:\Omega\times R\to R$ by

$$\gamma^1(x,r)=\begin{cases}w_-(x)+r+M & (r<-M)\\ \gamma(x,r) & (|r|\leq M)\\ w_+(x)+r-M & (r>M).\end{cases}$$

Then γ^1 satisfies (H2) and (H3). Define an operator B of $L^2(\Omega)$ by

$$B=\{(v,w)\varepsilon L^2(\Omega)\times L^2(\Omega)\,|\,w(x)=\alpha\gamma^1(x,v(x))\text{ a.e.}x\varepsilon\Omega\}.$$

Then B is everywhere defined, maximal monotone in $L^2(\Omega)$. Thus A_2+B is maximal monotone in $L^2(\Omega)$. There exists $v\varepsilon L^2(\Omega)$ such that $v+(A_2+B)v\ni f$. By Corollary 4, $v\varepsilon L^\infty(\Omega)$, $|v|_\infty\leq|f|_\infty$ and $v+(A+\alpha\gamma^1)v\ni f$. Since $|v|_\infty\leq M$, $\gamma^1v=\gamma v$. Put $u=v+\alpha\gamma v$. Then $u\varepsilon L^\infty(\Omega)$ and $u=(I+\alpha\gamma)^{-1}v$. Thus $u+A(+\alpha\gamma)^{-1}u\ni f$ which completes the proof.

<u>Corollary 6:</u> Let $\varepsilon>0$. Under the hypotheses of Proposition 5, $A(\varepsilon I+\alpha\gamma)^{-1}$ is m-accretive in $L^\infty(\Omega)$ for sufficiently small $\alpha>0$.

<u>Proof of Theorem 1:</u> Put $\gamma=\beta^{-1}$. Let $f\varepsilon L^\infty(\Omega)$. Take $M>0$ such that $|\alpha f|_\infty\leq M$. By (H3), there exist w_+, $w_-\varepsilon L^\infty(\Omega)$ such that a.e.$x\varepsilon\Omega$, $M\varepsilon\gamma(x,w_+(x))$, $-M\varepsilon\gamma(x,w_-(x))$. Define $\gamma^2:\Omega\times R\to2^R$ by

$$\gamma^2(x,r)=\begin{cases}-M+r-w_-(x) & (r<w_-(x))\\ \gamma(x,r)\cap[-M,M] & (w_-(x)\leq r\leq w_+(x))\\ M+r-w_+(x) & (r>w_+(x)).\end{cases}$$

Then γ^2 satisfies (H2) and (H3). By $A\beta\neq\Phi$, $A(\varepsilon I+\alpha\gamma^2)^{-1}\neq\Phi$ for every ε, $\alpha>0$. By Corollary 6, $A(\varepsilon I+\alpha\gamma^2)^{-1}$ is m-accretive in

$L^{\infty}(\Omega)$ for sufficiently small $\alpha>0$. Thus for every $\mu>0$ and f_1, $f_2 \in L^{\infty}(\Omega)$,

$$|(I+\mu A(\varepsilon I+\alpha\gamma^2)^{-1})^{-1}(\alpha^2 f_1)-(I+\mu A(\varepsilon I+\alpha\gamma^2)^{-1})^{-1}(\alpha^2 f_2)|_{\infty}$$
$$\leq \alpha^2 |f_1-f_2|_{\infty}.$$

Put for $f \in L^{\infty}(\Omega)$, $u_{\varepsilon,\nu}=(I+\mu A(\varepsilon I+\alpha\gamma_{\nu}^2)^{-1})^{-1}(\alpha^2 f)$ and $v_{\varepsilon,\nu}=(\varepsilon I+\alpha\gamma_{\nu}^2)^{-1}u_{\varepsilon,\nu}$. Then $v_{\varepsilon,\nu}+(\mu A_2+\alpha\gamma_{\nu}^2)v_{\varepsilon,\nu} \ni \alpha^2 f$. Using [8], [7], there exist $\{\varepsilon_n\}$, $\{\nu_n\}$, ε_n, $\nu_n>0$, $\varepsilon_n \to 0$, $\nu_n \to 0$ such that

$$\underset{\varepsilon_n\to 0+}{\text{w-lim}}\ \underset{\nu_n\to 0+}{\text{w-lim}}\ u_{\varepsilon_n,\nu_n} = u \text{ in } L^{\infty}(\Omega),$$

$$\lim_{\varepsilon_n\to 0+}\ \lim_{\nu_n\to 0+} v_{\varepsilon_n,\nu_n} = v \text{ in } L^2(\Omega)$$

and $u \in \alpha\gamma v$, $\frac{1}{\alpha}u + \frac{\mu}{\alpha^2}A\beta(\frac{1}{\alpha}u) \ni f$ by the definition of γ^2. By assumption, $f \in R(I+\frac{\mu}{\alpha^2}A\beta)$ and $u=\alpha(I+\frac{\mu}{\alpha^2}A\beta)^{-1}f$. Thus

$$|(I+\frac{\mu}{\alpha^2}A\beta)^{-1}f_1-(I+\frac{\mu}{\alpha^2}A\beta)^{-1}f_2|_{\infty}\leq\alpha|f_1-f_2|_{\infty}\leq|f_1-f_2|_{\infty}$$

and $R(I+\frac{\mu}{\alpha^2}A\beta)=L^{\infty}(\Omega)$. Hence $A\beta$ is m-accretive in $L^{\infty}(\Omega)$.

For the examples see [11] with [1], [6] and [9].

It is expected that detailed proofs of the results announced here, as well as other results on the m-accretiveness of $A\beta$ and the evolution equation associated with $A\beta$ in $L^{\infty}(\Omega)$, will appear elsewhere.

ACKNOWLEDGMENT

This paper is the first part of the research supported by the Korea Science and Engineering Foundation under the title "Evolution equation of type $(du/dt) + A\beta u \ni 0$ in $L^{\infty}(\Omega)$", 1981-1982. The author wishes to express his sincere thank to the Foundation.

REFERENCES

[1] H. Attouch et A. Damlamian, Convexité et monotonie dans certaines équations aux dérivées partielles non linéaires, Préprint.

[2] Ph. Bénilan, Equations d'évolution dans un espace de Banach quelquonque et applications, Thèse, 1972.

[3] Ph. Bénilan, Principe du maximum et perturbation d'opérateurs accrétifs dans $L^1(\Omega)$, Préprint.

[4] Ph. Bénilan, Semi-groups invariantes par un cône de fonctions convexes, Analyse Convexe et ses Applications, Lecture Notes in Economics Math. System 102 (1974).

[5] Ph. Bénilan et K. S. Ha, Equation d'évolution du type $(du/dt)+\beta\partial\phi(u)\ni 0$ dans $L^\infty(\Omega)$, C. R. Acad. Sci. Paris 281 (1974), Série A, 947-950.

[6] H. Brézis, Monotonicity methods in Hilbert spaces and some applications to nonlinear partial differential equations, Contributions to Nonlinear Functional Analysis, Academic Press, New York, 1971, 101-156.

[7] H. Brézis, Monotone operators, nonlinear semi-groups and applications, Proc. Int. Congress Math. Vancouver (1974).

[8] H. Brézis, Opérateurs maximaux monotones et semi-groupes de contractions dans les espaces de Hilbert, Math. Studies 5, North Holland, 1973.

[9] H. Brézis, Problèmes unilateraux, Jour. Math. Pures Appl. 51(1972), 1-168.

[10] M. G. Crandall and T. M. Liggett, Generation of nonlinear transformations on general Banach spaces, Amer. Jour. Math. 93(1971), 265-298.

[11] K. S. Ha, Sur des semi-groupes non linéaires dans les espaces $L^\infty(\Omega)$, Jour. Math. Soc. Jap. 31(1979), 593-622.

ENERGY BOUNDS FOR THE SLOW CAPTURE BY A CENTER*

Richard Haberman

Department of Mathematics
Southern Methodist University
Dallas, Texas

ABSTRACT

Critical trajectories are analyzed for an autonomous nonlinear differential equation perturbed by slowly varying coefficients. The method of matched asymptotic expansions is used to calculate initial conditions for which a previously inaccessible center is captured. The results are significantly simplified using the energy dissipated along the critical trajectories.

INTRODUCTION

In a series of papers [4]-[6], [9], summarized in [7], Kevorkian has studied systems of nonlinear ordinary differential equations with slowly varying coefficients with the intent of understanding the phenomenon of sustained reentry roll resonance. Simplified equations were derived. It was shown that sustained resonance corresponds to the capture of solutions by a previously inaccessible center in the phase plane due to the slow variations. The initial conditions that produced sustained resonance were obtained by numerical integrations of the appropriate differential equation.

Here, we consider a second-order nonlinear differential equation with slowly varying coefficients:

$$\frac{d^2y}{dt^2} = f(y,T) + \varepsilon^{\frac{1}{2}} f_1(y, \frac{dy}{dt}, T) , \tag{1}$$

where $T = \varepsilon^{\frac{1}{2}} t$. This is more general than the problem analyzed by Kevorkian, since our method and results have a wider validity. We

*The details will appear in [2].

assume, as in Kevorkian's model, that the phase plane for the conservative system with $\varepsilon = 0$ has a largest closed trajectory (separatrix) connecting to a saddle point encircling a center as illustrated in Figure 1.

SUMMARY OF RESULTS

In [2] we calculate the perturbed separatrix which results from the slowly varying coefficients. We show, using the method of matched asymptotic expansions, that solutions may be captured by the center. There are two separatrices forming the boundary of the captured set (see Figure 2), one corresponding to the smallest initial energy (E_{min}) and the other to the largest (E_{max}). Time-dependent solutions are qualitatively sketched in Figures 3-5. For t large but εt small, these critical solutions approach the unstable equilibrium nearly following the separatrix corresponding to $\varepsilon = 0$. For certain large times with εt small, the one with largest energy may be approximated by the solitary wave of the conservative system. For larger t such that $\varepsilon t = \mathcal{O}(1)$, these solutions drift with the slowly varying equilibrium (see also [1]). For initial energy between these extremes (Figure 4), solutions are captured by the center: first (t large but εt small) they approach the solitary wave, and then ($\varepsilon t = \mathcal{O}(1)$) they become slowly varying nonlinear oscillators decaying towards the slowly varying equilibrium, requiring Kuzmak's analysis [8].

In [2] we calculate, by the method of matched asymptotic expansions, the initial conditions necessary to produce the extreme solutions. There it is also shown that the initial energy approximately equals the energy dissipated along the critical trajectories. Thus, the details developed by the method of matched asymptotic expansions are not needed for the energy bounds. Instead, we define the slowly varying potential energy g:

$$g(y,T) \equiv - \int_{y_e^*}^{y} f(\bar{y},T)\,d\bar{y} \tag{2}$$

where y_e^* is the "initial" slowly varying equilibrium, $f(y_e^*,0) = 0$, and we define the slowly varying total energy:

$$E(T) \equiv \frac{1}{2}\left(\frac{dy}{dt}\right)^2 + g. \tag{3}$$

In [2], we show that the minimum initial energy bound ($E(0) = E_{min}$) satisfies to leading order

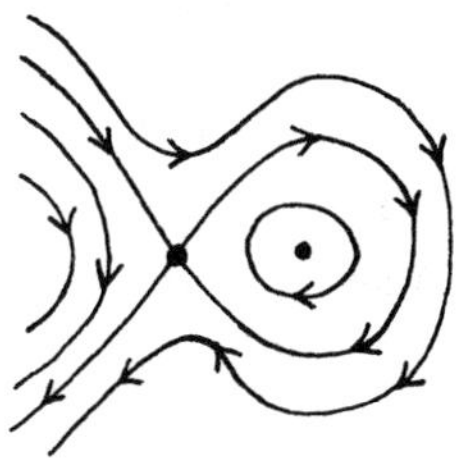

FIG. 1

Autonomous

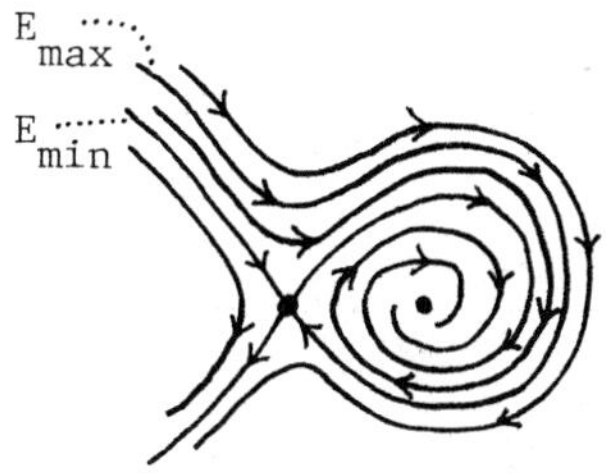

FIG. 2

Slowly Varying

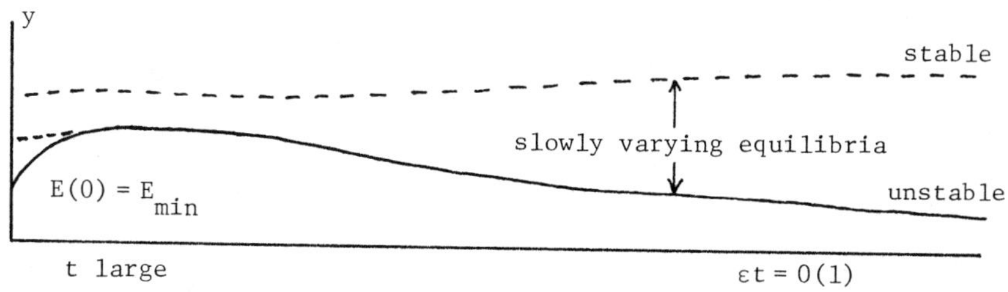

FIG. 3

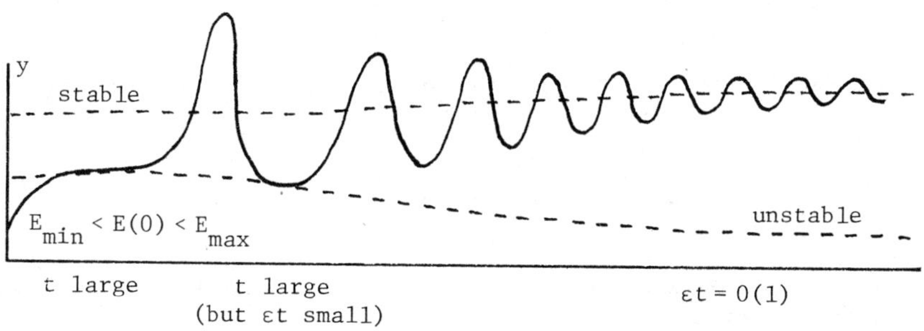

FIG. 4

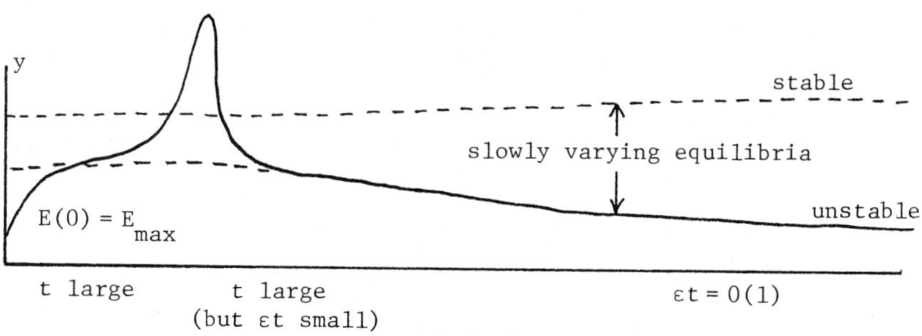

FIG. 5

$$E_{min} \sim -\varepsilon^{\frac{1}{2}} \int_{y_{ic}}^{y_e^*} \left[\frac{\frac{\partial g}{\partial T}(T=0)}{\sqrt{-2g(y,0)}} + f_1(y, \sqrt{-2g(y,0)}, 0 \right] dy , \qquad (4)$$

where $y(0) = y_{ic}$ assuming $y_e^* > y_{ic}$. A similar expression is derived in [2] for $E_{max} - E_{min}$. Solutions are captured by the center if $E_{max} > E_{min}$; see also Kath [3]. Robinson [10] has also obtained these energy bounds by deriving the Melnikov integral for systems with slowly varying coefficients.

REFERENCES

[1] R. Haberman, Slowly varying jump and transition phenomena associated with algebraic bifurcation problems, SIAM J. Appl. Math. 37 (1979), 69-106.

[2] R. Haberman, Energy bounds for the slow capture by a center in sustained resonance, SIAM J. Appl. Math., to appear.

[3] W. L. Kath, Conditions for sustained resonance, II, preprint.

[4] J. Kevorkian, Passage through resonance for a one-dimensional oscillator with slowly varying frequency, SIAM J. Appl. Math. 20 (1971), 364-373.

[5] J. Kevorkian, Errata: Passage through resonance for a one-dimensional oscillator with slowly varying frequency, SIAM J. Appl. Math. 26 (1974), 686.

[6] J. Kevorkian, On a model for reentry roll resonance, SIAM J. Appl. Math. 26 (1974), 638-669.

[7] J. Kevorkian, Passage through resonance, in Singular Perturbations and Asymptotics, R. Meyer and S. Parter, eds., Academic Press: New York, 1980, 191-222.

[8] G. E. Kuzmak, Asymptotic solution of nonlinear second order differential equations with variable coefficients, J. Appl. Math. Mech. 23 (1959), 730-744.

[9] L. Lewin and J. Kevorkian, On the problem of sustained resonance, SIAM J. Appl. Math. 35 (1978), 738-754.

[10] C. Robinson, Sustained resonance for a nonlinear system with slowly varying coefficients, preprint.

AN INTEGRAL EQUATION FOR HELMINTHIC INFECTIONS: STABILITY OF THE NON-INFECTED POPULATION

K. P. Hadeler
Lehrstuhl für Biomathematik
Universität Tübingen
Auf der Morgenstelle
Tübingen, West Germany

SUMMARY

For helminthic infections, where the host carries only a small number of adult parasites, it is appropriate to describe the state of the population by a generating function which for each age class and each number of parasites gives the corresponding fraction of the host population. The generating function satisfies a partial differential equation with integral terms. Global existence of solutions, existence and uniqueness of non-trivial stationary solutions and linear stability of the trivial stationary solution are shown under biologically reasonable assumptions.

In the classical epidemic model (the Kermack-McKendrick model) and its extensions the population is subdivided into several classes such as susceptibles, infectious, or recovered, and differential equations for the size of these classes are derived from a mass action law. Such equations are well-suited to model bacterial or viral infections where the number of parasites per host is very large and is not related to the number of parasites initially acquired. On the contrary, in typical helminthic infections such as in onchocerciasis (river-blindness) the host carries a small number of adult parasites which are acquired at different times and each of which contributes to the host's mortality. Such diseases should be described by semi-stochastic (hybrid) models which take account of the individual parasite load.

Let $n(t,a,r)$ be the number of hosts at time t which carry r parasites and whose age is in $(a,a+da)$. The state of the total population can be described by the generating function

$$u(t,a,z) = \sum_{r=0}^{\infty} n(t,a,r)\, z^r \qquad (1)$$

where z is a formal variable. In [1] a first order partial differential equation

$$u_t + u_a + [\alpha z + \sigma(z-1)]u_z - [(z-1)\varphi(t) - \mu(a)]u = 0 \qquad (2)$$

has been derived for the function u. The subscripts indicate partial derivatives with respect to the variables t,a,z. The function μ describes the age dependent death rate of hosts in the absence of parasites, the parameter α is the differential fertility due to the presence of one parasite, σ is the mortality of parasites within the hosts. The time-dependent function φ describes the rate at which parasites are acquired by the hosts.

In [1] we have assumed that φ is a function of the parasite load of the total population, i. e.

$$\varphi(t) = \beta f(\bar{m}(t)), \qquad (3)$$

where

$$\bar{m}(t) = \frac{\int_0^\infty u_z(t,a,1)\,da}{\int_0^\infty u(t,a,1)\,da}. \qquad (4)$$

Thus, with f and β given, the equation (2) becomes an integro-differential equation for the function u.

Equation (2) has been considered together with an initial condition at time zero

$$u(0,a,z) = u_0(a,z), \quad 0 < a < \infty, \quad 0 \leq z \leq 1 \qquad (5)$$

and a boundary condition at age zero

$$u(t,0,z) = N(t,z), \quad 0 < t < \infty, \quad 0 \leq z \leq 1. \tag{6}$$

In [1] we have chosen the following approach to the existence problem: Suppose the function φ in equation (2) were known. Then equation (2) is just a linear first order equation. Using the side conditions (5)(6), it can be solved in the domains $a > t$ and $a < t$ by the method of characteristics. The explicit representation, which of course depends on the function φ, can be inserted into equation (3). Then one obtains a Volterra integral equation for the function $\varphi = \varphi(t)$,

$$\varphi(t) = (\mathfrak{F}\varphi)(t) \tag{7}$$

where

$$(\mathfrak{F}\varphi)(t) = \beta f\left(\frac{\mathcal{N}}{\mathfrak{D}}\right), \tag{8}$$

$$\mathcal{N} = \int_0^t e^{A\varphi}\left\{\bar{N}_1 + \bar{N}B\varphi\right\} da + e^{C\varphi}\left\{F_1 + FD\varphi\right\}, \tag{9}$$

$$\mathfrak{D} = \int_0^t e^{A\varphi}\,\bar{N}\,da + e^{C\varphi}F. \tag{10}$$

Here the notation is very much condensed, the symbols stand for certain integral operators and functions.

Let

$$\varkappa = \alpha + \sigma, \tag{11}$$

$$L(z,t) = 1 - (1-z)e^{-\varkappa t} - \frac{\alpha}{\varkappa}(1-e^{-\varkappa t}), \tag{12}$$

$$L_1(t) = L(1,t) - 1 = -\frac{\alpha}{\varkappa}(1-e^{-\varkappa t}), \tag{13}$$

$$M(t) = \int_0^t \mu(s)\,ds, \tag{14}$$

$$A\varphi = (A\varphi)(t,a) = \int_{t-a}^t L_1(t-s)\varphi(s)\,ds, \tag{15a}$$

$$B\varphi = (B\varphi)(t,a) = \int_{t-a}^t e^{-\varkappa(t-s)}\varphi(s)\,ds, \tag{15b}$$

$$C\varphi = (C\varphi)(t) = \int_0^t L_1(t-s)\varphi(s)\,ds, \tag{15c}$$

$$D\varphi = (D\varphi)(t) = \int_0^t e^{-\varkappa(t-s)}\varphi(s)\,ds, \tag{15d}$$

$$\bar{N} = \bar{N}(t,a) = e^{-M(a)} N(t-a, L(1,a)) , \tag{16a}$$

$$\bar{N}_1 = \bar{N}_1(t,a) = e^{-M(a)-\varkappa a} N_z(t-a, L(1,a)) , \tag{16b}$$

$$F = F(t) = \int_t^\infty e^{-M(a)+M(a-t)} u_o(a-t, L(1,t))da , \tag{16c}$$

$$F_1 = F_1(t) = \int_t^\infty e^{-M(a)+M(a-t)} u_{oz}(a-t, L(1,t))da \, e^{-\varkappa t} . \tag{16d}$$

In [1] a first local existence result has been obtained under the assumption that u_o, u_{oz} are exponentially decaying functions of a, and that

$$\int_0^\infty u_o(a,1)da > 0. \tag{17}$$

In the following existence result we take advantage of the fact that u is a generating function. In particular we assume that

$$u_o(a,z) = \sum_{r=0}^\infty n_o(a,r)z^r , \tag{18}$$

$$N(t,z) = \sum_{r=0}^\infty N_o(t,r)z^r . \tag{19}$$

are (not necessarily normalized) generating functions.

<u>Theorem 1:</u> Let $f: [0,\infty) \to [0,\infty)$ be continuously differentiable, $f(0) = 0$, $f'(0) = 1$, and $0 < f(u) \leq f_o u$ for $u > 0$, where f_o is a constant. Let $\beta > 0$.
Let $\mu: [0,\infty) \to [0,\infty)$ be continuous, and such that for the function

$$M(a) = \int_0^a \mu(s)ds$$

holds

$$\int_0^\infty \exp\,[-M(a)]\,da < \infty. \tag{20}$$

Let u_o and N be continuously differentiable in both variables, let

$$0 < \int_0^\infty u_o(a,1)da < \infty \qquad (21)$$

and

$$\sup \left\{ \frac{u_{oz}(a,1)}{u_o(a,1)} : u_o(a,1) > 0, \quad 0 \leq a < \infty \right\} < \infty \; . \qquad (22)$$

Then the integral equation (7) has a unique solution for all $t > 0$.

For a proof see [2].

Stable stationary solutions are of great significance.

Stationary solutions of equation (2)(3) can occur if the boundary condition is independent of t,

$$u(t,0,z) = N(z). \qquad (23)$$

If u(a,z) is a stationary solution of (2)(3) then

$$\varphi = \beta f \left(\frac{\int_0^\infty u_z(a,1)da}{\int_0^\infty u(a,1)da} \right) \qquad (24)$$

is a stationary solution of equation (7). For stationary solutions equation (7) assumes the following form. Let

$$q(a) = \frac{1}{\varkappa}(1-e^{-\varkappa a}) = -\frac{1}{\alpha} L_1(a) \qquad (25)$$

$$Q(a) = \int_0^a q(s)ds = \frac{1}{\varkappa}\left(a - \frac{1}{\varkappa}(1 - e^{-\varkappa a})\right) \qquad (26)$$

Then

$$\varphi = \beta f\left(\frac{\int_0^\infty e^{-\alpha Q(a)\varphi - M(a)} \left\{ N_z(L(1,a))e^{-\varkappa a} + N(L(1,a))q(a)\varphi \right\} da}{\int_0^\infty e^{-\alpha Q(a)\varphi - M(a)} N(L(1,a))da} \right). \qquad (27)$$

In the most realistic special case newborns ar not infected. Then N is a positive constant, and $N_z = 0$. Define

$$I_1(\varphi) = \int_0^\infty e^{-\alpha Q(a)\varphi - M(a)} q(a)da , \tag{28}$$

$$I_0(\varphi) = \int_0^\infty e^{-\alpha Q(a)\varphi - M(a)} da , \tag{29}$$

$$W(\varphi) = I_1(\varphi)/I_0(\varphi) . \tag{30}$$

Then the equation for stationary solutions reads

$$\varphi = \beta f(W(\varphi)\varphi) . \tag{31}$$

In view of $f(0) = 0$ there is a trivial solution $\varphi = 0$ for all β. In (31) one can express β as a function of φ; hence there is always a branch of positive stationary solutions. Since $f'(0) = 0$, this branch bifurcates from the trivial solution.

Theorem 2: There is a branch of nontrivial solutions (β,φ) with $\varphi > 0$ starting from $(\beta_0, 0)$, where

$$\beta_0 = W(0)^{-1} . \tag{32}$$

If the function f satisfies

$$u^{-2} f(u) \to 0 \quad \text{for } u \to \infty \tag{33}$$

then the branch exists for all $\beta > \beta_0$.
In general, along the branch β is not a monotone function of φ. However, if the function f is concave in the sense of Krasnoselskij, i. e.

$$\frac{df(u)}{du} \geq 0 , \quad \frac{d}{du}\left(\frac{f(u)}{u}\right) \leq 0 \quad \text{for } u > 0 \tag{34}$$

then for each $\beta > \beta_0$ there is exactly one $\varphi > 0$.

The proof has been given in [1]. The conditions (34) are most natural, e. g. they are satisfied for $f(u) = u$, $f(u) = \log(1+u)$, $f(u) = 1-\exp(-u)$, $f(u) = u/(1+u)$.

The stability analysis of equation (2)(3) can be reduced to a stability analysis of the Volterra equation (7). Variations in the initial datum u_o enter the integral equation via the functions F and F_1 and lead to variations in the function $\varphi = \varphi(t)$.
Let again N be constant and let $\bar{\varphi}$ be a stationary solution. Let $\bar{u}(a,z)$ be the corresponding solution of equation (1),

$$\bar{u}(a,z) = N \cdot \exp[-((1-z)q(a)+\alpha Q(a))\varphi - M(a)], \tag{35}$$

Then equation (7) is satisfied with $u_o = \bar{u}, u_{oz} = \bar{u}_z$, $N = \text{const.}$, $N_z = 0$, $\varphi(t) = \bar{\varphi}$. Let $\bar{u}(a,z) + v_o(a,z)$ be a perturbed initial datum and let $\bar{\varphi} + \psi(t)$ be the corresponding solution of the integral equation (7) (which is uniquely determined).

With some rather tedious calculations [1] one can obtain the linearization of $\mathfrak{F}$ with respect to independent variations in u_o and φ. Then one obtains a linear integral equation for small variations v of the form

$$\psi(t) - (\mathcal{L}\psi)(t) = S(t) \tag{36}$$

where the linear integral operator $\mathcal{L}$ depends only on $\bar{\varphi}$ ($\bar{u}(a,z)$) depends only on φ) and the function S is a linear operator in v and v_z. Of course S depends on $\bar{\varphi}$.

In the special case $\bar{\varphi} = 0$ the integral operator is the convolution operator

$$(\mathcal{L}\psi)(t) = \int_0^t K(t-a)\psi(a)\,da, \tag{37}$$

$$K(t) = \beta \frac{\int_t^\infty e^{-M(s)}ds\; e^{-\varkappa t}}{\int_0^\infty e^{-M(s)}ds} \tag{38}$$

The function S has the general form (see [1])

$$S(t) = \beta f'(W(\bar{\varphi})\bar{\varphi})\frac{1}{\mathfrak{D}^2}(\mathfrak{D}\tilde{\mathcal{N}}-\mathcal{N}\tilde{\mathfrak{D}})$$

where $\mathfrak{D}$, $\mathcal{N}$ are given by (9)(10) with (16c)(16d)(35) and $\tilde{\mathcal{N}}$, $\tilde{\mathfrak{D}}$ are operators in v_o, v_{oz}. In the special case $\varphi = 0$ one has $f'(0) = 1$ and

$$\mathfrak{D} = N\int_0^\infty e^{-M(a)}da,$$

$$\mathcal{N} = 0,$$

$$\tilde{\mathfrak{D}} = \int_t^\infty e^{-M(a)+M(a-t)}v_o(a-t,L(1,t))da,$$

$$\tilde{\mathcal{N}} = \int_t^\infty e^{-M(a)+M(a-t)}e^{-\varkappa t}v_{oz}(a-t,L(1,t))da.$$

In particular $S(t) \equiv 0$ if v_o does not depend on z, as it should be.

In view of the biological interpretation one expects that the bifurcation at $(\beta_o,0)$ is related to a change in stability, i. e. that the stationary solution $\varphi = 0$ is stable up to the bifurcation point β_o. In fact, generalizing a result in [1], one can show the following theorem.

<u>Theorem 3</u>: For $\beta < \beta_o$ the solution of equation (2)(3) which is given by $\varphi = 0$ and (35) is linearly asymptotically stable with respect to perturbations v_o which satisfy

$$\int_0^\infty |v(a,z)|da < \infty, \quad 0 \leq z \leq 1.$$

Proof: In view of $0 \leq K(t) \leq \text{const}\cdot\exp(-\varkappa t)$ the Laplace transform

$$\hat{K}(\lambda) = \int_0^\infty e^{-\lambda t} K(t)\,dt$$

exists for $\operatorname{Re}\lambda > -\varkappa$. By partial integration one arrives at

$$\hat{K}(\lambda) = \beta \frac{\int_0^\infty e^{-M(t)} \frac{1}{\lambda+\varkappa}(1-e^{-(\lambda+\varkappa)t})\,dt}{\int_0^\infty e^{-M(t)}\,dt}.$$

A simple calculation shows

$$\left|\frac{1}{\eta}(1-e^{-\eta t})\right| \leq \frac{1}{|\eta|}(1-e^{-|\eta|t}), \quad \operatorname{Re}\eta \geq 0.$$

Choose $\varepsilon \in (0,\varkappa)$. Then, for $\operatorname{Re}\lambda \geq -\varepsilon$,

$$\left|\frac{1}{\lambda+\varkappa}(1-e^{-(\lambda+\varkappa)t})\right| \leq \frac{1}{\varkappa-\varepsilon}(1-e^{-(\varkappa-\varepsilon)t})$$

and

$$|\hat{K}(\lambda)| \leq \beta \frac{\int_0^\infty e^{-M(t)} \frac{1}{\varkappa-\varepsilon}(1-e^{-(\varkappa-\varepsilon)t})\,dt}{\int_0^\infty e^{-M(t)}\,dt}, \quad \operatorname{Re}\lambda \geq -\varepsilon.$$

The right hand side depends continuously on β and on ε. For $\beta = \beta_0$ and $\varepsilon = 0$ it is equal to 1. For any $\beta < \beta_0$, one can choose $\varepsilon = \varepsilon(\beta) > 0$ sufficiently small such that the right hand side is less than 1.

Let the resolvent equation be given by

$$R(t) = K(t) + \int_0^t K(t-s)R(s)\,ds.$$

The transform $\hat{R}(\lambda) = \hat{K}(\lambda)/(1-\hat{K}(\lambda))$ exists for $\operatorname{Re}\lambda \geq -\varepsilon$. Thus the resolvent is exponentially bounded,

$$|R(t)| \leq c \cdot e^{-\varepsilon t}$$

with some constant c. The function ψ has a representation

$$\psi(t) = S(t) + \int_0^\infty R(t-s)S(s)\,ds$$

$$|\psi(t)| \leq |S(t)| + c \int_0^t e^{-\varepsilon(t-s)} |S(s)| ds$$

Thus from $S(t) \to 0$ it follows that $\psi(t) \to 0$ for $t \to \infty$.

REFERENCES

[1] K.P.Hadeler, K.Dietz, Nonlinear hyperbolic partial differential equations for the dynamics of parasite populations. Computers and Applied Mathematics: An International Journal (Special volume, Editor M.Witten) Pergamon Press 1982/83.

[2] K.P.Hadeler, Nonlinear differential equations from biology. In: John, Kufner (Editors): Nonlinear Analysis, Function Spaces and Applications. Teubner Texte zur Mathematik Vol.49, Leipzig 1982.

[3] K.P.Hadeler, K.Dietz, An integral equation for helminthic infections: Global existence of solutions.
In: Conference Proceedings "Recent Trends in Mathematics" Teubner Texte zur Mathematik Vol.50, Leipzig 1982/83.

ASYMPTOTIC ANALYSIS OF RUDIMENTARY MODELS OF TOXICANT-POPULATION INTERACTIONS

Thomas G. Hallam

Department of Mathematics and Program in Ecology
University of Tennessee
Knoxville, Tennessee

ABSTRACT

Asymptotic behaviors of two models of a population-toxicant interaction, distinguished by the manners in which the environment toxicant concentrations are represented, are discussed. One model considers environmental concentration to be a function of population density only. Multiple equilibria that are asymptotically stable are demonstrated to be feasible behavior modes for this model. The second model focuses on effects of first order kinetics and a linear dose-response function. Persistence and extinction results are found for the cases of chronic and acute toxicant exposures.

INTRODUCTION

The fate, effects, and risks associated with the exposure of a population to a toxicant have been studied by mathematical models (e.g. [1], [7], [8]). There are many processes involved in the stress situation caused by a toxicant. Consequently, these interactions are not well understood and predictive capability is not presently at tolerable levels. Such difficulties are not surprising considering that the range of ecotoxicological effects covers the spectrum from the physiological to the biospherical levels.

Many past works have been devoted to simulation studies of ecotoxicological models but few have been analytical in character. The purpose of this note is to synthesize aspects of some recent investigations that employ analytical techniques to study toxicant-population interactions. It is tacitly assumed that the only source of toxicant is the environmental pathway. While the food chain pathway is important for some chemicals, there are many chemicals for which it is insignificant [9].

An objective of models that predict the fate of a toxic substance is to determine the concentration of toxicant in an environment. The following models specify the role of the environmental concentration of toxicant and then examine the effects (that is, any change in the nominal operating behavior of the system) of the toxicant on a population. Both models employ as state variables the population density, $x = x(t)$, and the concentration of toxicant in the organism at time t, $C_0 = C_0(t)$. The concentration of toxicant in the environment, C_E, is postulated to be a density dependent parameter in the first model. In the second model, the role of first order kinetics in toxicant dynamics is explored in conjunction with the view of C_E as a state variable.

DENSITY DEPENDENT ENVIRONMENTAL CONCENTRATION

Many classical population models, such as the logistic equation, ultimately allow only a single population size. When toxicant is present, it is possible for these models to possess multiple attracting states. A mechanism for generating multiple equilibria is indicated in this section.

The environmental concentration of toxicant, C_E, is regarded as a prescribed function of the population density, x. An example where this situation might be realized occurs for chemicals that are very persistent in an environment.

The model is

$$\frac{dx}{dt} = x[r_0 - r_1(C_0) - fx] \tag{1.a}$$

$$\frac{dC_0}{dt} = kC_E - gC_0 \tag{1.b}$$

where r_0, f, k, and g are positive parameters. The function r_1 in the dose-response term is assumed to be nonnegative for $0 \leq C_0 \leq 1$. The dynamics of the population, (1.a), are based upon a logistic equation with the toxic effects reflected in the birth, b, and death, d, rates by the dose-response function, $b - d = r_0 - r_1(C_0)$. The equation for the body burden, C_0, is determined by first order kinetics except for the hypothesis that the concentration of toxicant in the environment is a function of population density.

A traditional form for C_E is a sigmoid function. With such a form for C_E, system (1) can be analyzed by employing an isocline analysis. The resulting behavior depends upon the form of the dose-response function. The simplest form for which there is some biological support is the linear dose-response; even this representation can lead to behavior that cannot occur in the absence of the toxicant (as there can exist two equilibria of (1) that are locally asymptotally stable). As may be seen by sketching the isoclines, more general nonlinear dose-response curves of sigmoid type can result in multiple equilibria that are asymptotically stable. Additional details may be found in [5].

FIRST ORDER KINETICS

A model which assumes that the environmental concentration of toxicant is a state variable is formulated here. The population is again governed by the logistic equation

$$\frac{dx}{dt} = x(b - d - fx) \qquad (2.a).$$

The concentration of the toxicant in the organism is required to satisfy

$$\frac{dC_0}{dt} = kC_E - gC_0 - mC_0 \qquad (2.b).$$

The first two terms on the right in (2.b) are the first order kinetic representation for net uptake of toxicant by the organisms. The last term represents depuration of toxicant by the organism.

A differential equation for C_E can be derived by employing total toxicant as a variable. The resulting equation is

$$\frac{dC_E}{dt} = -k_1C_Ex + g_1C_0x - hC_E + u \qquad (2.c)$$

where the first two terms on the right in (2.c) represent net toxicant uptake by the population; the third term represents the totality of losses of toxicant from the environment; and the last term represents input of toxicant into the environment.

The coupling of the population and the toxicant is by a dose-response representation of the net growth rate b - d. It is required that $b - d = r_0 - r_1C_0$; that is, b - d is a linear function of body burden. A final hypothesis concerns the allowable exogeneous input u; it is assumed that there exist constants V_1 and V_2 such that $0 \le V_1 \le u(t) \le V_2$ for all $t \in [0, \infty)$.

It is readily established by using differential inequalities [6] that all populations, x = x(t), as modelled by (2.a) are bounded by $X_0 \equiv \max[x(0), r_0/f]$.

Studies of persistence or extinction often focus on identifying appropriate system level parameters that are functions of the model parameters [2], [3]. The next result for toxicant-population interactions is of the same character in that a single parameter is found which is sufficient for persistence or extinction.

Theorem 1. The population as represented by model (2) is persistent, that is, $\limsup_{t\to\infty} x(t) > 0$ for any solution x of (2.a), provided $\rho_V = r_0 - \frac{r_1 k g_1 X_0}{h(g + m)} - \frac{r_1 k V_1}{h(g+m)} > 0$. The population goes to extinction provided $\rho_L = r_0 - \frac{r_1 k V_0}{(h + k_1 X_0)(g + m)} < 0$.

Indication of the Proof. An integral equation representation of (1) is obtained by using the linearity of equations (2.b) and (2.c), and the integrability of the logistic equation. The resulting equation is of the form

$$x(t) = \frac{x(0)\ H(t)}{1 + x(0) f \int_0^t H(s) ds} \tag{3}$$

where $H(t) = \exp[r_0 t - \int_0^t r_1 C_0(s) ds]$, and

$$C_0(t) = ke^{-(g + m)t} \int_0^t C_E(0)\{\exp[-hs - \int_0^s k_1 x(s_1) ds_1]$$

$$+ \exp[-hs - \int_0^s k_1 x(s_1) ds_1] \int_0^s \exp[hs_2 + \int_0^{s_2} k_1 x(s_3) ds_3] \cdot$$

$$\cdot [g_1 C_0(s_2) + u(s_2)] ds_2\} e^{(g + m)s} ds.$$

Estimates can be found in (3) to establish $\liminf_{t\to\infty} fx(t) \geq \rho_L$ and $x(t) \leq x(0) \exp(\rho_V t)$. These bounds imply the conclusion of Theorem 1.

Theorem 1 requires that the population be exposed to a chronic dose of toxicant for extinction to occur. The next result demonstrates that extinction of some populations and persistence of others can occur simultaneously, at least in a special case of (2), when only an acute dose of toxicant is administered.

Theorem 2. In model 2, suppose that $g_1 = 0$, $h = 0$, $u = 0$, $x(0) > 0$, $C_0(0) = 0$, and $C_E(0) > 0$. Then, there exist solutions of (2) where the population component goes to extinction and there exist solutions of (2) that represent persistent populations with the behavior $\lim_{t\to\infty} x(t) = r_0/f$.

The existence of persistent populations can be demonstrated by a linearization of (2) about the equilibrium (r_0/f, 0, 0). The extinction result requires finding an appropriate integral operator and demonstrating the existence of a fixed point by employing a modification of the classical Schauder-Tychonoff Theorem. The details of the proof may be found in [4].

ACKNOWLEDGMENT

This research was supported in part by the U.S. Environmental Protection Agency under cooperative agreement R806727030.

REFERENCES

[1] S. Bartell, SCUM, Oak Ridge National Laboratory, personal communication.

[2] T.C. Gard, Top Predator Persistence in Differential Equation Models of Food Chains: The Effects of Omnivory and External Forcing of Lower Trophic Levels. J. Math. Biology 14(1982), 285-299.

[3] T.C. Gard and T.G. Hallam, Persistence in Food Webs I. Lotka-Volterra Food Chains. Bull. Math. Biol. 41(1979), 877-891.

[4] T.G. Hallam, C.E. Clark, and G.S. Jordan, Effects of Toxicants on Populations: A Qualitative Approach II. First Order Kinetics, to appear.

[5] T.G. Hallam, C.E. Clark, and R.R. Lassiter, Effects of Toxicants on Populations: A Qualitative Approach I. Equilibrium Environmental Exposure. Ecolog. Modelling, in proof.

[6] V. Lakshmikantham and S. Leela, Differential and Integral Inequalities, Vol. 1. New York: Academic Press (1969).

[7] R.R. Lassiter, G.L. Baughman, and L.A. Burns, Fate of Toxic Organic Substances in the Aquatic Environment. In: State-of-the-Art in Ecological Modelling (S.E. Jorgensen, ed.) Int. Soc. Ecol. Modelling. Copenhagen. Vol. 7 (1978), 219-245.

[8] R.A. Park, C.I. Connolly, J.R. Albanese, L.S. Clesceri, G.W. Heitzman, H.H. Herbrandson, B.H. Indyke, J.R. Loche, S. Ross, D.D. Sharma and W.W. Shuster, Modeling Transport and Behavior of Pesticides and Other Toxic Organic Materials in Aquatic Environments. Report No. 7, Center for Ecological Modeling, Rensselaer Polytechnic Institute, Troy, New York (1980).

[9] D.M. Rosenberg. Food Chain Concentration of Chlorinated Hydrocarbon Pesticides in Invertebrate Communities: a re-evaluation. Quaestiones Entomologicae 11(1975), 97-110.

ON THE ASYMPTOTIC STABILITY BY LYAPUNOV FUNCTION WITH SEMIDEFINITE DERIVATIVE

L. Hatvani

Bolyai Institute
Aradi vértanúk t. 1
Szeged, Hungary

ABSTRACT

Sufficient conditions for the asymptotic stability and partial asymptotic stability of the zero solution of a differential system are given. In the nonautonomous case the derivative is assumed to be estimated by the product of two semidefinite functions depending only the time and the phase variable, respectively. By the help of the method of limiting equation a sufficient condition for the partial asymptotic stability of the zero solution of autonomous system is found without requiring boundedness of the uncontrolled coordinates. The results are applied to the study of stability properties of equilibria in mechanical systems being under the action of dissipative forces.

1. INTRODUCTION

In classical theorems of Lyapunov's direct method the asymptotic stability of the zero solution of a differential system is guaranteed by the existence of an appropriate Lyapunov function with negative definite derivative. One finds, however, that in numerous applications it is very complicated to construct such a function. For example, the derivative of the total mechanical energy of a holonomic mechanical system under the action of dissipative forces is negative definite with respect to velocities only, thus it cannot be used in the basic theorems to establish asymptotic stability with respect to the generalized coordinates. We are going to give some generalizations of the basic theorems from which one can derive sufficient conditions for the asymptotic stability or partial asymptotic stability with respect to the generalized coordinates as well.

Consider the differential system

$$\dot{y} = Y(y,z,t), \quad \dot{z} = Z(y,z,t), \tag{1.1}$$

where $t \in R_+ = [0,\infty)$, $y \in R^m$, $z \in R^n$ $(m \geq 1,\ n \geq 0)$, the norm of y is defined by

$|y| = (y_1^2+\ldots+y_m^2)^{1/2}$. Denote by $B_k(r)$ the open ball in R^k with center at the origin and radius $r>0$. Assume that functions Y, Z are defined on the set $D_y = G_y \times R_+$ $(G_y = B_m(h)\times R^n;\ 0<h\leqq\infty)$, they are continuous in $x = (y,z)$, measurable in t, and satisfy the Carathéodory condition locally. We denote by $x(t) = x(t;x_o,t_o)$ any solution of (1.1) with $x(t_o) = x_o$, and always assume that solutions are z-continuable, i.e. if $x(t) = (y(t),z(t))$ is any solutions of (1.1) and $|y(t)|\leqq h'<h$ for $[t_o,T)$, then $x(t)$ can be continued to the segment $[t_o,T]$. Let $Y(0,0,t) \equiv 0$, $Z(0,0,t) \equiv 0$, i.e. let $x = 0$ be a solution of (1.1).

The zero solution of (1.1) is said to be:

1) y-stable if for every $\varepsilon>0$, $t_o\in R_+$ there exists a $\delta(\varepsilon,t_o)>0$ such that $|x_o|<\delta(\varepsilon,t_o)$ implies $|y(t;x_o,t_o)|<\varepsilon$ for $t\geqq t_o$; 2) asymptotically y-stable if it is y-stable and for every $t_o\in R_+$ there exists a $\sigma(t_o)>0$ such that $|x_o|< <\sigma(t_o)$ implies $|y(t;x_o,t_o)|\to 0$ as $t\to\infty$; 3) equi-asymptotically y-stable if it is asymptotically y-stable and $|y(t;x_o,t_o)|\to 0$ as $t\to\infty$ uniformly in $x_o\in B_{m+n}(\sigma(t_o))$ (see [1]).

We say that a function $a:R_+\to R_+$ belongs to the class K if it is continuous, strictly increasing and $a(0) = 0$. A locally Lipschitzian function $V:D'_y\to R_+$, $D'_y = G'_y\times R_+$ $(G'_y = B_m(h')\times R^n;\ 0<h'<h)$ is a Lyapunov function of (1.1) with respect to y if $V(0,t) \equiv 0$, V is positive y-definite (i.e. there exists an $a\in K$ such that $V(y,z,t)\geqq a(|y|)$ in the set D'_y) and $\dot V(x,t)\leqq 0$, where $\dot V$ denotes the derivative of V with respect to (1.1) (see [1]).

2. NONAUTONOMOUS SYSTEM

A measurable function $e:R_+\to R_-$ is said to be integrally negative [2] if

$$\left(I = \bigcup_{i=1}^{\infty} [a_i,b_i],\ a_i<b_i<a_{i+1},\ b_i-a_i\geqq c>0\right)\Rightarrow \int_I e = -\infty.$$

If in this implication also $a_{i+1}-b_i\leqq C$ is supposed on the set I, then e is called weakly integrally negative. Obviously, function $-\sin^2 t$ is integrally negative; $-1/t$, $-(1/t)\sin^2 t$ are weakly integrally negative but not integrally negative. Finally, for a $K\subset R^m$ introduce the notation $S(K,r) = \{y\in R^m:\mathrm{dist}\ (y,K)<r\}$.

Combining a generalization of LaSalle's invariance principle [3] with the method of several Lyapunov functions [2] we can prove the following

Theorem 2.1: Suppose that there exist locally Lipschitzian functions $V,A:D'_y\to R$ satisfying the following conditions:

(i) there are functions $a,b\in K$ such that

$$a(|y|)\leq V(x,t)\leq b(|y|);$$

(ii) $\dot{V}(x,t)\leq e(t)U(y)$, where $e:R_+\to R_-$, $U:B_m(h')\to R_+$ are continuous, e is weakly integrally negative;

(iii) if F denotes the zero set of U, then for every compact set $K\subset B_m(h')\setminus F$ there is an $r>0$ so that $\overline{S(K,r)}\cap F = \emptyset$ and the function $\int_0^t Y(u(s),s)ds$ is uniformly continuous on $(0,\infty)$ for each continuous function $u:R_+\to S(K,r)\times R^n$;

(iv) for every c_1,c_2 $(0<c_1<c_2<h')$ there exist $d_1,d_2,d_3>0$ such that $|A(x,t)|<d_1$, $\dot{A}(x,t)\geq d_2$ on the set

$$\{(y,z,t): c_1\leq|y|\leq c_2,\quad y\in\overline{S(F,d_3)},\quad z\in R^n,\quad t\in R_+\}.$$

Then the zero solution of (1.1) is equi-asymptotically y-stable.

In [4] the reader can find the proof and some modifications of this theorem. Namely: a) if e is integrally negative then (iv) can be considerably weakened; b) in condition (iii) the right-hand side Y of (1.1) can be replaced by the derivative $\dot{W}$ of an appropriate vector function W, possibly different from y.

As an example, consider the differential equation

$$\ddot{x}+a(t)\dot{x}+q(t)f(x) = 0, \qquad (x\in R) \tag{2.2}$$

where $q(t)>0$,

$$xf(x)>0 \ (x\neq 0);\ F(x) = \int_0^x f(s)ds\to\infty \ (|x|\to\infty).$$

Introducing the variables $x_1 = x$, $x_2 = \dot{x}$ and choosing the functions

$$V(x_1,x_2,t) = x_2^2/q(t)+2F(x_1),\ A(x_1,x_2) = -x_2f(x_1),$$

one can derive from Theorem 2.1 the following assertion: if $0<c_1\leq q(t)\leq C_1$, $|a(t)|\leq C_2$ for $t\in R_+$, and the function $-(\dot{q}+2aq)$ is weakly integrally negative, then the trivial solution $x = \dot{x} = 0$ is equi-asymptotically stable. This sharpens and generalizes a theorem of R.J. Ballieu and K. Peiffer [5] which reads as follows: the origin $(x,\dot{x}) = (0,0)$ is a global attractor for the differential equation $\ddot{x}+a(t)\dot{x}+f(x) = 0$ provided that $a:R_+\to R_+$ is bounded, non-increasing and $\int^\infty a = \infty$.

3. AUTONOMOUS SYSTEMS

Let $V:G'_y\to R$ be a Lyapunov function of the autonomous system

$$\dot{y} = Y(y,z),\quad \dot{z} = Z(y,z) \tag{3.1}$$

with respect to y. Following LaSalle's notation [3], for $c\in R$ we denote by

$V_m^{-1}[c,\infty]$ the set of the points $y \in R^m$ for which there is a sequence $\{(y_i,z_i)\}$ such that $y_i \to y$, $|z_i| \to \infty$, $V(y_i,z_i) \to c$ as $i \to \infty$. $V_m^{-1}[c,\infty]_o$ denotes the set of those points $y \in V_m^{-1}[c,\infty]$ for which the sequence $\{(y_i,z_i)\}$ satisfies, in addition, the condition $\dot{V}(y_i,z_i) \to 0$ as $i \to \infty$. If $W:G_y' \to R$, then $W^{-1}(c)$ denotes the inverse under W of c as usual.

Suppose that

(*) for every $c>0$ if the set $\dot{V}^{-1}(0) \cap V^{-1}(c)$ contains a complete trajectory of (3.1) then this trajectory is contained in the set $\{(y,z): y = 0\}$.

If the vector z is empty (i.e. $n = 0$ and all coordinates are controlled) then by LaSalle's invariance principle condition (*) implies asymptotic stability of the zero solution. However, in the case $n \geq 1$ one can conclude asymptotic y-stability of the zero solution only with knowledge of the boundedness of all uncontrolled z-coordinates along every motion [6, 7]. Now we are going to replace this boundedness condition by such one which can be checked directly i.e. without a priori knowledge of the solutions. While in the first theorem it is the Lyapunov function V on which additional condition is imposed, in the second one system (3.1) is restricted.

Theorem 3.1: Let $V:G_y' \to R$ be a Lyapunov function of (3.1) with respect to y, and let (*) be satisfied. If for every $c>0$ we have $V_m^{-1}[c,\infty] \subset \{0\}$, then the zero solution of (3.1) is asymptotically y-stable.

Theorem 3.2: Let condition (*) be satisfied for the Lyapunov function V. Suppose, in addition, that $Y(y,z) \to Y_*(y)$ uniformly in $y \in B_m(h')$ as $|z| \to \infty$, and the function Z in (3.1) is bounded in G_y'. If for every $c>0$ the set $V_m^{-1}[c,\infty]_o$ contains no complete trajectory of the limiting system $\dot{y} = Y_*(y)$ except the origin of R^m, then the zero solution of (3.1) is asymptotically y-stable.

4. APPLICATIONS TO MECHANICAL SYSTEMS

Let us now consider a holonomic mechanical system of r degrees of freedom described by the Lagrangian equation

$$\frac{d}{dt}\frac{\partial T}{\partial \dot{q}} - \frac{\partial T}{\partial q} = -\frac{\partial P}{\partial q} + Q, \tag{4.1}$$

where the following notations are used: the column vectors $q, \dot{q} \in R^r$ consist of generalized coordinates and velocities, respectively (v^T denotes the transposed of $v \in R^r$); $P: q \to P(q) \in R$ is the potential energy ($P(0) = 0$);

$T = T(q,\dot{q}) = (1/2)\dot{q}^T A(q)\dot{q}$ is the kinetic energy, where $A: q \to A(q) \in R^{r\times r}$ is symmetric; $Q: (q,\dot{q},t) \to Q(q,\dot{q},t) \in R^r$ denotes the resultant of gyroscopic and dissipative forces, i.e.

$$Q^T(q,\dot{q},t)\dot{q} \leqq e(t)c(|\dot{q}|) \qquad (t \in R_+,\ q,\dot{q} \in R^r)$$

with some $e: R_+ \to R_-$ and $c \in K$. Assume that $q = \dot{q} = 0$ is an equilibrium of (4.1).

L. Salvadori [9] proved that the equilibrium is asymptotically stable in the presence of time-indepenent dissipative and gyroscopic forces ($Q = Q(q,\dot{q})$) provided that the equilibrium is isolated, the potential energy has a minimum at $q = 0$ and the dissipation is complete (i.e. $e(t) = $ = const.<0). Applying our results we can give sufficient conditions for partial asymptotic stability of the equilibrium at $q = 0$ possibly non--isolated also in the case of time-dependent dissipative and gyroscopic forces [4, 8].

Let $q = \text{col}\ (q_1, q_2)$ be a partition of the vector of the generalized coordinates, where $q_1 \in R^s$, $q_2 \in R^{r-s}$ ($0 \leqq s \leqq r$; if $s = 0$ then $q_2 = q$, and the conditions and statements concerning q_1 are to be dropped). The following theorem concerns the stationary case ($Q = Q(q,\dot{q})$, $e(t) \equiv -1$).

<u>Theorem 4.1:</u> If (i) P is positive q_1-definite;

(ii) the set $\{q: q_1 \neq 0\}$ contains no equilibria;

(iii) for every $q_1^o \neq 0$

$$\lim_{q_1 \to q_1^o,\ |q_2| \to \infty} P(q_1, q_2) = \infty,$$

then the equilibrium $q = \dot{q} = 0$ is asymptotically q_1-stable.

We say that (4.1) is <u>asymptotically q_2-independent</u> if for some constant $h'>0$ and for every compact set $K \subset R^r$

(a) there is a $\lambda>0$ such that $T(q,\dot{q}) \geqq \lambda|\dot{q}|^2$ for all $q_1 \in \overline{B_s(h')}$, $q_2 \in R^{r-s}$, $\dot{q} \in R^r$;

(b) $A(q_1,q_2) \to A_*(q_1)$, $P(q_1,q_2) \to P_*(q_1)$ as $|q_2| \to \infty$; in addition, $Q(q_1,q_2,\dot{q}) \to Q_*(q_1,\dot{q})$ uniformly in $q_1 \in \overline{B_s(h')}$, $\dot{q} \in K$ as $|q_2| \to \infty$, as well as $\partial A/\partial q$, $\partial P/\partial q$ converge uniformly in $q_1 \in \overline{B_s(h')}$ as $|q_2| \to \infty$.

<u>Theorem 4.2:</u> Suppose that (4.1) is asymptotically q_2-independent. If (i) P is positive q_1-definite; (ii) (4.1) has no equilibrium position in the region $\{(q_1,q_2): P(q_1,q_2)>0,\ q_1 \neq 0\}$, and (iii) the equality $\partial P_*(q_1)/\partial q_1 = 0$ implies either $q_1 = 0$ or $P_*(q_1) = 0$, then the equilibrium $q = \dot{q} = 0$ of

(4.1) is asymptotically $(q_1,\dot{q})$-stable.

We can derive these results from the general theorems as follows. By the aid of the Hamiltonian variables q, $p = A(q)\dot{q}$ system (4.1) can be rewritten into the normal form (3.1). As is known, the derivative of the total mechanical energy H = T+P along motions is

$$\dot{H}(q,p) = Q^T(q,A^{-1}(q)p)A^{-1}(q)p \leq -d(|p|)$$

with some $d \in K$. Consequently, for arbitrary $c \in R$ we have

$$\dot{H}^{-1}(0) \cap H^{-1}(c) = \{(q,p): P(q) = c,\ p = 0\},$$

so the trajectories contained in this set are the equilibria $q = q^0$, $p = 0$ for which $P(q^0) = c$. By assumption (ii) condition (*) is satisfied. Now Theorem 4.1 is a simple corollary of Theorem 3.1. To deduce Theorem 4.2 from Theorem 3.2 we have to observe

$$H^{-1}_{s+r}[c,\infty]_0 \subset \{(q_1,p): P_*(q_1) = c,\quad p = 0\},$$

and the trajectories of the limiting equation contained in this set are of the form $q_1 = q_1^0$, $p = 0$ with

$$P_*(q_1^0) = c, \quad \frac{\partial P_*}{\partial q_1}(q_1^0) = 0.$$

To illustrate these results consider a material point moving on a surface in a constant field of gravity in the inertial frame Oxyz (Oz directed vertically upward) and subject to viscous friction. In the cases of the surfaces $z = y^2(1+x^2)$ and $z = y^2(1+1/(1+x^2))$ the equilibrium position $y = z = 0$ is y-stable.

Finally, applying Theorem 2.1 to the instationary case $Q = Q(q,\dot{q},t)$ we obtain the following sufficient condition for the asymptotic $(q_1,\dot{q})$--stability of the equilibrium $q = \dot{q} = 0$.

Theorem 4.3: Suppose that for every compact set $K_1 \subset R^s$ the following conditions are satisfied for all $(q_1,q_2)\in K_1\times R^{r-s}$:

(i) there exist functions $a_1,b_1\in K$ such that $a_1(|q_1|)\leq P(q_1,q_2)\leq \leq b_1(|q_1|)$;

(ii) for every α_1,α_2 $(0<\alpha_1<\alpha_2)$ there exists an $\eta>0$ such that $\alpha_1\leq|q_1|\leq\alpha_2$ implies $|\text{grad } P(q_1,q_2)|\geq\eta$;

(iii) there are positive constants λ,Λ so that $\lambda|\dot{q}|^2\leq T(q,\dot{q})\leq\Lambda|\dot{q}|^2$ for all $\dot{q}\in R^r$;

(iv) all the first partial derivatives of $A^{-1}(q)$ and all the first and second partial derivatives of P(q) are bounded;

(v) function e is weakly integrally negative;

(vi) for every compact set $K \subset R^r$ function Q is bounded on the set $K_1 \times R^{r-s} \times K \times R_+$.

Then the equilibrium $q = \dot{q} = 0$ is equi-asymptotically $(q_1, \dot{q})$-stable.

In the special case when all the generalized coordinates are controlled (i.e. $q_1 = q$, $s = r$) Theorem 4.3 gives the following generalization of Salvadori's result to nonstationary system: if the potential energy has a minimum at the isolated equilibrium position $q = 0$, function e is weakly integrally negative and $Q(q, \dot{q}, t)$ is bounded for all t when $(q, \dot{q})$ belongs to an arbitrary compact set, then the equilibrium $q = \dot{q} = 0$ is asymptotically stable.

REFERENCES

[1] V. Lakshmikantham and S. Leela, Differential and Integral Inequalities, Theory and Applications, Academic Press, New York-London, 1969.

[2] V.M. Matrosov, On the stability of motion (in Russian), Prikl. Mat. Meh., 26(1962), 885-895.

[3] J.P. LaSalle, Stability of nonautonomous systems, Nonlinear Anal., 1(1976), 83-91.

[4] L. Hatvani, A generalization of the Barbashin-Krasovskij theorems to the partial stability in nonautonomous systems, Colloquia Mathematica Societatis János Bolyai, 30. Qualitative Theory of Differential Equations, Szeged (Hungary), 1979, 381-409.

[5] R.J. Ballieu and K. Peiffer, Attractivity of the origin for the equation $\ddot{x}+f(t,x,\dot{x})|\dot{x}|^{\alpha}\dot{x}+g(x) = 0$, J. Math. Anal. Appl., 65(1978), 321-332.

[6] C. Risito, Sulla stabilità assintotica parziale, Ann. Mat. Pura. Appl. (4), 84(1970), 279-292.

[7] A.C. Oziraner, On the asymptotic stability and instability with respect to a part of variables (in Russian), Prikl. Mat. Meh., 37(1973), 659-665.

[8] L. Hatvani, On partial asymptotic stability and instability I-II, Acta Sci. Math. (Szeged) (to appear).

[9] L. Salvadori, Sull' estenzione ai sistemi dissipativi del criterio di stabilitá del Routh, Ricerche Mat., 15(1966), 162-167.

ERGODIC THEOREMS FOR NONLINEAR VOLTERRA INTEGRAL EQUATIONS

Douglas S. Hulbert and Simeon Reich

Department of Mathematics
The University of Southern California
Los Angeles, California

ABSTRACT

We present two mean ergodic theorems for the solutions of nonlinear Volterra integral equations in Hilbert space.

Our purpose in this paper is to present two mean ergodic theorems for the solutions of nonlinear Volterra integral equations in Hilbert space (Theorems 1 and 2). The first theorem deals with weak almost convergence and the second with strong almost convergence (see the definitions below). These results improve upon those of Baillon and Clément [1] and include the known results for nonlinear semigroups [7]. In addition, our results hold for a broader class of integral equations than the one treated in [1]. It is important to establish almost convergence because it then follows that the solution is summed by every strongly regular kernel (Corollaries 2 and 3). A criterion for weak convergence is also obtained (Corollary 1). These results can be applied, for example, to a problem of nonlinear heat flow in materials with memory [1].

Consider the nonlinear Volterra integral equation

(1) $\quad u(t) + b*Au(t) \ni F(t) \quad , \quad 0 \leq t < \infty,$

where A is a closed accretive operator in a real Hilbert space H, F maps R^+ into H, the kernel b is real-valued, and the convolution $b*w(t) = \int_0^t b(t-s)w(s)ds$. In previous papers A was always assumed to be maximal monotone. We shall assume instead that A satisfies the following weaker cone condition: there is a closed convex cone C that contains the domain $D(A)$ of A and is contained in the range $R(I+rA)$ of $I+rA$ for all positive r. In addition, we shall make the following standard assumptions on the kernel b and the function F: for each $T > 0$ $b:[0,T] \to R$ is absolutely continuous with a derivative that is of essentially bounded variation, $F \in W^{1,1}(0,T;H)$ with $f(t) = F'(t)$ in C for almost all t, $b(0) = 1$, and $x = F(0) \in cl(D(A))$, the closure of the domain of A. (The case $b(0) > 0$ can be reduced to the case $b(0) = 1$.) Recall that the kernel b is said to be completely positive [3] on $[0,T]$ if there is a non-negative and non-increasing $k \in L^1(0,T;R)$ such that $b(t) + k*b(t) = 1$, $0 \leq t \leq T$. (This equation actually defines $k \in L^1(0,T)$.) Since $b' \in BV([0,T];R)$, $k \in BV([0,T];R)$ too. Throughout this paper we also assume that b is completely positive on $[0,T]$ for all $T > 0$.

Consider now the initial value problem

$$(2) \quad \begin{cases} u'(t)+(A+k(0)I)u(t) \ni g(t)+k(t)x-u*k'(t), & 0 \leq t < \infty, \\ u(0) = x, \end{cases}$$

where I denotes the identity operator, $g(t) = f(t) + k*f(t)$, and $u*k'(t)$ denotes $\int_0^t u(t-s)dk(s)$. A strong solution to (1) on $[0,T]$ is a function $u \in L^1(0,T;H)$ for which there exists w in $L^1(0,T;H)$ with $w(t) \in Au(t)$ and $u(t) + b*w(t) = F(t)$ almost everywhere on $[0,T]$. A strong solution to (2) is a function $u \in W^{1,1}(0,T;H) \cap C([0,T];\, cl(D(A)))$ with $u(0) = x$ satisfying (2) a.e. on $[0,T]$. It is a fact [4] that a function u is a strong solution of (1) if and only if it is a strong solution of (2).

If A satisfies the cone condition, $h \in L^1(0,T;H)$, and $h(t) \in C$ a.e. on $[0,T]$, then there is (cf. [2]) a unique integral solution $v \in C([0,T];\ cl(D(A)))$ of the problem

$$(3)\quad \begin{cases} v'(t)+(A+k(0)I)v(t) \ni h(t), & 0 \leq t \leq T, \\ v(0) = x. \end{cases}$$

We denote this solution by Qh. Let
$G: C([0,T];\ cl(D(A))) \to L^1(0,T;H)$ be defined by
$G(u)(t) = g(t) + k(t)x - u*k'(t)$. Since $G(u)(t) \in C$ for almost all t, we can define a self-mapping K of $C([0,T];\ cl(D(A)))$ by $K(u) = Q(G(u))$. It can be shown that some iterate of K is a strict contraction. Hence K has a unique fixed point. We define this fixed point to be the unique generalized solution of (2) and (1) on $[0,T]$. Clearly a strong solution is also a generalized solution. In any case, we see that (1) has a unique (generalized) solution $u \in C([0,\infty);\ H)$. This is the solution that is studied in the present paper.

For positive r, let J_r denote the resolvent $(I+rA)^{-1}$ of A, and let $A_r = (I-J_r)/r$ be the Yosida

approximation of A. Consider the problems

$$(4)\quad u_r(t) + b*A_r u_r(t) = F(t),\ 0 \leq t < \infty,$$

and

$$(5)\quad \begin{cases} u_r'(t)+(A_r+k(0)I)u_r(t)=g(t)+k(t)x-u_r*k'(t), & 0 \leq t < \infty, \\ u_r(0) = x. \end{cases}$$

It can be shown that these two (equivalent) equations have unique strong solutions $u_r: [0,\infty) \to C$. Moreover, $\lim_{r\to 0+} u_r = u$ in $C([0,T];\ H)$ for all $T > 0$, where u is

the solution of (1).

We now turn our attention to the ergodic theory of the nonlinear Volterra integral equation (1). Recall that a locally integrable $v: [0,\infty) \to H$ is said to be (weakly) almost convergent [6] (as $T \to \infty$) to $z \in H$ if the (weak) $\lim_{T\to\infty} \frac{1}{T}\int_c^{T+c} v(t)dt = z$ uniformly in $c \geq 0$.

Theorem 1. Let $u:[0,\infty) \to H$ be the solution of the nonlinear Volterra integral equation (1) (and the initial value problem (2)). If $0 \in R(A)$, $k \in L^1(0,\infty;R)$, and $g \in L^1(0,\infty;H)$, then u is weakly almost convergent to a zero of A.

Proof. It is known [1] that u is bounded and that if f_1 and f_2 are two zeros of A, then $\lim_{t\to\infty} (u(t), f_1 - f_2)$ exists. Therefore it suffices to show that if

$c_n \geq 0$, $T_n \to \infty$, and $\{ \frac{1}{T_n}\int_{c_n}^{T_n+c_n} u(t)dt)\}$ converges

weakly to z, then z is a zero of A. To this end, let u_r, $r > 0$, be the solution of (4) (or (5)) with $u_r(0) = u(0)$, B a maximal monotone extension of A with $D(B) \subset \operatorname{clco}(D(A)) \subset C$, and y any point in D(B). Modifying the method of [1], we estimate $(B_r y, y - \frac{1}{T}\int_c^{T+c} u_r(t)dt)$ from below and then let $r \to 0+$. Since $B_r y \to B^o y$ (the canonical restriction of B), we obtain

$$(B^0 y, y - \frac{1}{T}\int_c^{T+c} u(t)dt) \geq \frac{1}{T} \int_c^{T+c}(g(t),y-u(t))dt$$

$$- \frac{1}{2T} |u(c) - y|^2 - \frac{1}{2T} (k*|u-x|^2)(c)$$

$$+ \frac{1}{T} (k*(u-x)(T+c) - k*(u-x)(c), x-y).$$

Since u is bounded, $g \in L^1(0,\infty;H)$, and $k \in L^1(0,\infty;R)$, the right-hand side of this inequality tends to 0 as $T \to \infty$, uniformly in $c \geq 0$. Hence $(B^o y, y-z) \geq 0$ for all $y \in D(B)$, so that $0 \in Bz$. If $w = (I + A)^{-1}z$, then $(z-w, w-z) \geq 0$. Therefore $z = w \in D(A)$ and $0 \in Az$, as required.

Theorem 1 has two immediate corollaries. Corollary 1 is known [1], but our proof is simpler and more general. Corollary 2 is new. However, in view of Theorem 1, its proof is essentially that of [8, Corollary 2] and therefore will be omitted.

Corollary 1. In the setting of Theorem 1, the solution $u(t)$ of the integral equation (1) converges weakly as $t \to \infty$ to a zero of A if and only if the weak $\lim_{t\to\infty} (u(t+h) - u(t)) = 0$ for all $h > 0$.

Proof. We may assume that the weak $\lim_{T\to\infty} \frac{1}{T}\int_c^{T+c} u(t)dt = 0$ uniformly in $c \geq 0$. Given a sequence $c_n \to \infty$, a point w in H, and a positive ε, we first find a T_o such that $|(w, \frac{1}{T_o}\int_c^{T_o+c} u(t)dt)| < \varepsilon$ for all $c \geq 0$. Since $|(w,u(c_n))| =$ $|(w, \frac{1}{T_o}\int_0^{T_o} u(c_n)dt)| \leq |(w, \frac{1}{T_o}\int_0^{T_o} u(c_n+t)dt)|$ $+|(w, \frac{1}{T_o}\int_0^{T_o} (u(c_n) - u(c_n+t))dt)|$, we can apply Lebesgue's convergence theorem to conclude that $\limsup_{n\to\infty} |(w,u(c_n))| \leq \varepsilon$. The result follows.

For each positive s let $K(s,t) : R^+ \to R$ be of bounded variation on R^+, and denote its total variation by $V(s)$. We say that the kernel K is strongly regular if $\int_0^\infty |K(s,t)|dt$ is bounded in s, $\lim_{s\to\infty} \int_0^\infty K(s,t)dt = 1$, and $\lim_{s\to\infty} V(s) = 0$. (It then follows that $\lim_{s\to\infty} \int_0^T |K(s,t)|dt = 0$ for each $T > 0$.)

Corollary 2. In the setting of Theorem 1, let $u: R^+ \to H$ be the solution of the integral equation (1). If the kernel K is strongly regular and $R(s) = \int_0^\infty K(s,t)u(t)dt$, then R(s) converges weakly as $s \to \infty$ to a zero of A.

Remark 1. Let clco(W) denote the closed convex hull of the weak omega-limit set of $\{u(t): t \geq 0\}$. The limit obtained in Theorem 1 can be identified [1] as the unique point in $A^{-1}(0) \cap \text{clco}(W)$.

Remark 2. Let $P: H \to A^{-1}(0)$ be the nearest point mapping onto the (closed and convex) zero set of A. We can show that the strong $\lim_{t\to\infty} Pu(t)$ exists and equals the (weak) limit of Theorem 1.

Remark 3. Using different assumptions, Hirano [5] has also proved a mean ergodic theorem for nonlinear Volterra integral equations. We can establish (weak) almost convergence in his setting too.

Under the assumptions of Theorem 1, strong almost convergence does not occur in general, even in the nonlinear semigroups case ($b(t) = 1$ and $F(t) = x$). However, when an additional assumption is made, strong almost convergence does take place.

Theorem 2. Let $u: [0,\infty) \to H$ be the solution of the nonlinear Volterra integral equation (1) (and the initial value problem (2)). Assume that $0 \in R(A)$, $k \in L^1(0,\infty;R)$, and $g \in L^1(0,\infty;H)$. If $\lim_{t\to\infty} |u(t+h) - u(t)|$ exists uniformly in $h \geq 0$, then u is almost convergent to a zero of A.

Proof. Let z be the weak limit obtained in Theorem 1. In order to show that z is, in fact, a strong limit, we compute $I(c,T) = \left| \frac{1}{T} \int_c^{T+c} u(t)dt - z \right|^2$.

Using the fact that $\lim_{t\to\infty} |u(t) - z|$ exists and the method of proof of [1, Theorem 2], we can indeed show that $\lim_{T\to\infty} I(c,T) = 0$, uniformly in $c \geq 0$. (Note that in [1], $\varepsilon(|t-s|)$ should be replaced by $\varepsilon(\min(t,s))$.)

We conclude this paper with an analog of Corollary 2.

Corollary 3. In the setting of Theorem 2, let $u:R^+ \to H$ be the solution of the integral equation (1). If the kernel K is strongly regular and $R(s) = \int_0^\infty K(s,t)u(t)dt$, then R(s) converges strongly as $s \to \infty$ to a zero of A.

It is expected that detailed proofs of the results announced here, as well as other results on the asymptotic behavior of solutions to nonlinear Volterra integral equations in Hilbert and Banach spaces, will appear elsewhere.

REFERENCES

[1] J.B. Baillon and Ph. Clément, Ergodic theorems for nonlinear Volterra equations in Hilbert space, Nonlinear Analysis 5(1981), 789-801.

[2] Ph. Bénilan, Solutions intégrales d'équations d'évolution dans un espace de Banach, C.R. Acad. Sci. Paris 274(1972), 47-50.

[3] Ph. Clément and J.A. Nohel, Asymptotic behavior of positive solutions of nonlinear Volterra equations for heat flow, Mathematics Research Center Report #2069, 1980.

[4] M.G. Crandall and J.A. Nohel, An abstract functional differential equation and a related nonlinear Volterra equation, Israel J. Math. 29(1978), 313-328.

[5] N. Hirano, Asymptotic behavior of solutions of nonlinear Volterra equations, preprint.

[6] G.G. Lorentz, A contribution to the theory of divergent series, Acta Math. 80(1948), 167-190.

[7] S. Reich, Almost convergence and nonlinear ergodic theorems, J. Approximation Theory 24(1978), 269-272.

[8] S. Reich, A note on the mean ergodic theorem for nonlinear semigroups, J. Math. Anal. Appl., to appear.

SOME REMARKS
ON
CONTINUOUS AND DISCRETE FLOWS

S.K. Kaul
Department of Mathematics and Statistics
University of Regina
Regina, Saskatchewan, Canada

ABSTRACT

Let (X,T,π) be a continuous or a discrete flow on a connected, locally compact metric phase space X. We consider flows which are of characteristic $0^{\pm}$ everywhere except on a non-empty set B whose closure is of dimension zero and does not disconnect the space X. Such flows are called flows of characteristic $0^{\pm}$ almost everywhere. The results here are related to results in a recent paper of Lam [9], and earlier results of Lam and other authors [see references in [8]] on transformation groups which are equicontinuous almost everywhere. We prove that (a) for continuous flows B consists of at most two points, and (b) for discrete flows, if $X-\bar{B}$ is connected im kleinen then B consists of at most two points.

1. Let (X,T,π) be a continuous or a discrete flow, where X is a locally compact connected metric space, T is either the group of Reals $\mathbb{R}$, or the group of integers Z, π: $X\times T\to X$ is a continuous function which has the following properties, (i) $((x,t)\pi,s)\pi=(x,t+s)\pi$, (ii) $(x,e)\pi=x$, where t, s are any elements of T and e is the identity in T. For convenience we shall write $(x,t)\pi$ simply as xt.

We shall discuss some results which are concerned with special cases of a general problem which may be described as follows: Suppose A and B are disjoint subsets of X whose union is X. Suppose T has some "nice" property on A and $B\neq\phi$ and T fails to have that property on B. We wish to study the nature of B in some special cases where the "nice" property has been specified.

To start with consider the case where the nice property T has on A is pointwise equicontinuity on A. Then the following theorem is known [8].

Theorem (1.1): Let (X,T) be a discrete or a continuous flow, and the following conditions be satisfied.

(1) $\bar{B}$ is zero dimensional

(2) for any $p\varepsilon B$, $\overline{pT}$ is compact

(3) A^O is connected.

Then B consists of at most two points and each is fixed under T.

This result can be improved by analysing points of non-equicontinuity in detail. We divide B into three mutually disjoint sects, B_1, B_2 and B_3. Suppose $x\varepsilon B$. Then

(1) $x\varepsilon B_1$ if there exists a sequence $\{t_n\}$ in T such that for some sequence $\{x_n\}$ in X converging to x, $x_nt_n\to p$ but $xt_n\not\to p$.

(2) $x\varepsilon B_2$ if there exists a sequence $\{t_n\}$ in T such that for some sequence $\{x_n\}$ in X, converging to x, $xt_n\to p$ but $x_nt_n\not\to p$.

(3) $B_3=B-(B_1\cup B_2)$.

Then we have the following theorem:

Theorem (1.2): Let (X,T) be a discrete or a continuous flow, $B=B_1\cup B_2$ and the following conditions be satisfied:

(1) $\bar{B}$ is zero dimensional

(2) A^O is connected.

Then B consists of at most two points each of which is fixed under T.

Remark (1.3): Note that if $\overline{xT}$ is compact for each $x\varepsilon B$, then $B_3=\phi$.

The crucial property of points in B obtained from the assumption that $\bar{B}$ be zero dimensional is:

Lemma (1.4): Given $p\varepsilon B$ there exists a point $x_0\varepsilon A^O$ and a sequence $\{t_n\}$ such that $x_0t_n\to p$, that is $\{x_0t_n\}$ converges to p.

A crucial step after having established the existence of one such point x_0 is to show that $xt_n\to p$ for all $x\varepsilon A^O$. This is the so called property of indivisibility [6, Def. (1.4), p.170]. We shall say that (X,T) is indivisible if given any point $p\varepsilon B$, whenever there is a point x_0 and a sequence $\{t_n\}$ in T such that, $x_0t_n\to p$, then $xt_n\to p$ for all $x\varepsilon A^O$.

A simple example below will show that without condition (3) in the theorem (1.1) that B does not disconnect the space, B need not consist of at most two points.

Example (1.5): Suppose I=[0,1] is the closed unit interval with the usual metric. Let h: $I\to I$ be the homeomorphism defined by $h(x)=x^2$ for any $x\varepsilon I$. The transformation group (X,Z,π) is defined by $(x,n)\pi=h^n(x)$. Then clearly $B=\{0,1\}$, and A is the open interval (0,1).

Now let X=[a,b] be any closed interval in the real line with integral end points or the real line itself. Consider the transformation group (X,Z,π) where, $\pi:X\times Z\to X$ is defined by

$$(x,n)\pi=h^n(x-i)+i \text{ where } i\le x\le i+1$$

for all integers i in [a,b]. Then B is the set of all integers in X, hence it is zero dimensional, but disconnects the space if X is not of unit length. Thus, B does not consist of at most two points.

In the above theorem $\bar{B}$ was assumed to be zero dimensional. Suppose we put no such condition on $\bar{B}$ and ask the question as to the number of components of $\bar{B}$ and their invariance with respect to T.

This problem has been studied by Lam [6] and Kaul [5] by identifying the components of $\bar{B}$ to points to get a quotient space $\tilde{X}$ so that $(\tilde{X},T)$ is again a transformation group. For the following discussion see [2].

Let $\mathcal{C}$ denote the components of $\bar{B}$ and assume that each component is compact. Then a topological result says that given any $C\in\mathcal{C}$ and an open set U containing C there exists an open set V, such that, $C\subset V\subset U$ and $\partial U\cap\bar{B}=\phi$, where ∂V is the boundary of V. Thus the decomposition space $\tilde{X}=\mathcal{C}\cup\{\{x\}: x\in A^O\}$ is upper semi-continuous, and hence the projection map, $\phi:X\to\tilde{X}$, where $\tilde{X}$ has the quotient topology, is closed. Hence, $\phi(\bar{B})$ is a closed and zero dimensional subset of $\tilde{X}$, $\tilde{X}$ is metrizable, and ϕ being a homeomorphism on A^O, points $\phi(x)$, $x\in A^O$, are points of equicontinuity in $(\tilde{X},T)$. $(\tilde{X},T)$ is a transformation group and for each $x\in X$, the action of T on $\tilde{X}$ is defined by, $(x\phi)t=(xt)\phi$. Now let $\tilde{A}$ denote the set of all points of equicontinuity in $(\tilde{X},T)$ and $\tilde{B}=\tilde{X}-\tilde{A}$. Then since $\tilde{B}\subset\phi(\bar{B})$, $cl(\tilde{B})$ is zero dimensional in $\tilde{X}$. Now we may apply Theorem (1.1) to $(\tilde{X},T)$ and from that draw conclusions about $\mathcal{C}$.

The problem with this approach is that we may lose a component C of $\bar{B}$ in that $\phi(C)$ may become a point of equicontinuity of T. This may happen even if we assume that $\bar{B}$ is compact. More precisely it may be that for a point $x\in C$, there is a sequence $\{t_n\}$ in T, such that $xt_n\to p\in\bar{B}$ but for some $\{x_n\}$ converging to x, $\{x_nt_n\}$ converges to some other point of C. In such a situation we do not have the crucial lemma (1.4) and the approach in general fails to identify such components of $\bar{B}$. We therefore look for irregular sets: e.g. $C\in\mathcal{C}$ is said to be irregular if there exists a point $x\in A^O$ and a sequence $\{t_n\}$ in T such that $x_0t_n\to p\in C$.

If $C\in\mathcal{C}$ is an irregular set, then because $\phi t=t\phi$, and $(x_0\phi)\in(\tilde{A})^O$, $(x_0\phi)t_n=(x_0t_n)\to p\phi=C\phi$, and therefore $C\phi\in\tilde{B}$. Thus, theorem (1.1) applied to $(\tilde{X},T)$ gives the following:

Theorem (1.6): If each component of $\bar{B}$ is compact, then $\bar{B}$ contains at most two irregular components, and each of them is invariant under T.

Here we do not assume that $\bar{B}$ be compact, as is the case in [8] and [5] because of the following indivisibility result.

Theorem (1.7): Let (X,T) be a continuous or a discrete flow, and A and B be as above, and let $\{t_n\}$ be any sequence in T. Then the set of points $\{x\in A^O: \limsup xt_n\subset\bar{B}\cup\{\infty\}\}$ is both open and closed in A^O.

Consequently we have the following:

Theorem (1.8): If $\bar{B}$ is compact and has only a finite number of components then it has at most two components and each is invariant under T.

2. We may now ask the question whether there are any other nice properties, of course weaker than pointwise equicontinuity, that lead to similar results.

Two properties studied in this regard are those of characteristic $0^{\pm}$ [7] and strong characteristic 0 [1]. Suppose $T^+=\{x\in T: x\geq 0\}$ and $T^-=\{x\in T: x\leq 0\}$. Then for a transformation group (X,T), the positive limit set $L^+(x)$ and the positive prolongation set $D^+(x)$ are defined as follows: $L^+(x)=\{y\in X: xt_n\to y$ for some divergent sequence $\{t_n\}$ in $T^+\}$, and $D^+(x)=\{y\in X: x_nt_n\to y$ for some sequence $\{t_n\}$ in T^+ and $\{x_n\}$ in X converging to $x\}$ $L^-(x)$ and $D^-(x)$ are similarly defined for T^-. A point $x\in X$ is said to be of characteristic 0^+ if $D^+(x)=\overline{xT^+}$ and of characteristic 0^- if $D^-(x)=\overline{xT^-}$; and of characteristic $0^{\pm}$ if both conditions are satisfied. Let A denote all the points of characteristic $0^{\pm}$ and B=X-A in (X,T).

Definition (2.1): A flow (X,T) is CAE, or said to be of characteristic $0^{\pm}$ almost everywhere, if $B\neq\phi$, $\bar{B}$ is zero dimensional, and A^O is connected.

Lam initiated the study of such flows in [9] under the additional condition that $\overline{xT}$ be compact for each $x\in\bar{B}$. Our results that follow do not make such an assumption. Furthermore, we do not make the assumption as in [9] that A^O contain a point which is not almost periodic in view of the following:

Theorem (2.2): Let (X,T) be a flow of characteristic $0^{\pm}$ at each point of X. Then the set $Y=\{y\in X$: T is recurrent at $x\}$ is both open and closed in X.

Definition: T is recurrent at x if $x\in L^+(x)$ and $x\in L^-(x)$.

Remark: Because each point is a point of characteristic $0^{\pm}$, one can show that T is recurrent at x if $x\in L^+(x)$.

The above theorem is proved in [1] for general transformation groups under the additional condition that X be locally connected. In case T=R or Z one does not need this condition.

We need the following crucial lemma:

Lemma (2.3): If (X,T) is a CAE flow, then for any $p \in B$ there exists a point $x_0 \in A^o$ and a sequence $\{t_n\}$ in T such that $x_0 t_n \to p$.

The following theorems on indivisibility are true.

Theorem (2.4): A continuous CAE transformation group is indivisible.

Theorem (2.5): If (X,Z) is CAE and A^o is a semi-continuum, then it is indivisible.

Definition (2.6): A space X is said to be a semi-continuum if any two points in X lie in a compact connected set.

The above results lead to the following:

Theorem (2.7): If (X,T) is CAE and indivisible, then B consists of at most two points and each is fixed under T. Furthermore, each point of A^o is a point of equicontinuity of T.

3. The property of strong characteristic 0 introduced in [1] is stronger than that of characteristic $0^{\pm}$ and weaker than equicontinuity.

Definition (3.1): A point $x \in X$ is of strong characteristic 0 if, whenever for some sequence $\{t_n\}$ in T and $x_n \to x$ in X, $x_n t_n \to p$ then $xt_n \to p$.

If now A and B refer to points of strong characteristic 0 and its complement in X, then (X,T) is SCAE or is said to be of strong characteristic 0 almost everywhere. We then have the following results.

Theorem (3.2): Let (X,Z) be SCAE and for each $x \in A^o$ let $L^+(x) \cap \bar{B}$ and $L^-(x) \cap \bar{B}$ be both non-empty and compact. Then (X,Z) is indivisible, and B consists of at most two points each of which is fixed under T.

Corollary (3.3): If (X,Z) is SCAE and X is compact, then B consists of at most two points, each of which is fixed under T.

Remark: For detailed proves of these results see [6] and [7].

REFERENCES

[1] Elaydi, S. and Kaul, S.K. "Flows of almost strong characteristic 0 with generative phase group", to appear in the Journal of Nonlinear Analysis.

[2] Gottschalk, W.H. and Hedlund, G.A., Topological Dynamics, Amer. Math. Soc. Colloq. Pub. Vol. 36, 1955.

[3] Grey, W.J. and Roberson, F.A., "On the near equicontinuity of transformation groups", Proc. Amer. Math. Soc. 23 (1969) p.59-63.

[4] Kaul, S.K., "On almost regular homeomorphisms", Can. Jour. Math. 20 (1968) p.1-6.

[5] Kaul, S.K., "On the irregular sets of a transformation group", Can. Math. Bull. 16 (1973) p.225-232.

[6] Kaul, S.K., "Indivisibility in Transformation Groups", to appear.

[7] Kaul, S.K., "Continuous and discrete flows of characteristic $0^{\pm}$ almost everywhere", to appear.

[8] Lam, P.F., "Almost equicontinuous transformation groups", Trans. Amer. Math. Soc. 195 (1974) p.165-200.

[9] Lam, P.F., "Discrete and continuous flows of characteristic $0^{\pm}$", Fund. Math. XCII (1976) p.142-147.

SOME STABILITY NOTIONS

Ronald A. Knight

Northeast Missouri State University
Kirksville, Missouri

ABSTRACT

Invariance and stability are defined for each fundamental dynamical system relation and the interrelationships are obtained.

INTRODUCTION

Ever since its inception dynamical system research has focused on stability theory. Classically, orbital stability, Poisson stability, and Lagrange stability have been the most useful stability concepts. The development of several dynamical relations over the past three decades has remarkably facilitated the study of orbital stability. Bhatia and Hajek defined stability relative to two of these relations in [3] and Elaydi defined weaker versions of these for transformation groups in [5]. The author used the pattern of the stronger version to define stability in terms of any dynamical function for transformation groups in [8].

Our objective in this paper is to determine the interrelationships among the various forms of invariance and stability. Throughout the paper we let (X,π) be a given dynamical system on a Hausdorff space X. The fundamental dynamical relations (functions) are the unilateral and bilateral versions of the orbit, orbit closure, limit, prolongation, prolongational limit, weak attraction, attraction, and strong attraction relations (functions) denoted, respectively, by C, K, L, D, J, A_w, A, and A_s with the unilateral versions each carrying the appropriate superscript. Our terminology and notations follow [1], [2], [6], [7], and [8]. The reader is referred to [3] and [4] for basic properties of dynamical system theory.

This section concludes with two important related results.

Theorem 1.1. (Ura [9]) Let X be locally compact and let M be a closed subset with compact boundary. Then, M is positively (bilaterally) stable if and only if $D^+(M) = M$ $(D(M) = M)$.

Theorem 1.2. (Bhatia [2]) A subset M of X is positively (bilaterally) stable if and only if $M \subset A_s^+(M)$ $(M = A_s(M))$.

INVARIANCE

Let $F: \mathcal{P}(X) \to \mathcal{P}(X)$ be a function on the power set $\mathcal{P}(X)$ of X. A subset M of X is said to be F-*invariant* if $F(M) = M$. When F is C^+, C^-, or C we simply say that M is *positively, negatively,* or *bilaterally invariant,* respectively. Often the adjective bilateral is omitted.

For brevity in our first theorem we adopt the following notation. For a given subset M of X and a function F we let (F) denote the statement "M is F-invariant." By $(F_1,\ldots,F_n) \to (G_1,\ldots,G_n)$ we mean (F_i) implies (G_i) for $i = 1, \ldots, n$.

The results of the following invariance interrelationship theorem are summarized in the diagram following Proposition 3.9.

Theorem 2.1. For a subset M of X we have the following:

(a) $(F) \to (F^+)$ for $F = D, K, C$, or A_s;

(b) $(F^+) \to (F)$ for $F = A_w$ or A;

(c) $(J,J^+,L,L^+,A_s^+) \to (D,D^+,K,K^+,A_w^+) \to (K,K^+,C,C^+,A^+)$;

(d) $(D,J^+,L^+,A_w,A,A^+,A_s^+) \to (A_w^+,A_w^-,C,A,C,C,A_w^-)$; and

(e) the dual of each implication in (a) through (d).

Proof: To prove part (a) let F be D, K, or C. If $F(M) = M$, then $F^+(M) \subset F(M) = M \subset F^+(M)$, and hence, $F^+(M) = M$. On the other hand, if $A_s(M) = M$, then M is bilaterally stable by Theorem 1.2 and $C(M) = M$. Thus, $A_w^+(M) = M$. Since $M \subset A_s^+(M) \subset A_w^+(M)$, $A_s^+(M) = M$.

Next, consider part (b). Let F be A_w or A. If $F^+(M) = M$, then $C(M) = M$ and $F^-(M)$ contains M. Consequently, $F(M) = F^+(M) \cap F^-(M) = M$.

To prove part (c) let (F,G) be any of the pairs (J,D), (J^+,D^+), (L,K), or (L^+,K^+) and let (F,H) be (D,K), (D^+,K^+), (K,C), or (K^+,C^+). If $F(M) = M$, then $C(M) = M$ so that $G(M) = C(M) \cup F(M) = M$ or else $M \subset H(M) \subset F(M)$, and hence, $H(M) = M$. Thus, $(F,F) \to (G,H)$. To complete the proof of part (c) we need to show $(A_s^+) \to (A_w^+) \to (A^+)$. Let $A_s^+(M) = M$. Then, M is stable, invariant, and $M = A_s^+(M) \subset A_w^+(M)$. Suppose that $x \in A_w^+(M)$ but not M. Given a neighborhood U of M there is an open neighborhood V of M such that $V = C^+(V) \subset U$. There is also a $t > 0$ such that $xt \in V$. The set Wt is contained in V for some neighborhood W of x, and hence, $C^+(Wt)$ is contained in U. This means that $x \in A_s^+(M)$ contradicting $A_s^+(M) = M$. Thus, $A_w^+(M) = M$ yielding $(A_s^+) \to (A_w^+)$. Next, let $A_w^+(M) = M$. Then, $A^+(M) \subset A_w^+(M) = M = C(M) \subset A^+(M)$ so that $A^+(M) = M$ and $(A_w^+) \to (A^+)$.

We consider part (d) by first letting F be L^+, A^+, or A. If $F(M) = M$, then $M \subset C(M) = C(F(M)) = F(M) = M$, and so, $(F) \to (C)$. The proof of $(A_w) \to (A)$ can be constructed in a manner similar to the proof of the unilateral version $(A_w^+) \to (A^+)$. Next, let $D(M) = M$. Then, $C(M) = M$ so that $M \subset A_w^+(M)$. Let $x \in A_w^+(M)$. Either $x \in M$ or else there is a point y in M belonging to $L^+(x) \subset J^+(x)$, and hence, $x \in J^-(y) \subset D(y) \subset M$. Therefore, $A_w^+(M) = M$ proving that $(D) \to (A_w^+)$. Now, let $J^+(M) = M$. Again, $C(M) = M$ yielding $M \subset A_w^-(M)$. For $x \in A_w^-(M)$ either $x \in M$ or else there is a point y in M belonging to $L^-(x)$ which means that $x \in J^-(y) \subset M$. Thus, $A_w^-(M) = M$ and $(J^+) \to (A_w^-)$. All we have left to show is $(A_s^+) \to (A_w^-)$. Let $A_s^+(M) = M$. Because $C(M) = M$ we have $M \subset A_w^-(M)$. Suppose that $x \in A_w^-(M)$ but not M. The set M is stable, and so, for any neighborhood U of M there is a neighborhood V of M excluding x such that $C^+(V) = V$. Since $x \in A_w^-(M)$, there is a point xt in V such that $t < 0$, and hence, $x \in C^+(xt) \subset V$ which is absurd because it leads to $x \in A_s^+(M) = M$ as in the proof of part (c).

Finally, part (e) follows by the dual arguments completing the proof.

Except for the orbital versions, each type of invariance will now be characterized. Four conditions used for this are stated first. A point x belongs to M if and only if there exist

(1) a point $y \in M$ and a net $t_i \geq 0$ such that $yt_i \to x$,

(2) a point $y \in M$ and a net $t_i \to +\infty$ such that $yt_i \to x$,

(3) a point $y \in M$, a net $y_i \to y$, and a net $t_i \geq 0$ such that $y_i t_i \to x$, and

(4) a point $y \in M$, a net $y_i \to y$, and a net $t_i \to +\infty$ such that $y_i t_i \to x$.

The proof of the following invariance characterization proposition is an easy consequence of the definitions.

<u>Proposition 2.2.</u> A subset M of X is

(a) K^+-invariant if and only if condition (1) above holds,

(b) L^+-invariant if and only if condition (2) above holds,

(c) D^+-invariant if and only if condition (3) above holds, and

(d) J^+-invariant if and only if condition (4) above holds.

The dual versions are characterized by the duals of (1) through (4), respectively. To characterize the bilateral versions we need only replace "net $t_i \geq 0$" by "net t_i" and replace "net $t_i \to +\infty$" by "net $t_i \to +\infty$ or $-\infty$" because $F(M) = F^+(M) \cup F^-(M)$ for F = K, L, D, or J.

The next three propositions characterize invariance relative to A_w^+

and A^+ and the concluding theorem of this section characterizes A_s^+-invariance. The dual versions are characterized by the respective duals and the bilateral versions are characterized by the conjunction of the unilateral versions because $F(M) = F^+(M) \cap F^-(M)$ for $F = A_w$, A, or A_s. The proofs of these results are omitted because straight forward applications of known results is all that is required.

Proposition 2.3. A subset M of X is A_w^+-invariant if and only if it is invariant and it satisfies the dual to Zubov's condition, $L^+(X - M) \cap M = \emptyset$.

Proposition 2.4. A subset M of X is A^+-invariant if and only if it is invariant and $C^+(x)$ is not ultimately contained in some neighborhood of $L^+(x) \cap M$ whenever $x \notin M$ and $L^+(x) \cap M \neq \emptyset$.

Proposition 2.5. Let X be locally compact and let M be a closed subset of X with compact boundary. The set M is A^+-invariant if and only if M is invariant and $L^+(x) \cap (X - M) \neq \emptyset$ for each positively nondivergent point x of X - M.

The following theorem identifies the A_s^+-invariant sets as the invariant stable sets not weakly attracting any exterior point in either direction.

Theorem 2.6. A subset M of X is A_s^+-invariant if and only if M is an A_w^+-invariant positively stable set.

Recall that the A_s-invariant sets are characterized in Theorem 1.2 as the bilaterally stable sets.

STABILITY

Let $F: P(X) \to P(X)$ be a function on the power set $P(X)$ of X. A subset M of X is said to be F-*stable* if given a neighborhood U of M there exists a neighborhood V of M contained in U such that $F(V) = V$. If $F = C^+$, C^-, or C, then F-stability is one of the classical orbital stability notions and we say that M is *stable*, *negatively stable*, or *bilaterally stable*, respectively, as usual. Also, for such trajectoral functions F we sometimes use the equivalent requirement "$F(V) \subset U$" in place of "$F(V) = V \subset U$." In general, however, we can only say that our version of F-stability which follows the pattern of [3] implies the version using "$F(V) \subset U$" which is the pattern followed in [5]. The converse is false for the fundamental dynamical functions as the following simple examples illustrate except for the various orbit and orbit closure functions (see Corollary 3.1.1). In fact, as Example 3.1 illustrates the alternate condition applies to many noninvariant sets whereas only C^+, C^-, and C-invariant sets are F-stable relative to a fundamental dynamical function F.

Example 3.1. Let (R^2,π) be a parallel flow and let M be any nonempty bounded subset of M. Then, M is not F-stable relative to any of the fundamental dynamical functions, however, it is stable relative to each of the L, J, A_w, A, and A_s functions with respect to the alternate definition.

Example 3.2. Consider the flow of Example 1, pp. 257-8, of reference [9]. Let (X,π) be the restriction to the y-axis and let M consist of just the origin. Then, M is not F-stable relative to the various D and J functions but is stable with respect to these functions relative to the alternate definition.

By examining the situation for the orbit relations we easily obtain the following proposition containing both necessary and sufficient conditions for the two stability versions to coincide, namely, $V \subset F(V)$ and $F(V) = F^2(V)$ for appropriate V. These conditions are stated so that the results of [5] can be related to ours.

Proposition 3.1. An F-stable subset M of X is also F-stable relative to the alternate definition if and only if for each neighborhood U of M there is a neighborhood V of M such that $V \subset F(V) = F^2(V) \subset U$.

Corollary 3.1.1. The alternate version of F-stability coincides with our version for a set M relative to

(a) each orbit and orbit closure function and

(b) each limit and attraction (prolongation, prolongational limit) function provided that for each neighborhood U of M there is a neighborhood V of M such that $V \subset F(V) \subset U$ $(F(V) = F^2(V) \subset U$, $V \subset F(V) = F^2(V) \subset U)$.

Throughout the remainder of this section we let F be a given fundamental dynamical function defined on $P(X)$.

Theorem 3.2. Any F-stable set is F-invariant.

Proof: Each bilateral version is proved in [8] and each unilateral version of C, K, and D is proved in [3]. Since dual results follow by dual arguments, we need only consider $F = L^+$, J^+, A_w^+, A^+, and A_s^+. Let M be an F-stable subset of X. Since $F(M) \subset F(V)$ for every neighborhood V of M, no point of X - M can belong to F(M), and hence, $F(M) \subset M$. Suppose that x belongs to M but not to F(M). Consider $F = L^+$ or J^+. Then, for any neighborhood U of M there must be a point x_U in U - M such that $x \in F(x_U)$. Let V be an F-invariant neighborhood of M contained in $U - \{x_U\}$. Since $F(x_U) \cap V$ is contained in the boundary of V, the point x cannot belong to $F(x_U)$ which is absurd. Hence, F(M) = M. On the other hand, let F be A_w^+, A^+, or A_s^+. Since any F-invariant neighborhood of M is positively invariant, M must be stable. According to Theorem 1.2, $M \subset A_s^+(M)$. Thus, $A_s^+(M) \subset A^+(M) \subset A_w^+(M)$ implies $F(M) = M$. The proof is complete.

The next two theorems completely determine the interrelationships amomg the various F-stability concepts. The first follows easily from Theorem 2.1.

Theorem 3.3. Each statement of Theorem 2.1 holds whenever (F) denotes "M is F-stable."

Corollary 3.3.1. Every F-stable set is either positively, negatively, or bilaterally stable.

Theorem 3.4. The following statements are pairwise equivalent for a subset M of X:

(a) M is bilaterally stable,

(b) M is A_w^+-stable,

(c) M is A^+-stable,

(d) M is A_s^+-stable,

(e) M is A_s-invariant, and

(f) the dual to (b), (c), or (d).

Proof: In view of Theorem 1.2 and the dual arguments it is sufficient to show that (a) through (d) are pairwise equivalent.

First, let M be bilaterally stable and let U be a neighborhood of M. Select an open invariant neighborhood V of M contained in U. Then, $A_w^+(V) = V$, and hence, M is A_w^+-stable.

Next, let M be A_w^+-stable. Any neighborhood U of M contains an A_w^+-invariant neighborhood V of M. Since $A^+(V) \subset A_w^+(V)$ and $C(V) = V$, we have $A^+(V) = A_w^+(V)$. Therefore, M is A^+-stable.

Now, let M be A^+-stable. Each neighborhood U of M contains an open neighborhood V such that $A_s^+(V) \subset A^+(V) = V$. Since V is open and invariant, $V \subset A_s^+(V)$, and hence, M is A_s^+-stable.

Finally, let M be A_s^+-stable. Any neighborhood U of M contains a neighborhood V of M such that $A_s^+(V) = V$. Since $C(V) = V$ and $V \subset U$, we have M bilaterally stable. The proof is complete.

We conclude the results of this paper with a series of necessary and sufficient conditions for converse implications of Theorems 2.1 and 3.3 to hold. Each proof is quite straight forward, and therefore, each is omitted.

Proposition 3.5. Let (E,G,H) be the triple (K,L,C), (K^+,L^+,C^+), (D,J,K), or (D^+,J^+,K^+) and let M be an H-stable (H-invariant) set. Then, M is E-stable (E-invariant) if and only if each neighborhood of M contains a neighborhood V of M such that $G(V) \subset V$ (that $G(M) \subset M$),

Proposition 3.6. Let (G,H) be the pair (L,K), (L^+,K^+), (J,D), or (J^+,D^+) and let M be an H-stable (H-invariant) set. Then, M is G-stable (G-invariant) if and only if each neighborhood of M contains a G-asymptotic neighborhood of M (the set M is G-asymptotic relative to $(M,\pi|M)$).

Proposition 3.7. Let (G,H) denote (K,L) or (D,J) and let M be a G^+-stable (G^+-invariant) set. Then, M is G-stable (G-invariant) if and only if M is bilaterally stable and each neighborhood of M contains a neighborhood V of M such that $H^-(V) \subset V$ (the set M is invariant and $H^-(M) \subset M$).

Proposition 3.8. Let G denote C, L, or J and let M be G^+-stable (G^+-invariant). Then, M is G-stable (G-invariant) if and only if each neighborhood of M contains a neighborhood V of M such that $G^-(V) \subset V$ ($G^-(M) \subset M$).

Proposition 3.9. Let G denote L or J and Let M be G-stable (G-invariant). Then, M is G^+-stable (G^+-invariant) if and only if each neighborhood of M contains a neighborhood V of M such that $G^-(V) \subset G^+(V)$ ($G^-(M) \subset G^+(M)$).

The following diagram summarizes the results of Theorems 1.1, 1.2, 2.1,

DIAGRAM

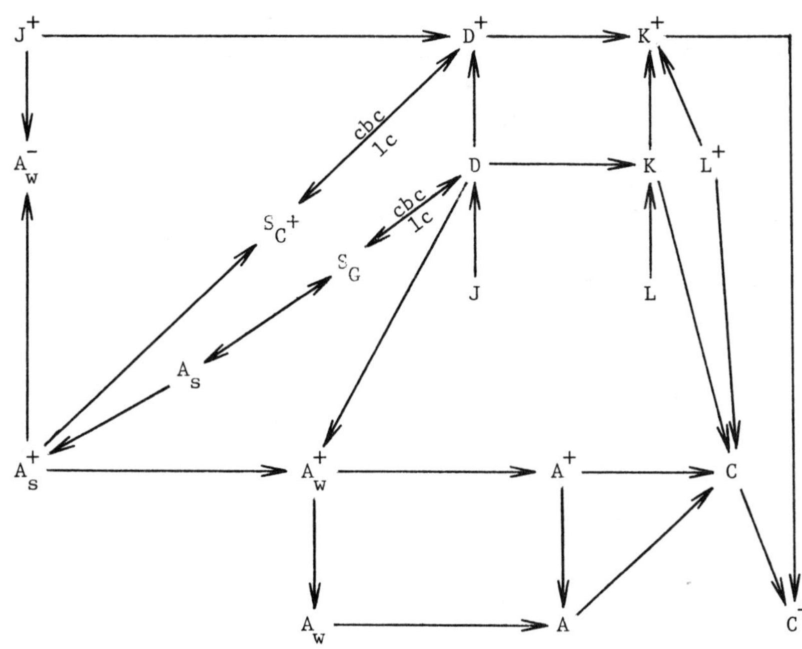

and 3.4 provided F denotes F-invariant, S_F denotes F-stable, cbc means that M is closed with compact boundary, lc means that X is locally compact, and $G = C$, A_s, A_s^+, A, A^+, A_w, or A_w^+. It summarizes the results of Theorem 3.3 provided F denotes F-stable and the other parts are interpreted as above.

REFERENCES

[1] S. Ahmad, Strong attraction and classification of certain continuous flows, Math. Systems Theory 5, 1970, 157-163.

[2] N. Bhatia, Attraction and nonsaddle sets in dynamical systems, J. Differential Equations 8, 1970, 229-249.

[3] N. Bhatia and O. Hajek, Theory of dynamical systems, Parts I and II, Tech. Notes BN-599 and BN-606, Univ. of Maryland, College Park, Md., 1969.

[4] N. Bhatia and G. Szego, Stability Theory of Dynamical Systems, Springer-Verlag, Berlin and New York, 1970.

[5] S. Elaydi, On some stability notions in topological dynamics, to appear in J. Differential Equations.

[6] R. Knight, Dynamically symmetric flows, J. Math. Phy. Sci. 15, 1981, 79-98.

[7] R. Knight, Characterizations of recurrent and stable transformation groups, Bull. Inst. Math. Acad. Sinica 10, 1982, 185-196.

[8] R. Knight, Attraction and stability in transformation groups, unpublished manuscript.

[9] T. Ura, On a flow outside a closed invariant set; stability, relative stability, and saddle sets, Contrib. Differential Equations 3, 1964, 249-294.

FUNCTIONAL DIFFERENTIAL INEQUALITIES AND EQUATIONS WITH OSCILLATING COEFFICIENTS

G. Ladas, Y. G. Sficas* and I. P. Stavroulakis**

Department of Mathematics
University of Rhode Island
Kingston, Rhode Island

ABSTRACT

In this paper we study the oscillatory character of the solutions of differential equations of the form $x'(t) + p(t)x(t-r) = 0$ with oscillating coefficient $p(t)$. It is noteworthy to observe that when the coefficient $p(t)$ oscillates then the derivative $x'(t)$ of the solution also oscillates. And this fact is true when r is zero, positive or negative, i.e., for ordinary, retarded or advanced differential equations. Now the question is under what additional hypotheses on $p(t)$ and r will the solution $x(t)$, itself oscillate? Clearly in the case of ordinary differential equations this is impossible for nontrivial solutions. However, for functional differential equations we present sufficient conditions under which all solutions oscillate.

1. INTRODUCTION

The work in the present paper has been motivated by our curiosity to learn the oscillatory character of the solutions of differential equations of the form

$$x'(t) + p(t)x(t-r) = 0$$

with oscillating coefficient $p(t)$.

In this paper we study differential inequalities and equations of the retarded

$$x'(t) + p(t)x(t-\tau) \leq 0 \tag{1}$$

$$x'(t) + p(t)x(t-\tau) \geq 0 \tag{2}$$

$$x'(t) + p(t)x(t-\tau) = 0 \tag{3}$$

*On sabbatical leave from the University of Ioannina, Greece.
**On leave from the University of Ioannina, Greece.

of the advanced

$$x'(t) - q(t)x(t+\sigma) \leq 0 \tag{1'}$$

$$x'(t) - q(t)x(t+\sigma) \geq 0 \tag{2'}$$

$$x'(t) - q(t)x(t+\sigma) = 0 \tag{3'}$$

and of the mixed type

$$x'(t) + p(t)x(t-\tau) + q(t)x(t+\sigma) = 0 \tag{4}$$

$$x'(t) - p(t)x(t-\tau) - q(t)x(t+\sigma) = 0 \tag{4'}$$

where τ,σ are positive constants and the coefficients $p(t)$, $q(t)$ are continuous functions which are allowed to oscillate.

As in the case where the coefficient $p(t)$ is positive we give conditions which involve the limit superior and conditions which involve the limit inferior of some integrals of $p(t)$.

Throughout this paper the solutions are assumed to be valid for all $t \geq 0$. As it is customary, a function is said to be oscillatory if it has arbitrarily large zeros.

2. MAIN RESULTS

The first theorem in this section generalizes a result by Ladas, Lakshmikantham and Papadakis [3, p. 222] for retarded equations with positive coefficients.

Theorem 2.1. *Consider the retarded differential inequalities*

$$x'(t) + p(t)x(t-\tau) \leq 0, \tag{1}$$

$$x'(t) + p(t)x(t-\tau) \geq 0 \tag{2}$$

and the equation

$$x'(t) + p(t)x(t-\tau) = 0 \tag{3}$$

and assume that $p(t) > 0$ *(at least) on a sequence of disjoint intervals* $\{(\xi_n,t_n)\}_{n=1}^{\infty}$ *with* $t_n - \xi_n = 2\tau$. *If*

$$\limsup_{n\to\infty} \int_{t_n-\tau}^{t_n} p(s)ds \geq 1, \tag{5}$$

then

(1) *has no eventually positive solutions,*

(2) *has no eventually negative solutions, and*

(3) *has oscillatory solutions only.*

Proof: First we prove that (1) has no eventually positive solutions. To this end suppose that $x(t)$ is a solution of (1) such that for t_0 sufficiently large

$$x(t) > 0, \quad t > t_0.$$

Then $x(t-\tau) > 0$ for $t > t_0+\tau$. Since $\xi_n \to \infty$ there exists an N such that for $n \geq N$, $\xi_n > t_0+\tau$ and, because $p(t) > 0$ on (ξ_n,t_n), $x'(t) < 0$ for $t \in (\xi_n,t_n)$, $n \geq N$. Hence $x(t-\tau) > x(t)$ for $t \in (\xi_n+\tau,t_n)$, $n \geq N$. Integrating (1) from $t_n-\tau$ to t_n, $n \geq N$, we obtain

$$x(t_n) - x(t_n-\tau) + \int_{t_n-\tau}^{t_n} p(s)x(s-\tau)ds \leq 0$$

and, because $x(t)$ is decreasing, it follows that

$$x(t_n) - x(t_n-\tau) + x(t_n-\tau) \int_{t_n-\tau}^{t_n} p(s)ds \leq 0$$

or

$$x(t_n) + x(t_n-\tau)\left[\int_{t_n-\tau}^{t_n} p(s)ds - 1\right] \leq 0$$

which, in view of (5), is a contradiction.

To prove that (2) has no eventually negative solutions it suffices to observe that if $x(t)$ is a solution of (2) then $-x(t)$ is a solution of (1).

From the above it follows that (3) has neither eventually positive nor eventually negative solutions and therefore every solution of (3) oscillates.

The proof of the theorem is complete.

In the next theorem we will assume that the coefficients are positive in intervals of arbitrarily large length but otherwise they may oscillate. The results in [2] and [4] can be viewed as special cases because the coefficients there are positive everywhere. The conditions here are "sharp" as those in [2] and [4].

Theorem 2.2. *Consider the retarded differential inequalities*

$$x'(t) + p(t)x(t-\tau) \leq 0 \tag{1}$$

$$x'(t) + p(t)x(t-\tau) \geq 0 \tag{2}$$

and the equation

$$x'(t) + p(t)x(t-\tau) = 0 \tag{3}$$

and assume that $p(t) > 0$ *(at least) on a sequence of disjoint intervals* $\{(\xi_n,t_n)\}_{n=1}^{\infty}$ *with* $t_n - \xi_n \geq 2\tau$ *and* $\lim_{n\to\infty}(t_n-\xi_n) = +\infty$. *If*

$$\liminf_{t\to\infty} \int_{t-\frac{\tau}{2}}^{t} p(s)ds > 0 \quad \text{for} \quad t \in \bigcup_{n=1}^{\infty} (\xi_n + \frac{\tau}{2}, t_n) \tag{6}$$

and

$$\liminf_{t\to\infty} \int_{t-\tau}^{t} p(s)ds > \frac{1}{e} \quad \text{for} \quad t \in \bigcup_{n=1}^{\infty} (\xi_n+\tau, t_n) \tag{7}$$

then

(1) *has no eventually positive solutions,*

(2) *has no eventually negative solutions, and*

(3) *has oscillatory solutions only.*

To prove this theorem, we first establish the following lemmas.

Lemma 2.1. *If condition (6) is satisfied, then for every eventually positive solution* $x(t)$ *of*

$$x'(t) + p(t)x(t-\tau) \leq 0 \tag{1}$$

where $p(t)$ *satisfies the assumption of Theorem 2.2, the function*

$$\frac{x(t-\tau)}{x(t)} \tag{8}$$

is bounded on $I = \bigcup_{n=N}^{\infty} (\xi_n + \frac{3\tau}{2}, t_n - \frac{\tau}{2})$ *for* N *sufficiently large.*

Proof: Suppose that $x(t)$ is a solution of (1) such that for t_0 sufficiently large

$$x(t) > 0, \quad t > t_0.$$

Then $x(t-\tau) > 0$ for $t > t_0+\tau$. Since $\xi_n \to \infty$ there exists an N such that for $n \geq N$, $\xi_n > t_0+\tau$ and therefore for $t \in \bigcup_{n=N}^{\infty} (\xi_n,t_n)$, $x(t) > 0$, $x(t-\tau) > 0$ and $x'(t) < 0$. Integrating (1) from $t-\frac{\tau}{2}$ to t for $t \in \bigcup_{n=N}^{\infty} (\xi_n+\frac{3\tau}{2},t_n)$, we obtain

$$x(t) - x(t-\frac{\tau}{2}) + \int_{t-\frac{\tau}{2}}^{t} p(s)x(s-\tau)ds \leq 0.$$

Since $s-\tau \geq t - \frac{3\tau}{2} \geq \xi_n$ it follows that $x(s-\tau) \geq x(t-\tau)$ and therefore

$$x(t) - x(t-\frac{\tau}{2}) + x(t-\tau)\int_{t-\frac{\tau}{2}}^{t} p(s)ds \leq 0, \quad t \in \bigcup_{n=N}^{\infty} (\xi_n + \frac{3\tau}{2}, t_n).$$

Dividing the last inequality first by $x(t)$ and then by $x(t-\frac{\tau}{2})$, we obtain respectively

$$1 - \frac{x(t-\frac{\tau}{2})}{x(t)} + \frac{x(t-\tau)}{x(t)} \int_{t-\frac{\tau}{2}}^{t} p(s)ds \le 0, \quad t \in \bigcup_{n=N}^{\infty} (\xi_n + \frac{3\tau}{2}, t_n) \tag{9}$$

and

$$\frac{x(t)}{x(t-\frac{\tau}{2})} - 1 + \frac{x(t-\tau)}{x(t-\frac{\tau}{2})} \int_{t-\frac{\tau}{2}}^{t} p(s)ds \le 0, \quad t \in \bigcup_{n=N}^{\infty} (\xi_n + \frac{3\tau}{2}, t_n) \tag{10}$$

If the assertion of the lemma were false, then there would exist a sequence $s_n \to \infty$ with $s_n \in \bigcup_{n=N}^{\infty} (\xi_n + \frac{3\tau}{2}, t_n - \frac{\tau}{2})$ such that

$$\lim_{n\to\infty} \frac{x(s_n-\tau)}{x(s_n)} = \infty.$$

In view of (6), inequality (9) implies

$$\lim_{n\to\infty} \frac{x(s_n - \frac{\tau}{2})}{x(s_n)} = \infty,$$

or setting $u_n = s_n + \frac{\tau}{2}$ $(\in \bigcup_{n=N}^{\infty} (\xi_n + \frac{3\tau}{2}, t_n))$

$$\lim_{n\to\infty} \frac{x(u_n - \tau)}{x(u_n - \frac{\tau}{2})} = \infty.$$

But this, for $t = u_n$, contradicts (10). The proof is complete.

Lemma 2.2. *If* $K > \frac{1}{e}$ *the sequence*

$$s_0 = e^K, \quad s_n = e^{Ks_{n-1}}, \quad n = 1,2,\ldots$$

diverges to $+\infty$.

Proof: It is easy to show by induction that the sequence s_n increases. Therefore if the assertion were false the sequence would converge to some finite limit $\ell > 0$, i.e.

$$\lim_{n\to\infty} s_n = \ell.$$

Thus

$$\ell = e^{K\ell}$$

which implies that

$$\frac{\log \ell}{\ell} = K.$$

But

$$\max_{\ell>0} \frac{\log \ell}{\ell} = \frac{1}{e}$$

and so

$$K \le \frac{1}{e}$$

which contradicts the hypothesis $K > \frac{1}{e}$. The proof is complete.

Proof of Theorem 2.2: First we prove that (1) has no eventually positive solutions. To this end suppose that $x(t)$ is a solution of (1) such that for t_0 sufficiently large

$$x(t) > 0, \quad t > t_0.$$

Then for N_1 sufficiently large

$$x(t) > 0, \quad x(t-\tau) > 0 \quad \text{and} \quad x'(t) < 0, \quad t \in \bigcup_{n=N_1}^{\infty} (\xi_n, t_n).$$

By Lemma 2.1 there exists a number M such that

$$\frac{x(t-\tau)}{x(t)} \le M, \quad t \in \bigcup_{n=N_1}^{\infty} (\xi_n + \frac{3\tau}{2}, t_n - \frac{\tau}{2}). \tag{11}$$

Next, we choose a number K such that

$$\liminf_{t\to\infty} \int_{t-\tau}^{t} p(s)ds > K > \frac{1}{e} \quad \text{for} \quad t \in \bigcup_{n=N_1}^{\infty} (\xi_n+\tau, t_n).$$

Thus, for N_2 sufficiently large and also $N_2 > N_1$,

$$\int_{t-\tau}^{t} p(s)ds \ge K > \frac{1}{e}, \quad t \in \bigcup_{n=N_2}^{\infty} (\xi_n + \tau, t_n - \frac{\tau}{2}).$$

Now consider the sequence

$$s_0 = e^K, \quad s_n = e^{Ks_{n-1}}, \quad n = 1,2,\ldots$$

By Lemma 2.2, the sequence s_n tends to $+\infty$ and therefore there exists an $N > N_2$ such that $s_N > M$. Since

$$\lim_{n\to\infty} (t_n - \xi_n) = +\infty$$

there exists an n such that

$$t_n - \xi_n \ge (N+2)\tau + \frac{\tau}{2}.$$

To complete the proof we will show that (11) is violated in the interval $(\xi_n + \frac{3\tau}{2}, t_n - \frac{\tau}{2})$.

Dividing (1) by x(t) and integrating from t-τ to t, we obtain

$$\log \frac{x(t-\tau)}{x(t)} \geq \int_{t-\tau}^{t} p(s)\,\frac{x(s-\tau)}{x(s)}\,ds, \quad t \in (\xi_n+\tau,\ t_n - \tfrac{\tau}{2}). \tag{12}$$

Since

$$\frac{x(t-\tau)}{x(t)} > 1, \quad t \in (\xi_n+\tau,\ t_n - \tfrac{\tau}{2})$$

inequality (12) yields

$$\log \frac{x(t-\tau)}{x(t)} > K, \quad t \in (\xi_n+2\tau,\ t_n - \tfrac{\tau}{2}).$$

That is,

$$\frac{x(t-\tau)}{x(t)} > e^K = s_0, \quad t \in (\xi_n+2\tau,\ t_n - \tfrac{\tau}{2}).$$

Similarly

$$\log \frac{x(t-\tau)}{x(t)} > Ke^K, \quad t \in (\xi_n+3\tau,\ t_n - \tfrac{\tau}{2})$$

that is,

$$\frac{x(t-\tau)}{x(t)} > e^{Ke^K} = s_1, \quad t \in (\xi_n+3\tau,\ t_n - \tfrac{\tau}{2}).$$

And inductively after N steps

$$\frac{x(t-\tau)}{x(t)} > s_N > M, \quad t \in (\xi_n+(N+2)\tau,\ t_n - \tfrac{\tau}{2})$$

which contradicts (11).

To prove that (2) has no eventually negative solutions it suffices to observe that if x(t) is a solution of (2) then -x(t) is a solution of (1).

From the above it follows that (3) has neither eventually positive nor eventually negative solutions and therefore every solution of (3) oscillates.

The proof of the theorem is complete.

By similar arguments we have established parallel results about the differential inequalities and equations (1)', (2)', (3)', (4) and (4)'. We omit the details for lack of space.

3. EXAMPLES

Consider the differential equations

$$x'(t) + \sin t\; x(t - \tfrac{\pi}{2}) = 0, \tag{13}$$

$$x'(t) - \sin t\, x(t + \tfrac{\pi}{2}) = 0, \tag{13'}$$

$$x'(t) + 2 \sin t\, x(t - \tfrac{\pi}{4}) + 2 \cos t\, x(t + \tfrac{\pi}{6}) = 0, \tag{14}$$

$$x'(t) - 2 \sin t\, x(t - \tfrac{\pi}{4}) - 2 \cos t\, x(t + \tfrac{\pi}{6}) = 0, \tag{14'}$$

$x'(t) + p(t)x(t-1) = 0$ where

$$p(t) = \begin{cases} p > 1/e & \text{for } t \in [2^n\pi, 2^{n+1}\pi], \quad n \text{ odd} \\ \cos t & \text{for } t \in (2^n\pi, 2^{n+1}\pi), \quad n \text{ even} \end{cases} \tag{15}$$

$$x'(t) + \sin t\, x(t - 2\pi) = 0, \quad \text{(see [1, p. 197])}, \tag{16}$$

and

$$x'(t) + \frac{\sin t}{2 + \sin t}\, x(t - \tfrac{\pi}{2}) = 0, \quad \text{([cf. 5])}. \tag{17}$$

From our results it follows that every solution of Eq. (13), (13)', (14), (14)' and (15) oscillates. However Eq. (16) and (17) have the non-oscillatory solutions $x_1(t) = e^{\cos t}$ and $x_2(t) = 2 + \cos t$ respectively. As expected the conditions of Theorem 2.1 and 2.2 are violated for the last two equations.

REFERENCES

[1] Hale, J. Theory of Functional Differential Equations, Springer-Verlag, New York, 1977.

[2] Ladas, G. Sharp Conditions for Oscillations Caused by Delays, Applicable Analysis, 9(1979), 93-98.

[3] Ladas, G., Lakshmikantham, V. and Papadakis, J. S. Oscillations of Higher-Order Retarded Differential Equations Generated by the Retarded Argument, Delay and Functional Differential Equations and Their Applications, Academic Press, Inc., New York, 1972, 219-231.

[4] Ladas, G. and Stavroulakis, I. P. On Delay Differential Inequalities of First Order, Funkcial. Ekvac., (in press).

[5] Travis, C. Oscillation theorems for second-order differential equations with functional arguments, Proc. Amer. Math. Soc., 31(1972), 199-202.

THE PRESENT STATE OF THE METHOD OF UPPER AND LOWER SOLUTIONS

V. Lakshmikantham

Department of Mathematics
The University of Texas at Arlington
Arlington, Texas

ABSTRACT

If we know the existence of upper and lower solutions w,v of a nonlinear problem satisfying $v \le w$, then we can prove existence of a solution u of the same problem such that $v \le u \le w$. This approach depends heavily on the theory of inequalities and is known as the method of upper and lower solutions. If we are further interested in establishing a constructive method which yields monotone sequences that converge to solutions, we need to employ monotone iterative technique coupled with the method of upper and lower solutions. This combined approach also offers existence of multiple solutions of the problem. Recently, because of the fruitful applications to real world problems, this method has been employed to various situations and in this paper, we give a survey of the present state of this method.

1. FIRST ORDER EQUATIONS

Let us begin by considering the Scalar IVP

$$u' = f(t,u),\ u(0) = u_0, \tag{1.1}$$

where $f \in C[J\times R,R]$ and $J = [0,T]$. If we know the existence of w and v (upper and lower solutions of (1.1)) such that $v \le w$ on J, then we can prove the existence of a solution of (1.1) such that $v(t) \le u(t) \le w(t)$ on J. Precisely, the following result is true.

Theorem 1.1. Assume that

(A_0) $v,w \in C^1[J,R]$, $v' \le f(t,v)$, $w' \ge f(t,w)$ and $v \le w$ on J.

Then, there exists a solution $u(t)$ of (1.1) satisfying $v(t) \le u(t) \le w(t)$ on J, provided $v(0) \le u(0) \le w(0)$.

The proof of this result depends crucially on the modified function approach and the theory of differential inequalities. We consider the modified IVP

$$u' = \tilde{f}(t,u), \quad u(0) = u_0, \tag{1.2}$$

where $\tilde{f}(t,u) = f(t,u) = f(t,p(t,u))$ and $p(t,u) = \max[v(t),\min(u,w(t))]$. Since $\tilde{f}$ is continuous and bounded on $J \times R$, there exists a solution $u(t)$ of (1.2) on J. If we show that this solution $u(t)$ is such that $v(t) \leq u(t) \leq w(t)$ on J, then it is clear that $u(t)$ is actually a solution of (1.1) and the theorem is proved.

To show that $v(t) \leq u(t) \leq w(t)$ on J, we consider $w_\varepsilon(t) = w(t) + \varepsilon(1+t)$, $v_\varepsilon(t) = v(t) - \varepsilon(1+t)$ for $\varepsilon > 0$ sufficiently small and prove by using the theory of differential inequalities that $v_\varepsilon(t) < u(t) < w_\varepsilon(t)$ on J. Letting $\varepsilon \longrightarrow 0$ gives the required conclusion.

If $f(t,u)$ is nonincreasing in u, then we can actually construct upper and lower solutions in a simple way and hence (A_0) becomes superfluous. This important special case is given in

<u>Theorem 1.2</u>. Suppose that $f(t,u)$ is nonincreasing in u for each $t \in J$. Then

(i) there exist lower and upper solutions v,w of (1.1) such that $v \leq w$ on J;

(ii) there exists a unique solution $u(t)$ of (1.1) such that $v(t) \leq u(t) \leq w(t)$ on J.

The proof is simple and instructive. Let $z(t)$ be a solution of $z' = f(t,0)$, $z(0) = u_0$ and set $v(t) = -R + z(t)$, $w(t) = R + z(t)$. Choose $R > 0$ so large that $v(t) \leq 0 \leq w(t)$ on J. Since f is nonincreasing, we have

$$v' = z' = f(t,0) \leq f(t,v),$$

and

$$w' = z' = f(t,0) \geq f(t,w).$$

By Theorem 1.1, it now follows that there exists a solution $u(t)$ of (1.1) such that $v(t) \leq u(t) \leq w(t)$ on J. Uniqueness of solution u of (1.1) is obvious because of nonincreasing nature of f.

The existence results given in Theorems 1.1 and 1.2 are based on theoretical considerations. We now describe a constructive method which yields monotone sequences that converge to extremal solutions of (1.1). Since the members of these monotone sequences are solutions of certain linear differential equations which can be solved explicitly, the advantage and importance of this technique needs no emphasis. See [29].

Theorem 1.3. Let the assumption (A_0) of Theorem 1.1 hold and let $v(0) \le u_0 \le w(0)$. Suppose further that

$$(A_1) \qquad f(t,u_1) - f(t,u_2) \ge -M(u_1-u_2)$$

for $v(t) \le u_2 \le u_1 \le w(t)$ and $M \ge 0$.

Then there exist monotone sequences $\{v_n\}$, $\{w_n\}$ such that $v_n \longrightarrow \rho$, $w_n \longrightarrow r$ uniformly and monotonically on J, ρ, r being the extremal solutions of (1.1).

Let us indicate the method of proof. For any $\eta \in C[J,R]$ such that $v \le \eta \le w$ on J, consider the linear differential equation

$$(1.3) \qquad u' = f(t,\eta) - M(u-\eta),\ u(0) = u_0,$$

and define a mapping A by $A\eta = u$ where $u = u(t)$ is the unique solution of (1.3) on J, which clearly exists for each η. It can be shown by using the theory of differential inequalities that

(i) $v \le Av$, $w \ge Aw$

and

(ii) A is a monotone operator on the segment

$$[v,w] = \{u \in C[J,R]: v \le u \le w\}.$$

Then it is easy to define the sequences $\{v_n\}$, $\{w_n\}$ by $v_n = Av_{n-1}$, $w_n = Aw_{n-1}$ with $v_0 = v$, $w_0 = w$ and conclude that

$$v \le v_1 \le \dots \le v_n \le w_n \le \dots \le w_1 \le w \quad \text{on} \quad J.$$

Standard arguments show that $v_n \longrightarrow \rho$, $w_n \longrightarrow r$ uniquely and monotonically. If u is any solution of (1.1) such that $v \le u \le w$ on J, then one can show using induction arguments that $v_n \le u \le w_n$ on J for all u which proves that ρ, r are minimal and maximal solutions of (1.1) such that $v \le \rho \le u \le r \le w$ on J.

Suppose that (A_0) holds and that f is nondecreasing in u. In this special case, (A_1) holds with $M = 0$. Consequently, the conclusion of Theorem 1.3 is trivially true. If f is nonincreasing in u, then we have seen in Theorem 1.2 that (A_0) holds. However (A_1) need not be satisfied. Nonetheless, we can proceed with the iteration scheme

$$v'_n = f(t, v_{n-1}),\ v_n(0) = u_0$$

which yields an alternating sequence consisting of two monotone sequences bounding the solution of (1.1), provided we assume that $v_2 \geq v$. The limits of these sequences are not necessarily solutions of (1.1). Similar conclusions follow from the other iteration scheme $w'_n = f(t, w_{n-1})$, $w_n(0) = u_0$, if we assume that $w_2 \leq w$.

The ideas involved in these two main results discussed above, namely Theorems 1.1 and 1.3, can be extended to several other situations. As an illustration, let us consider the terminal value problem (TVP)

$$(1.4) \qquad u' = f(t,u),\ u(\infty) = u_\infty,$$

where $f \in C[R_+ \times R, R]$. Extensions of Theorems 1.1 and 1.3 relative to TVP (1.4) are as follows.

Theorem 1.4. Assume that

(i) $v, w \in C^1[R_+, R]$, $v' \geq f(t,v)$, $w' \leq f(t,w)$ and $v \leq w$ on R_+;

(ii) $|f(t,u)| \leq \lambda(t)$ on $\Omega = \{(t,u): v(t) \leq u \leq w(t),\ t \in R_+\}$ and $\lambda \in L^1[0,\infty)$.

Then the TVP (1.4) has a solution $u(t)$ defined on $[a,\infty)$ for some $a \geq 0$, such that $v(t) \leq u(t) \leq w(t)$ on $[a,\infty)$, provided $v(\infty)$, $w(\infty)$ exist and satisfy $v(\infty) \leq u_\infty \leq w(\infty)$.

Theorem 1.5. Let the assumption (i) of Theorem 1.4 hold. Suppose further that $|f(t,u)| \leq \lambda(t)\,|u|$ with $\lambda \in L^1[0,\infty)$ and

$$f(t,u_1) - f(t,u_2) \leq M(t)(u_1 - u_2)$$

whenever $v(t) \leq u_2 \leq u_1 \leq w(t)$ on R_+ and $M \in L^1[0,\infty)$.

Then there exist monotone sequences $\{v_n\}$, $\{w_n\}$ such that $v_n \longrightarrow \rho$, $w_n \longrightarrow r$ as $n \longrightarrow \infty$ uniquely and monotonically on $[a,\infty)$, ρ, r being extremal solutions of TVP (1.4).

Since the terminal value problems are more complicated than the initial value problems, the proofs of Theorems 1.4 and 1.5 require additional tools. See [1] for details.

As another illustration, let us consider the periodic boundary value problem (PBVP)

$$(1.5) \qquad u' = f(t,u),\ u(0) = u(2\pi),$$

where $f \in C[[0,2\pi] \times R, R]$. In order to extend Theorems 1.1 and 1.3 to PBVP

(1.5), we need to utilize Lyapunov-Schmidt method together with an abstract set-up.

Theorem 1.6. Assume that

(i) $v,w \in C^1[[0,2\pi],R]$, $v' \leq f(t,v)$, $w' \geq f(t,w)$ for $0 \leq t \leq 2\pi$ and $v(t) \leq w(t)$ on $[0,2\pi]$;

(ii) $v(0) \leq v(2\pi)$ and $w(0) \geq w(2\pi)$.

Then there exists a solution $u(t)$ of PBVP (1.5) such that $v(t) \leq u(t) \leq w(t)$ on $[0,2\pi]$.

Assumption (ii) implies the integral inequalities

$$\int_0^{2\pi} f(s,v(s))ds \geq 0 \quad \text{and} \quad \int_0^{2\pi} f(s,w(s))ds \leq 0,$$

which are essential to prove existence of a solution of (1.5) when f is bounded. This condition is also employed in showing $v \leq u \leq w$ on $[0,2\pi]$.

Theorem 1.7. If in addition to the assumptions (i) and (ii) of Theorem 1.6, we suppose that

(iii) $f(t,u_1) - f(t,u_2) \geq -M(u_1-u_2)$ whenever $v(t) \leq u_2 \leq u_1 \leq w(t)$ for some $M \geq 0$,

then there exist monotone sequences $\{v_n\}$, $\{w_n\}$ such that $v_n \longrightarrow \rho$, $w_n \longrightarrow r$ uniformly and monotonically on $[0,2\pi]$, ρ, r being extremal solutions of PBVP (1.5).

In case of PBVP (1.5), the definition of extremal solutions is different compared to IVP (1.1). Also, in this case, we do not know a priori that $v(0) \leq u(0) \leq w(0)$ as in the case of IVP (1.1) and this creates problems. Furthermore, problem (1.5) falls under the category called "problems at resonance" and consequently, the proofs need more technical tools. For details, see [34].

Let us next consider the extension of foregoing results to finite systems. Consider first

$$u' = f(t,u), \quad u(0) = u_0, \tag{1.6}$$

where $f \in C[J\times R^n, R^n]$ and suppose that the inequalities between vectors are componentwise. Then in order for Theorem 1.1 to be valid, we need a further assumption on f called quasimonotone nondecreasing propery of f, that is

$$v \leq u,\ v_i = u_i,\ 1 \leq i < n \ \text{ implies } \ f_i(t,v) \leq f_i(t,u).$$

If f is not known to be quasimonotone, the definition of upper and lower solutions has to be strengthened for the validity of Theorem 1.1 as follows:

for each i, $1 \le i \le N$,

$$v_i' \le f_i(t,\sigma) \quad \text{for all } \sigma \text{ such that } v(t) \le \sigma \le w(t) \text{ and } \sigma_i = v_i(t) \tag{1.7}$$
$$w_i' \ge f_i(t,\sigma) \quad \text{for all } \sigma \text{ such that } v(t) \le \sigma \le w(t) \text{ and } \sigma_i = w_i(t)$$

Then the conclusion is the classical result of Müller [15,18,37]. The monotone iterative technique for (1.6) when f satisfies a mixed monotone condition can be developed which leads to the notion of quasi-solutions. See [14,28].

The corresponding developments for systems concerning TVP is open and seems to present several difficulties. However parallel results for PBVP of systems have been proved recently [46].

Consider next the IVP

$$u' = f(t,u), \; u(0) = u_0, \tag{1.8}$$

where $f \in C[J\times E,E]$, E being a real banach space. Let $K \subset E$ be a cone and by means of K, we induce a partial order defined by $v \le u$ if $u - v \in K$. Let $K^* = \{\phi \in E^*: \phi(u) \ge 0 \text{ for all } u \in K\}$, E^* being the dual of E. A function f is said to be quasimonotone relative to K, if

$$v \le u \text{ and } \phi(v-u) = 0, \; \phi \in K^* \text{ implies } \phi(f(t,v)) \le \phi(f(t,u)).$$

If $E = R^n$ and $K = R^n_+$, the inequalities induced by K are componentwise. The modified function approach which was an effective tool in the proof of Theorem 1.1 does no longer make sense since $p(t,u)$ cannot be defined. If the two ideas contained in Theorems 1.1 and 1.3 are suitably unified, then one can prove the following result [17].

Theorem 1.8. Assume that

(i) $v,w \in C^1[J,E]$, $v' \le f(t,v)$, $w' \ge f(t,w)$ and $v(t) \le w(t)$ on J;

(ii) $f(t,u_1) - f(t,u_2) \ge - M(u_1-u_2)$ whenever $v(t) \le u_2 \le u_1 \le w(t)$ for some $M \ge 0$;

(iii) the cone K is normal and for any bounded set $A \subset E$, $\alpha(f(J\times A)) \le L\alpha(A)$, $L > 0$, where α denotes the measure of noncompactness of Kuratowski.

Then there exist monotone sequences $\{v_n\}$, $\{w_n\}$ which converge uniformly and monotonically to minimal and maximal solutions ρ, r respectively of (1.8) such that for any solution u of (1.8) with $v(0) \le u_0 \le w(0)$ we have $v \le \rho \le u \le r \le w$ on J.

When K is an arbitrary cone, the inequalities (1.7) can be formulated in terms of linear functionals of K^*, namely, for $\phi \in K^*$,

$$\phi(v'-f(t,\sigma)) \leq 0 \quad \text{for all} \quad \sigma \quad \text{such that} \quad v(t) \leq \sigma \leq w(t),\ \phi(v(t)-\sigma) = 0$$
$$\phi(w'-f(t,\sigma)) \geq 0 \quad \text{for all} \quad \sigma \quad \text{such that} \quad v(t) \leq \sigma \leq w(t),\ \phi(w(t)-\sigma) = 0.$$

For arbitrary cones, the extension of classical Müller's result need not hold even in R^n [47]. For, in R^n, consider $K = \{u \in R^3 : (u_1^2+u_2^2)^{1/2} \leq u_3\}$. Choose $v = (0,0,0)$, $w = (2,0,2)$, $f_1 \equiv f_3 \equiv 0$ and

$$f_2 = \begin{cases} u, & u \in [0,1] \\ 2-u, & u \in [1,2] \\ 0 & \text{otherwise}. \end{cases}$$

But the solution through $u_0 = (1,0,1)$ is $u(t) = (1,t,1)$ which does not remain in the sector $[v,w]$. Note also that f is Lipschitzian. Thus f being Lipschitzian in addition is not sufficient to prove the extension of Müller's result in arbitrary cones. However, if the notion of upper and lower solutions is further strengthened, one can prove such a result [27].

Theorem 1.9. Assume that

(i) $v,w \in C^1[J,E]$ with $v \leq w$ on J such that there exists an $M > 0$ satisfying

$$\phi(v'-f(t,\sigma) + M(v-\sigma)) \leq 0,$$
$$\phi(w'-f(t,\sigma) + M(w-\sigma)) \geq 0,$$

for all σ such that $v(t) \leq \sigma \leq w(t)$, $t \in J$ and $\phi \in K^*$;

(ii) f is Lipschitzian and condition (iii) of Theorem 1.8 holds.

Then there exists a unique solution $u(t)$ of (1.8) such that $v(t) \leq u(t) \leq w(t)$ on J, provided $v(0) \leq u(0) \leq w(0)$.

If f admits a decomposition of the form $f = f_0 + f_1 + f_2$ where f_1 is nondecreasing, f_2 is nonincreasing and f_2 satisfies (ii) of Theorem 1.8, it is possible to prove [30] that there exists a solution $u(t)$ of (1.8) such that $v \leq u \leq w$ on J and such a result includes several special cases of importance.

Consider the PBVP

$$u' = f(t,u),\ u(0) = u(2\pi), \tag{1.9}$$

where $f \in C[[0,2\pi]\times E,E]$. Recall that when $E = R^n$, Lyapunov-Schmidt method is employed to solve the problem (1.9). However when E is a Banach space, this approach is no longer useful since the null space of $u' = 0$, $u(0) = u(2\pi)$ is not finite dimensional. As a result, the methods employed for finite dimensional problems are not suitable. Nonetheless, the monotone iterative technique used in Theorem 1.8 is appropriate in the present case since the corresponding linear PBVP can be solved directly. However, in the present setup we do not know $v(0) \leq u(0) \leq w(0)$ and this creates further problems. For further details, see [18a].

Finally consider the IVP for delay differential equation

$$u'(t) = f(t,u(t),u_t), \quad u_{t_0} = \phi_0, \tag{1.10}$$

where $f \in C[[t_0,t_0+a]\times E\times \mathcal{C},E]$, $\mathcal{C} = c[[-\tau,0],E]$ and $\phi_0 \in \mathcal{C}$. Here $u_t = u_t(s) = u(t+s)$, $s \in [-\tau,0]$ and $u_t \in \mathcal{C}$ for each $t \in [t_0,t_0+a]$. Under suitable conditions, monotone method has been successfully developed for the IVP (1.10). For technical details, see [18].

2. SECOND ORDER EQUATIONS

Let us consider the scalar BVP

$$\begin{cases} -u'' = f(t,u,u'), \quad 0 < t < 1, \\ B^i u = \alpha_i u(i) + (-1)^{i+1}\beta_i u'(i) = b_i, \quad i = 0,1, \end{cases} \tag{2.1}$$

where $f \in C[I\times R\times R,R]$, $I = [0,1]$, $\alpha_i,\beta_i \geq 0$ and $\alpha_i^2 + \beta_i^2 \neq 0$. The following classical result is well known; se [8a].

Theorem 2.1. Assume that

(H_0) $v,w \in C^2[I,R]$ with $v \leq w$ on I and $-v'' \leq f(t,v,v')$, $-w'' \geq f(t,w,w')$;

(H_1) f satisfies a Nagumo condition.

Then there exists a solution $u(t)$ of (2.1) such that $v(t) \leq u(t) \leq w(t)$ on I, provided $B^i v \leq b_i \leq B^i w$.

The proof depends on the modified function approach and the theory of differential inequalities as before. However, construction of the modified function is different. For example, the modified function $\tilde{f}$ is defined as follows:

$$\tilde{f}(t,u,u') = f(t,p(t,u),q(u')) + r(u),$$

where $p(t,u) = \max[v, \min(u,w)]$, $q(u') = \max[-c, \min(u',c)]$ and

$$r(u) = \begin{cases} \dfrac{u-w}{1+u^2} & \text{if } u > w, \\ 0 & \text{if } v \le u \le w, \\ \dfrac{u-v}{1+u^2} & \text{if } u < v, \end{cases}$$

where $c > \max[N, |v'|, |w'|]$, N bieng the constant obtained from Nagumo condition (H_1).

Since $\tilde{f}$ is bounded, there exists a solution $u(t)$ for BVP

$$- u'' = \tilde{f}(t,u,u'), \ B^i u = b_i$$

and we show that this $u(t)$ is actually a solution of (2.1) by showing that $v \le u \le w$ and $|u'| \le N$ on I.

When $f(t,u,u')$ is nonincreasing in u, we can actually construct v,w such that (H_0) holds. We let $z(t)$ be a solution of

$$- z''(t) = f(t,0,z'), \ B_i z = b_i$$

and set $R + z$, $w = -R + z$. Choosing $R > 0$ sufficiently large so that $v \le 0 \le w$, noting that $v' = z' = w'$ and using nonincreasing nature of f, we can easily verify that (H_0) holds.

To develop the monotone method for BVP (2.1), one can proceed using ideas of Theorem 1.3. Since the gradient term in f offers difficulties, this development was achieved in successive stages; see [12,8]. The best result known may be stated as follows.

<u>Theorem 2.2.</u> Assume that (H_0) and (H_1) hold. Suppose further that (H_2) $f(t,u_1,u') - f(t,u_2,u') \ge - M(u_1-u_2)$ whenever $v(t) \le u_2 \le u_1 \le w(t)$ and $|u'| \le c$ for some $M \ge 0$.
Then there exist monotone sequences $\{v_n\}$, $\{w_n\}$ such that $v_n \longrightarrow \rho$, $w_n \longrightarrow r$ as $n \longrightarrow \infty$ uniformly and monotonically, ρ, r being the minimal and maximal solutions of BVP (2.1) respectively, provided $B^i v \le B^i u \le B^i w$.

The results of Theorems 2.1 and 2.2 can be extended to finite or countably infinite systems which are weakly coupled, namely,

$$\text{(2.2)} \qquad \begin{cases} - u_i'' = f_i(t,u,u_i'), \ B^i u = b_i, \\ i = 1,2,\ldots, \end{cases}$$

by appropriately modifying the notion of upper and lower solutions and incorporating the needed mixed quasimonotone conditions. See for details [10,31,13].

There are several dificulties in extending these results to BVP's in a Banach space. When f is independent of u' or when f depends on u' linearly, it is possible to obtain the existence of extremal solutions by combining the method of upper and lower solutions and the monotone technique. In this case, the iterates are solutions of simple linear BVP's and consequently are computable. See [7,48] for necessary discussion. The general problem remains open.

Consider next the scalar PBVP

$$-u'' = f(t,u,u'),\ u(0) = u(2\pi),\ u'(0) = u'(2\pi),$$

where $f \in C[[0,2\pi]\times R\times R,R]$. By combining the method of upper and lower solutions and Lyapunov-Schmidt method, one can prove the following results [22,23,35].

Theorem 2.3. Assume (H_0) and (H_1) hold. Suppose further that

(H_4) $v(0) = v(2\pi)$, $w(0) = w(2\pi)$, $v'(0) \geq v'(2\pi)$ and $w'(0) \leq w'(2\pi)$.

Then there exists a solution $u(t)$ of (2.3) such that $v(t) \leq u(t) \leq w(t)$ on $[0,2\pi]$.

Theorem 2.4. If (H_3) holds, in addition to the assumptions of Theorem 2.3, then there exist monotone sequences $\{v_n\}$, $\{w_n\}$ such that $v_n \longrightarrow \rho$, $w_n \longrightarrow r$ uniformly and monotonically as $n \longrightarrow \infty$ and ρ, r are extremal solutions of PBVP (2.3).

The extension of the foregoing results to finite systems remains open. Furthermore, extending PBVP (2.3) to an arbitrary Banach space is very difficult. However, if f is independent of u', then it appears that the extension is possible. This is also an open problem.

Special boundary value problems occurring in transport phenomena are considered in [33,39] and a syetematic study of stochastic scalar BVP is given in [11].

3. PARTIAL DIFFERENTIAL EQUATIONS

Consider the IVP for first order partial differential equation

$$u_t + f(t,x)u_x = g(t,x,u),\ u(0,x) = \phi(x), \tag{3.1}$$

where $f \in C[\Omega,R^n]$, $g \in C[\Omega\times R,R]$, $fu_x = \sum^n f_i(t,x)u_{x_i}$, $\phi \in C^1[[a,b],R]$ and $\Omega = \{(t,x): 0 \leq t \leq T$ and $a \leq x \leq b,\ a,b \in R^n\}$. Recently monotone technique is developed for problem (3.1) under suitable conditions [32]. The monotone sequences converge to functions which are not solutions of (3.1) since ρ,r need not be continuously differentiable. If additional assumptions are imposed so that they are solutions of (3.1), then it also follows that the solution is unique and hence $\rho = u = r$. The problem of obtaining extremal solutions seems to be difficult and is open. For detailed discussion, see [32].

In [5] monotone technique is employed for semilinear elliptic BVP and in [6] the method is extended to systems when f does not depend on the gradient term. General results for systems remain open as well as problems at resonance in this setup. For special cases, see [21,42,43]. For results corresponding to Theorem 2.1, see [28,49,5b,3,2,43]. For the theory of inequalities at resonance see [44].

Concerning parabolic initial boundary value problems, various results are known [2,9.19,24,26,40-43] for scalar case as well as for systems. Again many papers deal with the case when f does not depend on the gradient term since this causes certain technical difficulties. General results for systems remain open. For the theory of inequalities at resonance see [45].

These ideas have not yet found their way into hyperbolic equations since the techniques need a totally different approach. For nonlinear operators in ordered Banach spaces, there are some interesting results incorporating these ideas. See [4,5a,36]. Also for Volterra integral equation, see [25].

REFERENCES

[1] A. R. Aftabizadeh and V. Lakshmikantham, "On the theory of terminal value problems for ordinary differential equations," J. Nonlinear Analysis, 5, (1981), 1173-1180.

[2] H. Amann, "Invariant sets and existence theorems for semilinear parabolic and elliptic systems," J. Math. Anal. and Appl., 65, (1978), 432-467.

[3] H. Amann, "On the existence of positive solutions of nonlinear elliptic boundary value problems," Indiana Univ. Math. J., 21, (1971), 124-146.

[5] H. Amann and M. G. Crandall, "On some existence theorems for semilinear elliptic equations," Indiana Univ. Math. J., 27, (1978), 779-790.

[5a] H. Amann, "Nonlinear operators in a Banach space and some applications to nonlinear boundary value problems, in nonlinear operators and calculus of variations," Lecture Notes in Math #543, Springer-Verlag, 1976.

[5b] H. Amann and P. Hess, "A multiplicity result for a class of elliptic boundary value problems," Proc. Royal Soc. Edinburgh, 84, (1979), 145-152.

[6] R. Aris, "Weakly coupled systems of nonlinear elliptic boundary value problems," J. Nonlinear Analysis, 6, (1982), 555-569.

[7] S. R. Bernfeld and V. Lakshmikantham, "Linear monotone method for nonlinear boundary value problems in Banach space," Rocky Mountain J., (to appear).

[8] S. R. Bernfeld and J. Chandra, "Minimal and maximal solutions of nonlinear boundary value problems," Pacific J. Math. 71, (1977), 13-20.

[8a] S. R. Bernfeld and V. Lakshmikantham, An introduction to nonlinear boundary value Problems, Academic Press, New York, 1974.

[9] J. Chandra, F. Norman and F. Dressel, "A monotone method for a system of nonlinear parabolic differential equations," Proc. of Royal Soc. Edinburgh, 87A, (1981), 209-217.

[10] J. Chandra, V. Lakshmikantham and S. Leela, "A monotone method for infinite systems of nonlinear boundary value problems," Arch. Rat. Mech. Anal., 68, (1978), 179-190.

[11] J. Chandra, G. S. Ladde and V. Lakshmikantham, "On the fundamental theory of nonlinear second order stochastic boundary value problems, Stochastic Anal. and Appl., 1, (1983)

[12] J. Chandra and P. W. Davis, "A monotone method for quasilinear boundary value problems," Arch. Rat. Mech. Anal., 54, (1974), 257-266.

[13] K. M. Das and M. P. Devasahayam, "Quasisolutions and monotone method for infinite systems of nonlinear boundary value problems," Nonlinear Analysis, (to appear).

[14] K. Deimling and V. Lakshmikantham, "Quasisolutions and their role in the qualitative theory of differential equations," J. Nonlinear Analysis, 4, (1980), 657-663.

[15] K. Deimling and V. Lakshmikantham, "Existence and comparison theorems for differential equations in a Banach space," J. Nonlinear Analysis, 3, (1979), 569-575.

[16] K. Deimling and V. Lakshmikantham, "On existence of extremal solutions of differential equations in Banach spaces," J. Nonlinear Analysis, 3, (1979), 563-568.

[17] S. W. Du and V. Lakshmikantham, "Monotone iterative technique for differential equations in a Banach space," J. Math. Anal. and Appl., 87, (1982), 454-459.

[18] S. W. Du, "Monotone iterative technique for delay differential equations in abstract cones," (to appear).

[18a] S. W. Du, V. Lakshmikantham and A. S. Vatsala, "Monotone iterative technique for first order differential equations with periodic boundary conditions in a Banach space," (to appear).

[19] J. Hernandez, "Some existence and stability results for solutions of reaction diffusion systems with nonlinear boundary conditions," Nonlinear Differential Equations: Invariance, Stability and Bifurcation, Academic Press, New York, 1981, pp. 161-173.

[20] F. Inkmann, "Existence and multiplicity theorems for semilinear elliptic equations with nonlinear boundary conditions." Indian Univ. Math. J., 31, (1982), 213-221.

[21] H. B. Keller, "Elliptic boundary value problems suggested by nonlinear diffusion processes," Arch. Rat. Mech. Anal., 35, (1969), 363-381.

[22] R. Kannan and V. Lakshmikantham, "Existence of periodic solutions of nonlinear boundary value problems," Tech. Rep. #173, U. T. Arlington.

[23] R. Kannan and V. Lakshmikantham, "Existence of periodic solution of semilinear parabolic equations and the method of upper and lower solutions," J. Math. Anal. and Appl., (to appear).

[24] G. S. Ladde, V. Lakshmikantham and A. S. Vatsala, "Existence and asymptotic behaviour of reaction diffusion systems in a coupled quasi-solution," Proc. of the Int'l. Conf. on Nonlinear Phenomena in Math. Sciences, Academic Press, New York, 1982, pp. 619-628.

[25] G. S. Ladde, V. Lakshmikantham and B. G. Pachpatte, "The method of upper, lower solutions and Volterra integral equations," J. Integral Equations, 4, (1982), 353-360.

[26] V. Lakshmikantham, "Comparison results for reaction-diffusion equations in a Banach space," Lecture notes of talks delivered at the Conference on "A survey of theoretical and numerical trends in nonlinear analysis," at Bari, Italy.

[27] V. Lakshmikantham and S. Leela, "The method of upper and lower solutions in abstract cones," Anal. Polon Math., (to appear).

[28] V. Lakshmikantham, S. Leela and M. N. Oguztoreli, "Quasi-solutions vector Lyapunov functions and monotone method," IEEE, Trans Autom. Control, 26, (1981).

[29] V. Lakshmikantham and S. Leela, Differential and integral inequalities, Vol. 1, Academic Press, New York, 1969.

[30] V. Lakshmikantham, S. Leela and A. S. Vatsala, "Method of quasi-upper and lower solutions in Abstract cones," J. Nonlinear Analysis, 6, (1982), 833-838.

[31] V. Lakshmikantham and A. S. Vatsala, "Quasisolutions and monotone method for systems of nonlinear boundary value problems," J. Math. Anal. Appl., 79, (1981), 38-47.

[32] V. Lakshmikantham, M. N. Oguztoreli and A. S. Vatsala, "Monotone iterative technique for semilinear partial differential equations of first order," (to appear in J. Math. Anal. Appl.).

[33] V. Lakshmikantham and B. G. Pachpatte, "Monotone method for nonlinear boundary value problems arising in transport processes," J. Appl. Math. and Comp., 9, (1981), 35-41.

[34] V. Lakshmikantham and S. Leela, "Existence and Monotone method for periodic solutions of first order differential equations," J. Math. Anal. and Appl., (to appear).

[35] S. Leela, "Monotone method for second order periodic boundary value problems," J. Nonlinear Analysis, (to appear).

[36] Jane Moore, "Existence of multiple quasi-fixed points of mixed monotone operators by iterative techniques," J. Appl. Math. and Comp., 9, (1981), 135-141.

[37] M. Müller, "Uber das fundamental theorem in der theorie der gewöhnlichen differential gleichungen," Math. Z., 26, (1926), 619-645.

[38] M. Nagumo, "On principally linear elliptic differential equations of the second order," Osaka Math. J., 6, (1954), 207-229.

[39] B. G. Pachpatte and A. S. Vatsala, "Monotone method for equations describing transport phenomena in a Banach space," Appl. Math. and Comp., 9, (1981), 165-174.

[40] C. V. Pao, "Asymptotic behaviour and nonexistence of global solutions of a class of nonlinear boundary value problems of parabolic type," J. Math. Anal. and Appl., 65, (1978), 616-637.

[41] C. V. Pao, "On nonlinear reaction-diffusion systems," J. Math. Anal. and Appl., 87, (1982), 165-198.

[42] D. H. Sattinger, "Monotone methods in nonlinear elliptic and parabolic boundary value problems," Indiana Univ. Math. J., 21, (1972), 979-1000.

[43] K. Schmitt, "Boundary value problems for quasilinear second order elliptic equations," J. Nonlinear Analysis: Theory, Methods and Appls., 3, (1978), 263-309.

[44] G. R. Shendge and A. S. Vatsala, "Comparison results for first and second order boundary value problems at resonance," J. Appl. Math and Comp., (1982), to appear).

[45] G. R. Shendge and A. S. Vatsala, "Comparison results for parabolic differential equations at resonance," J. Appl. Math. and Comp., (to appear).

[46] A. S. Vatsala, "On the existence of periodic quasi-solutions for first order systems," J. Nonlinear Analysis, (to appear).

[47] P. Volkman, "Ausdehnungeiner Satzer von maxmüller auf unendliche systeme von gewöhnlichen differentialgleichungen," Funkcialaj Ekvacioj, 21, (1978), 81-96.

[48] G. F. Von Harten, "Dissertation," Padeborn, (1979).

[49] F. Tomi, Über semilineare elliplische Differentialgleichungen zweiter ordnung," Math. z., III, (1969), 350-366.

ON A CONJECTURE ON THE NUMBER OF SOLUTIONS OF A NONLINEAR DIRICHLET PROBLEM WITH JUMPING NONLINEARITY

A. C. Lazer

Department of Mathematics
University of Miami
Coral Gables, Florida

P. J. McKenna

Department of Mathematics
University of Florida
Gainesville, Florida

ABSTRACT

We consider nonlinear Dirichlet problems in which the range of the derivative of the nonlinearity includes at least n eigenvalues. We show that for the ordinary differential equation, there exist at least 2n solutions for a described part of the range of the nonlinear operator. Some remarks are made on the conjecture for the partial differential equation.

1. INTRODUCTION

In this paper we review some recent progress on the study of semilinear elliptic equations of the form

$$\begin{aligned} \Delta u + f(u) &= h(x) && \text{in } \Omega \\ u &= 0 && \text{on } \partial\Omega \end{aligned}$$

where the range of the derivative of f contains at least the first two eigenvalues of the Laplacian.

Let $\Omega \subset \mathbb{R}^N$ $(N\geq 1)$ be a bounded domain with boundary $\partial\Omega$ of class $C^{2+\alpha}$ $(0 < \alpha < 1)$. Let $h \in C^{\alpha}(\bar{\Omega})$ and let f be a C^1 function defined on $(-\infty,\infty)$. It follows from a classical result on nonlinear integral equations due to Hammerstein [5] that if λ_1 denotes the smallest eigenvalue of the problem

(1) $\Delta u + \lambda u = 0$ in Ω, $u|\partial\Omega = 0$

and there exists a number γ such that $f'(\xi) \leq \gamma < \lambda_1$ for all $\xi \in (-\infty,\infty)$, then the problem

(2) $\Delta u + f(u) = h(x)$, $u|\partial\Omega = 0$

has a unique solution in $C^{2+\alpha}(\bar{\Omega})$. Subsequent work of Dolph [4] showed that if there exist two numbers γ and γ' such that $\gamma \leq f'(\xi) \leq \gamma'$ for all $\xi \in (-\infty,\infty)$ and the interval $[\gamma,\gamma']$ contains no eigenvalue of the problem (1), then (2) is uniquely solvable in $C^{2+\alpha}(\bar{\Omega})$. In [2] Ambrosetti and Prodi considered the case in which the range of f' contains an eigenvalue of (1). They showed that the conditions, $-\infty < f'(-\infty) < \lambda_1 < f'(\infty) < \lambda_2$ and $f'' > 0$ on $(-\infty,\infty)$, where $f'(\pm\infty) = \lim_{\xi\to\pm\infty} f'(\xi)$, and λ_2 is the second eigenvalue of (1), imply the existence of a closed connected C^2 manifold M of codimension one in the Banach space $C^{\alpha}(\bar{\Omega})$ such that $C^{\alpha}(\bar{\Omega}) - M$ has exactly two components U_0 and U_2 with the property that (2) has no solution if $h \in U_0$, exactly two solutions if $h \in U_2$, and exactly one solution if $h \in M$. Subsequently, Manes and Micheletti [10] were able to weaken the condition $-\infty < \lim_{\xi\to-\infty} f'(\xi) < \lambda_1$ to $-\infty \leq \lim_{\xi\to-\infty} f'(\xi) < \lambda_1$. In [7] Kazdan and Warner showed that less precise results of the Ambrosetti-Prodi type could be established under much less restrictive grwoth conditions on f. It follows from the results of [7] that if the C^1 function f satisfies the growth conditions

(3) $$\limsup_{\xi\to-\infty} \frac{f(\xi)}{\xi} < \lambda_1 < \liminf_{\xi\to+\infty} \frac{f(\xi)}{\xi},$$

if θ denotes a normalized eigenfunction of (1) corresponding to λ_1, and if $h_1 \in C^{\alpha}(\bar{\Omega})$ is orthogonal to θ with respect the usual $L^2(\Omega)$ inner-product then there exists a number $S_0 = S_0(h_1)$ such that the problem

(4) $\Delta u + f(u) = h_1(x) + S\theta(x)$ in Ω $u|\partial\Omega = 0$

has no solution if $S < S_0$ and at least one solution if $S > S_0$. This result, which elucidated the structure of the sets U_0 and U_2 in the Ambrosetti-Prodi theorem, was obtained using the method of upper and lower solutions. Simultaneously, Dancer [3] and Amann and Hess [1] used degree theoretic methods to show that if $f(\xi)$ satisfies a certain growth condition as $\xi \to \infty$ then (4) has at least two solutions for $S > S_0(h_1)$ and is solvable for $S = S_0(h_1)$. Amann and Hess require $f(\xi)$ to satisfy a linear growth condition as $\xi \to \infty$ where as Dancer has a stronger result which allows $f(\xi)$ to grow no faster than ξ^r as $\xi \to \infty$ where $r > 1$ depends on the dimension of Ω.

Recently the present authors showed in [9] that if λ_2 is a simple eigenvalue, if $\lambda_2 < \lim_{\xi\to\infty} f'(\xi) < \lambda_2$, and if (3) holds then there exists a number $S_1 = S_1(h_1)$ such that for $S > S_1(h_1)$ the problem (4) has at least three solutions. The proof of this result was based on the fact that for large S the problem (2) has a solution Z_S such that Z_S/S is close to the solution of the linear problem

$$\Delta u + f'(\infty)u = \theta, \qquad u|\partial\Omega = 0,$$

and on degree theory. The proof showed that "generically" there should be four solutions for sufficiently large S.

This led us to conjecture in [9] that if the eigenvalues $\{\lambda_m\}_1^\infty$ of (1) are arranged in increasing order, each counted as often as its multiplicity, $\lambda_n < \lambda_{n+1}$, and $\lambda_n < f'(\infty) < \lambda_{n+1}$ then there exists $S_1 = S_1(h_1)$ such that if $S > S_1$ then (4) has at least 2n solutions.

The purpose of this paper is to show that the conjecture is true in the one-dimensional case. For convenience we take Ω to be the interval $(0,\pi)$ since the boundary value problem $u''(t) + \lambda u(t) = 0$, $u(0) = u(\pi) = 0$ has a nontrivial solution if and only if $\lambda = n^2$, $n = 1,2,\ldots$. We prove the following result:

<u>If</u> $f \in C^1(-\infty,\infty)$, $\liminf_{\xi\to-\infty} f(\xi)/\xi < 1$, <u>and</u> $n^2 < \lim_{\xi\to\infty} f'(\xi) < (n+1)^2$ <u>for some integer</u> $n \geq 1$, <u>then for any</u> $h_1 \in C[0,\pi]$ <u>satisfying</u>

$$\int_0^\pi h_1(t)\sin t\, dt = 0$$

<u>there exists a number</u> $S_1 = S_1(h_1)$ <u>such that for</u> $S > S_1(h_1)$ <u>the boundary value problem</u>

$$\text{(5)}\quad \begin{aligned} &u''(t) + f(u(t)) = h_1(t) + S\sin t \\ &u(0) = u(\pi) = 0 \end{aligned}$$

<u>has at least</u> 2n <u>solutions</u>. Actually the result is true without the orthogonality condition but since every function in $C[0,\pi]$ can be represented in the form of right hand side of (5), we have chosen to state the result as above. Theorem 2, below, will be a more precise form of the above result which will give a qualitative description of 2n solutions.

For a specific nonlinearity f which is asymptotically linear in one direction D. Hart and the second author [6] have established the exact number of solutions of the autonomous Neumann problem

$$u''(t) + f(u(t)) = S,\ u'(0) = u'(\pi) = 0$$

where S is a large parameter. The principal tool used in [6] is phase plane analysis which is not possible here. Our principal tools are fixed point theorems, the method of upper and lower solutions and "shooting methods".

2. It will be convenient to set $f(u) = \lambda u + g(u)$ where $\lambda = \lim_{u\to\infty} f'(u)$.

From our assumptions on f it follows that

(6) $g'(u) \to 0,\ g(u)/u \to 0$ as $u \to \infty$

and

(7) $\lambda + \liminf_{u\to-\infty} g(u)/u < \lambda_1$.

A version of our first theorem was established in [9] by means of Leray-Schauder degree. Here we establish a more precise result, which includes uniqueness as well as existence, using the contraction mapping principle.

Theorem 1. There exists a number δ_0 such that for $0 < \delta \le \delta_0$ there exists a number $\hat{S}(\delta)$ such that for $S \ge \hat{S}(\delta)$ the boundary value problem

(8) $u'' + \lambda u + g(u) = S \sin t + h_1(t)$

$u(0) = u(\pi) = 0$

has a unique solution $Z_S(t)$ which satisfies $||Z_S - SX_0||_1 < S\delta$ where $X_0(t) = \dfrac{\sin t}{\lambda - 1}$. Moreover $||(Z_S/S) - X_0||_1 \to 0$ as $S \to \infty$.

Here, for $F \in C^1[0,\pi]$, we set $||F||_1 = \max(\max_{[0,\pi]} |F(t)|, \max_{[0,\pi]} |F'(t)|)$.

Proof: Setting $u(t) = Sv(t)$ in (8) for $S > 0$ we obtain the boundary-value problem

(9) $v'' + \lambda v + g\dfrac{(Sv)}{S} + \sin t + \dfrac{h_1(t)}{S}$

$v(0) = v(\pi) = 0.$

We view this problem as a perturbation of the linear boundary value problem

$$x''(t) + \lambda x(t) = \sin t,\ x(0) = x(\pi) = 0$$

which has the explicit solution $X_0(t)$. The problem (9) is equivalent to the integral equation

$$V(t) = X_0(t) - \int_0^{\pi} G(t,r)\left[\frac{h_1(r)}{S} - \frac{g(SV(r))}{S}\right]dr$$

where G is the Green's function for the boundary value problem $y'' + \lambda y = f \quad y(0) = y(\pi) = 0$.

Let us choose $\delta_0 > 0$ such that $\|V - X_0\|_1 \le \delta_0$ implies $V(t) > 0$ for $0 < t < \pi$. For $0 < \delta \le \delta_0$ let $B(\delta)$ denote the functions in $C^1[0,\pi]$ which satisfy $\|V - X_0\|_1 \le \delta$. If I is a compact subinterval of $(0,\pi)$ then $C(I) \cap B(\delta)$ is a compact subset of $C(I)$ and $\min_I V(t)$ is a continuous function on $C(I)$. Therefore, there exists $\alpha(I,\delta)$, independent of $V \in B(\delta)$, such that $V \in B(\delta)$ implies that

(10) $V(t) \ge \alpha(I,\delta) > 0$ if $t \in I$.

For $V \in B(\delta)$ define $T_S V = V^*$ where for $0 \le t \le \pi$

$$V^*(t) = X_0(t) - \int_0^{\pi} G(t,r)\left[\frac{h_1(r)}{S} - \frac{g(sv(r))}{S}\right]dr.$$

Let L_1 be a bound for $|g'|$ on $[0,\infty)$ and let L_2 be a bound for both $|G|$ and $\left|\frac{\partial G}{\partial t}\right|$ on $[0,\pi] \times [0,\pi]$. Let $\epsilon > 0$ be chosen so $2\epsilon < \pi$ and $2\epsilon L_1 L_2 < 1/4$. Since $g'(x) \to 0$ as $x \to \infty$ there exists $S^*(\delta) > 0$ defined for $0 < \delta \le \delta_0$ such that $x \ge S^*(\delta)\alpha([\epsilon,\pi-\epsilon],\delta)$ implies $|g'(x)| \le 1/4L_2\pi$. It then follows that if $V_1, V_2 \in B(\delta)$ and $S \ge S^*(\delta)$ then for all $t \in [\epsilon,\pi-\epsilon]$

$$|g(SV_1(t)) - g(SV_2(t))| \le \frac{1}{4L_2\pi} S\,|V_1(t) - V_2(t)|$$

and consequently

$$|V_1^*(t) - V_2^*(t)| \le \int_0^{\pi} |G(t,r)|\,\frac{|g(SV_1(r) - g(SV_2(r))|}{S}\,dr \le$$

$$L_1L_2\int_0^{\epsilon} |V_1(r) - V_2(r)|dr + L_2\int_{\epsilon}^{\pi-\epsilon} \frac{1}{4L_2\pi}|V_1(r) - V_2(r)|dr$$

$$+ L_1L_2\int_{\pi-\epsilon}^{\pi} |V_1(r) - V_2(r)|dr \le (2\epsilon L_1L_2 + \tfrac{1}{4})\|V_1 - V_2\|_1$$

$$< \tfrac{1}{2}\|V_1 - V_2\|_1.$$

In exactly the same way it follows that for $0 \le t \le \pi$

$$\left|\frac{d}{dt}(V_2^*(t) - V_2^*(t)\right| \le \frac{1}{2}\|V_1 - V_2\|_1.$$

Therefore, $||T_sV_1 - T_sV_2||_1 \le \frac{1}{2} ||V_1 - V_2||_1$

for all $V_1, V_2 \in B(\delta)$.

Since $g(x)/x \to 0$ as $x \to \infty$ there exists a number $\hat{S}(\delta) \ge S^*(\delta)$ such that if $S \ge \hat{S}(\delta)$ then

$$L_2 \int_0^\pi \left| \frac{g(Sx_0(r))}{S} - \frac{h_1(r)}{S} \right| dr \le \frac{\delta}{2}.$$

It follows that for $S \ge \hat{S}(\delta)$, $t \in [0,\pi]$, and $V \in B(\delta)$,

$$|V^*(t) - X_0(t)| \le \int_0^\pi |G(t,r)| \frac{|g(SV(r)) - g(SX_0(r)|}{S} dr$$

$$+ \int_0^\pi |G(t,r)| \left| \frac{g(SX_0(r))}{S} - \frac{h_1(r)}{S} \right| dr.$$

The exact same argument as before shows that the first term in the right-hand side of the last inequality is less than $\frac{1}{2} ||V - X_0||_1$. Hence for $S \ge \hat{S}(\delta)$, $t \in [0,\pi]$

$$|V^*(t) - X_0(t)| \le \frac{1}{2} ||V - X||_1 + \frac{\delta}{2} \le \delta.$$

In the same way it follows that for $S \ge \hat{S}(\delta)$, and $t \in [0,\pi]$, $\left|\frac{dV^*}{dt}(t) - \frac{dX_0}{dt}(t)\right| \le \delta$. Thus $T_s(B(\delta)) \subseteq B(\delta)$ for $S \ge \hat{S}(\delta)$ so by the contraction mapping principle it follows that for $S \ge \hat{S}(\delta)$ there exists a unique $V_s \in B(\delta)$ such that $T_s V_s = V_s$. By earlier remarks if $Z_s(t) \equiv S\, V_s(t)$ then Z_s is a solution of (8).

The last statement follows from the uniqueness of Z_s satisfying the theorem, the condition

$$||(Z_s/s) - X_0||_1 < \delta \quad \text{for } S \ge \hat{S}(\delta),$$

and the fact that $\delta > 0$ can be made arbitrarily small. This proves the theorem.

In the differential equation $y'' + \lambda y + g(y) = S \sin t + h(t)$, we set $y(t) = u(t) + Z_s(t)$ and we arrive at the following differential equation for u

(11) $u'' + \lambda u + g(u + Z_s) - g(Z_s) = 0.$

The proof of our main theorem uses three lemmas which we now state. The proofs of these lemmas will be defered until after the proof of the main theorem.

Lemma 2.1. *There exists a constant* S_0 *such that for* $S \ge S_0$ (11) *has a solution* $U_0(t)$ *such* that $U_0(0) = U_0(\pi) = 0$, $U_0'(0) < 0$, $U_0'(\pi) > 0$, *and* $U(t) < 0$ for $0 < t < \pi$.

Lemma 2.2 *There exists a constant* $S_1 \geq S_0$ *such that for* $S \geq S_1$ (11) *has solutions* V_1, V_2, *and* W, *defined on* $[0,\pi]$ *such that* $V_1(0) = V_2(0) = W(\pi) = 0$, $V_1'(0) < 0$, $V_2'(0) > 0$, $W'(\pi) > 0$, *and all three solutions have* n *simple zeros on the open interval* $(0,\pi)$.

Lemma 2.3 *Let* $p \geq 0$ *be an integer and suppose that for some* $S \geq S_1$ *there exists a solution* V *of* (11) *satisfying the boundary conditions* $V(0) = V(\pi) = 0$ *such that* V *has exactly* p *simple zeros on the open interval* $(0,\pi)$. *Suppose that there exists a solution* y *of* (11) *such that*

(12) $y(0) = 0$, $\operatorname{sgn} y'(0) = \operatorname{sgn} v'(0)$, and y has $q > p+1$ zeros in $(0,\pi)$.

Then for any integer k *with* $p < k < q$ *there exists a solution* U of (11) *such that* $U(0) = U(\pi) = 0$, $\operatorname{sgn} U'(0) = \operatorname{sgn} y'(0)$, *and* U *has exactly* k *simple zeros* on $(0,\pi)$. *If* (12) *is changed to*

(13) $y(\pi) = 0$, $\operatorname{sgn} y'(\pi) = \operatorname{sgn} V'(\pi)$

then there exists a solution U of (11) *with exactly* k *simple zeros on* $(0,\pi)$ *satisfying* $U(0) = U(\pi) = 0$, $\operatorname{sgn} U'(\pi) = \operatorname{sgn} V'(\pi)$.

Remark Since the function $U \equiv 0$ is a solution of (11), it follows by the uniqueness theorem that the zeros of a nontrival solution are all simple.

Our main result is

Theorem 2. *There exists a number* S_1 *such that for* $S > S_1$ (11) *has* 2n-1 *distinct solutions* $U_0, U_1, \ldots, U_{n-1}, \Theta_1, \Theta_2, \ldots, \Theta_{n-1}$ *such that* $U_j(0) = U_j(\pi) = 0$ $j = 0,\ldots,n-1$, $\Theta_j(0) = \Theta_j(\pi) = 0$ $j = 1,\ldots n-1$. *These solutions are distinguished by the* conditions $U_j'(0) < 0$ *and* $U_j(t)$ *has exactly* j *simple zeros on* $(0,\pi)$ *for* $j = 0,\ldots,n-1$, *where as* $\Theta_j'(0) > 0$ *and* $\Theta_j(t)$ *has exactly* j *simple zeros on* $(0,\pi)$ *for* $j = 1,\ldots,n-1$.

Consequently, for $S > S_1$ *the boundary value problem* (5) *has the* 2n *distinct solutions* Z_s, $Z_s + U_j$ for $j=0,\ldots,n-1$, *and* $Z_s + \Theta_j$ for $j=1,\ldots,n-1$.

Proof: Let U_0 be as the statement of Lemma 1.2 which exists for $S \geq S_0$. Let $S \geq S_1 \geq S_0$ and let V_1 be as in the statement of Lemma 2.2. Since $V_1(0) = 0$, $V_1'(0) < 0$, $U_0'(0) < 0$, U_0 has no zeros on $(0,\pi)$ and V_1 has n simple zeros on $(0,\pi)$, it follows from Lemma 2.3 that for each integer j with $1 \leq j \leq n-1$, there exists a solution U_j of (11) such that U_j has exactly j simple zeros on the interval $(0,\pi)$ and $U_j(0) = U_j(\pi) = 0$.

Now consider U_0 and the solution W of Lemma 2.2. Since $W(\pi) = 0$, $W'(\pi) > 0$, $W(t)$ has $n \geq 2$ simple zeros on $(0,\pi)$, where as $U_0(0) = U_0(\pi) = 0$ and U_0 has no zeros on $(0,\pi)$, Lemma 2.3 (second-part) implies the existence of a solution Θ_1 of (11) satisfying $\Theta_1(0) = \Theta_1(\pi) = 0$ and

$\Theta_1'(\pi) > 0$ such that Θ_1 has exactly one simple zero on $(0,\pi)$. Since Θ_1 is a nontrivial solution it follows that $\Theta_1'(0) > 0$.

We now apply Lemma 2.3 to the two solutions $\Theta_1(t)$ and the solution $V_2(t)$ of Lemma 2.2. Since $V_2(0) = 0$, $V_2'(0) > 0$ and V_2 has n simple zeros on $(0,\pi)$ it follows from Lemma 2.3 that if $n > 2$, then for each integer j with $1 < j < n-1$ there exists a solution $\Theta_j(t)$ of (11) such that $\Theta_j(0) = \Theta_j(\pi) = 0$, $\Theta_j'(0) > 0$ and Θ_j has exactly j simple zeros on $(0,\pi)$. This completes the proof of the theorem.

Proof of Lemma 2.1: We prove the lemma by the method of upper and lower solutions (see [11]). Since $n^2 < \lambda < (n+1)^2$, there exists a unique function $\psi_1(t)$ such that

$$\psi_1'' + \lambda\psi_1 = -1, \ \psi_1(0) = \psi_1(\pi) = 0.$$

Since the function sin t is positive on $(0,\pi)$ and has simple zeros at the end points of this interval there exists small constant $k > 0$ such that

$$(14) \quad \psi_2(t) \equiv k\psi_1(t) - \sin t < 0, \ t \in (0,\pi).$$

We claim that there exists a number $S_0 > 0$ such that for $S > S_0$ ψ_2 is an upper solution for the boundary value problem

$$(15) \quad U'' + \lambda U + g(Z_s(t) + U) - g(Z_s(t)) = 0 \qquad U(0) = U(\pi) = 0.$$

We choose S_o so large that $S \geq S_0$ implies $Z_s(t) \geq 0$ on $[0,\pi]$. If r_2 is a bound for $|\psi_2(t)|$ on $[0,\pi]$ and M is a bound for $|g'(\xi)|$ when $-r_2 \leq \xi < \infty$ then $|g(Z_s(t) + \psi_2(t)) - g(Z_s(t))| \leq M\,|\psi_2(t)|$. Since $\psi_2(0) = \psi_2(\pi) = 0$ there exists a number $\alpha > 0$ such that $|\psi_2(t)| \leq k/2M$ for $0 \leq t \leq \alpha$ and for $\pi - \alpha \leq t \leq \pi$. It follows that for $t \in [0,\alpha]$ or $t \in [\pi - \alpha, \pi]$

$$(16) \quad \psi_2''(t) + \lambda\psi_2(t) + g(Z_s(t) + \psi_2(t)) - g(Z_s(t))$$

$$\leq -(\lambda-1) \sin t - k + M (k/2M \leq 0.$$

Now consider the interval $[\alpha, \pi-\alpha]$. Since as $S \to \infty, Z_s(t) \to \infty$ uniformly with respect to $t \in [\alpha, \pi-\alpha]$, and $g'(\xi) \to 0$ as $\xi \to \infty$ we may assume S_0 is so large that if $t \in [\alpha,\pi-\alpha]$ then $|g(Z_s(t) + \psi_2(t)) - g(Z_s(t))| \leq k/2$ for all $S \geq S_0$. Consequently, for $S \geq S_0$ and $\alpha \leq t \leq \pi-\alpha$

$$(17) \quad \psi_2''(t) + \lambda\psi_2(t) + g(Z_s(t) + \psi_2(t)) - g(Z_s(t))$$

$$\leq -(\lambda - 1) \sin t - k + k/2 \leq 0.$$

Since $\psi_2(0) = \psi_2(\pi) = 0$, it follows from (16) and (17) that ψ_2 is an upper solution for the boundary value problem (15).

The conditions (6) and (7) imply the existence of constants b_1 and $b_2 \geq 0$ such that $0 < b_1 < 1$ and

(18) $g(\xi) + \lambda\xi \geq b_1\,\xi - b_2$ for all $\xi \in (-\infty,\infty)$.

In fact, since $\lambda > n \geq 1$, it follows from (6) and (7) that there exists a number $C_0 > 0$ and $b_1 \in (0,1)$ such that $\frac{g(\xi)}{\xi} + \lambda > b_1$ if $\xi \geq C_0$ and $\frac{g(\xi)}{\xi} + \lambda < b_1$ if $\xi \leq -C_0$.

If $b_2 \geq 0$ is chosen so that $g(\xi) - \lambda\xi - b_1\,\xi \geq -b_2$ for $-C_0 \leq \xi \leq C_0$ then (18) follows.

Let ψ_3 be the unique solution of the boundary value problem

$$\psi_3''(t) + b_1\,\psi_3(t) = b_2 + g(Z_s(t)) + (\lambda - b_1)\,Z_s(t) \qquad \psi_3(0) = \psi_3(\pi) = 0$$

where $S \geq S_0$ is fixed. Let ψ_2 be as above and let $C > 0$ be so large that

(19) $\psi_4(t) \equiv \psi_3(t) - c\sin t \leq \psi_2(t)$ on $[0,\pi]$.

We have

$$\psi_4''(t) + \lambda\psi_4(t) + g(\psi_4(t) + Z_s(t)) - g(Z_s(t)) =$$

$$\psi_4''(t) + \lambda(\psi_4(t) + Z_s(t)) + g(\psi_4(t)) + Z_s(t)) - g(Z_s(t)) - \lambda Z_s(t) \geq$$

$$\psi_4''(t) + b_1(\psi_4(t) + Z_s(t) - b_2 - g(Z_s(t)) - \lambda Z_s(t) = (1 - b_1)\,c\sin t \geq 0$$

for $0 \leq t \leq \pi$. Since $\psi_4(0) = \psi_4(\pi) = 0$, $\psi_4(t)$ is a lower solution for the boundary value problem (15) which, according to (19), is less than the upper solution $\psi_2(t)$. From this we infer the existence of a solution U_0 of the problem (15) such that $\psi_4(t) \leq U_o(t) \leq \psi_2(t) < 0$ on $(0,\pi)$. The conditions $U_0'(0) < 0$ and $U_0'(\pi) > 0$ follow from the fact that the zeros of $U_0(t)$ must be simple. This proves the lemma.

<u>Proof of Lemma 2.2:</u> We prove the existence of the solution $V_2(t)$ - - the proofs of the existence of $V_1(t)$ and $W(t)$ are based on the same type of argument and will only be indicated.

Since $n^2 < \lambda < (n+1)^2$ the function $\Theta(t) = \sin\sqrt{\lambda}\,t$ has exactly n simple zeros on $(0,\pi)$. It follows that there exists a number $\varepsilon > 0$ such that if $U \in C^1[0,\pi]$ satisfies $U(0) = \Theta(0) = 0$, $U'(0) = \Theta'(0)$ and $||\Theta(t) - U(t)||_1 \leq \varepsilon$ then U has exactly n simple zeros on $(0,\pi)$. We may also suppose that $\varepsilon < 1$. From the variation of parameters formula, or as may be verified directly, if $\psi \in C^1[0,\pi]$ satisfies the integral equation

(20) $$\psi(t) = \int_0^t \frac{1}{\sqrt{\lambda}}\sin\sqrt{\lambda}(t-r)[g(Z_s(r) + \psi(r) + \Theta(r)) - g(Z_s(r))]dr$$

then $\psi(0) = \psi'(0) = 0$, $\psi \in C^2$ and

$\psi''(t) + \lambda\psi(t) + g(Z_s(t) + \Theta(t) + \psi(t)) - g(Z_s(t)) = 0$. We show that there exists a number $S_1 \geq S_0$ such that for $S \geq S_1$, (20) has a solution ψ such that $||\psi||_1 < \varepsilon$. Since $\Theta'' + \lambda\Theta = 0$ it will then follow that $V_2 \equiv \Theta + \psi$ is a solution of (11) which satisfies the assertion of Lemma 2.2.

Let M be a bound for $|g'(X)|$ if $X \in [-2, \infty)$ and let $\delta > 0$ be a number such that $2\delta < \pi$ and

(21) $4 M \delta < \varepsilon/3$.

Since $g'(X) \to 0$ as $X \to \infty$ there exists a positive number X_0 such that $|g'(X)| < \varepsilon/6\pi$ for $X \ge X_0$. By Theorem 1, there exists a number $S_1 \ge S_0$ such that for $S \ge S_1$ the solution $Z_s(t)$ satisfies $Z_s(t) \ge X_0 + 2$ for $\delta \le t \le \pi-\delta$. If $|C| \le 2$ and $S \ge S_1$ it follows from the mean value theorem that for $\delta \le t \le \pi-\delta$

(22) $$|g(Z_s(t) + C) - g(Z_s(t)| \le \frac{\varepsilon|C|}{6\pi} \le \frac{\varepsilon}{3\pi}.$$

Let A be the closed, convex and bounded subset of $C^1[0,\pi]$ such that $\phi \in A$ iff $\phi(0) = \phi'(0) = 0$ and $||\phi||_1 \le \varepsilon$. For fixed $S \ge S_1$ we define a mapping $N_s : A \to C^1[0,\pi]$ by

$$(N_s\phi)(t) = \frac{1}{\sqrt{\lambda}} \int_0^t (\sin \sqrt{\lambda}(t-r))(g(Z_s(r) + \Theta(r) + \phi(r)) - g(Z_s(r)))dr$$

for $\phi \in A$. Since $|\Theta(t) + \phi(t)| \le |\Theta(t)| + |\phi(t)| \le 1 + \varepsilon \le 2$ it follows from (21), (22), and the mean-value theorem that for all $t \in [0,\pi]$

$$N_s\phi(t)| \le \int_0^\pi |g(Z_s(r) + \Theta(r) + \phi(r)) - g(Z_s(r))|\, dr \le$$

$$\int_0^\delta 2M\, dr + \int_\delta^{\pi-\delta} \frac{\varepsilon}{3\pi}\, dr + \int_{\pi-\delta}^\pi 2M\, dr < 4\, M\, \delta + \frac{\varepsilon}{3} < \frac{2\varepsilon}{3}.$$ Since

$$(N_s\phi)'(t) = \int_0^t (\cos \sqrt{\lambda}\,(t-r))[g(Z_s(r) + \Theta(r) + \phi(r)) - g(Z_s(r))]\, dr$$

the same reasoning shows that $|(N_s\phi)'(t)| \le 2\varepsilon/3$ for all $t \in [0,\pi]$. Therefore, since $(N_s\phi)(0) = (N_s\phi)'(0) = 0$, it follows that $N_s(A) \subseteq A$. If $\phi \in A$ and $U = N_s\phi$ then $U''(t) = -\lambda U(t) - g(Z_s(t) + \Theta(t) + \phi(t)) + g(Z_s(t))$ so $|U''(t)| \le k$ for all $t \in [0,\pi]$ where the constant k does not depend on ϕ. Therefore N_s is compact. Since N_s is obviously continuous it follows by Schauder's fixed point theorem that there exists $\psi \in A$ such that $N_s\psi = \psi$. Since $||\psi||_1 \le \varepsilon$ and ψ is a solution of the integral equation (20), it follows by earlier remarks that $V_2(t) = \psi(t) + \Theta(t)$ satisfies the assertion of Lemma 2.2.

The proofs of the existence of $V_1(t)$ and $W(t)$ follow along the same lines by considering the solutions of $y'' + \lambda\, y = 0$ given by $-\sin \sqrt{\lambda}\, t$ and $\sin \sqrt{\lambda}\,(t-\pi)$ respectively instead of Θ.

Proof of Lemma 2.3: We first show that for fixed $S \ge S_1$ there exists a number r_1 such that every solution of the boundary value problem

(15) $U''(t) + \lambda U(t) + g(Z_s(t) + U(t)) - g(Z_s(t)) = 0, U(0) = U(\pi) = 0$

satisfies $U(t) \geq r_1$ on $[0,\pi]$. According to (18) $\lambda(U(t) + Z_s(t)) + g(U(t) + Z_s(t)) \geq b_1(U(t) + Z_s(t)) - b_2$ where $b_1 < 1$, so if $U(t)$ is a solution of (15) then $U''(t) + b_1U(t) \leq (\lambda - b_1)Z_s(t) + g(Z_s(t)) + b_2$ for $0 \leq t \leq \pi$. If ψ is the unique solution of $\psi''(t) + b_1\psi(t) = (\lambda - b_1)Z_s(t) + g(Z_s(t)) + b_2$ $\psi(0) = \psi(\pi) = 0$, and $\psi_1(t) = U(t) - \psi(t)$, then $-\psi_1'' - b_1\psi_1 \geq 0$, $\psi_1(0) = \psi_1(\pi) = 0$. Since $b_1 < 1$, an explicit calculation shows that the Green's function of the problem $-y'' - b_1 y = f$ $y(0) = y(\pi) = 0$ is non negative. (For another proof see [8 p.222]). Consequently $\psi_1(t) \geq 0$ on $[0,\pi]$ so $U(t) \geq r_1 \equiv \min_{[0,\pi]} \psi(t)$ for $0 \leq t \leq \pi$.

Now assume that V and y satisfy the hypothesis of Lemma 2.3. We choose r_2 such that $r_2 \leq y(t)$, $r_2 \leq V(t)$ on $[0,\pi]$, $r_2 < 0$ and $r_2 \leq r_1$. We define $F(t,U)$ for $0 \leq t \leq \pi$, $-\infty < U < \infty$ by
$F(t,U) = \lambda(U + Z_s(t)) + g(U + Z_s(t)) - \lambda Z_s(t) + g(Z_s(t))$ if $r_2 \leq U$
and by
$F(t,U) = b_1(U + Z_s(t)) - b_2 - \lambda Z_s(t) + g(Z_s(t)) + h(t)$ if $U \leq r_2$ where $h(t) = \lambda(r_2 + Z_s(t)) + g(r_2 + Z_s(t)) - [b_1(r_2 + Z_s(t)) - b_2]$. According to (18) $h(t) \geq 0$ on $[0,\pi]$. If U is a solution of the boundary value problem

(23) $U''(t) + F(t,U(t)) = 0, \quad U(0) = U(\pi) = 0$

then $U''(t) + b_1U(t) \leq (\lambda - b_1)Z_s(t) + g(Z_s(t) + b_2$ so, by the argument given above, it follows that $U(t) \geq r_1 \geq r_2$ on $[0,\pi]$. Hence, <u>any solution</u> of (23) <u>is a solution</u> of (15). Moreover, by the definition of F, y and V are solutions of the differential equation $U'' + F(t,u) = 0$. Since $g(u)$ grows linearly as $U \to +\infty$ the function F is continuous and satisfies a uniform Lipshitz condition with respect to the second variable. It follows that if $U(t,a)$ denotes the solution of the initial value problem $U'' + F(t,U) = 0$, $U(0,a) = 0$, $U'(0,a) = a$ then $U(t,a)$ is defined and continuous for $0 \leq t \leq \pi$ and $-\infty < a < \infty$. Since $r_2 < 0$, $F(t,0) \equiv 0$ so the uniqueness theorem for differential equations implies that <u>the zeros of a nontrival solution of</u> $U'' + F(t,U) = 0$ <u>on</u> $[0,\pi]$ <u>are all simple</u>.

Since $V'(0)\, y'(0) > 0$, if <u>a</u> is between $V'(0)$ and $y'(0)$ then $U(t,a)$ is a nontrival solution of $U'' + F(t,U) = 0$. Let k be an integer with $p < k < q$. Let $a_1 = V'(0)$ and $a_2 = y'(0)$ so that $U(t,a_1)$ has $p < k$ simple zeros on $(0,\pi)$ and $U(t,a_2)$ has $q > k$ simple zeros on $(0,\pi)$. For definiteness we assume $a_1 < a_2$ -- the case $a_2 < a_1$ is treated similarly. If $\bar{a}$

is the infimum of the numbers $a > a_1$ such that $U(t,a)$ has $k + 1$ simple zeros on $(0,\pi)$ then $a_1 < \bar{a} \leq a_2$. The equality $a_1 = \bar{a}$ is ruled out by the conditions $U(0,a_1) = 0$, $U'(0,a_1) \neq 0$, and $U(t,a_1)$ has fewer than $k + 1$ simple zeros on the half-open interval $(0,\pi]$. Indeed, the continuity of $U(t,a)$ and $U'(t,a)$ in $\underline{a}$ and t implies that for $\underline{a}$ close to a_1, $U(t,a)$ has fewer than $k + 1$ zeros on the half-open interval $(0,\pi]$. The same reasoning shows that $U(t,\bar{a})$ must have at least $k + 1$ zeros on the half open-interval $(0,\pi]$. Otherwise, since all the zeros of $U(t,\bar{a})$ are simple, $U(t,a)$ would have fewer than $k + 1$ zeros on $(0,\pi]$ for a slightly greater than $\bar{a}$, which contradicts the definition of $\bar{a}$. If $U(t,\bar{a})$ had at least $k + 1$ simple zeros on the open interval $(0,\pi)$ then so would $U(t,a)$ for a slightly less than $\bar{a}$, which contradicts the definition of $\bar{a}$. It follows that $U(t,\bar{a})$ has exactly k simple zeros on $(0,\pi)$ and $U(\pi,\bar{a}) = 0$. Since $U(0,\bar{a}) = 0$ and $\operatorname{sgn} U'(0,\bar{a}) = \operatorname{sgn} y'(0)$, the conclusion of the first part of Lemma 2.3 follows. The proof of the second part follows in the same way.

REFERENCES

[1.] Amann, H and Hess, P. - "A multiplicity result for a class of elliptic boundary value problems," Proc. Royal Soc. Edinburgh 84A (1979) 145-151.

[2.] Ambrosetti, A. and Prodi G. "On the inversion of some differentiable mappings with singularities between Banach spaces," Ann. Math. Pura Appl. 93 (1972), 231-247.

[3.] Pancer, E. N. "On the ranges of certain weakly nonlinear elliptic partial differential equations," J. Math. Pures et Appl. 57 (1978), 351-366.

[4.] Dolph, C. L., "Nonlinear Integral equations of the Hammersteing type," Trans. Amer. Math. Soc. 60 (1949), 289-307.

[5.] Hammerstein A., "Nichtlineare Integralgleichungen nebst anwendungen, Acta Math. 54 (1929), 117-176.

[6.] Hart, D. C. and McKenna, P. J., "On the number of solutions of a generalized Ambrosetti-Prodi problem," to appear.

[7.] Kazdan, J. L., and Warner F. W., "Remarks on some quasilinear elliptic equations," Comm. Pure and Appl. Math. 28 (1975), 837-846.

[8.] Keller, H. B., "Some positone problems suggested by nonlinear heat generaration," In Bifurcation and Nonlinear Eigenvalue Problems (J.B. Keller and S. Antman Eds.) Benhamin, New York, 1969.

[9.] Lazer, A. C. and McKenna P. J. "On the number of solutions of a nonlinear Dirichlet problem," J. Math. Anal. Appl. 84 (1981).

[10.] Manes A., and Micheletti A.M., "Un estensione della teoria variazionale classica degli autovalori per operatori ellittici del secundo ordine," Boll. U. M. I. 7 (1973), 285-301.

[11.] Sattinger D. H. "Topics in Stability and Bifurcation Theory," Lecture Notes in Mathematics, No. 309, Springer-Verlag, Berlin, Heidelberg, New York, 1973.

OPERATOR PARTS OF ORDINARY DIFFERENTIAL SUBSPACES WITH APPLICATION TO NONLINEAR PROBLEMS

Sung J. Lee
Department of Mathematics
University of South Florida
Tampa, Florida

ABSTRACT

We compute all possible operator parts of the inverse relation of a regular ordinary differential operator in a Hilbert space. The use of an operator part to a nonlinear problem is discussed and is also compared with the alternative method.

§1. NOTATION AND PRELIMARIES

Most of this work is taken from a joint work of Lee and Nashed [6]. Let [a,b] be a closed bounded interval and τ be the nth order of expression:

$$\tau y: = \sum_{0}^{n} p_j(x)y^{(j)}, \quad a \leq x \leq b.$$

Here $p_m \varepsilon C^j\left([a.b]\right)$ and $p_n(x) \neq 0$ on [a,b]. Let $\tau^{\dagger}$ denote the adjoint of τ. Let X denote the Hilbert space $L_2[a.b]$ with the usual inner product < , >. If $F = \left\{f_{ij}\right\}$, $G = \left\{g_{ij}\right\}$ are m x n and n x p matrices, then < F,G > will denote the m x p matrices whose (i,j) -entry is $\sum_{k=1}^{n} < f_{ik}, g_{kj} >$. Thus

$$< F,G > = \int F\bar{G}dx,$$

$$< F,G >^t = < G^*,F^* >,$$

$$\overline{< F,G >} = < \bar{F},\bar{G} >.$$

Let T_1 $(T_1^{\dagger})$ be the graph of the maximal closed operator in X generated by $\tau(\tau^{\dagger})$, respectively. For y ε Dom T_1, or Dom $T_1^{\dagger}$, $\tilde{y}(x)$ will denote the n x 1 vector $[y(x), y(x), \ldots, y^{(n-1)}(x)]^t$. Let $C_1(a)$ and $C_1(b)$ be the n x n invertible matrices such that

$$\int_a^b \left[(\tau y)\bar{z} - y\,\overline{(\tau^{\dagger}z)}\right] dx = \tilde{y}(b)C_1(b)\tilde{z}(b) - \tilde{y}(a)C_1(a)\tilde{z}(a)$$

for all yε Dom T_1, z ε Dom $T_1^{\dagger}$. Define boundary operators B on T_1,

$B^\dagger$ on $T_1^\dagger$ by

$$B(\{y,\tau\}: = [\tilde{y}^t(a),\ \tilde{y}^t(b)],\quad B^\dagger(\{z,\tau^\dagger z\}): = [\tilde{z}^t(a),\ \tilde{z}^t(b)]$$

for $y \in \text{Dom } T_1$, $z \in \text{Dom } T_1^\dagger$. Then

$$\int_a^b \left[(\tau y)\bar{z} - y\,\overline{(\tau^\dagger z)} \right] dx = i\, B(\{y,\ \tau y\})C\left(B^\dagger(\{z,\tau^\dagger z\})\right)^*$$

for all $y \in \text{Dom } T_1$, $z \in \text{Dom } T_1^\dagger$.

Here

$$C: = \left(\begin{array}{c|c} {}_1C_1(a) & 0 \\ \hline 0 & -{}_1C_1(b) \end{array}\right).$$

It is well-known that T_1, $T_1^\dagger$ are closed and that $(T_1^\dagger)^* \subset T_1$.

§ 2. OPERATOR PARTS

Suppose that $M \subset X \oplus X$ is a closed vector space. Then $R \subset X \oplus X$ is called an operator part of M^{-1} if R is the graph of a closed linear operator such that M^{-1} is the direct sum of R and $\{0\} \oplus$ Null M. For a complete development, see Lee and Nashed [6]. The main problem is the following: Let T be an arbitrary, but fixed closed linear operator such that $(T^\dagger)^* \subset T \subset T_1$. Find all possible operator parts of T^{-1} and $(T^*)^{-1}$. It is shown in an abstract theorem of Lee and Nashed [6] that such operator parts are necessarily compact integral operators. Thus the main purpose here is to compute them explicitly. Now, as is well-known, we can write

$$T: = \left\{\{y,\tau y\} \in T_1 \mid P\left(B(\{y,\tau y\})\right)^* = 0_{m \times 1}\right\},$$

where P is a m x 2n ($m \leq 2n$) constant matrix of rank m, and hence T^* can be written as

$$T^*: = \left\{\{z,\tau^\dagger z\} \in T_1^\dagger \mid \tilde{P}\left(B^\dagger(\{z,\tau^\dagger z\})\right)^* = 0_{(2n-m) \times 1}\right\}$$

where P is any (2n − m) x 2n constant matrix of rank 2n − m such that

$$PC_1^{*-1}\tilde{P}^* = 0_{m \times (2n - m)}.$$

Let Θ be an 1 x n fundamental matrix solution of $\tau y = 0$ and $\Delta^\dagger$ be the one for $\tau^\dagger y = 0$.

Define functions K_0, $K_0^\dagger$ [a,b] $\oplus$ [a,b] by

$$K_0(x,t) := \begin{cases} \theta(x) \begin{pmatrix} \theta(t) \\ \theta'(t) \\ \vdots \\ \theta^{(n-1)}(t) \end{pmatrix} \begin{pmatrix} 0_{(n-1)\times 1} \\ P_n^{-1}(t) \end{pmatrix}, & a \le t \le x \le b, \\ 0 & , a \le x \le t \le b, \end{cases}$$

$$K_0^{\dagger}(x,t) := \begin{cases} \Delta^{\dagger}(x) \begin{pmatrix} \Delta^{\dagger}(t) \\ (\Delta^{\dagger})'(t) \\ \vdots \\ (\Delta^{\dagger})^{(n-1)}(t) \end{pmatrix} \begin{pmatrix} 0_{(n-1)\times 1} \\ P_n^{\dagger}(t) \end{pmatrix}, & a \le t \le x \le b, \\ 0 & , a \le x \le t \le b. \end{cases}$$

Here $P_n^{\dagger}$ is the leading coefficient of $\tau^{\dagger}$.

An operator part R of T^{-1} is called the orthogonal operator part if T^{-1} is the orthogonal sum of R and $\{0\} \oplus \text{Null } T$.

Proposition 2.1. Let R_1 be the orthogonal operator part of T_1^{-1}, and $R_1^{\dagger}$ be the one for $(T_1^{\dagger})^{-1}$. Then they are given as follows:

Dom R_1 = Dom R_1 = X,

$$\begin{cases} R_1(g) := \mathcal{K}_0(g) - \theta\,(<\theta^*, \overline{\theta}>)^{-1} < \mathcal{K}_0(g), \theta >^t, \\ R_1^{\dagger}(g) := \mathcal{K}_0^{\dagger}(g) - \Delta^{\dagger}\Big(<(\Delta^{\dagger})^*, \overline{\Delta}^{\dagger}>\Big)^{-1} \quad \mathcal{K}_0(g), \Delta^{\dagger} >^t, \end{cases}$$

for $g \in X$.

Here

$$\mathcal{K}_0(g) := \int_a^b K_0(x,t)g(t)dt, \quad \mathcal{K}_0^{\dagger}(g) := \int_a^b K_0^{\dagger}(x,t)g(t)dt.$$

Proof: R_1 is the orthogonal operator part of T_1^{-1} if and only if

$$R_1 := \Big\{ \{\tau y, (I-\mathcal{P})(y)\} \mid y \in \text{Dom } T_1 \Big\}.$$

Here $\mathcal{P}$ is the orthogonal projector from X onto Null T_1 (see Lee and Nashed [6]. Now, letting $g = \tau y$ and using the variation of -constant formula and simplifying the above R_1, we get the expression for R_1 in the theorem. Similarly, we can obtain the representation for $R_1^{\dagger}$.

Let k: = dim Null T. Then

$$\dim \text{Null } T^* = k + m - n.$$

Let

$$D: = \begin{pmatrix} \tilde{\theta}\ (a) \\ \tilde{\theta}\ (b) \end{pmatrix}^t, \quad D^{\dagger}: = \begin{pmatrix} \tilde{\Delta}^{\dagger}\ (a) \\ \tilde{\Delta}^{\dagger}\ (b) \end{pmatrix}^t$$

Let E (k x n) be the matrix such that

$$\text{rank } E = k, \ EDP^* = 0_{k \times m}.$$

Let $E^{\dagger}\left((k + m - n) \times n\right)$ be the matrix such that

$$\text{rank } E^{\dagger} = k + m - n, \ E^{\dagger} D^{\dagger} P^* = 0_{(k + m - n) \times (2n - m)}.$$

Finally, if F is a p x q matrix, then $(.F)^{\#}$ will denote the q x p matrix such that if we treat $(.F)^{\#}$ as an operator from $\mathbb{C}^q$ into $\mathbb{C}^p$, then it is a generalized inverse of the operator $\alpha \longmapsto \alpha F$.

<u>Theorem 2.2.</u> Let R_1 be the orthogonal operator part of T_1^{-1}. Then R is an operator part of T^{-1} if and only if R is defined as follows:

$$\text{Dom } R = \text{Range } T = \left\{ g \varepsilon X \mid \begin{pmatrix} \tilde{R}_1(g)(a) \\ \tilde{R}_1(g)(b) \end{pmatrix}^t P^* \varepsilon \text{ Range } (.DP^*) \right\},$$

$$R(g): = \eta(g) - \theta E^t \langle \overline{E\theta}^*, \overline{\theta E}^* \rangle^{-1} \langle \overline{E\theta}^*, \overline{\eta(g)} \rangle - A\left(\{g, \eta(g)\}\right)$$

for all g ε Dom R.

Here A is a bounded linear operator from X ⊕ X into Null T with the property that

$$A\left(\{0, y\}\right) = 0, \text{ all } y \in \text{Null } T,$$

and η (g) denotes the function:

$$\eta(g): = R_1(g) - \begin{pmatrix} \tilde{R}_1(g)(a) \\ \tilde{R}_1(g)(b) \end{pmatrix}^t P^* \left(\cdot DP^*\right)^{\#} \theta^t.$$

In particular, the orthogonal operator part of T^{-1} is given as R in the above with $A \equiv 0$.

Proof. Let P be the orthogonal projector from X onto Null T. Then the orthogonal operator part of T^{-1} is given by

$$R_0: = \left\{ \{\tau y, (I - P)(y)\} \mid y \varepsilon \text{ Dom } T \right\}.$$

Letting $g = \tau y$ and considering the boundary conditions for T, we can compute R_0. Now it follows from Theorem 1.9 of Lee and Nashed [6], R is an operator part of T^{-1} if and only if

$$R: = \left\{\{ g, R_0(g) - A\left(\{g, R_0(g) \}\right)\} \mid g \in \text{Range } T\right\}$$

for same continuous linear operator A from $X \oplus X$ into Null T such that

$$A\left(\{0,y\}\right) = 0, \text{ all } y \in \text{Null } T.$$

This R, after simplifying, will coincide with the result in the theorem.

Theorem 2.3. Let $R_1^{\dagger}$ be the orthogonal operator part of $(T_1^{\dagger})^{-1}$. Then $R^{\dagger}$ is an operator part of $(T^*)^{-1}$ if and only if $R^{\dagger}$ is given as in Theorem 2.2 where in this case R_1 is replaced by $R_1^{\dagger}$, P by $\tilde{P}$, D by $D^{\dagger}$, Θ by $\Delta^{\dagger}$, E by $E^{\dagger}$. In particular, the orthogonal operator part of $(T^*)^{-1}$ is given as the above $R^{\dagger}$ with $A \equiv 0$.

Example of Operator part.

Let

$$T: = \left\{ \{y,y''\} \mid y' \in A\,C\,[0,2\pi],\ y'' \in L_2\,[0,2\pi],\ y(0) = y(2\pi),\ y'(0) = y'(2\pi)\right\}.$$

We consider $T \subset L_2[0,2\pi] \oplus L_2\,[0,2\pi]$ and wish to find all operator parts of T^{-1}. Define operators U and V (the latter depends on u and v) by

$$U(g): = \int_0^x (x-s)g(s)ds - \frac{1}{4\pi}\int_0^{2\pi} s^2 g(s)ds + \frac{\pi + x}{2\pi}\int_0^{2\pi} sg(s)ds,$$

for $g \in$ Range T,

$$V(g): = \int_0^{2\pi} g(s)v(s)ds + \int_0^{2\pi}\left\{\int_0^t (t-s)g(s)ds + \frac{\pi + t}{2\pi}\int_0^{2\pi} sg(s)ds\right\} u(t)\,dt$$

for $g \in$ Range T.

Here u is an element in $L_2\,[0,2\pi]$ such that $\int_0^{2\pi} u(s)ds = 0$, and v is an element in $L_2\,[0,2\pi]$. Then U is the orthogonal operator part, and R is an operator part if and only if $R = U + V$.

We can check that

$$U = \left(T|_{(\text{Null } T)^{\perp}}\right)^{-1},$$

and

$$\text{Range } T = \left\{ g \in L_2\,[0,2\pi] \;\middle|\; \int^{2\pi} g(s)ds = 0 \right\}$$

§ 3 Nonlinear Problem

Suppose that H_1, H_2 are Hilbert spaces. Let $L \subset H_1 \oplus H_2$ be a closed linear operator and $N \subset H_1 \oplus H_2$ a nonlinear operator. Let R be an arbitrary, but fixed operator part of L^{-1}, and define a nonlinear operator S_k($k \in$ Null L) by

$$S_k(y) = RN(y) + k$$

for $y \in \text{Dom } S_k := \text{Dom } RN$.

Proposition 3.1

$L(y) = N(y)$ if and only if $y = S_k(y)$ for same $k \in$ Null L.

Proof: We know that $L^{-1} = R \dotplus (\{0\} \oplus \text{Null } L)$, direct sum. Suppose $L(y) = N(y)$. Then $\{y, N(y)\} \in L$, or $\{N(y), y \in L^{-1}$. Thus $N(y) = u$, $y = R(u) + k$ for some $u \in$ Dom R, $k \in$ Null L. This can be written as

$$y = RN(y) + k =: S_k(y).$$

Suppose now $y = S_k(Y)$. Then $y = RN(y) + k$. Since $k \in$ Null L Range $R \subset$ Dom L, it follows that $y \in$ Dom L. Now $\{y - k, N(y)\} \in R^{-1} \subset L$. Hence $L(y - k) = N(y)$, or $L(y) = N(y)$.

Remark: The strength of using S_k is that Dom S_k does not depend on boundary conditions. Note that Range $S_k \subset$ Dom L, as Range $R \subset L$.

The above proposition resembles the alternative method. Thus we give some differences between them. The basic differences are:

(i) The class of "pseudoinverses" of L via operator parts of L^{-1} is much larger than that obtained by considering projectors from H_1 onto Null L and those from H_2 onto (Range L)$^{\perp}$.

(ii) The operator part makes use of only one projector, while the alternative method makes use of two projectors by using generalized inverses.

To be more specific about (ii), let $L^{\#}$ denote a generalized inverse defined by

$$L^{\#} := \left\{ \{g, (I-P)(y)\} \;\middle|\; y \in \text{Dom } L,\ Ly = g - (P^{+})^{*}(g) \right\}$$

where P is a continuous projector from H_1 onto Null L, and P^{+} is a continuous projector from H_2 onto (Range $L)^{\perp}$. For the definition of $L^{\#}$, see Lee and Nashed [6] for a multi-valued case, and Nashed [8] for a single-valued case.

Then because $L^{\#}$ is an extension of an operator part R of L (by enlarging the domain of R without changing its value), the following holds:
Ly = Ny if and only if

(Auxiliary Equation) $y = L^{\#}(I-P^{+})\,N(Y) + k,$

(Bifurcation Equation) $(I-LL^{\#})\,N(y) = 0$

for same k ε Null L.

REFERENCES

[1] L. Cesari, Functional analysis, nonlinear differential equations and the alternative method, Nonlinear Functional Analysis and Differential Equations (Cesari, Kannan and Schuur, ed). 1-199, Marcel Dekker, New York (1976).

[2] E. A. Coddington and A. Dijksma, Adjoing subspaces in Banach spaces with application to ordinary differential subspaces, Anal. Mat. Pura Appl. 4 (118) (1978), 1-118.

[3] A. Dijksma, The Generalized Green's functions for regular ordinary differential subspaces in $L^2[a,b] \oplus L^2[a,b]$, Differential Equations and Applications (Eckhaus and de Jager, ed), Narth-Holland, Amsterdam (1978), 199-221.

[4] S. J. Lee, Boundary conditions for linear manifolds, I, J. Math. Anal. Appl. 73 (1980), 366-380.

[5] S. J. Lee, Nonhomegeneous boundary-value problems for linear manifolds, J. Math. Anal. Appl. (to appear)

[6] S. J. Lee, and M. Z. Nashed, Operator parts and generalized inverses of multi-valued operators with applications to ordinary differential subspaces (to appear).

[7] J. Locker, The generalized Green's functions for an nth order linear differential operator, Trans. Amer. Math. Soc. 228 (1977), 243-268.

[8] M. Z. Nashed, Aspects of generalized inverses in analysis and regularization, Generalized Inverses and Applications (Nashed, ed) (1976), Academic Press, New York, 193-244.

LARGE-SCALE SYSTEMS, CONE-VALUED LYAPUNOV FUNCTIONS AND QUASI-SOLUTIONS

S. Leela

State University of New York
College at Geneseo
Geneseo, New York

1. INTRODUCTION

It is well known that the method of vector Lyapunov functions offers an effective way to study the qualitative behavior of large-scale interconnected dynamical and control systems [1,3,5,12,14,15]. Several Lyapunov functions result in a natural way in the investigation of such systems by (i) the decomposition and aggregation method and (ii) the inclusion principle [5,14,15]. The inclusion principle allows one to use a suitable transformation to expand the given system and then decompose the expanded system into disjoint subsystems for which appropriate vector Lyapunov functions are constructed. The stability of the original system is determined by the stability of the expanded system.

The classical vector Lyapunov function method requires the comparison system to be quasi-monotone nondecreasing [2,9,11]. However, this stringent condition can be relaxed by making use of appropriate cone-valued Lyapunov functions and/or quasi-solutions [4,6-8,10]. In this brief article, we note that a judicious combination of these various ideas may indeed be a creative as well as an effective way for the study of large-scale systems.

2. CONE-VALUED LYAPUNOV FUNCTIONS AND QUASI-SOLUTIONS

Consider the large-scale system

$$x' = F(t,x), \quad x(t_0) = x_0 \tag{2.1}$$

where $f \in C[I\times R^N, R^N]$, $I = [t_0, t_0+T]$, $t_0 \geq 0$. Let us suppose that the system (2.1) is decomposable into isolated subsystems

$$z'_\eta = g_\eta(t,z_\eta) + R_\eta(t,x), z_\eta(t_0) = z_{\eta 0} \tag{2.2}$$

where $z_\eta \in R^{N}$ such that $\sum_{\eta=1}^{n} N_\eta = N$ and the interaction terms $R_\eta(t,x)$ are usually of the form $R_\eta(t,x) = \sum_{j=1}^{n} R_{\eta j}(t,z_j)$. If there exist functions V_η, d_η and w_η such that

$$D^+V_\eta(t,z_\eta) \leq - d_\eta(t,z_\eta) + w_\eta(t,z_\eta,x) \tag{2.3}$$

where D^+V_η denotes the Dini-derivative of V_η with respect to (2.2) and if the right-hand side of (2.3) can be majorized so that (2.3) takes the form

$$D^+V(t,z) \leq f(t,V(t,z)) \tag{2.4}$$

with $f(t,u)$ being quasi-monotone nondecreasing in u (that is, $f_i(t,u)$ is monotone nondecreasing in $u_j, j \neq i$), the method of vector Lyapunov functions can be effectively employed with appropriate assumptions on V and f to obtain the stability analysis of the null solution of (2.1). See [3,5,12,14,15].

However, the requirement of quasi-monotone property on the comparison system in the context of comparison principle and differential inequalities theory poses an unpleasant situation since in many practical problems, it is possible to have a system that is stable without being necessarily quasi-monotone. In order to overcome this difficulty, two approaches have been suggested and employed. They are (a) using the cone-valued Lyapunov functions and finding an appropriate cone K other than R^n_+ (R^n_+ is the standard cone with respect to which the comparison system f is not quasi-monotone) relative to which f will be quasi-monotone; and (b) using the notion of quasi-solutions. See [6-8,4,10] for details.

The following result [7] is a compariosn theorem in the context of approach in (a).

<u>Theorem 1</u>. Let K be a cone in $R^n (n \leq N)$. Let $V \in C[R_+ \times S(\rho), K]$ be locally Lipschitzian in z relative to K and satisfy the differential inequality

$$D^+V(t,z) \underset{K}{\leq} g(t,V(t,z)) \tag{2.5}$$

where $S(\rho) = \{z \in R^n: \|z-z_0\| \leq \rho\}$, $z_0 = (z_{10},\dots,z_{n0})$ and $\underset{K}{\leq}$ denotes the partial ordering induced by the cone K and the function $g(t,u)$ is quasi-monotone relative to K. Then

$$V(t,z(t,t_0,z_0)) \underset{K}{\leq} r(t,t_0,u_0)$$

whenever $V(t_0,z_0) \underset{K}{\leq} u_0, r(t)$ being the maximal solution of

$$u' = g(t,u),\ u(t_0) = u_0. \tag{2.6}$$

Once such an estimate is available for $V(t,z(t))$, the stability analysis of the null solution $z(t) \equiv 0$ can be carried out, knowing the corresponding stability behavior of system (2.6). See [7] for details.

Remark 1. Let P,Q be two cones in R^n, $P \subset Q$. If V is Q-valued, Lipschitzian relative to P and (2.5) holds relative to P, then it can be shown that

$$V(t,z(t,t_0,z_0)) \underset{Q}{\leq} r(t,t_0,u_0)$$

whenever $V(t_0 z_0) \underset{P}{\leq} u_0$. If $Q = R^n_+$, we get the usual componentwise estimates.

Example. Consider the system

$$\left.\begin{aligned} u_1' &= a_{11}u_1 + a_{12}u_2 \equiv g_1(t,u) \\ u_2' &= a_{21}u_1 + a_{22}u_2 \equiv g_2(t,u) \end{aligned}\right\}$$

with $u(t_0) = u_0$, where $a_{11},a_{22} < 0$ and a_{21},a_{12} are not necessarily non-negative. If we choose $Q = R^2_+$, the function $g(t,u)$ is not quasi-monotone relative to Q. Now choose $P \subset Q$ so that $P = \{u \in R^2_+: \alpha u_1 \leq u_2 \leq \beta u_1, \beta > \alpha$. Using $\phi_1 = (-\alpha,1), \phi_2 = (\beta,-1)$ as positive linear functional supporting P, we can show that if the coefficients $a_{11},a_{12},a_{21},a_{22}$ and α,β satisfy the inequalities

$$a_{21} + \alpha a_{22} \geq \alpha a_{11} + \alpha^2 a_{12},\ \beta a_{11} + \beta^2 a_{12} \geq a_{21} + \beta a_{22},$$

which in turn imply $a_{21} + \beta\alpha a_{12} > 0$. This last inequality can hold with a_{12} or a_{21} negative. Relative to the cone P, the function g will be quasimonotone and Theorem 1 can be applied to obtain componentwise estimates on V and stability analysis can be made using the cone-valued Lyapunov function V [7].

The concept of quasi-solutions provides another approach to deal with the lack of quasi-monotonicity of the comparison system [4,10]. Let $a \in C[I,R^n]$. A function $u \in C^1[I,R^n]$ is a quasi-solution of the system (2.6) relative to this choice function a if, for every $i = 1,2,\ldots,n$,

$$u_i' = g_i(t,u_i,[u]_{pi},[a]_{qi}),\ u_i(t_0) = u_{i0}, \tag{2.7}$$

where the vector $u \in R^n$ is split into $u = (u_i, [u]_{pi}, [u]_{qi})$ such that $p_i + q_i = n - 1$. A maximal quasi-solution of (2.6) with respect to a is the maximal solution of (2.7). Suppose now that g is not quasimonotone nondecreasing but it is mixed quasi-monotone. That is to say that for every $i = 1,2,\ldots,n, g_i(t,u_i,[u]_{pi},[u]_{qi})$ is monotone nondecreasing in $[u]_{pi}$ and monotone nonincreasing in $[u]_{qi}$. Since the Lyapunov function V is suth that $V_i \geq 0$ for every i, we choose $a(t) \equiv 0$ and if $r(t)$ is the maximal solution of the system

$$r_i' = g_i(t,r_i,[r]_{pi},[r]_{pi},[0]_{qi}),\ r_i(t_0) = V_i(t_0,z_0),$$

the comparison principle (Theorem 1 with $K = R^n_+$) yields the estimate $V(t,z(t)) \leq r(t)$. The stability of the null solution of (2.1) can be derived from the stability behavior of the quasi-solution $r(t)$.

For example, consider the linear system $u' = Au$ with $a_{ii} < 0$, $a_{ij} \leq 0$ for $j < i$ and $a_{ij} \geq 0$ for $j > i$. The quasi-solutions relative to $a(t) \equiv 0$ are determined by $v' = \tilde{A}v$ where $\tilde{a}_{ii} = a_{ii}$, $\tilde{a}_{ij} = 0$ for $j < i$ and $\tilde{a}_{ij} = a_{ij}$ for $j > i$. This matrix $\tilde{A}$ is stable since its eigenvalues $a_{11}, a_{22}, \ldots, a_{nn}$ are all negative.

Quasi-solutions are also useful in decomposing (2.1) into suitable isolated subsystems of the form (2.2) in a natural way. Let (2.1) be written as

$$z_\mu' = F_\mu(t,z_{\mu 1},z_{\mu 2},\ldots,z_{\mu n}),\ z_\mu(t_0) = x_{\mu 0},$$

where $x = (z_{\mu 1},z_{\mu 2},\ldots,z_{\mu n})$, $z_{\mu i} \in R^{N_i}$ and $\sum_{i=1}^{n} N_i = N$. Let $a \in C[I,R^N]$ and consider the subsystems

$$z_{\mu i}' = F_{\mu i}(t,a_{\mu 1},\ldots,z_{\mu i},\ldots,a_{\mu n}) \equiv g_{\mu i}(tz_{\mu i})$$

so that the large-scale system (2.1) can be written as

$$z' = g_\mu(t,z_\mu) + R_\mu(t,z_\mu,x)$$

where $R_\mu(t,z_\mu,x) = F_\mu(t,x) - g_\mu(t,z_\mu)$.

3. GENERAL INCLUSION PRINCIPLE

Large-scale systems have been extensively studied by using the decomposition and aggregation method [1,3,4,15] and a generalized decomposition [5,15]. A generalized decomposition is a disjoint decomposition of a system which is obtained by expanding the state-space of the given system. The inclusion principle is a framework which allows the expansion of a system by means of a suitable transformation such that the expanded system includes the solutions of the original system and its stability implies the stability of the original system.

Consider the differential system (2.1) and let $y = h(t,x)$, $h \in R^m$ be a continuously differentiable transformation which has an inverse given by $x = \tilde{h}(t,y)$ (for details about inverses and generalized inverse of a mapping, see [13]). Using the transformation, (2.1) can be written as

$$\left.\begin{aligned} y' &= h_t(t,\tilde{h}(t,y)) + h_x(t,\tilde{h}(t,y)) \cdot F(t,\tilde{h}(t,y)) \equiv f(t,y) \\ y(t_0) &= h(t_0,x_0). \end{aligned}\right\} \tag{3.1}$$

The solutions $x(t,t_0,x_0)$, $y(t,t_0,y_0)$ (of (2.1), (3.1) respectively) are related by $y(t,t_0,y_0) = h(t,x(t,t_0,x_0))$ and $x(t,t_0,x_0) = \tilde{h}(t,y(t,t_0,y_0))$ for all $t \geq t_0$ and $y_0 = h(t_0,x_0)$. The choice of the transformation can yield an expanded system $(m > N)$ or a contracted system $(m < N)$.

The original system may be subjected to (i) a disjoint decomposition directly or (ii) an overlapping decomposition, i.e., disjoint decomposition of an expanded system obtained by repeating the states of the given system or (iii) a generalized decomposition which deals with situations in which the original and the expanded states are linear combinations of each other. Suitable subsystems which are obtained by any of the above decomposition procedures are then studied by using the classical vector Lyapunov function method. In [5], it has been shown that (ii) may work while (i) fails and that (iii) may be useful even in cases when both (i) and (ii) fail to provide any useful information about the stability behavior of the original system.

Let us consider a few examples to illustrate the above ideas.

Example 1. Let $x = (x_1\ x_2,x_3)$ and $y = (y_1,y_2,y_3,y_4)$ that are related by the transformation $y = Tx$ with

$$T = \begin{pmatrix} 1 & 0 & 0 \\ 0 & \frac{1}{2} & 0 \\ 0 & \frac{1}{2} & 0 \\ 0 & 0 & 1 \end{pmatrix}.$$

This matrix has the generalized inverse T^I given by

$$T^I = \begin{pmatrix} 1 & 0 & 0 & 0 \\ 0 & 1 & 1 & 0 \\ 0 & 0 & 0 & 1 \end{pmatrix}.$$

Let the given system be

$$\left.\begin{aligned} x_1' &= f_{11}(x_1) + f_{12}(x_2) + f_{13}(x), \\ x_2' &= f_{21}(x_1) + f_{22}(x_2) + f_{23}(x_3), \\ x_3' &= f_{31}(x) + f_{32}(x_2) + f_{33}(x_3), \end{aligned}\right\}$$

where f_{12}, f_{22} and f_{32} are linear. The transformed system is then

$$\left.\begin{aligned} \begin{pmatrix} y_1' \\ y_2' \end{pmatrix} &= z_1' = \begin{pmatrix} f_{11}(y_1)+f_{12}(y_2) \\ \frac{1}{2}[f_{21}(y_1)+f_{22}(y_2)] \end{pmatrix} + R_1(z_1,z_2), \\ \begin{pmatrix} y_3' \\ y_4' \end{pmatrix} &= z_2' = \begin{pmatrix} \frac{1}{2}[f_{22}(y_3)+f_{23}(y_4)] \\ f_{32}(y_3)+f_{33}(y_4) \end{pmatrix} + R_2(z_1,z_2), \end{aligned}\right\}$$

which can be expressed as

$$\left.\begin{aligned} z_1' &= g_1(z_1) + R_1(z_1,z_2), \\ z_2' &= g_2(z_2) + R_2(z_1,z_2). \end{aligned}\right\} \tag{3.2}$$

The system (3.2) can then be analyzed by using either the classical vector Lyapunov function or an appropriate cone-valued Lyapunov function constructed for the subsystems.

To illustrate the idea that one way of decomposition may be preferred to another, let us consider the following example [5].

Example 2. The system

$$x' = \begin{pmatrix} -3 & 2 & 0 \\ -6 & -5 & -6 \\ 0 & 2 & -3 \end{pmatrix} x$$

may be transformed into the expanded system

$$y' = \left(\begin{array}{cc|cc} -3 & 2 & 0 & 0 \\ -6 & -5 & 0 & -6 \\ \hline -6 & 0 & -5 & -6 \\ 0 & 0 & 2 & -3 \end{array}\right) y \tag{3.3}$$

by means of the transformation

$$T = \begin{pmatrix} 1 & 0 & 0 \\ 0 & 1 & 0 \\ 0 & 1 & 0 \\ 0 & 0 & 1 \end{pmatrix},$$

whose generalized

inverse is

$$T^I = \begin{pmatrix} 1 & 0 & 0 & 0 \\ 0 & \frac{1}{2} & \frac{1}{2} & 0 \\ 0 & 0 & 0 & 1 \end{pmatrix}.$$

It is shown in [5] that stability of (3.3) cannot be established in the context of vector Lyapunov functions. Also, none of the four possible disjoint decompositions of (3.3) can be used to obtain the stability information. Now choosing T and T^I as in Example 1, we get the expanded system

$$y' = \left(\begin{array}{cc|cc} -3 & 2 & 2 & 0 \\ -6 & -5 & 0 & 0 \\ \hline 0 & 0 & -5 & -6 \\ 0 & 2 & 2 & -3 \end{array}\right) y \tag{3.4}$$

which can be shown to be stable by means of vector Lyapunov functions.

In disjoint and overlapping decompositions as in (3.3) the influence of interconnection is not incorporated into subsystems whereas the generalized decomposition as in (3.4) includes the interconnection in a beneficial way. It seems that the ability to study an overall large-scale system depends on the appropriate inclusion principle and the use of vector (or cone-valued) Lyapunov functions.

A judicious combination of the ideas discussed in Sections 2 and 3 might offer an even greater flexibility in studying (2.1), and utilizing the structural properties of the interconnected subsystems profitably.

REFERENCES

[1] M. Araki , "Stability of large scale nonlinear systems," IEEE Trans. Automat. Control, Vol. Ac-23, (1978), 129-142.

[2] R. Bellman, "Vector Lyapunov functions," SIAM J. Control, Vol. 1, (1962), 32-34.

[3] F. N. Bailey, "The application of Lyapunov's second method to interconnected systems," SIAM J. Control, Vol. 27, (1971), 543-548.

[4] K. Deimling and V. Lakshmikantham, "Quasi-solutions and their role in the qualitative theory of differential equations," J. Nonlinear Analysis, 4,(1980), 657-663.

[5] M. Ikeda and D. D. Šiljak, Generalized decompositions of dynamic systems and vector Lyapunov functions," IEEE Trans. Automat. Control, Vol. Ac-26, (1981), 1118-1125.

[6] V. Lakshmikantham, "On the method of vector Lyapunov functions," Proc. 12th Annual Allerton Conf., Circuit and System Theory, (1974), 71-76.

[7] V. Lakshmikantham and S. Leela, "Cone-valued Lyapunov functions," J. Nonlinear Analysis, Vol. 1, (1977), 215-222.

[8] ------------------------------, "Cone-valued Lyapunov functions and large scale systems, IEEE Trans. Automat. Control, Vol. Ac-74, (1979)

[9] ------------------------------, Differential and integral inequalities, Vol. I. Academic Press (1969).

[10] V. Lakshmikantham, S. Leela and M. N. Oǧuztöreli, "Quasi-solutions, vector Lyapunov functions and monotone method," IEEE Trans. Automat Control, Vol. Ac-26, (1981), 1149-1153.

[11] V. M. Matrosov, "On the theory of stability of motion," Prikl. Mat. ech., Vol. 26, (1962), 992-1002.

[12] A. N. Michel and R. K. Miller, Qualitative analysis of large scale dynamical systems, Academic Press, New York (1977).

[13] M. Z. Nashed, "Generalized inverse mapping theorems and related applications of generalized inverses in nonlinear analysis," Proc. of Intern. Symp. on Nonlinear Equations in Abstract Spaces, held at the University of Texas at Arlington, Arlington, Texas (1977), Academic Press, New York, 217-252.

[14] N. R. Sandell, P. Varaiya, M. Athans and M. G. Safanov, "Survey of decentralized control methods for large scale systems, IEEE Trans. Automat. Control, Vol. Ac-23, (1978), 108-128.

[15] D. D. Šiljak, Large Scale Dynamical Systems, North Holland, New York, Academic Press (1978).

A NONLINEAR VOLTERRA INTEGRODIFFERENTIAL EQUATION OCCURRING IN HEAT FLOW*

S.-O. Londen

Helsinki University of Technology
Espoo, Finland

J. A. Nohel

Mathematics Research Center
University of Wisconsin-Madison
Madison, Wisconsin

ABSTRACT

We study the nonlinear Volterra equation

$$(1)\qquad \begin{aligned} &u'(t) + Bu(t) + a * Au(t) + \frac{d}{dt}(b * u(t)) = F(t), \quad t \geq 0, \\ &u(0) = u_0 . \end{aligned}$$

Under various conditions on the nonlinear operators A,B and on the given functions a,b,F we give existence and asymptotic results for (1). An application to a problem in nonlinear heat flow is presented. Details of proofs will appear elsewhere

In this work we discuss the abstract nonlinear Volterra integro-differential equation

$$(1)\qquad \begin{aligned} &u'(t) + Bu(t) + a * Au(t) + \frac{d}{dt}(b * u(t)) = F(t), \quad t \geq 0, \\ &u(0) = u_0 . \end{aligned}$$

Here A and B are nonlinear operators, a,b are given real kernels defined on R^+, $*$ denotes the convolution (thus $a * Au(t) = \int_0^t a(t - s)Au(s)ds$), and F is a given function defined on R^+. Details and additional results concerning asymptotic behavior will appear in [3].

The equation (1) with $b(t) \equiv 0$ was studied at some length in a paper by Crandall, Londen, Nohel [2] (see also an article by Barbu [1]) and as regards existence the results that will be given here are extensions of the results which appeared in [2]. The situation is different when it comes to asymptotics; the results which will be stated here are new, even if $b(t)$ happens to vanish identically.

* Sponsored by the United States Army under Contract No. DAAG29-80-C-0041.

One may reasonably ask: why study (1) and in particular why add the specific linear term involving $b(t)$? The answer is that this equation occurs in nonlinear heat flow in materials with memory and consequently there is a specific application of (1) that we have in mind. In particular, one should note that (1) with $b(t) \not\equiv 0$ represents a much better model of the physics than the corresponding equation with $b(t) \equiv 0$. We will return to the application later. Of course, if both $a(t)$ and $b(t)$ vanish identically then (1) becomes the ordinary nonlinear evolution equation.

We begin by stating two existence results. To do that we need the following general assumptions that will be used in both.

Let H be a real Hilbert space, W a real reflexive Banach space satisfying $W \subset H \subset W'$ where W' is the dual of W. We assume that the injections are continuous and dense and denote the inner product and norm in H by $(\, , \,)$ and $| \; |$ respectively. Then let $\psi : W \to (-\infty,\infty]$ and $\phi : H \to (-\infty,\infty]$ be convex, lower semicontinuous (proper) functions and assume

$$A = \partial\psi, \qquad B = \partial\phi$$

where $\partial\psi, \partial\phi$ are the subdifferentials of ψ and ϕ. Then A and B will be maximal monotone operators from W and H to W' and H respectively. Define $\psi_H : H \to (-\infty,\infty]$ by

$$\psi_H(u) = \lim_{r \downarrow 0} \inf\{\psi(v) \mid v \in W \text{ and } |v - u| < r\}$$

and assume that $\psi_H(u) = \psi(u)$ for $u \in W$.

Let A_λ be the Yosida-approximation of A. We assume

$$(Bu, A_\lambda u) \geq -\beta(|Bu|^2 + |u|^2 + 1), \qquad u \in W, \quad \lambda \in (0,1]$$

and some constant $\beta \geq 0$. For simplicity we take A and B to be singlevalued; most of the results can, however be reformulated for the case when one or sometimes both operators A,B are multivalued.

We also need a compactness assumption:

$$\begin{cases} \text{For every } K > 0, \ \{u \in H \,|\, |\phi(u)| + |u| \leq K\} \text{ is} \\ \text{precompact in } W. \end{cases}$$

Regarding the kernel $a(t)$ we assume that it satisfies either (i) or (ii) where

$$\text{(i)} \qquad \begin{cases} a, a' \in L^1_{loc}(R^+, R) \\ a(0) > 0 \\ a' \in BV_{loc}(R^+) \end{cases}$$

and

$$\text{(ii)} \qquad \begin{cases} a(t) \text{ is nonnegative, decreasing and convex on } R^+ \\ a(0) > 0 \\ a \in C^2(0,\infty) \cap C[0,\infty) . \end{cases}$$

One can also have $a = a_1 + a_2$ where a_1 satisfies (i) and a_2 satisfies (ii). Finally concerning b, F we require that

$$b \in LAC(R^+, R) \qquad F \in W^{1,1}_{loc}(R^+, H) .$$

We can now state the following existence result:

Theorem 1. Let the general assumptions above be satisfied. In addition suppose that $D(A) = W$ and that $u_0 \in D(\phi)$. Then there exists u such that

$$u \in C(R^+, W); \qquad u', Bu \in L^2_{loc}(R^+, H) ,$$

$$\int_0^t Au(s)ds \in L^\infty_{loc}(R^+, H); \qquad \text{(1) is satisfied a.e. on } R^+ .$$

The way to prove this result is to add a regularizing term $\varepsilon Au(t)$, $\varepsilon > 0$, outside the integrals and replace A by the Yosida-approximation A_λ. Thus one looks at

$$u_\lambda'(t) + Bu_\lambda(t) = F(t) - \varepsilon A_\lambda u_\lambda(t) - a * A_\lambda u_\lambda(t) - \frac{d}{dt}(b * u_\lambda(t)) .$$

For this regularized equation one obtains existence quite easily as everything on the right side is Lipschitz-continuous in u_λ. The next step is to let $\lambda \downarrow 0$ with $\varepsilon > 0$ fixed. To do that one of course needs some a priori estimates on the functions involved which is where the assumption regarding $(Bu, A_\lambda u)$ is used. Finally one lets $\varepsilon \downarrow 0$, uses the compactness assumption, together with relevant a priori bounds.

What one cannot do under the assumptions of Theorem 1 is to obtain $Au \in L^2_{loc}(R^+, H)$. For that one needs a set of different conditions which are formulated next.

Theorem 2. Let the general assumptions be satisfied with $W = H = W'$. Also suppose that for every $r > 0$ there exists a number $k(r)$ such that

$$k(r)(1 + |Bu|) \geq |Au|, \qquad |u| \leq r ,$$

and let $u_0 \in D(\psi) \cap D(\phi)$. Then there exists a solution u of (1) such that $u, u', Au, Bu \in L^2_{loc}(R^+, H)$.

This second simpler theorem is in fact in the spirit of the asymptotic results. Common to both existence results is that there is no claim concerning uniqueness.

The first asymptotic result which we wish to formulate concerns conditions under which solutions u of (1) satisfy $u \in L^2(R^+, H)$. Recall that $a(t)$ strongly positive definite means that there exists $\alpha > 0$ such that $a(t) - \alpha e^{-t}$ is positive definite.

Theorem 3. Let

$$(2)\quad \begin{cases} a, a' \in L^1(R^+);\ a \text{ be strongly positive definite}, \\ (-1)^k b^{(k)}(t) \geq 0, \quad t \in R^+, \quad k = 0,1; \\ A = \partial\psi \text{ where } \psi : H \to (-\infty, \infty] \text{ is convex, l.s.c. and proper}, \\ B : D(B) \subset H \to H \text{ with } (u, Bu) \geq \lambda|u|^2 \text{ for some } \lambda > 0, \\ \lambda_1 |Au|^2 \leq \lambda_0^2 \mu |u|^2 + (Au, Bu) - 2^{-1} b(0)|Au - u|^2, \ \forall u \in D(A) \cap D(B) \\ \qquad \text{for some } \lambda_0, \lambda_1 \text{ such that } \lambda_1 > 0,\ \lambda_0 < \lambda. \\ \text{There exists } u \in LAC(R^+, D(A) \cap D(B)) \text{ satisfying (1) on } R^+ \\ \qquad \text{and such that } Au, Bu \in L^2_{loc}(R^+, H). \end{cases}$$

If

$$F \in L^2(R^+, H),$$

then

$$u, Au \in L^2(R^+, H), \qquad u \in L^\infty(R^+, H).$$

The key idea of the proof is to obtain both upper and lower bounds for the quadratic form $\int_0^t (Au(\tau), \int_0^\tau a(\tau - s)Au(s)ds)d\tau$. Then one combines these estimates and uses the key inequality above involving λ_1, λ_0 to arrive at the desired conclusions. Note that in Theorem 3 the existence of u has to be assumed as the remaining conditions do not imply that a solution exists.

In the next two theorems we let the nonhomogeneous term F be "large". (Below we use the notation $g \in L^2_\infty(R^+, H)$ iff $\sup_{1 \leq t < \infty} \int_{t-1}^t |g(\tau)|^2 d\tau < \infty$).

Theorem 4. Let the assumptions (2) hold. Also assume that for some $\gamma > 3/2$ and some constant K

$$|a(t)| \leq Kt^{-\gamma}, \qquad |b'(t)| \leq Kt^{-\gamma}, \qquad t \geq 1 .$$

Suppose

$$F \in L^2_\infty(R^+,H) \tag{3}$$

and let there exist $\rho > 0$ satisfying

$$|u| \leq \rho|Au|, \qquad u \in D(A) .$$

Then $u \in L_\infty(R^+,H)$, $Au \in L^2_\infty(R^+,H)$. If in addition $B = \partial\phi$ where $\phi : H \to (-\infty,\infty]$ is convex, l.s.c. and proper, then $u',Bu \in L^2_\infty(R^+,H)$.

This theorem is our main asymptotic result. The key difficulty when proving Theorem 4 is that F is not small at infinity but only bounded on the average; this forces us to assume that the kernels a and b decay sufficiently rapidly for large t.

Under the assumptions of Theorem 4 a reasonable guess is that if $F \to 0$ in some sense as $t \to \infty$ then so will the solution u:

Theorem 5. If in Theorem 4 the assumption (3) is strengthened to $\lim_{t\to\infty} \int_{t-1}^{t} |F(\tau)|^2 d\tau = 0$, with the remaining assumptions unchanged, then

$$\lim_{t\to\infty} |u(t)| = 0, \qquad \lim_{t\to\infty} \int_{t-1}^{t} |Au(\tau)|^2 d\tau = 0 .$$

The proofs of these two last theorems are quite lengthy and technical.

We end this report with the example which has motivated our work.

Example. Consider the scalar equation (subscripts denote partial derivatives)

$$\begin{aligned} &u_t(t,x) - (\sigma_1(u_x(t,x)))_x - \int_0^t a(t-s)(\sigma_2(u_x(s,x)))_x ds \\ &\quad + b(0)u(t,x) + \int_0^t b'(t-s)u(s,x)ds = F(t,x), \quad t > 0, \quad 0 < x < 1 \end{aligned} \tag{4}$$

with the boundary and initial conditions

$$u(t,0) = u(t,1) = 0, \quad t > 0; \qquad u(0,x) = u_0(x), \quad 0 \leq x \leq 1 .$$

In (4) assume that for $i = 1,2$ and for some constants M, c_i,

$$\sigma_i \in C^1(R); \qquad 0 \leq \sigma_i' \leq M < \infty, \qquad x \in \mathbf{R} ;$$
$$x^{-1}\sigma_i(x) \geq c_i > 0, \qquad x \neq 0 ;$$

and suppose there exists $\beta > 0$ such that

$$\sigma_2'(x) \leq \beta\sigma_1'(x), \qquad x \in \mathbf{R} .$$

Take $W = H = L^2(0,1)$ and define $\phi_i : L^2(0,1) \to (-\infty,\infty]$ by

$$\phi_i(u) = \begin{cases} \int_0^1 \Sigma_i(\frac{du}{dx})dx, & u \in H_0^1(0,1) , \\ \infty, & \text{otherwise} , \end{cases}$$

where $\Sigma_i(r) \overset{\text{def}}{=} \int_0^r \sigma_i(s)ds$, $i = 1,2$. Then the functions ϕ_i are convex, l.s.c. and

$$\partial\phi_i(u) = -\frac{d}{dx}\left(\sigma_i(\frac{du}{dx})\right); \quad u \in D(\partial\phi_i) = \left\{u \in H_0^1 \mid \frac{d}{dx}\left(\sigma_i(\frac{du}{dx})\right) \in L^2(0,1)\right\} .$$

Denote $Bu = \partial\phi_1(u)$, $Au = \partial\phi_2(u)$. By the growth condition on σ_1 one of course has that $\{u \mid |\phi_1(u)| + |u| \leq K\}$ is compact in $L^2(0,1)$ for every K. Also,

$$|Au|^2 = \int_0^1 \left(\sigma_2'(\frac{du}{dx})\right)^2 \left(\frac{d^2u}{dx^2}\right)^2 dx \leq \beta^2 \int_0^1 \left(\sigma_1'(\frac{du}{dx})\right)^2 \left(\frac{d^2u}{dx^2}\right)^2 dx = \beta^2|Bu|^2$$

and so $|Au| \leq \beta|Bu|$. Therefore,

$$(A_\lambda u,Bu) \geq -|A_\lambda u|\,|Bu| \geq -|Au|\,|Bu| \geq -\beta|Bu|^2$$

which gives one of the needed general assumptions. From $|Au| \leq \beta|Bu|$ we trivially have $k(r)(1 + |Bu|) \geq |Au|$ with $k(r) \equiv \beta$. So, if $u_0 \in H_0^1(0,1)$ and $t \to F(t,\cdot) \in W_{loc}^{1,1}(R^+,L^2(0,1))$, and if a,b satisfy the general assumptions, then we have existence of a solution $u(t,x)$ of (4) such that $u,u_t,u_{xx} \in L_{loc}^2(R^+,L^2(0,1))$. The kernels a,b are frequently decaying exponentials and consequently the size assumptions on a,b are not very crucial in practice.

As regards asymptotics we observe that

$$(u,Bu) = -\int_0^1 u(x)\frac{d}{dx}\left(\sigma_1(\frac{du}{dx})\right)dx \geq c_1\int_0^1 \left(\frac{du}{dx}\right)^2 dx \geq c_1\pi^2\int_0^1 |u(x)|^2dx$$
$$= c_1\pi^2|u|^2$$

and analogously

(5) $$(u,Au) \geq c_2\pi^2|u|^2 .$$

From (5) and as $(Au,Bu) \geq \beta^{-1}|Au|^2$ we have that the key inequality of Theorem 3 is satisfied if

$$b(0) < 2\beta^{-1}, \quad \text{and} \quad b(0) < 2(\pi^2 c_1)^2\mu + 2b(0)c_2\pi^2 .$$

Finally, concerning the application of Theorems 4 and 5 we note that $|u| \leq \rho|Au|$ holds provided $\sigma_2'(x) \geq \varepsilon > 0$, for all x and some $\varepsilon > 0$. Hence, under the above assumptions and of course requiring the appropriate size conditions on F to hold one may apply Theorems 3-5.

REFERENCES

[1] V. Barbu, Integro-differential equations in Hilbert spaces, An. Sti. Univ. "Al. I. Cuza" Iasi Sect. Ia Mat. 19 (1973), 365-383.

[2] M. G. Crandall, S.-O. Londen and J. A. Nohel, An abstract nonlinear Volterra integrodifferential equation, J. Math. Anal. Appl. 64 (1978), 701-735.

[3] S.-O. Londen and J. A. Nohel, A nonlinear Volterra integro-differential equation occurring in heat flow, MRC Tech. Summary Report (to appear).

ON EXACT CONTROLLABILITY IN HILBERT SPACES

J.-Cl. Louis and D. Wexler

Department of Mathematics
Facultés Universitaires Notre-Dame de la Paix,
Rempart de la Vierge,
Namur, Belgium

ABSTRACT

We discuss the limitations of the exact controllability concept for evolution equations in Hilbert spaces.

§1. INTRODUCTION AND STATEMENT OF MAIN RESULTS

Let X and U be Hilbert spaces (both real or both complex) with inner product $<. , .>$ and norm $|.|$ (when confusion is possible we use the subscript X, respectively U). Let B be an element of the space $\mathcal{L}(U, X)$ of bounded linear operators from U to X and A be the generator of a linear C_o-semigroup S on X with domain $D(A)$; for C_o-semigroups, we refer to reader to [4, 5]. Consider the Cauchy problem on the interval $[t_o , t_1]$ for the evolution equation

$$(1.1) \qquad \frac{dx}{dt}(.) = A\, x(.) + B\, u(.) \quad , \quad x(t_o) = a \quad ,$$

where $a \in X$ and $u(.) \in L^1([t_o, t_1], U)$. The (mild) solution of (1.1) is

$$x(t) = S(t - t_o)\, a + \int_{t_o}^{t} S(t - s)\, B\, u(s)\, ds \quad , \quad t \in [t_o, t_1] .$$

As usual, exact controllability of the pair (A, B) on some interval $[0, T]$, $T > 0$, (with L^2-controls) means that for each $(a, a') \in X^2$, there exists $u(.) \in L^2([0, T], U$ such that

$$S(T)\, a + \int_0^T S(T - s)\, B\, u(s)\, ds = a' \quad ;$$

this is equivalent to saying that the operator

(1.2) $u(.) \mapsto \int_0^T S(T-s)\, Bu(s)\, ds$ from $L^2([0, T], U)$ to X

is surjective. The above concept, which is basic in finite-dimensional control theory, has also some relevance to the infinite-dimensional control since exact controllability of (A, B) holds for certain significant systems as, for instance, some controlled wave equations (see Curtain and Pritchard [3, chapter 2], Russel [9] and systems in which A generates a C_o-group and B is surjective. See also Zabczyk [17] and Leigh [6, § 10.10] for a simple exemple in which (A, B) is exactly controllable and A does not generate a C_o-group but merely a C_o-semigroup (the semigroup of left translations in $L^2(\mathbb{R}^+)$).
However it is now well-known that in infinite dimensions, the exact controllability concept is rather restricitve, for it never applies for important classes of evolution equations. Notable results of general scope have been obtained in this direction by Triggiani. He has proved that, if $\dim X = \infty$, the pair (A, B) is never exactly controllable when : 1. B is compact [10] ; or 2. $S(t)$ is compact for all $t > 0$ [11, 12]. Case 1. occurs in significant control problems for evolution equations associated, say, to retarded functional differential equations; case 2. occurs for C_o-semigroups associated to parabolic partial differential equations in bounded domains. In Triggiani's proofs, the lack of surjectivity of (1.2) follows from its compacity for all $T > 0$.

The purpose of this paper is to reveal some further restrictive features of the exact controllability concept in the setting of evolution equations. Our main result is the following :

<u>Theorem 1.</u> Assume that A generates a C_o-semigroup S on the Hilbert space X. Then the following conditions are equivalent :

(i) there exists a Hilbert space U and an operator $B \in \mathcal{L}(U, X)$ such that the pair (A, B) is exactly controllable;

(ii) S admits a right-inverse C_o-semigroup $\hat{S}$ on X i.e. $S(t)\,\hat{S}(t) = I_X$ (the identity on X) for all $t \geq 0$;

(iii) $S(t)$ is surjective for all $t \geq 0$; and

(iv) there exists $t_o > 0$ such that $S(t)$ is surjective.

To a certain extent, the above theorem explains why exact controllability occurs somewhat exceptionnally in the evolution equations setting. So condition (ii) says that the solutions of the homogeneous equation

$$\frac{dx}{dt}(.) = A\,x(.)$$

admit backward continuation (possibly without uniqueness). Indeed, in infinite dimensions, the above condition fairly narrows the class of generators A for which we can find B such that (A, B) is exactly controllable.

Theorem 1 also allows to sharpen some earlier results in [11, 12]. So it follows that, if $\dim X = \infty$, exact controllability of (A, B) can never arise when 1. there exists $t_o > 0$ such that $S(t_o)$ is compact, which occurs for retarded functional differential equations: or 2. S is differentiable (i.e. $S(t)\,X \subset D(A)$ for all $t > 0$) and $D(A) \neq X$, which occurs for parabolic partial differential equations in bounded and unbounded domains.

Note that although retarded functional differential equations are never exactly controllable in the associated to them abstract setting (1.1), by a slight modification one can obtain a setting in which they possess some exact controllability properties [2]; the price to be paid is then the loss of the evolution equations dynamics.

From Theorem 1 we also obtain some new information about the unique selfadjoint solution $K \in \mathcal{L}(X)$ of the Liapunov equation

(1.3) $\quad 2\,\mathrm{Re}\,\langle A\,x\,,\,K\,x\rangle = -|x|^2 \quad , \quad \forall\; x \in D(A)\,,$

where A generates an exponentially stable C_o-semigroups S. It is wellknown that K is defined by $Kx = \int_o^{+\infty} S^\star(t)\,S(t)\,x\,dt$ for all $x \in X$; the star stands for the adjoint operator. Hence $\langle K\,x\,,\,x\rangle > 0$ for all $x \neq 0$, but, if $\dim X = \infty$, the above does not imply coercivity of K.

We may use the relation between exact controllability and Liapounov equations to derive from Theorem 1 the following :

<u>Corollary</u>. Assume that A generates an exponentially stable C_o-semigroup S on the Hilbert space X. Then the following conditions are equivalent:

(i) the unique selfadjoint solution K of the Liapunov equation (1.3) is coercive;

(ii) S admits a left-inverse C_o-semigroup on X;

(iii) there exists $t_o > 0$ such that $S(t_o)$ admits a bounded linear left-inverse.

We note that the above equivalence (i) ⇔ (iii) is not new. In fact it follows from an earlier result by Pazy in the Banach setting [8] .

§2. PROOFS

The implications (ii) ⇒ (iii) ⇒ (iv) of Theorem 1 are obvious.

(iv) ⇒ (i). Clearly there exists $M \geq 1$ such that $|S^{\star}(t)| \leq M$ on $[0,t_o]$. By [4, chpt. VI, §6], (iv) implies the existence of a $k > 0$ such that $|S^{\star}(t_o)x| \geq k\, x$ for all $x \in X$, hence

$$k\, x \leq |S^{\star}(t_o - t)\, S^{\star}(t)x\,| \leq M|S^{\star}(t)x|, \quad \forall\, t \in [0,t_o],\ x \in X.$$

It follows that for some $\gamma > 0$ we have

$$|S^{\star}(t)x| \geq \gamma|x|, \quad \forall\, t \in [0,\, t_o],\quad x \in X\,.$$

Use then [3, thm. 3.7] to see that $(A,\, I_X)$ is exaclty controllable.

In the proof of (i) ⇒ (ii) we consider an optimal control (in fact a minimum norm) problem on the negative half-axis $\mathbb{R}^-$. In finite-dimensional spaces the contol problem on $\mathbb{R}^-$ with cost function wich is not sign definite has been related to the minimal solution of the algebraic Riccati equation, see Willems [14]. Our proof makes use of some ideas in dynamic programming. We are also inspired by Yakubovish's approach to the regulator problem on $\mathbb{R}^+$ with <u>bounded</u> Hilbert space operators [16]. For more generality on this point see [13] .

If (A,B) is exactly controllable, so is $(A + \alpha I_X,\, I_X)$ for all $\alpha \in \mathbb{R}$. Hence without loss of generality we may prove (i) ⇒ (ii) under the additional assumptions that $U = X$, $B = I_X$ and S is exponentially stable, wich we assume in the remaining of this contribution.

For any $u(.) \in L^2(\mathbb{R}^-,\, X)$ the function $t \to \int_{-\infty}^{t} S(t - s)\, u(s)ds$ is well-defined on $\mathbb{R}^-$ and its restriction on each compact interval $I \subset \mathbb{R}^-$ is a mild solution of (1.1) on I. By exact controllability, for each $a \in X$ the set

$$N_a^- \overset{\text{def}}{=} \{u(.) \in L^2(\mathbb{R}^-,\, X)\ :\ \int_{-\infty}^{0} S(-s)\, u(s)ds = a\}$$

is $\neq \emptyset$; N_a^- is the translation of the closed vector space N_o^- by some vector in N_a^-. Denote then by $u^-(.,a)$ the unique element of minimal norm in N_a^- and put

$$x^-(t,a) \overset{\text{def}}{=} \int_{-\infty}^{t} S(t-s)\, u^-(s,a)ds, \quad \forall \ t \in \mathbb{R}^- .$$

Note that the application $a \mapsto u^-(.,a)$ is injective.

Lemma 1. The functions $a \mapsto u^-(.,a)$ and $a \mapsto x^-(.,a)$ from X to $L^2(\mathbb{R}^-,X)$ are linear and continuous, so that $a \mapsto V^-(a) \overset{\text{def}}{=} \|u^-(.,a)\|^2_{L^2(R^-,X)}$ is a continuous Hermitian form on X .

The proof follows easily by the fact that $N_a^- + N_{\hat{a}}^- \subset N_{a+\hat{a}}^-$ and $\lambda N_a^- \subset N_{\lambda a}^-$ for all $\lambda \in \mathbb{C}$, $a,\hat{a} \in X$; use also the projection theorem and the closed graph theorem.

For each $a \in X$ and $t \geqslant 0$ we put $S'(t)a \overset{\text{def}}{=} x^-(-t,a)$.

Lemma 2. (i) S' is a C_o-semigroup on X; (ii) for each a in the domain $D(A')$ of the generator A' of S' we have $x^-(.,a) \in C^1(\mathbb{R}^-,X)$ and

(2.1) $\qquad x^-(t,a) \in D(A)$ and $\dfrac{dx^-}{dt}(t,a) = Ax^-(t,a) + u^-(t,a)$

a.e.; and (iii) for each $a \in X$ and $s \in \mathbb{R}^-$ we have

(2.2) $\qquad u^-(t+s,\ a) = u^-(t,x^-(s,a))$ for almost all $t \in \mathbb{R}^-$.

Proof. $S'(0) = I_X$ and continuity of $S'(.)a$ for each $a \in X$ are obvious. Lemma 1 implies the linearity of $a \mapsto S'(.)a$ from X to $L^2(\mathbb{R}^-,X)$ which combined with continuity of $S'(.)a$ for each $a \in X$ yields the linearity of $S'(t)$ from X to X for each $t \geqslant 0$.
The continuity of $S'(t) : X \to X$ for each $t \geqslant 0$ follows by the continuity of $a \mapsto u^-(.,a)$ from X to $L^2(\mathbb{R}^-,X)$.

To prove (iii) it sufficies to observe that for each $a \in X$ and $s \leqslant 0$ we have $u^-(.+s,\ a) \in N^-_{x^-(s,a)}$ and $u^-(.+s,a) \perp N_o^-$ and to apply the projection theorem. The semigroup property of S' follows by (iii).

To prove chaim (ii) consider an arbitrary $a \in D(A')$.
Since S' is a C_o-semigroup, we have $x^-(.,a) \in C^1(\mathbb{R}^-,X)$, which combined with [1, Thm. 4.8.3] implies the existence of a set $N \subset \mathbb{R}^-$ such that for each $y \in D(A^\star)$ we have

$$<\frac{dx^-}{dt}(t,a)\ ,\ y> = <x^-(t,a)\ ,\ A^\star y> + <u^-(t,a)\ ,\ y> \quad ,$$

for all $t \in \mathbb{R}^- \setminus N$. Then (2.1) follows by observing that $A^{\star\star} = A$.

Lemma 3. For each $a \in X$ and each $u(.) \in N_a^-$ we have

(2.3) $\qquad V^-(x(s)) - V^-(x(\sigma)) + \int_\sigma^s |u(s)|^2 ds \geqslant 0,\ \forall\ s,\ \sigma \in \mathbb{R}^-,\ \sigma < s,$

where $x(t) = \int_{-\infty}^{t} S(t-s)u(s)ds$. If, moreover, $u(.) = u^-(.,a)$, so that $x(.) = x^-(.,a)$, we may replace in (2.3) the inequality by equality.

Proof. Define $\hat{u}(.)$ by $\hat{u}(t) = u^-(t - \sigma + s, x(\sigma))$ a.e. in $]-\infty, \sigma-s[$, $\hat{u}(t) = u(t + s)$ a.e. in $[\sigma - s, 0]$ and put $\hat{x}(t) = \int_{-\infty}^{t} S(t-\lambda)\hat{u}(\lambda)d\lambda$. It follows $\hat{x}(t) = x^-(t - \sigma + s, x(\sigma))$ on $]-\infty, \tau - s[$, $\hat{x}(t) = x(t + s)$ on $[\sigma - s, 0]$, so that $\hat{u}(.) \in N^-_{x(s)}$. Then to see that (2.3) holds observe that

$$-V^-(x(s)) \leqslant \|\hat{u}(.)\|^2_{L^2(\mathbb{R}^-,X)} = \int_\sigma^s |u(t)|^2 dt - V^-(x(\sigma)).$$

Lemma 4. Let $H^- \in \mathcal{L}(X)$ be the selfadjoint operator of V^- ($V^-(a) = \langle H^- a, a\rangle$ for all $a \in X$) and define

$$L^- : D(A) \times X \to \mathbb{R} \text{ by } L^-(x,v) = 2 \text{ Re } \langle Ax + v, H^- x\rangle + |v|^2 .$$

Then for each $a \in D(A^\star)$ there exists a set $N \subset \mathbb{R}^-$ with mes $N = 0$ such that : (i) $x^-(t,a) \in D(A)$ and $L^-(x^-(t,a), u^-(t,a)) = 0$ for all $t \in \mathbb{R}\setminus N$; and (ii) $L^-(x^-(t,a),v) \geqslant 0$ for all $v \in X$ and $t \in]-\infty, 0[\setminus N$.

Proof. By (ii) of Lemma 2 there exists $N_1 \subset \mathbb{R}^-$ with mes $N_1 = 0$ such that (2.1) holds on $\mathbb{R}^- \setminus N_1$. By (2.3) with equality the function $t \mapsto V^-(x^-(t,a))$ is locally absolutely continuous, so that there exists $N_2 \subset \mathbb{R}^-$ with mes $N_2 = 0$ such that the above function is differentiable on $\mathbb{R}^- \setminus N_2$. Then (i) follows with $N = N_1 \cup N_2$.

Fix now $t_o \in]-\infty, 0[\setminus N$ and $v \in X$. Let $x_v(.)$ be the solution of (1.1) on $[t_o, 0]$ with initial condition $x^-(t_o,a)$ and constant control v on $[t_o,0]$. By the choice of N we have $x^-(t_o,a) \in D(A)$, so that $x_v(.)$ is a strong solution of (1.1). It follows that the function $t \mapsto V^-(x_v(t))$ is differentiable on the right at t_o and its right derivative satisfies

$$(2.4) \qquad \frac{d^+}{dt} V^-(x_v(t)) \Big|_{t = t_o} = 2 \text{ Re } \langle Ax^-(t_o,a) + v, H^- x^-(t_o,a)\rangle$$

Define now $u(.)$ by $u(t) = u^-(t,a)$ a.e. in $]-\infty, t_o[$, $u(t) = v$ on $[t_o,0]$ and put $x(t) = \int_{-\infty}^{t} S(t - \lambda)u(\lambda)d\lambda$. It follows $x(t) = x^-(t,a)$ on $]-\infty, t_o[$, $x(t) = x_v(t)$ on $[t_o, 0]$, so that $u(.) \in N^-_{x_v(0)}$. Apply then Lemma 3 with a and σ replaced by $x_v(0)$ and t_o respectively to see that

$$\frac{d^+}{dt} V^-(X_v(t)) \Big|_{t=t_o} + |v|^2 \geqslant 0 ,$$

which combined with (2.4) implies claim (ii).

Lemma 5. The generator A' of S' satisfies $D(A') \subset D(A)$ and

$A' = -A + H^-$ on $D(A')$.

Proof. By applying Lemma 4 with $a \in D(A')$ and $t \in \mathbb{R}^- \setminus N$ and [15, Lemma 8] with $\Gamma = I_X$, $g = H^- x^-(t,a)$ and $\gamma = \text{Re} \langle 2H^- Ax^-(t,a), x^-(t,a)\rangle$ we see that

$$(2.5) \qquad u^-(t,a) = -H^- x^-(t,a) \ , \quad \forall \ t \in \mathbb{R}^- \text{ and } a \in X \ .$$

Observe that in this way we obtain equality in (2.5) first for each $a \in D(A')$ and $t \in \mathbb{R}^- \setminus N$; since $x^-(.,a)$ is continuous we may assume it holds for all $a \in D(A')$ and $t \in \mathbb{R}^-$ (modify, if necessary, $u^-(.,a)$ ou N); then to see that (2.5) holds use the fact that $D(A')$ is dense in X combined with the fact that by Lemma 1 the operators $a \mapsto u^-(.,a)$ and $a \mapsto x^-(.a)$ from X to $L^2(\mathbb{R}^-, X)$ are continuous.

According to [5, chpt.X, §2] the operator $A - H^-$ with domain $D(A)$ generates a C_o-semigroup too. So we may apply (2.5) and [1, thm. 4.8.3] to see that $x^-(.,a)$ is for each $a \in X$ the mild solution of

$$\frac{dx^-}{dt}(.,a) = (A - H^-)x^-(.,a)$$

on all compact intervals in $\mathbb{R}^-$. By Lemma 2, $a \in D(A')$ implies $x^-(.,a) \in C^1(\mathbb{R}^-, X)$ so that it follows

$$x^-(t,a) \in D(A) \text{ and } \frac{dx^-}{dt}(t,a) = (A-H^-)x^-(t,a) \ , \ \forall \ t\in\mathbb{R}^- \ , \ a\in D(A').$$

Since the generator of S' is A' it follows $D(A') \subset D(A)$ and $(A-H^-)a = -A'a$ for all $a \in D(A')$.

Proof of the implication (i) ⇒ (ii) of Theorem 1. Since $H^- \in \mathcal{L}(X)$, $\hat{A} \overset{\text{def}}{=} A' - H^-$ generates a C_o-semigroup $\hat{S}$. By Lemma 5 we have $D(\hat{A}) \subset D(A)$ and $\hat{A} = -A$ on $D(\hat{A})$. Hence we see easily that for each $x \in D(\hat{A})$ we have $d[S(t)\,\hat{S}(t)x]/dt = 0$ for all $t \geqslant 0$, which combined with $S(0)\,\hat{S}(0)x = x$ implies $S(t)\,\hat{S}(t)x = x$ for all $x \in D(\hat{A})$ and $t \geqslant 0$. Then (ii) of Theorem 1 follows by the denseness of $D(\hat{A})$ in X and the continuity of $t \mapsto S(t)\,\hat{S}(t)$.

Proof of Corollary of Theorem 1. For each $T > 0$ we define $K_T \in \mathcal{L}(X)$ by

$$K_T\, x = \int_0^T S^\star(t)\, S(t)\, x \, dt \quad .$$

Use then the exponential stability of S and [3, Theorem 3.7] to see that

K is coercive $\Leftrightarrow$ there exists $T > 0$ such that K_T is coercive $\Leftrightarrow$

$\Leftrightarrow$ the pair $(A^{\star}, I_X)$ is exactly controllable.

The above combined with Theorem 1 imply the equivalences in the Corollary.

REFERENCES

[1] A.V. BALAKRISHNAN, Applied Functional Analysis, Springer-Verlag, 1976.

[2] H.T. BANKS, M.Q. JACOBS, C.E. LANGENHOP, Characterization of the Controlled States in $W_2^{(1)}$ of Linear Hereditary Systems", SIAM J. Control 13 (1975), pp. 611-649.

[3] R.F. CURTAIN, A.J. PRITCHARD, Infinite Dimensional Linear Systems Theory, Springer-Verlag, 1978.

[4] N. DUNFORD, J.T. SCHWARTZ, Linear Operators, Part I, Interscience Publishers, 1963.

[5] T. KATO, Perturbation Theory for Linear Operators, Springer-Verlag, 1966.

[6] J.R. LEIGH, Functional Analysis and Linear Control Theory, Academic Press, 1980.

[7] J.L. LIONS, Optimal Control of Systems Governed by Partial Differential Equations, Springer-Verlag, 1971.

[8] A. PAZY, On the applicability of Lyapunov's theorem in Hilbert spaces, SIAM J. Math. Anal. 3 (1972), pp. 291-294.

[9] D.L. RUSSEL, Controllability and stabilization theory for linear partial differential equations : recent progress and open questions, SIAM Review 20 (1978), pp. 639-739.

[10] R. TRIGGIANI, On the lack of exact controllability for mild solutions in Banach spaces, J. Math. Anal. Appl. 50 (1975), pp. 438-446.

[11] R. TRIGGIANI, A note on the lack of exact controllability for mild solutions in Banach spaces, SIAM J. Control and Optimization 15 (1977), pp. 407-411.

[12] R. TRIGGIANI, Addendum : A note on the lack of exact controllability for mild solutions in Banash spaces, SIAM J. Control and Optimization 18 (1980), pp. 98-99.

[13] J.-Cl. LOUIS, D. WEXLER, On exact controllability and Liapunov equations for evolution equations, J. Diff. Eq (to appear).

[14] J.C. WILLEMS, Least squares stationary optimal control and the algebraic Riccati equation, IEEE Trans. Autom. Control AC-16 N°6 (1971), pp. 621-634.

[15] V.A. YAKUBOVICH, A frequency theorem for the case in which the state and control spaces are Hilbert spaces with an application to some problems in the synthesis of optimal controls I, Siberian Math. J. 15 (1974), pp. 457-476.

[16] V.A. YAKUBOVICH, A frequency theorem for the case in which the state and control spaces are Hilbert spaces with an application to some problems in the synthesis of optimal controls II, Siberian Math. J. 16 (1975), pp. 828-845.

[17] J. ZABCZYK, Remarks on the algebraic Riccati equation in Hilbert spaces, Applied Math. and Optimization 2 (1975/1976), pp. 251-258.

CONTINUOUS DEPENDENCE FOR DEMICOMPACT MAPS

Rafiq Makhluf and Zdenek Vorel

University of Southern California
Los Angeles, California

ABSTRACT

If $\{T_k\}_{k=0}^{\infty}$ is a sequence of operators in a Banach space, $x_k = T_k x_k$ is a fixed point of T_k in a closed and bounded set B for $k = 0,1,\ldots$ and if T_0 satisfies some regularity conditions, the usual problem of continuous dependence of fixed points on parameters is to give conditions on T_k which guarantee that $x_k \to x_0$ as $k \to \infty$. To obtain the weakest conditions one can either try to find the weakest topology in the set of operators or to restrict the sets on which T_k should be close to T_0. The last approach is used to obtain sufficient and necessary conditions for $x_k \to x$.

INTRODUCTION

Usual formulations of continuous dependence theorems involve functions $T:G \times \Lambda \to X$ where G is an open set in a Banach space X and Λ is a metric space of parameters with a limit point λ_0. If for $\lambda \in \Lambda$, x_λ is a fixed point of $T(\cdot,\lambda)$ (not necessarily unique) and if $x_{\lambda_0} \in G$ is the unique fixed point of $T(\cdot,\lambda_0)$, what conditions on T guarantee that $x_\lambda \to x_{\lambda_0}$ with $\lambda \to \lambda_0$ and which of these conditions are necessary? No conditions of closeness of $T(\cdot,\lambda)$ and $T(\cdot,\lambda_0)$ on the whole domain G can be necessary for closeness of x_λ and x_{λ_0}. This is easy to see if one assumes that $T(\cdot,\lambda_0)$ is continuous at x_{λ_0}; then $x_\lambda \to x_{\lambda_0}$ implies that $|T(x_\lambda,\lambda) - T(x_\lambda,\lambda_0)| \leq |x_\lambda - x_{\lambda_0}| + |T(x_{\lambda_0},\lambda_0) - T(x_\lambda,\lambda_0)| \to 0$ with $\lambda \to \lambda_0$. E.g. if $r > 0$, $x_\lambda \varepsilon B(x_{\lambda_0},r) = \{x \in X \mid |x - x_{\lambda_0}| < r\}$ for $\lambda \in \Lambda$, $x_\lambda \to x_{\lambda_0}$ may hold although $|T(x,\lambda) - T(x,\lambda_0)|$ is large in the complement $B^c(x_{\lambda_0},r)$. In [1] this difficulty was avoided by narrowing the definition of continuous dependence in the following sense: Instead of the equation $x = T(x,\lambda)$ one considers $x = T(x,\lambda) + y$, where y is an arbitrary forcing term, and a characterization is given for the upper semicontinuity of the solution set $s(\lambda,y)$ (as a set-valued function of two variables λ and y) under the assumption that $T(\cdot,\lambda)$ is collectively α-condensing: $s(\lambda,y)$ is upper semicontinuous iff T is (jointly) continuous in x and λ. The latter condition is not satisfied in the following

Example 1: $X = G = C([0,1],R^1)$, $\Lambda = [0,1]$, $\lambda_0 = 0$,

$T(x,\lambda)(t) = \lambda + \int_0^t x(s)\,\lambda^{-\frac{1}{2}}\cos(\lambda^{-1}s)ds$ for $\lambda > 0$, $T(x,0) = 0$.
To see that T is not continuous at $x = 0$, $\lambda = 0$ take
$x(t) = \lambda^{1/4}\cos\lambda^{-1}t$, $t \in [0,1]$.

CONTINUOUS DEPENDENCE OF FIXED POINTS

It is natural, in this approach to continuous dependence to work with demicompact operators: Let X be a Banach space, let D be set in X, $V:D \to X$. V is called demicompact if for every bounded sequence $x_n \in D$ such that $x_n - V(x_n)$ converges there is a convergent subsequence $\{x_{n_k}\}$ (cfr. [2]). It is not difficult to show that if V is α-condensing then V is demicompact [9]. Thus the results obtained will be applicable to a class of neutral functional differential equations introduced in [3].

Proposition 1. Let $T(\cdot,\lambda_0)$ be continuous and demicompact in an open set $G \subset X$. Assume $T(\cdot,\lambda_0)$ has a unique fixed point $x(\lambda_0) \in G$. Let $\{x_n\}_{n=1}^{\infty} \subset \overline{B}(x(\lambda_0),r) \subset G$, $x_n - T(x_n,\lambda_0) \to 0$ as $n \to \infty$. Then $x_n \to x(\lambda_0)$.

Proof: It is sufficient to prove that every subsequence $\{y_n\}$ of $\{x_n\}$ has a subsequence converging to $x(\lambda_0)$. As $T(\cdot,\lambda_0)$ is demicompact and as $x_n - T(x_n,\lambda_0) \to 0$, y_n has a subsequence $z_n \to z$. By the same hypothesis, $z_n - T(z_n,\lambda_0) \to 0$, which implies $T(z_n,\lambda_0) \to z$. Continuity of $T(\cdot,\lambda_0)$ and the uniqueness of $x(\lambda_0)$ forces $z = x(\lambda_0)$.

Note that the full force of the demicompactness of $T(\cdot,\lambda_0)$ was not used in the proof of Proposition 1. It is sufficient to assume that $\{x_n \in \overline{B}(x(\lambda_0),r),\ x_n - T(x_n,\lambda_0) \to 0\}$ implies that there is a convergent subsequence of x_n. Also, if the latter condition is not satisfied the conclusion of Proposition 1 does not hold, i.e. this condition cannot be relaxed any more. Further, it is not necessary to assume the existence of the fixed point $x(\lambda_0)$ in G. This is a consequence of the completeness of X.

<u>Proposition 2.</u> If $T(\cdot,\lambda_0)$ is continuous at $x(\lambda_0)$, which is the unique fixed point of $T(\cdot,\lambda_0)$ and if $x_n \to x(\lambda_0)$, then $x_n - T(x_n,\lambda_0) \to 0$.

<u>Proof:</u> Follows from
$|x_n - T(x_n,\lambda_0)| \leq |x_n - x(\lambda_0)| + |T[x(\lambda_0), \lambda_0] - T(x_n,\lambda_0)|$.

<u>Theorem 1.</u> Let B be a closed and bounded set in X, $T(\cdot,\lambda_0)$ continuous and demicompact in $G \supset B$, $T(B,\lambda) \subset B$ for every $\lambda \in \Lambda$, $T[x(\lambda),\lambda] = x(\lambda) \in B$, $x(\lambda_0)$ the unique fixed point of $T(\cdot,\lambda_0)$ in B. If there exists a function $p : \Lambda \to \{1,2,\ldots\}$ such that

(1) $T^{p(\lambda)}(x,\lambda) - T[T^{p(\lambda)}(x,\lambda), \lambda_0] \to 0$ uniformly

for $x \in B$ as $\lambda \to \lambda_0$, where $T^{p(\lambda)}(\cdot,\lambda)$ is the $p(\lambda)^{\underline{th}}$ iterate of $T(\cdot,\lambda)$, then $x(\lambda) \to x(\lambda_0)$.

<u>Proof:</u> Let $\lambda_n \to \lambda_0$, $\lambda_n \in \Lambda$ for $n = 1,2, \ldots$. (1) implies $x(\lambda_n) - T[x(\lambda_n),\lambda_0] \to 0$, as $x(\lambda_n)$ is a fixed point for every iterate of $T(\cdot,\lambda_n)$. By Proposition 1, $x(\lambda_n) \to x(\lambda_0)$ as $n \to \infty$.

A simple example in [5] shows that the uniformity of convergence in (1) is essential. Further, in [5] condition (1) ensures, in terms of numerical analysis, convergence of approximate solutions of $x(\lambda_n) = T[x(\lambda_n),\lambda_n]$ to the fixed point $x(\lambda_0)$.

Condition (1) is the best possible, if for every $\lambda \in \Lambda, x(\lambda)$ is a fixed point of a contraction mapping. In this case a partial converse of Theorem 1 is true.

<u>Theorem 2.</u> Let B be a closed and bounded set in X, $T(\cdot, \lambda_0)$ continuous at $x(\lambda_0)$, $T(B,\lambda) \subset B$ for every $\lambda \in \Lambda$. Let $x(\lambda) = T[x(\lambda),\lambda]$ for $\lambda \in \Lambda$ be a (unique) fixed point of $T(\cdot,\lambda)$ in B, let for every $\lambda \neq \lambda_0$

(2) $\sup\{|x(\lambda) - y| : y \in T^k(B,\lambda)\} \to 0$ as $k \to \infty$.

If $x(\lambda) \to x(\lambda_0)$ as $\lambda \to \lambda_0$, then there exists a function $p: \Lambda \to \{0,1,2,\ldots\}$ such that

$$T^{p(\lambda)}(x,\lambda) - T[T^{p(\lambda)}(x,\lambda),\lambda_0] \to 0 \text{ uniformly for } x \in B \text{ as } \lambda \to \lambda_0.$$

Proof: For every $\lambda \neq \lambda_0$ let $p(\lambda)$ be such that $|x(\lambda) - z| < d(\lambda,\lambda_0)$ whenever $z \in T^{p(\lambda)}(B,\lambda)$ ($p^{(\lambda)}$ exists by (2)). Let $\varepsilon > 0$; by hypotheses there is a $\delta(\varepsilon)$ such that $|x - x(\lambda_0)| < \delta$ implies $|T(x,\lambda_0) - T[x(\lambda_0),\lambda_0]| < \frac{\varepsilon}{3}$; then there exists an $\eta(\varepsilon,\delta) < \min(\frac{\varepsilon}{3}, \frac{\delta}{2})$ such that $d(\lambda,\lambda_0) < \eta$ implies $|x(\lambda) - x(\lambda_0)| < \min(\frac{\varepsilon}{3}, \frac{\delta}{2})$. Now, for $d(\lambda,\lambda_0) < \eta$ and $x \in B$ one has $|T^{p(\lambda)}(x,\lambda) - T[T^{p(\lambda)}(x,\lambda), \lambda_0]| \leq$

$$|T^{p(\lambda)}(x,\lambda) - x(\lambda)| + |x(\lambda) - x(\lambda_0)| + |T[x(\lambda_0),\lambda_0] - T[T^{p(\lambda)}(x,\lambda),\lambda_0]| < \varepsilon$$

as $|x(\lambda_0) - T^{p(\lambda)}(x,\lambda)| \leq |x(\lambda_0) - x(\lambda)| + |x(\lambda) - T^{p(\lambda)}(x,\lambda)| < \delta$.

Corollary: Let B be a closed and bounded set in X, $T(B,\lambda) \subset B$ for $\lambda \in \Lambda$, $T(\cdot,\lambda_0)$ continuous and demicompact in $G \supset B$; assume further that for every $\lambda \in \Lambda$ there exists a fixed point $x(\lambda)$ of $T(\cdot,\lambda)$ such that (2) holds. Then the following two conditions are equivalent:

a) For each $\lambda \in \Lambda$, $x(\lambda)$ is the unique fixed point of $T(\cdot,\lambda)$ in B and $x(\lambda) \to x(\lambda_0)$.

b) there exists a function $p: \Lambda \to \{1,2,3,...\}$ such that (1) holds uniformly for $x \in B$.

Remark: Theorem 1 may be applied to Example 1 with $p(\lambda) = 3$ for every λ as there exists $\alpha > 0$ such that $|T^3(x,\lambda)| < \alpha\lambda^{1/2}$ for every $x \in B(0,1)$. This shows that for Example 1 the full strength of Theorem 1 (i.e. the dependence of p on λ) is not needed. Simple examples in ODE's show that condition (1) with p constant is not necessary for continuous dependence [4] and also that Theorem 1 may be applied with p variable but not with p constant [9].

NEUSTADT'S OPERATOR EQUATIONS

Theorem 1 has two drawbacks: First, it assumes the existence of the fixed points $x(\lambda)$ in B and second, it requires $T(B,\lambda) \subset B$. Both assumptions can be relaxed if we assume local existence and continuation of solutions of causal integral-like operator equations [6]. There the Banach space $X = C([a,b],Y)$, the space of continuous functions from a compact interval $[a,b]$ into a Banach space Y with the supremum norm . The operator $T : G \times \Lambda \to X$, where G is an open set in X and Λ is as before, is assumed to be causal (i.e. $x,y \in G$, $\lambda \in \Lambda$, $t \in [a,b]$, $x = y$ in $[a,t]$

imply $T(x,\lambda)(t) = T(y,\lambda)(t))$ and of fixed initial value (i.e. $T(x,\lambda)(a) = T(y,\lambda)(a)$ for all $x,y \in G$, $\lambda \in \Lambda$). Let $B = \overline{B(x(\lambda_0),r)} = \{x \in G: |x - x(\lambda_0)| \leq r\}$, where $x(\lambda_0)$ is the unique fixed point of $T(\cdot, \lambda_0)$ in B. Let $[T(x,\lambda)]\ (a) \to [x(\lambda_0)](a)$ as $\lambda \to \lambda_0$ and for every $\lambda \in \Lambda$ let there exist a local solution of the equation

$$x = T(x,\lambda) \tag{3}$$

Assume that every maximal continuation y of a local solution of (3) is either a global solution of (3) on [a,b] or else for every r such that $B[x(\lambda_0),r] \subset G$ and $|y(a) - x(\lambda_0)(a)| < r$ there is a $t_1 \in (a,b)$ such that $|y(t_1) - x(\lambda_0)(t_1)| = r$.

Now let $r > 0$ be fixed so that $B[x(\lambda_0),r] \subset G$. For this r, a truncation operator Z on $C([a,b]),Y)$ is defined as follows: a) $Zx = x$ if $|x(a) - [x(\lambda_0)](a)| \geq r$ or if $x \in B[x(\lambda_0), r]$; b) if $|x(a) - x(\lambda_0)(a)| < r$ and if there exists a first $t_x > a$ such that $|x(t_x) - x(\lambda_0)(t_x)| = r$, one defines $Zx(t) = x(t)$ for $t \in [a,t_x]$, $Zx(t) = [x(\lambda_0)](t) + x(t_x) - [x(\lambda_0)](t_x)$ for $t \in [t_x,b]$. For iterates of the composition ZT the following notation will be used: $(ZT)^0(x,\lambda) = x$, $(ZT)^1(x,\lambda) = (ZT)(x,\lambda)$; $(ZT)^{j+1}(x,\lambda) = (ZT)[(ZT)^j(x,\lambda),\lambda]$ for $j = 1,2,3,\ldots$, $x \in G$, $\lambda \in \Lambda$.

If y is any maximal continuation of a local solution of equation (3) for which $|y(a) - [x(\lambda_0)](a)| < r$, define t_y as the first $t \in (a,b]$ such that $|y(t) - [x(\lambda_0)](t)| = r$. If such t does not exist it means that $y \in B[x(\lambda_0), r]$ and in this case one defines $t_y = b$.

Next, for $\lambda \in \Lambda$ such that $|T[x(\lambda_0),\lambda](a) - x(\lambda_0)(a)| < r$ define $S(\lambda) = \{ x \in \overline{B[x(\lambda_0), r]}$: $x(t) = [T(x,\lambda)](t)$ for every $t \in [a,t_x]$ and $x(t) = [x(\lambda_0)](t) + x(t_x) - [x(\lambda_0)](t_x)$ for every $t \in [t_x,b]\}$, i.e. $S(\lambda)$ is the set of all truncations of maximal continuations y of equation (3) for which $|y(a) - [x(\lambda_0)](a)| < r$.

The next theorem [9] gives a sufficient and necessary condition for existence of global solutions of (3) and for their convergence to the unique fixed point $x(\lambda_0)$ of $T(\cdot,\lambda_0)$.

<u>Theorem 3.</u> Let $T(\cdot,\lambda_0)$ be continuous and demicompact in G. Then the following two conditions are equivalent:

(a) Every maximal continuation of (3) is a (global) solution for $d(\lambda,\lambda_0)$ sufficiently small, and every solution $x(\lambda) \to x(\lambda_0)$ as $\lambda \to \lambda_0$.

(b) $|x - T(x,\lambda_0)| \to 0$ uniformly for $x \in S(\lambda)$ as $\lambda \to \lambda_0$.

Proof: If (b) holds then (a) follows from Proposition 1, as $S(\lambda) \subset \overline{B[x(\lambda_0),r]}$ for λ sufficiently close to λ_0. Condition (a) implies (b) by Proposition 2.

Condition (b) in Theorem 3 is stated in terms of sets $S(\lambda)$ which are generally unknown. Condition (b) is replaced by a mere sufficient but verifiable condition in the following result.

Theorem 4. Let $T(\cdot,\lambda_0)$ be continuous and demicompact on G. Let p be a function from Λ into the set of nonnegative integers such that

$$(4)\quad (ZT)^{p(\lambda)}(x,\lambda) \ - \ T[(ZT)^{p(\lambda)}(x,\lambda),\ \lambda_0] \to 0$$

uniformly for $x \in \overline{B[x(\lambda_0),r]}$ as $\lambda \to \lambda_0$. Then condition (a) of Theorem 3 holds.

Proof: It follows from the definition of $S(\lambda)$ that $x \in S(\lambda)$ implies $(ZT)^{p(\lambda)}(x,\lambda) = x$. Thus condition (b) of Theorem 3 is satisfied and (a) holds.

The next theorem shows that condition (4) is not only sufficient but also necessary for (a) if $x(\lambda)$ is the fixed point of a contraction mapping in some neighborhood of $x(\lambda_0)$ for every $\lambda \in \Lambda$.

Theorem 5. Let $T(\cdot,\lambda_0)$ be continuous at $x(\lambda_0)$ and let for every $\lambda \in \Lambda$ $x(\lambda)$ be a fixed point of $T(\cdot,\lambda)$ in $B = \overline{B[x(\lambda_0),r]}$. Assume $\sup\{|x(\lambda) - y|: y \in (ZT)^j(B,\lambda)\} \to 0$ as $j \to \infty$ for every $\lambda \in \Lambda$ and $x(\lambda) \to x(\lambda_0)$ as $\lambda \to \lambda_0$. Then condition (4) holds for some function p.

Proof: The hypotheses of Theorem 2 are satisfied with T replaced by ZT, as $ZT(B,\lambda) \subset B$. Thus condition (1) is satisfied and this implies (4).

Note that if condition (4) is satisfied for some r then it holds for every r such that $\overline{B(x(\lambda_0)r)} \subset G$.

REFERENCES

[1] Z. Artstein, On continuous dependence of fixed points of condensing maps, Dynamical Systems, Vol.III, An International Symposium, Academic Press; New York, (1976), pp. 73-75.

[2] Z. Artstein, Continuous dependence of operator equations, Trans. Amer. Math. Soc., 231, (1977), pp. 143-165.

[3] J. Hale, Theory of functional differential equations, Springer Verlag, New York, Heidelberg, Berlin, (1977).

[4] F. Flores, and Z. Vorel, Necessary and sufficient conditions for continuous dependence for Volterra operator equations, Bol. Soc. Mat. Mexicana, 23(2), (1978), pp. 34-35.

[5] R. Kannan, and Z. Vorel, Continuous dependence of approximate solutions of operator equations, (in preparation).

[6] L.W. Neustadt, On the solution of certain integral like operator equations, Arch. Rational Mech. Anal., 38 (2), (1970), pp. 131-160.

[7] Z. Vorel, Continuous dependence on parameters, J. Nonlinear Anal., Theory, Meth. & Appl., 5, (1981), 373-380.

[8] Z. Vorel, Necessary and sufficient conditions for continuous dependence of fixed points of α-condensing maps, Proceedings of Intern. Conference on Nonlinear Phenomena, UTA, (to appear).

[9] R. Makhluf, Ph.D Dissertation, University of Southern California (1982).

VECTOR LYAPUNOV FUNCTIONS METHOD IN THE ANALYSIS OF DYNAMICAL PROPERTIES OF NONLINEAR DIFFERENTIAL EQUATIONS

V.M.Matrosov
Irkutsk, Computer Centre, 1,USSR

ABSTRACT

The objective of the paper is the description of the development of principal ideas and present-day state of the vector Lyapunov functions method (VLF), and also of technology of method's application to the applied problems of stability and other dynamical properties analysis of non-linear differential equations. In general, the survey of investigations, realized by the staff of the Irkutsk computer centre of the Siberian Branch of USSR Academy of Sciences, is presented.

INTRODUCTION

Lyapunov functions method was founded in 1892 by the great Russian mathematician and mechanician, whose 125-th anniversary is celebrated on the 7 June this year, as the main method of stability investigations for ordinary differential equations. In many outstanding works in our century this method was essentially developed in the stability theory, extended to the various dynamical properties analysis of differential equations solutions. It became the principal method of qualitative analysis of nonlinear systems.

Combination with theory of differential inequalities of Chapligin type begun with the works of 1956 by M.A.Krasnoselsky, S.G.Krein, P.Conti, O.Borouvka, G.I.Melnikov. Investigations of this trend led H.Antosiewicz, K.Corduneanu, M.A.Krasnoselsky, V.Lakshmikantham and other authors to the formation at the beginning of 60-th the comparison method with the functional of Lyapunov type; by using this method one can easily obtain many classical and modern theorems of ordinary differential equations theory, the stability theory.

The main difficulty for method application is Lyapunov function construction. In this situation, although effective means of Lyapunov functions construction were supposed for a series of classes of nonlinear systems (see,survey [1]), the further development of this method to weakening the Lyapunov functions requirements, expansion of the class of used functions is of great interest.

For this purpose in 1962 R.Bellman [2] and the author [3] suggested the vector Lyapunov function (VLF), to satisfy the finite-dimensional inequalities of Chapligin-Wazewski type, the first theorems on stability with VLF [3-5] were obtained.

There along with the initial system the auxiliary comparison system (CS) is introduced, described by finite-dimensional differential equation. It was proved that upper semi-stability in the same sense of CS imply the stability of the initial system. Now we call such theorems the comparison theorems (CT) with VLF. In 1965 by V.Lakshmikantham and author CT with VLF were proved for the conditional stability, boundedness [6] and partial stability [7] . In 60-th F.Hartman [8] , V.Lakshmikantham with co-workers [9,18-20] ,N.Rouche, K.Peiffer [10,21] and Russian mathematicians [11-15] obtained CT with VLF for various types of stability and boundedness for different classes of systems. In [15] CT with VLF were obtained for the solutions existence and uniqueness of non-linear differential equations in Banach space with unbounded operators, their continuous dependence on perturbations, for attraction, dissipativeness, convergence and etc. This trend continued in 70-th [16,17,22-37] and at present it is being successfully developed [38-43] . In the works by S.Bernfeld,V. Lakshmikantham and others [44,45] the sufficient conditions of boundary problems solvability in the form of CT with VLF with values in the cone of Banach space are given; in the Soviet Union the comparison theorems with VLF for the stabilizability [46,29,40] , controllability, exponential invariance and boundedness, pursuit and evasion [40,46] ,optimal control existence and etc. for control systems are obtained [47,48] . The conditions of every CT depending essentially on the studied property and requiring the creative approach to their formulation.

The results seemed to be possible to unite as the intuitive idea of comparison principle: If exists VLF, satisfying some conditions, then the various dynamical properties of differential equation in Banach space follow from corresponding dynamical properties of comparison system [13]. Their development is in the works of V.Hahn and others [49].

Thus, the important step in the development of qualitative theory of differential equations of Lyapunov and Poincare was made.

The sphere of applicability of comparison method was considerably expanded; by extension to general systems [18], to polysystems [50], especially to the systems of processes, abstract controlled systems and systems in the medium. This sphere covers the majority of axiomatic constructions of mathematical systems theory, presented in literature [51-57].

We shall limit yourselves to the composition of differential equations.

I. COMPARISON VECTOR FUNCTIONS AND SYSTEMS. DYNAMICAL PROPERTIES.

Let a differential equation

$$\frac{dx(t)}{dt} = X(t, x(t)) \tag{1}$$

$$(t \in T = [0, \tau),\ 0 \leq \tau \leq +\infty,\ t_0 \in \overset{\circ}{T} \subseteq T,\ H \subseteq E,\ \Omega = T \times H,$$

$$H \subseteq H^* \subseteq \bar{H},\ H\text{-dense in domain } H^*,\ \Omega^* = T \times H^*)$$

be given in a real Banach space E with nonlinear unbounded (generally discontinuous) operator $X : \Omega \longrightarrow E$ satisfying the existence conditions of Caratheodory solutions $x : T(x) \longrightarrow E$ of Cauchy problem (I) for any $x(t_0) = X_0 \in H^0$ [15]. For initial data $h = (t_0, X_0) \in \overset{\circ}{T} \times H^0 = \Omega^0$ the solution set is denoted by τh [55].

The real measurable functionals v^s and vector function $v = (v^1, \ldots, v^K) : \Omega^* \longrightarrow R^K,\ (t, x) \longmapsto v(t, x)$ are being considered. The norm, semi-orderliness

$$\|y\| = \max_{s=\overline{1,K}} |y^s|;\ y_1 \leq y_2 : y_1^s \leq y_2^s;\ y_1 < y_2 : y_1^s < y_2^s\ (s = \overline{1, K})$$

are introduced in R^K; $\bar{v}(t,x) \triangleq \max_S v^S(t,x)$ $(s=\overline{1,\ell},\ \ell \le k)$. v is called vector comparison function (VCF) for equation (I), if a domain $A \subseteq T \times R^K$ $(v(t,H^*) \subseteq A(t))$ and a measurable function $f: A \longrightarrow R^K$ exist, such that :

I. $\bar{D}^+ v(t,x(t)) \le f(t, v(t,x(t)))$ for $t \dot{\in} T$, $x \in \tau\overset{o}{\Omega}$, $x(t) \in H$.

II. f in every bounded region $\bar{B} \subset A$ is bounded on the norm by summable function, it is measurable on t for fixed v and satisfies the quasimonotony condition by T.Wazewsky in A:

$$y_1^S = y_2^S,\ (\forall \nu \ne s)\ y_1^\nu \le y_2^\nu \Longrightarrow f^S(t,y_1) \le f^S(t,y_2).$$

III. For any absolutely continuous abstract function x the vector-function $v(\cdot, x(\cdot))$ is upper left semi-continuous and either an upper absolutely semi-continuous in V.M.Alekseev's sense, or upper right semi-continuous and the differential inequality (I) holds for $t \in T \setminus \Sigma$, where Σ is account.

Using f we form the comparison system τ_c (CS) as the system of upper solutions $\bar{x}_c$ of the equation

$$\frac{dx_c(t)}{dt} = f(t, x_c(t)). \tag{2}$$

Their generalized solutions of II kind absolutely continuous function $x_c: T(x_c) \longrightarrow R^K$ satisfying inequalities [14]

$$\underline{\lim}_{y \to x_c(t)} f(t,y) \dot{\le} \frac{dx_c(t)}{dt} \dot{\le} \overline{\lim}_{y \to x_c(t)} f(t,y).$$

<u>Lemma</u>. Let v is VCF for (I), $v(t_o, X_o) \le X_{co}$, $x \in \tau(t_o, X_o)$. Then upper solution $\bar{x}_c : T(\bar{x}_c) \to R^K$ exists, continuable to the right up to the $\bar{B} \setminus B$, and $v(t,x(t)) \le \bar{x}_c(t)$ $(t \in T(x) \cap \cap T(\bar{x}_c))$ [15,17].

The estimators $\rho_q : \overset{*}{\Omega} \longrightarrow R^{m_q}$, $\overset{o}{\rho}_q : \overset{o}{\Omega} \longrightarrow R^{\overset{o}{m}_q}$ and the families of estimation sets $(q=0,1,2)$

$$P_q(\varepsilon_q, t) \triangleq \{x \in H^* : \rho_q(t,x) < \varepsilon_q\},\ \varepsilon_q \in R_+^{m_q},$$

$$\overset{o}{P}_q(\delta_q, t_o) \triangleq \{x_o \in \overset{o}{H} : \overset{o}{\rho}_q(t_o,x_o) < \delta_q\},\ \delta_q \in R_+^{\overset{o}{m}_q},$$

$$P_q^*(\delta_q, \delta, t_o) \triangleq \overset{o}{P}_q(\delta_q, t_o) \cap \overset{o}{P}(\delta, t_o),\ R_+^{m_q} \triangleq \{y > 0\}$$

are considered $(\delta = \delta_o)$.

Here is introduced the unified representation of dynamical properties using the quantifier formulae $\mathfrak{P} \triangleq \mathfrak{S}_o(\mathfrak{S}_1 \mathcal{R}_1 \wedge \mathfrak{S}_2 \mathcal{R}_2)$, where $\mathcal{R}_q \triangleq (x \in P_q(\varepsilon_q, \varepsilon, t))$, $\mathfrak{S}_q$ are the words from quantifiers $\mathfrak{M}_q^{\mathcal{M}}$ (universal $\hat{\mathfrak{M}}_q^{\mathcal{M}} \triangleq (\forall z_q^{\mathcal{M}} \in \mathcal{Z}_q^{\mathcal{M}})$ or existential $\check{\mathfrak{M}}_q^{\mathcal{M}} \triangleq (\exists z_q^{\mathcal{M}} \in \mathcal{Z}_q^{\mathcal{M}})$ by the variables $z_q^{\mathcal{M}}: t_0, x_0, x, t, x$ (main variables), $\varepsilon_q, \delta_q, t_{*q}$ (index $q=0$ is omitted) (see [60]). The typed quantifiers on the main variables t_0, x_0, x, t, x are denoted by $\mathfrak{T}_0, \mathfrak{X}_0, \mathfrak{X}, \mathfrak{T}, \mathfrak{M}_x$. The replacement A by B in the formula $\mathcal{F}$ is denoted by $\sigma_B^A \; \mathcal{F} \,|$.

The dynamical properties of main type is introduced by some additional conditions [52]. For example, asymptotic stability

$$(\forall t_0 \in T^0 \, \exists \delta \in R_+^{m^0} \, \forall \varepsilon \in R_+^{m})[(\exists \delta_1 \in R_+^{m_1^0} \, \forall x_0 \in P_1^*(\delta_1, \delta, t_0)$$
$$\forall x \in \tau(t_0, x_0) \forall t \in T(x) \forall x = x(t))(x \in P(\varepsilon, t)) \wedge$$
$$\wedge (\forall x_0 \in P^0(\delta, t_0) \, \exists t_{*2} \in T_{t_0} \, \forall x \in \tau(t_0, x_0)$$
$$\forall t \in T(x) \cap [t_{*2}, \tau) \, \forall x = x(t))(x \in P(\varepsilon, t))].$$

Here are $T_{t_{*q}} \triangleq \{t \in T_{*q} : t_{*q} \le t\}$, $T_{*q} \triangleq [t_0, t_*]$ $(t_{*q} \prec x)$,

$$P_q(\varepsilon_q, \varepsilon, t) \triangleq \begin{cases} P(\varepsilon, t) \cap P_q(\varepsilon_q, t) \,|\, \varepsilon, \varepsilon_q \in \mathfrak{P} & (q=1,2). \\ P(\varepsilon, t) \,|\, \varepsilon \in \mathfrak{P}, \varepsilon_q \notin \mathfrak{P} \\ P_q(\varepsilon_q, t) \,|\, \varepsilon_q \in \mathfrak{P}, \varepsilon \notin \mathfrak{P} \end{cases}$$

For CS the corresponding comparison properties $\mathfrak{P}$ obtained by the formal index addition to all letters in the formula $\mathfrak{P}$, by the introduction of corresponding estimation sets $P_{qc}(\varepsilon_c, t)$, $P_{qc}^0(\delta_c, t_c)$ usually by cone type into $R^{m_{qc}}$, $R^{m_{qc}^0}$ and etc. are introduced. For example, the upper asymptotic semi-stability of CS:

$$\mathfrak{P}_c = \mathfrak{S}_{0c}(\mathfrak{S}_{1c} \mathcal{R}_{1c} \wedge \mathfrak{S}_{2c} \mathcal{R}_{2c}) \triangleq$$
$$(\forall t_0 \in T^0 \, \exists \delta_c \in R_+^K \, \forall \varepsilon_c \in R_+^e)[(\exists \delta_{1c} \in R_+^K \, \forall x_{c0} \in R_+^K :$$
$$x_{c0} < \min[\delta_{1c}, \delta_c] \, \forall \bar{x}_c \in \tau_c(t_0, x_{0c}) \, \forall t_c \in T_c(\bar{x}_c) \, \forall x_c =$$
$$= \bar{x}_c(t))(x_c < \varepsilon_c) \wedge (\forall x_{c0} \in R_+^K : x_{c0} < \delta_c \, \exists t_{*2c} \in T_{t_0}$$
$$\forall \bar{x}_c \in \tau_c(t_0, x_{c0}) \, \forall t \in [t_{*2c}, \tau) \cap T_c(x_c) \, \forall x_c = x_c(t))$$
$$(x_c < \varepsilon_c)].$$

2. COMPARISON PRINCIPLE WITH VLF

As an analysis result of formulations and proofs of hundreds theorems of Lyapunov function method and comparison method their unique structure (in the simplest case by Habets, K.Peiffer [58]), the connections between formulae of studied properties and conditions on VLF in comparison theorems were revealed, a series of heuristics were obtained the main of which [52]

$$\mathfrak{C}_q^{z_q^\mu} \triangleq \begin{cases} \hat{\mathfrak{M}}_q^\mu \check{\mathfrak{M}}_{qc}^\mu \mid \mathfrak{M}_q^\mu = \hat{\mathfrak{M}}_q^\mu \\ \hat{\mathfrak{M}}_{qc}^\mu \check{\mathfrak{M}}_q^\mu \mid \mathfrak{M}_q^\mu = \check{\mathfrak{M}}_q^\mu \end{cases} \qquad (q=0,1,2),$$

is the transformation of typed quantifiers $\mathfrak{M}_q^\mu$ into subformulae of requirements on VLF.

Comparison principle with VLF was proved. Here is denoted:

$$\hat{\mathfrak{X}}_{oq}\hat{\mathfrak{X}}_{oqc}^{v} \triangleq (\forall x_o \in P_q^*(\delta_q, \delta, t_o) \cap v(t_o, \cdot)^{-1} P_{qc}^*(\delta_{qc}, \delta_c, t_o)\, \forall x_o = v(t_o, x_o)),$$

$$\hat{\mathfrak{t}}_{*q} \triangleq (\forall t_{*q} \in T_{*q} \cap T_{*qc}),\ \hat{\mathfrak{t}}_{t*q} \triangleq (\forall t \in T_{t*q} \cap T_{t*qc}) \quad (q=0,1,2),$$

$$\mathfrak{C}_{t_o t_*}^{\varepsilon} \triangleq \begin{cases} \hat{\mathfrak{t}}_o\, \hat{\mathfrak{t}}_*\, \mathfrak{C}^\varepsilon \mid t_o \prec t_* \prec \varepsilon \\ \hat{\mathfrak{t}}_o\, \mathfrak{C}^\varepsilon\, \hat{\mathfrak{t}}_* \mid t_o \prec \varepsilon \prec t_* \\ \mathfrak{C}^\varepsilon\, \hat{\mathfrak{t}}_o\, \hat{\mathfrak{t}}_* \mid \varepsilon \prec t_o \prec t_* . \end{cases}$$

Here and further $\hat{\mathfrak{t}}_o$ ($\hat{\mathfrak{t}}_*$ or $\mathfrak{C}^\varepsilon, \ldots,$ respectively) are replaced by the empty formula Λ , if $t_o \notin \mathfrak{P}$ (t_* or $\varepsilon \notin \mathfrak{P}$). Analogously $\mathfrak{C}_{t*qt}^{\varepsilon_q}$ is determined. The formulae are introduced ($q=1,2$): $\mathfrak{P}_q \triangleq \mathfrak{C}_o \mathfrak{C}_q \mathfrak{R}_q$,

$$\mathfrak{D}_{tq} \triangleq \hat{\mathfrak{t}}_o \begin{Bmatrix} \mathfrak{C}^{t_*} \mathfrak{C}^{t_{*q}} & \mid t_*, t_{*q} \in \mathfrak{P}_q \\ \mathfrak{C}^{t_*} & \mid t_* \in \mathfrak{P}_q, t_{*q} \notin \mathfrak{P}_q \\ \mathfrak{C}_q^{t_{*q}} & \mid t_{*q} \in \mathfrak{P}_q, t_* \notin \mathfrak{P}_q \end{Bmatrix}$$

$$\left\{\begin{matrix} \sigma\, P_q^*(\delta_q,\delta,t_0), & P_{qc}^*(\delta_{qc},\delta_c,t_0) & \hat{\mathfrak{X}}_{oq}\,\hat{\mathfrak{X}}_{oqc}^{v} \mid \delta_q \vee \delta \in \mathfrak{P}_q \\ \cup\, P_q^*(\delta_q,\delta,t_0), & \cup\, P_{qc}^*(\delta_{qc},\delta_c,t_0) & \\ \delta_q,\delta & \delta_{qc},\delta_c & \hat{\mathfrak{X}}_{oq}\,\hat{\mathfrak{X}}_{oqc}^{v} \mid \delta_q,\delta\in\mathfrak{P}_q, x_0\in\mathfrak{P}_q \end{matrix}\right\}$$

$$\left\{\begin{matrix} \mathfrak{S}_q^x & \mid \mathfrak{X}_q=\hat{\mathfrak{X}} \\ \hat{\mathfrak{X}}\,\hat{\mathfrak{X}}_c & \mid \mathfrak{X}_q=\check{\mathfrak{X}} \end{matrix}\right\}\left\{\begin{matrix} T_q(x)\subseteq T_c(x_c) & \mid \mathfrak{f}_q=\hat{\mathfrak{f}} \\ T_c(x_c)\subseteq T_q(x) & \mid \mathfrak{f}_q=\check{\mathfrak{f}} \end{matrix}\right\},$$

$$\mathfrak{D}_{xq}\triangleq \mathfrak{S}_{t_0 t_*}^{\varepsilon}\,\mathfrak{S}_{t_{*}qt}^{\varepsilon_q}\left\{\begin{matrix} (\forall x\in\bar{X}\setminus P_q(\varepsilon_q,\varepsilon,t)\,\forall x_c\in P_{qc}(\varepsilon_{qc},\varepsilon_c,t))\,v(t,x)\notin X_c & \mid \mathcal{R}_q\neq\Lambda \\ \Lambda & \mid \mathcal{R}_q=\Lambda \end{matrix}\right\},$$

$$\mathfrak{D}_{x_0 q}\triangleq\left\{\begin{matrix} \mathfrak{S}^{\delta}\,\hat{\mathfrak{f}}_0 & \mid \delta<t_0 \\ \hat{\mathfrak{f}}\,\mathfrak{S}^{\delta} & \mid t_0<\delta \end{matrix}\right\}\left\{\begin{matrix} \mathfrak{S}^{\delta_q} & \mid \delta_q\in\mathfrak{P}_q \\ \Lambda & \mid \delta_q\notin\mathfrak{P}_q \end{matrix}\right\}$$

$$\left\{\begin{matrix} v(t_0,P_q^*(\delta_q,\delta,t_0))\subseteq P_{qc}^*(\delta_{qc},\delta_c,t_c) & \mid \mathfrak{X}_{oq}=\hat{\mathfrak{X}}_0 \\ P_{qc}^*(\delta_{qc},\delta_c,t_0)\subseteq v(t_0,P_q^*(\delta_q,\delta,t_0)) & \mid \mathfrak{X}_{oq}=\mathfrak{X}_0 \\ v(t_0,x_0)=X_c & \mid x_0\in\mathfrak{P}_q \end{matrix}\right\}.$$

VCF v, satisfying (with CS) the conditions $\mathfrak{D}_{tq},\mathfrak{D}_{xq},\mathfrak{D}_{x_0q}$ ($q=1,2$) is called vector Lyapunov function (VLF) for equation (I) with respect to $\mathfrak{P}$.

Comparison Principle. Let VLF v for (I) with respect to $\mathfrak{P}$ and CS τ_c exist. Then existence of dynamical property $\mathfrak{P}$ in the system (I) follows from existence of corresponding property $\mathfrak{P}_c$ in CS τ_c.

This explicit form of comparison principle in the theory of differential equations gives the explicit scheme of comparison theorems:

$$(\exists v_1 f)\ \mathrm{I},\mathrm{II},\mathrm{III},\ \mathfrak{D}_{xq},\mathfrak{D}_{tq},\mathfrak{D}_{x_0q}\mapsto\mathfrak{P}_c\Longrightarrow\mathfrak{P}.$$

VLF for equation (I) with respect to asymptotic stability is called VCF v, satisfying conditions:

$$(\forall t_0\in T^0\,\forall\varepsilon\in R_+^m\,\exists\varepsilon_c\in R_+^{\ell}\,\forall t\in T_{t_0}\,\forall x\in\bar{X}\setminus P(\varepsilon,t))\,v(t,x)\not< X_c$$

(of vector definite non-negativity type with respect to $\{P(\varepsilon, t)\}$);

$$(\forall t_0 \in T^0\, \forall \delta_c \in R_+^K\, \exists \delta \in R_+^{m^0}\, \forall x_0 \in P^0(\delta, t_0))\, v(t_0, x_0) < \delta_c \quad ,$$

$$(\forall t_0 \in T^0\, \forall \delta_{1c} \in R_+^K\, \exists \delta_1 \in R_+^{m_0}\, \forall x_0 \in P_1^0(\delta_1, t_0))\, v(t_0, x_0) < \delta_{1c}$$

(of upper semi-continuity type with respect to $\{P^0(\delta, t_0)\}$, $\{P_1^0(\delta_1, t_0)\}$);

$$(\forall t_0 \in T^0\, \forall x_0 \in (\bigcup_{\substack{\delta_1 \in R_+^{m_1^0} \\ \delta \in R_+^{m^0}}} P_1^*(\delta_1, \delta, t_0)) \cap v(t_0, \cdot)^{-1} (\bigcup_{\delta_{1c}, \delta_c \in R_+^K} P_c^0(\delta_{1c}, \delta_c, t_0)) \forall x_{c0} =$$

$$= v(t_0, x_0)\, \forall x \in \tau h\, \exists \bar{x}_c \in \tau_c h_c))\; (T(x) \subseteq T_c(\bar{x}_c));$$

$$(\forall t_0 \in T^0\, \forall t_{*2c} \in T_{t_0}\, \exists t_{*2} \in T_{t_0}\, \forall x_0 \in (\bigcup_{\delta \in R_+^{m^0}} P^0(\delta, t_0)) \cap v(t_0, \cdot)^{-1} \cdot$$

$$\cdot (\bigcup_{\delta_c \in R_+^K} P_c^0(\delta_c, t_0))\, \forall x_{c0} = v(t_0, x_0) \forall x \in \tau h \exists \bar{x}_c \in \tau_c h_c) \cdot$$

$$(T(x) \cap [t_{*2}, \tau) \subseteq T_c(\bar{x}_c) \cap [t_{*2c}, \tau))$$

(continuability condition of CS solutions).

<u>Comparison Theorem</u>. Let exists VLF v for (I) with respect to asymptotic stability. Then, if CS τ_c is upper asymptotically semistable, then system (I) is asymptotically stable (compar. [3,7,15,55]).

The usage of comparison theorems reduces the dynamical properties analysis to VLF, CS construction and to essentially simpler analysis of CS corresponding properties. The later one is realized separately in sufficiently general case by R.I.Kozlov [60] , A.A.Martinyuk, A.Ju.Obolenski [59] .Having them into comparison theorem $\mathfrak{p}_c$, there is obtained more effective theorem on dynamical property $\mathfrak{p}$ of the system (I), in the text of which CS is not mentioned. Some algorithms [60] , based on comparison principle ideas, can be aplied to CS analysis.

<u>Theorem 1 [17]</u>. For the exponential stability of the set M of the system (I) it is sufficient, and for $E = R^n$, $M = 0$,

$X \in C^{(1)}$ it is necessary, VLF existence possessing the following conditions in some neighbourhood $M^{\alpha} \triangleq \{(t,x) \in T \times E : d(x, M(t)) < \alpha\}$ $(\alpha > 0)$

1°. $\|v(t_0, x_0)\| \leq \gamma d(x_0, M_0(t_0))^{\xi}$, $\bar{v}(t,x) \geq \delta d(x, M(t))$ $(\gamma, \delta > 0)$

where $d(x, M(t))$ is distance between x, $M(t)$.

2°. $f(t, y) = Py + \tilde{f}(t, y)$, where $\tilde{f}$ is quasimonotone,

$p_{sj} \geq 0$ $(s \neq j)$, $(-1)^s |p_{ij}|_1^s > 0$ $(s = \overline{1,k})$, $\|\tilde{f}(t,y)\| / \|y\| \underset{t \in T}{\rightrightarrows} 0$ $y \longrightarrow 0$.

Applications of VLF method, its high effectiveness showed considerable advantages over other known methods, required its development in more general form and its program realization at computers. VLF method turned out possible to extend on the analysis of the various dynamical and other complicated properties, the definitions of which are described in some quantifier graphs. The comparison principle is stated in the abstract systems dynamics, control theory and mathematical systems theory, the main concepts of which covering many classes of real systems models, developed by our collective [55]. Comparison method appeared to be the first strict and universal method of the analysis of different properties. Comparison principle is the basic result, defining the algorithms of formulations and proofs derivation of comparison theorems with VLF. The new direction in artificial intelligence, called the algorithmized approach to theorems derivation is opened. It is realized by a computer program [54,56,57], using which more than two hundred theorems were obtained, being the new or the modifications of known ones. As for their pitheness, they are quite comparable with the theorems, published by authors at present time, as the new results. The majority of known comparison theorems can be now obtained by the algorithms of comparison principle.

Supplemented with the algorithms of choice of models [57] descriptions, of derivation of theorems on dynamical properties with VLF from comparison theorems, and with the algorithms of derivation and conversion of theorems on dynamical properties with VLF (S.N.Vassiljev [57]) they represented the foundation of dialoque applied program package (APP), created in Irkutsk computer centre of the Siberian Branch of

USSR Academy of Sciences.

3. VLF, CS CONSTRUCTION, APPLICATIONS.

We consider only the problem of exponential stability of zero solution (theorem I.) for finite-dimensional systems (I) ($E = R^n$, $M = 0$). The developed means are divided into three groups.

In the first group in VLF construction $v=(v^1,\dots,v^K)$ ascending to one Lyapunov - Poincare lemma for autonomous systems, CS $\dot{v}^s(x)=\varkappa_s v^s(x)(\varkappa_s\text{-const}<0,\ \tilde{f}=0)$ [1] is being given. For the linear autonomous system $\dot{x}=Ax$ the quadratic VLF ($v^s(x)=x^T V^s x \geq 0$ $((V^s)^T=V^s)$, $\bar{v}(x)=x^T V x$ ($\ell=K$)-positive-definite) is constructed. If all the eigenvalues $\lambda_s(A)$ are different, into [61] the construction algorithm of such quadratic VLF is implemented. The approach of holomorphic VLF construction in the form of $v^s(x)=z^s(x)\bar{z}^s(x)$, satisfying the precise exponential estimations $v^s(x(t))=v^s(x_0)\exp\varkappa_s(t-t_0)$ for the autonomous system (I) with holomorphic right side $X(x)=Ax+\tilde{X}^{(2)}(x)$, based on the solution of equation

$$\nabla z^s(x)X(x)=\lambda_s z^s(x)\quad (s=\overline{1,n})$$

$$(\mathrm{Re}\lambda_s<0,\quad \lambda_s\neq\sum_{j=1}^{n} m_j\lambda_j \text{ for } m_j=0,1,2,\dots,\sum_{j=1}^{n} m_j>0,\ s=\overline{1,n})$$

are given [1,61].

The means of VLF construction, satisfying the non-linear differential comparison equations, may be based on the results of K.P.Persidski [62], K.G.Waleev and G.S.Finin [63]. There is a common feature for this group of means: the accuracy of estimations, but the difficulties of the realization of construction algorithms of VLF.

The second group of means - they are called now the decomposition and aggregation means-are based on R.Bellman's and F.N.Bailey's ideas. For the complicated (large scale) systems

$$\frac{dx_i}{dt}=X_i^0(t,x_i)+\sum_{j\neq}X_{ij}(t,x)x_j,\quad x\in R^n\ (i=\overline{1,m}),\tag{3}$$

the Bailey's approach is [64] based on the decomposition of complicated system (3) to subsystems

$$\frac{dx_i}{dt}=X_i^0(t,x_i),\quad x_i\in R^{n_i}\quad (\sum_{i=1}^{m} n_i=n),$$

on construction for isolated subsystems (4) of Lyapunov - Krassovski scalar functions

$$c_{i1}\|x_i\|^{\xi} \leq v^i(t,x_i) \leq c_{i2}\|x_i\|^{\xi},\quad \dot{v}^i_{(4)}(t,x_i) \leq -c_{i3}\|x_i\|^{\xi} \quad (c_{ij}>0) \qquad (5)$$
$$\|\partial v^i/\partial x_i\| \leq c_{i4}\|x_i\|$$

with $\xi = 2$, on forming VLF from them $v=(v^1,\dots,v^m)$, on the consideration of its derivative $\dot{v}(t,x)$ by the complete system (3) taking into account the interconnections $X_{ij}(t,x)\,x_j$ and on the linear CS construction $\dot{x}_c = Px_c$, estimating the interconnections between subsystems (aggregation). The first applications [64] of the method are given to multiconnected electropower systems (EPS) and to the nonlinear control (NCS) [65] systems. The method is modified by A.S.Zemlyakov [66] in the part of matrix P choice.

The main advantage is exceptional simplicity. But the "too sufficient" conditions of stability [65] . It may be connected by the rough choice of P , of Lyapunov functions v^s and estimations (5), by the failure of system's decomposition. For elimination of the first defect the optimization of the choice of the matrix P was suggested, originating from the criterion $\lambda_{max}(P) \xrightarrow[P]{} min$ or $|\det P| \xrightarrow[P]{} max$ under the conditions $p_{ij} \geq 0$ $(j \neq i)$, $Pv - \dot{v} \geq 0$. The algorithm of solution of such non-linear programming problem in the local case was realized by a program. After finding P by linear approximation, are chosen $\tilde{f}^s(v) = v^T Q^s v$ $(q^s_{ij} \geq 0$ for $i \vee j \neq s)$. CS represents the vector Riccati equation [I ,68] .

The greatest development of decomposition-aggregation means is completed by D.D.Siljak with co-authors [23,25,34,26, 41] ; they were assumed as a basis of large-scale systems dynamics, and are being accompanied by the applications to power, space, ecological, economic and other systems. It should be noted also the interesting works by A.N.Mitchel with collaborators [16,33] , and V.D.Furassov [67,32,42] and other authors [17,26,29,35,39,43,70] . R.I.Kozlow with collaborators developed a means of VLF construction with components $\vartheta^s = |V^s \cdot X|$, for the non-linear systems. It gives more precise estimations, realized by a program for the stability analysis with perturbations, for the synthesis of parameters of complicated NCS and was used for EPS, gyroscopic stabilizer and orbital telescopes, accounting the non-flexibilities [37-39,43].

For complicated systems with linear and polynomial right sides R.Z.Abdullin developed [38] the method of VLF and CS construction with optimization of strings p_i of the matrix P by the criteria $f^i(t, \vartheta(x)) - \dot{\vartheta}^i(x) \longrightarrow min$. For multiconnected NCS A.I.Malikov developed the algorithms [38,39] of VLF construction with components of Lur'e type and the algorithms of estimations of attraction regions, of the most important indices of stability and of functioning accuracy. They created separate APP and used for the analysis of spaceship orientation, of absolute stability of NCS, for the estimations of the attraction regions of EPS.

The mixed methods, representing the finite iterative processes of improvement of the complicated system decomposition and VLF construction, are seemed to be more effective for applications [1,17,68].

As the first approximation in them the linearized subsystems (4), VLF (5) are chosen, linear CS is constructed obtainable as the result of optimization by one from the called criteria or CS of vector Riccati equation type.

If i-th subsystem, being isolated, is autonomous, and has an analytical right side, then by the $\lambda_i^s(A_i)$ roots of characteristic equation $\varkappa_i^s = \lambda_i^s + \bar{\lambda}_i^s$, their neighbourhood to ε_i in the inequalities $\dot{\vartheta}^i(x_i) \leq -\varepsilon_i \vartheta^i(x_i)$ is being calculated and estimated. There is being cleared up the expediency of replacement (5) of the latter inequality by closer to precise exponential estimations with increasing may be a number of subsystems, of VLF components, connected with it (systems' decomposition). We form VLF second approximation from them. If we cannot obtain the satisfactory majorization accuracy of interconnections, then by p_{ij} $(j \neq i)$ their intensity and influence on the comparison system is being estimated. The second level subsystems are being formed, some of them unite "strongly connected" subsystems of the first level(subsystems' aggregation). VLF of corresponding approximations are constructed for them.

With usage this mean dynamics of the first Soviet stratospheric astronomic observatory and of other stratospheric and orbital astronomic observatories worked out in USSR were investigated [69] . The principal problem of their creation is just the stabilizability of telescopes with high accuracy on vibrating bases.

Finally, about VLF usage for optimal control of natural systems, describing by the differential equations with partial derivatives, in particular, the model of spatial-age structure of forest [48] . By multiple CT usage A.I.Moskalenko managed to simplify essentially, but in a series of interesting cases, to solve the problem.

The most actual problems for the further development and applications of this method is the elaboration of algorithms of VLF and CS construction for the analysis of properties, different from the stability for non-linear systems, their program realization; the investigation of new dynamical problems of current technical, socio-economic and other systems with usage of VLF method.

CONCLUSIONS

Thus, the present state of VLF method enables us to speak about the creation of the new general and effective method of various dynamical properties analysis of non-linear differential equations solutions. Its usage led to the solution of a series important applied problems.

REFERENCES

[1] Anapolski, L. Ju., Irtegov V.D., Matrosov V.M. Summary of Sciences. Gen. Mechan., V.2, M., Viniti, 1975, 53-112 (Russian).

[2] Bellman, R. Journ. Soc. Industr. and Appl. Math, Ser. AI, 1962, pp.32-34.

[3] Matrosov, V. M. Applied Mat. and Mechanics, XXVI,N6, 992-1002 (Russian).

[4] Matrosov, V. M. Proceedings of KAI, 1963, 80, 22-23 (Russian).

[5] Matrosov, V. M. Proceedings of interuniversity conf. on appl. stability theory of motion and anal. mechan. Kazan, KAI, 1964, 103-109 (Russian).

[6] Lakshmikantham, V. Jour. Math. Anal. & Appl., 1956, 10, N2, 368-377.

[7] Matrosov, V. M. Proceedings of the II-nd All-Union Congress on the theory and appl. mechan. 1964,vl, Publish. Science, M., 1965, 112-125 (Russian).

[8] Hartman Ph. Ann. Math. Pura ed Appl., 1965, 69, 305-320; Duke Math. Jour. 1966, 33 N 2. 281-290.

[9] Lakshmikantham, V., Leela, S. Differential and integral inequalities. Theory and Applications, vol. 1,2, New-York-London, Acad. Press 1969.

[10] Rouche, N., Peiffer, K., Journ. Mech., 8, N 2, 1969, 323-334.

[11] Yeghorov, G. A. Proceedings of KAI. Math. and Mech., 1968, 107, 31-33 (Russian).

[12] Persidski, S. K. Diff. Equa., 1969, 5, N 4, 678-687 (Russian).

[13] Matrosov, V. M. ICM, section 6, Abstracts, M., 1966; 35-36; Proceedings of KAI, 1968, 97, 41-53 (Russian).

[14] Matrosov, V. M. Diff. Equa., 1967, v. III, N 3, 395-409, N 5,, 839-848 (Russian).

[15] Matrosov, V. M. Diff. Equa., 1968, v. 4, N 8, 1374-1386; N10, 173-1752; 1969, v. 5, N 7, 1171-1185; N 12, 2129-2143 (Russian).

[16] Michel, A. N. IEEE Trans. Circuit Theory, 1970, CT-17, N3, 292-301.

[17] Matrosov, V. M. Symp. Math., VI, London-New York, 1971, 209-242.

[18] Kayand, A. A., Lakshmikantham, V., Jour. Math. Anal. and Appl., 1966, 14, N 2, 285-293.

[19] Lakshmikantham, V., Leela, S., Sastry, T., Jour. Math. Analysis and Appl., 1967, 19, N 3, 444-456.

[20] Lakshmikantham, V., Verma, G. R., Bull. Math. Soc. Sci. Math. RSR, 1967/1968.

[21] Rouche, N., Int. J. Non-Linear Mechanics, 3, 1968, 295-306; 4, 1969, 301-309; Ann. Soc. Sci. de Bruxelles, ser. 1, 83, 1969, 5-17.

[22] Matrosov, V. M. Automatics and Telemechanics, 9, 1972, 62-75; 1, 1973, 5-22 (Russian).

[23] Siljak, D. D. Proc. Fifth IFAC Congress Comment. paper sess. Theory-Nonlin. Syst. Paris 1972, C 32, 1-11.

[24] Loloy, M. Institut math. pure appl. univ. Louvain rap. N 49, 54, 1972.

[25] Grujic, L. T., Siljak, D. D. I CC Conf., Stanford Univ. 1972, 550-555. IEEE trans. aut. control. vol. AC-18, 1973, N 5, 522-524; N 6, 636-645.

[26] Grujic, L. T., Stabilnost velikih sistema. Doct. diss. Beograd Univ., 1974.

[27] Hatvani, L., Vestnik of MU., Math., Mech. N 3, 1975, 83-89 (Russian).

[28] Lakshmikantham, V., Mitchell, A. R., Mitchell, R. W., Univ. of Texas at Arlington, 1975, TR-27.

[29] Martinyuk, A. A. Stability of motion of complicated systems. Kiev, "Naukova dumka", 1975 (Russian).

[30] Problems of analytical mechanics, stability and control theories, Ed. P. A. Kuz'min, M., "Science", 1975, v. 2, Kazan, KAI, 1976 (Russian) .

[31] Rouche,N.,Habets, P., Laloy, M., Stability Theory by Lyapunov's Direct Method. New York-Heidelberg-Berlin, "Springer", 1977.

[32] Furassov, V. D. Stability of motion, estimation and stabilization. M., "Science", 1977 Russian .

[33] Michel, A. N., Miller, R. K. Qualitative analysis of Large Scale Dynamical Systems. Acad. Press, New York-San Francisco-London, 1977.

[34] Siljak, D. D., Large-scale dynamic systems. Stability and structure. New York-Amsterdam-Oxford, North-Holland, 1978.

[35] Bitsoris, G. S., Grade Doct. d'etat Univ. Toulous, 1978.

[36] Martinyuk, A. A., Gutovski, R., Integral inequalities and stability of motion, "Naukova dumka", Kiev, 1979. (Russian).

[37] Stability theory and its applications. Ed. V. M. Matrosov, V. D. Irtegov, Novosibirsk, "Science", 1979 (Russian).

[38] Vector Lyapunov functions and its constructions. Ed. V. M. Matrosov, L. Ju. Anapolski. Novosibirsk, "Science", 1980 (Russian).

[39] Direct method in the stability theory and its applications. Ed. V. M. Matrosov, L. Ju. Anapolski. Novosibirsk, "Science", 1981 (Russian).

[40] Stability of motion. Analytical mechanics. Control of motion. Ed. V. M. Matrosov, V. G. Dyomin, M., "Science", 1981 (Russian).

[41] Ikeda, M., Siljak, D. D., IEEE Trans. aut. control, v. 26, N 5, 1981.

[42] Ivanova, E. I., Firassov, V. D. Diff. equa., XVIII, N 3, 1982, 409 418 (Russian).

[43] Lyapunov functions method in the dynamics of non-linear systems. Ed. V. M. Matrosov, R. I. Kozlov, Novosobirsk, "Science", 1982 (Russian).

[44] Bernfeld, S., Lakshmikantham, V. An introduction to non-linear boundary value problems. New York Academic Press, 1974.

[45] Lakshmikantham, V. Abstract boundary value problems. Non-linear equations in abstract spaces. New York-San Francisco-London, Academic Press, 1978, 117-123.

[46] Matrosov, V. M. VIII Intern. conf. on nonlin. oscilat. 1, Acad. Praque, 1979, 447-474.

[47] Moskalenko, A. I., PAS USSR, 1977, v. 232, N 3, 524-527 (Russian).

[48] Moskalenko, A. I., In book: Models of control of natural resources. M., "Science", 1981, 61-73, 184-198, 206-225, 231-243. (Russian).

[49] Hahn, W., Jour. reine und angewandte Mathe., 1967, 228, 189-192.

[50] Bushaw, D., Proc. United States-Japan seminar on Diff. Func. Equa., New York, Benjamin Inc., 1967.

[51] Anapolski, L. Ju., Matrosov, V. M., IFAC International symposium on problems of organization control and heir-

archic systems (Baku), (1971), Abstr. papers, p. I, M., 1972, p. 43-46 (Russian).

[52] Matrosov, V. M. Equations Differenti lles et fonctionnelles non lineaires. Actes conf. internat., "Equa-Diff. 73" Paris, Herman, 1973, 407-445.

[53] Matrosov, V. M. Diff. Equa., N 9, 1974, 1547-1559; II, N 3, 1975, 403-417 (Russian).

[54] Vassilyev, S. N., Matrosov, V. M., Proceedings of the IV-th International united conference on artificial intelligence. Suppl. materials. M., 1975 (Russian). 166-175.

[55] Matrosov, V. M., Anapolski, L. Yu., Vassilyev, S. N. Comparison method in mathematical systems theory, Novosibirsk, "Science", 1980 (Russian).

[56] Matrosov, V. M., Vassilyev, S. N., Comparison principle for automatic theorems derivation, 1978, N 2, 60-69; N 4, 40-49 (Russian).

[57] Matrosov, V. M., Vassilyev, S. N., Kozlov, R. I. and oth. Algorithms of theorems derivation of vector Lyapunov functions method, Novosibirsk, "Science", 1981 (Russian).

[58] Habets, P., Peiffer, K. Jour. Math. Anal. Appl. 1973, 43, 537-570.

[59] Martinyuk, A. A., Obolenski, A. Ju. Diff. Equa. 1980, 16, N 8, 1392-1407 (Russian).

[60] Kozlov, R. I., Matrosov, V. M., Proc. of AS USSR, 1980, 253, N 4, 807-811, 255, N 3, 521-525 (Russian).

[61] Vachonin, G. S., Zemlyakov, A. S., Matrosov, V. M. Automatics and telemechanics, 1973, N 2, 5-16 (Russian).

[62] Persidski, K. P. Selected proceedings, v. 2, Alma-Ata, "Science"; 1976, 134-136, 154-161 (Russian).

[63] Weleev, K. G., Finin, G., Lyapunov function construction. Kiev, "Naukova dumka", 1981 (Russian).

[64] Bailey, F. N., Journ. SIAM, ser. A., Control, 1965-1966, 3, N 3, 443-462.

[65] Piontkovski, A. A., and others, Automatics and telemechanics, 1967, 10, 23-31; 1970, 11, 179-181 (Russian).

[66] Zemlyakov, A. S., Proc. of KAI, 1972, issue 144, 46-54 (Russian).

[67] Furassov, V. D. Achievements and perspectives of technical cybernetics development. Proceedings of the II-nd Interuniversity scientific-technical conference. M., 1972 (Russian).

[68] Zemlyakov, A. S., Matrosov, V. M., Optimal and adaptive control., Saratov, S. S., 1977, 202-215 (Russian).

[69] Danilov, A. M., Dul'kin, L. Z., Zemlyakov, A. S., Matrosov, V. M., Strezhnev, V. A., Control space. M., "Science" v. 1, 1975, 208-228; 1976, 153-171 (Russian).

[70] Voronov, A. A., Stability, controllability, "Science", M. 1979 (Russian). Automatics and telemechanics, 1982, N 2, 5-20.

SOME RECENT TOPICS ON NONLINEAR SEMI-GROUPS

Isao Miyadera

Waseda University
Shinjuku-ku, Tokyo, Japan

ABSTRACT

Let X be a Banach space. In the section 1 we show that if the dual of X has Fréchet differentiable norm and A is a closed dissipative operator in X satisfying range condition then the canonical restriction of A is the infinitesimal generator of the contraction semi-group generated by A. In the section 2 we study the behavior of "the inverse of (I-(r/h)(T(h)-I))" as $h \to 0+$, where T(t), $t \geq 0$, is a contraction semi-group on a closed and convex subset of X.

INTRODUCTION

In this paper we are concerned with the infinitesimal generators of contraction semi-groups in a Banach space X with norm $|\ |$. A family S = $\{T(t);t \geq 0\}$ of mappings from $X_0 (\subset X)$ into itself is called a contraction semi-group on X_0 if T(0)=I (the identity), T(t+s)=T(t)T(s), $|T(t)x - T(t)y| \leq |x - y|$ for $t,s \geq 0$ and $x,y \in X_0$, and $\lim_{t \to 0+} T(t)x = x$ for $x \in X_0$. The infinitesimal generator A_0 of a contraction semi-group S is defined by $A_0 x = \lim_{h \to 0+}(T(h)x - x)/h$ whenever the limit exists.

Let A: $D(A)(\subset X) \to X$ be a dissipative operator satisfying the following range condition

(R) $R(I - rA) \supset cl(D(A))$ (the closure of D(A)) for r>0.

The Crandall-Liggett theorem [2] asserts that for $t \geq 0$ and $x \in cl(D(A))$ $T(t)x = \lim_{r \to 0+}(I - rA)^{-[t/r]}x$ exists and $\{T(t);t \geq 0\}$ is a contraction semi-group on cl(D(A)), where [ ] denotes the Gaussian bracket. In this case, $\{T(t);t \geq 0\}$ is called the (contraction) semi-group generated by A.

If X is a Hilbert space then the following (a) and (b) hold true [7, 3] :

(a) Let A be an m-dissipative operator in X. If S is the semi-group generated by A, then the canonical restriction A^o of A is the infinitesimal generator of S.

(b) Let X_0 be a closed convex subset of X. If S is a contraction semi-group on X_0, then its infinitesimal generator A_0 is densely defined in X_0, and there exists a dissipative operator A with (R) such that $A_0 \subset A$ and S is generated by A. (Furthermore it is known that there is a unique m-dissipative operator A such that $A^o = A_0$.)

Unfortunately, (a) and (b) both do not hold in general Banach spaxe X [2]. However, under some assumptions on X, it is expected that these hold outside Hilbert space. Our purpose is to report some recent results in this direction. In the section 1 we give some new results about (a). In the section 2 we discuss the behavior of $(I - (r/h)(T(h)-I))^{-1}x$ as $h \to 0+$ in connection with (b).

SECTION 1

X is said to be smooth if the norm of X is Gâteaux differentiable. We say that X is (F) (or (UG)) if the norm of X is Fréchet differentiable (or uniformly Gâteaux differentiable). It is known that X is reflexive and (UG) if its dual X* is uniformly convex [4], and X* is (F) if and only if X is reflexive and strictly convex, and has the following property (A) [5].

(A) $\text{w-lim}_{n\to\infty} x_n = x$ and $\lim_{n\to\infty} |x_n| = |x|$ imply $\lim_{n\to\infty} x_n = x$.

Here $\text{w-lim}_{n\to\infty} x_n$ denotes the weak limit of x_n.

The duality map $F: X \to X^*$ is defined by $F(x) = \{x^* \in X^* ; (x,x^*) = |x|^2 = |x^*|^2\}$, where (x,x^*) denotes the value of $x^* \in X^*$ at $x \in X$. In general, F is a multi-valued map, and it is single-valued if and only if X is smooth.

The main result in this section is as follows.

Theorem 1.1 ([9]). Let $A: D(A)(\subset X) \to X$ be a dissipative operator satisfying (R) and let S be the semi-group generated by A. If X* is (F), then the canonical restriction $(\bar{A})^o$ of $\bar{A}$ is the infinitesimal generator of S and $D((\bar{A})^o) = D(\bar{A}) = \hat{D}(A)$, where $\bar{A}$ denotes the closure of A and $\hat{D}(A) = \{x \in \mathrm{cl}(D(A)); \lim_{r\to 0+} |(I-rA)^{-1}x - x|/r < \infty\}$.

This theorem improves on Plant's result [10, Theorem 5] and Reich's result [12, Corollary 6.2]. For the proof see [9]. As an immediate consequence of Theorem 1.1 we have

Corollary 1.2. Let A be m-dissipative in X and S the semi-group generated by A. If X* is (F), then the canonical restriction A^o of A is the infinitesimal generator of S.

The weak infinitesimal generator A' of a contraction semi-group S is defined by $A'x = \text{w-lim}_{h\to 0+}(T(h)x - x)/h$ whenever the weak limit exists.

Theorem 1.3. Let A, S and $\hat{D}(A)$ be as in Theorem 1.1. Suppose that X is reflexive and strictly convex. Then for every maximal dissipative operator $\tilde{A}$ in cl(D(A)) such that $\tilde{A} \supset A$, the canonical restriction $(\tilde{A})^o$ of $\tilde{A}$ is the weak infinitesimal generator of S and $D((\tilde{A})^o) = D(\tilde{A}) = \hat{D}(A)$. In particular, if A is m-dissipative in X then A^o is the weak infinitesimal generator of S.

Proof. It has been proved in [9, 12] that $A'x = \text{w-lim}_{t\to 0+}(T(t)x - x)/t$ exists and $\lim_{t\to 0+}|T(t)x - x|/t = |A'x| = \lim_{t\to 0+}|J_t x - x|/t$ for $x \in \hat{D}(A)$, where $J_t = (I - tA)^{-1}$. It is easy to see that the above A' defind on $\hat{D}(A)$ is the weak infinitesimal generator of S.

Let $\tilde{A}$ be a maximal dissipative operator in cl(D(A)) such that $\tilde{A} \supset A$. Note that S is generated by $\tilde{A}$, because $R(I - t\tilde{A}) \supset R(I - tA) \supset \text{cl}(D(A)) = \text{cl}(D(\tilde{A}))$ for $t>0$. By [8, Corollary 1] we have $\hat{D}(A) = D((\tilde{A})^o) = D(\tilde{A})$ and $A' \subset (\tilde{A})^o$. To show that $(\tilde{A})^o = A'$, let $x \in D((\tilde{A})^o)$ and $v \in (\tilde{A})^o x$. Since $(J_t x - x)/t \in \tilde{A}J_t x$ and $\tilde{A}$ is dissipative, for each $t>0$ there is an $f(t) \in F((J_t x-x)/t)$ such that $\text{Re}(v,f(t)) \geqq |J_t x - x|^2/t^2$. As shown in [9]

$$\text{Re}((T(s)x - x)/s,\ f(t)) \geqq |J_t x - x|^2/t^2 - (2/t^2)|J_t x - x|(1/s)\int_0^s |T(r)x - x|\, dr$$

for $t,s>0$. Letting $s \to 0+$, $\text{Re}(A'x,\ f(t)) \geqq |J_t x - x|^2/t^2$ for $t>0$. So that $\text{Re}(v + A'x,\ f(t)) \geqq 2|J_t x - x|^2/t^2$ and hence $|v + A'x| \geqq 2|J_t x - x|/t$ for $t>0$. Letting $t \to 0+$, $|v + A'x| \geqq 2|A'x|$. Noting that $|A'x| = |v|$, it follows from the strict convexity of X that $v = A'x$. This completes the proof.

SECTION 2

Throughout this section, let X_0 be a closed convex subset of X, $S = \{T(t); t\geqq 0\}$ a contraction semi-group on X_0, $J_{r,h} = (I - (r/h)(T(h)-I))^{-1}$ for $r,h>0$.

In Hilbert space, to prove (b) in the introduction Kōmura [7] and Kato [6] established the existence of $\lim_{h\to 0+} J_{r,h}x$. Later, Crandall and Liggett [2] proved that if for some sequence $h_n \to 0+$ $\lim_{n\to\infty} J_{r,h_n} x$ exists for $x \in X_0$ and $r>0$, then there exists a dissipative operator A such that $D(A) \subset \{x \in X_0;\ T(t)x$ is Lipschitz continuous in $t \geqq 0\}$, $cl(D(A)) = X_0$, $R(I - rA) \supset X_0$ for $r>0$ and S is generated by A (and hence (b) holds whenever X is reflexive). In 1978, Baillon [1] established that (b) holds if X* is uniformly convex, and, in 1980, Reich [11] succeeded in extending this result to the case that X is reflexive and (UG). Their proofs are also based on the existence of $\lim_{h\to 0+} J_{r,h}x$. So in this section we study the convergence of $J_{r,h}x$ as $h \to 0+$.

Let us define $(\ ,\ )_s : X \times X \to R = (-\infty,\infty)$ by $(x,y)_s = \sup\{Re(x,y^*);\ y^* \in F(y)\}$. Here $Re(x,y^*)$ denotes the real part of (x,y^*). Fix $x \in X_0$ and $r>0$, and set $y(h) = J_{r,h}x$ for $h>0$. The following are known [2] :

(2.1) $y(h)$ is continuous in $h>0$ and bounded as $h \to 0+$.

(2.2) $$|y(h) - z|^2 \geqq (2h/r) \sum_{k=1}^{n} (y(h)-x,\ y(h)-T(kh)z)_s + |y(h) - T(nh)z|^2 \quad \text{for } z \in X_0,\ n \geqq 1 \text{ and } h>0.$$

By (2.2) and $|y(h) - u|^2 - (y(h)-x,\ y(h)-T(kh)u)_s + (u-x,\ y(h)-T(kh)u)_s \leqq (|y(h) - u| + 2|y(h) - T(kh)u|)|T(kh)u - u|$, we have

(2.3) $$|y(h) - u|^2 \leqq (r/2nh)(|y(h) - u|^2 - |y(h) - T(nh)u|^2) + (1/n) \sum_{k=1}^{n} [(|y(h) - u| + 2|y(h) - T(kh)u|)|T(kh)u - u| - (u-x,\ y(h)-T(kh)u)_s] \quad \text{for } h>0,\ n \geqq 1 \text{ and } u \in X_0.$$

<u>Theorem 2.1.</u> Let X be reflexive, and put $E(y) = \{y(h) - T(t)y;\ 0<h\leqq 1,\ 0\leqq t\leqq 1\}$ for $y \in X_0$. Then the following are equivalent :

(i) $y(h)$ is convergent as $h \to 0+$ and $(y-x,\ )_s : cl(E(y)) \to R$ is continuous for every $y \in X_0$.

(ii) $\{y(h);\ 0<h\leqq 1\}$ is sequentially compact and $(y-x,\ )_s : cl(E(y)) \to R$ is continuous for every $y \in X_0$.

(iii) For each $y \in X_0$ $\lim_{a\to 0+}(|z + a(y-x)|^2 - |z|^2)/a$ exists uniformly in $z \in E(y)$.

<u>Proof.</u> By (2.1), (i) implies (ii). Let $y \in X_0$ and put $g_a(z) = (|z + a(y-x)|^2 - |z|^2)/a$. Then for $z \in X$, $g_a(z) \leqq g_b(z)$ if $0<a\leqq b$ and $\lim_{a\to 0+} g_a(z) = 2(y-x,z)_s$. So, if (ii) is satisfied then by Dini's theorem $\lim_{a\to 0+} g_a(z)$

$= 2(y-x,z)_s$ uniformly in $z \in cl(E(y))$. To prove that (iii) implies (i), let $h_{m'}$ be a sequence such that $h_{m'} \in (0,1]$ and $h_{m'} \to 0$ as $m' \to \infty$. We use the idea due to Baillon and Reich [1, 11]. Set $y_{m'} = y(h_{m'})$. We show that $y_{m'}$ has a convergent subsequence. By [11] there exists a closed convex and separable subset C of X_0 such that $T(t)C \subset C$ and $x, y(h) \in C$, and then by (2.1) there is a subsequence h_m of $h_{m'}$ such that $|y_m - z| (= |y(h_m) - z|)$ converges for each $z \in C$. Since X is reflexive and C is closed convex, there exists a $u \in C$ such that

$$(2.4) \quad \lim_{m\to\infty} |y_m - u| \leqq \lim_{m\to\infty} |y_m - z| \quad \text{for } z \in C.$$

We claim that $y_m \to u$. Let $\varepsilon>0$. By (iii) there is an $a_0 \in (0,1]$ such that for every $h \in (0,1]$ and $t \in [0,1]$

$$(2.5) \quad (|y(h) - T(t)u + a_0(u-x)|^2 - |y(h) - T(t)u|^2)/2a_0 < (u - x, y(h) - T(t)u)_s + \varepsilon.$$

Choose a $t_0 \in (0,1]$ such that $|T(t)u - u| < a_0\varepsilon$ for $t \in [0,t_0]$, and put $n_m = [t_0/h_m]$. By (2.5), $-(u - x, y_m - T(kh_m)u)_s < (|y_m - u|^2 - |y_m - u + a_0(u-x)|^2)/2a_0 + (2M + |u-x| + 1)\varepsilon$ for $k = 1, 2, \ldots, n_m$, where $M = \sup\{|y(h) - T(t)u|; 0<h\leq 1, 0\leq t\leq 1\}$. Combining this with (2.3) we have that $|y_m - u|^2 \leqq (r/2n_m h_m)(|y_m - u|^2 - |y_m - T(n_m h_m)u|^2) + (|y_m - u|^2 - |y_m - u + a_0(u-x)|^2)/2a_0 + (5M + |u-x| + 1)\varepsilon$. Hence by (2.4) $\lim_{m\to\infty}|y_m - u|^2 \leqq (5M + |u-x| + 1)\varepsilon$.

Next, by (iii), for $y \in X_0$ $\lim_{a\to 0+}(|z + a(y-x)|^2 - |z|^2)/a = 2(y-x, z)_s$ uniformly in $z \in cl(E(y))$. Thus we have

$$(2.6) \quad (y-x, \ )_s : cl(E(y)) \to R \text{ is continuous for } y \in X_0.$$

Now the existence of $\lim_{h\to 0+} y(h)$ follows from the next lemma.

<u>Lemma 2.2.</u> Suppose that (2.6) is satisfied. Let $h_m \to 0+$ and $a_p \to 0+$ as $m,p \to \infty$. If $y(h_m) \to u$ and $y(a_p) \to v$ then $u = v$.

<u>Proof.</u> Put $y_m = y(h_m)$ and $z_p = y(a_p)$. Let $\varepsilon>0$. Since $y_m - T(t)z_p \to u - v$ as $m,p \to \infty$ and $t \to 0+$, by the upper semi-continuity of $(\ , \ )_s$ and $-(y,z)_s \leqq (y,-z)_s$ there exist positive numbers m_0, p_0 and t_0 such that for $m \geqq m_0$, $p \geqq p_0$ and $0 \leqq t \leqq t_0$

$$(2.7) \quad -(y_m - x, y_m - T(t)z_p)_s < (u - x, v - u)_s + \varepsilon.$$

Let $0<t\leqq t_0$, $m\geqq m_0$, $p\geqq p_0$ and put $n_m = [t/h_m]$. By (2.2) and (2.7), $|y_m - T(n_m h_m)z_p|^2 - |y_m - z_p|^2 \leqq (2h_m n_m/r)((u-x,v-u)_s + \varepsilon)$, and hence $|T(t)z_p - u|^2 - |z_p - u|^2 \leqq (2t/r)((u-x,v-u)_s + \varepsilon)$ for $0<t\leqq t_0$ and $p\geqq p_0$. So

that for sufficiently large p,

$$|z_p - u|^2 \leqq (u - x, v - u)_s - (u - x, z_p - u)_s + \varepsilon.$$

Here we have used that $T(a_p)z_p = z_p + (a_p/r)(z_p - x)$. Letting $p \to \infty$, $|v - u|^2 \leqq \varepsilon$ by (2.6). This completes the proof.

Remark. It is easy to see that (2.6) is satisfied if X is smooth.

As an immediate consequence of Theorem 2.1 we have

Theorem 2.3 ([11]). If X is reflexive and (UG), then for every $x \in X_0$ and $r>0$, $J_{r,h}x$ is convergent as $h \to 0+$.

Proof. Let $x \in X_0$ and $r>0$, and let y(h) and E(y) be as in Theorem 2.1. Note that X is (UG) if and only if for every $v \in X$ and every bounded set $B \subset X$ $\lim_{a\to 0}(|z + av|^2 - |z|^2)/a$ exists uniformly in $z \in B$. Therefore (iii) in Theorem 2.1 is satisfied, because E(y) is bounded for every $y \in X_0$. This completes the proof.

By a selection of F we mean a single-valued map $f: X \to X^*$ such that $f(x) \in F(x)$ for all $x \in X$. Let f be a selection of F. We say that f is weakly sequentially continuous at 0, if $w^*\text{-}\lim_{n\to\infty} f(x_n) = 0$ whenever $w\text{-}\lim_{n\to\infty} x_n = 0$. Here w*-lim denotes the weak-star limit.

Theorem 2.4. If X is reflexive and F has a selection f which is weakly sequentially continuous at 0, then $\{J_{r,h}x;\ h \in (0,1]\}$ is sequentially compact for every $x \in X_0$ and $r>0$.

Proof. Fix $x \in X_0$ and $r>0$, and set $y(h) = J_{r,h}x$ for $h>0$. By (2.1) it suffices to show that for any sequence $h_{m'} \to 0+$, $y(h_{m'})$ has a convergent subsequence. To this end, let $h_{m'}$, C and M be as in the proof of Theorem 2.1. By (2.1) and the reflexivity of X there exists a $u \in C$ and a subsequence h_m of $h_{m'}$ such that $w\text{-}\lim_{m\to\infty} y_m = u$ and $|y_m - z|$ is convergent for every $z \in C$, where $y_m = y(h_m)$. We claim that y_m converges to u. By $|y_m - z|^2 \geqq |y_m - u|^2 + 2\mathrm{Re}(u-z,\ f(y_m-u))$ for $z \in C$, we have

$$\text{(2.8)}\qquad \lim_{m\to\infty} |y_m - z| \geqq \lim_{m\to\infty} |y_m - u| \quad \text{for } z \in C.$$

By (2.3), for h>0 and $n \geqq 1$

$$\text{(2.9)}\qquad |y(h) - u|^2 \leqq (r/2nh)(|y(h) - u|^2 - |y(h) - T(nh)u|^2) + n^{-1}\sum_{k=1}^{n} [(|y(h) - u| + 2|y(h) - T(kh)u|)|T(kh)u - u| - \mathrm{Re}(u - x,\ f(y(h) - T(kh)u))].$$

Let $\varepsilon>0$. Since $w^*\text{-}\lim_{m\to\infty, t\to 0+} f(y_m - T(t)u) = 0$, there exists $m_0 \geqq 1$ and $t_0 \in (0,1]$ such that $|T(t)u - u| < \varepsilon$ and $|(u-x, f(y_m - T(t)u))| < \varepsilon$ for $m \geqq m_0$ and $t \in [0,t_0]$. Using (2.9) with $h = h_m$ and $n = [t_0/h_m]$ and letting $m \to \infty$, it follows from (2.8) that $\lim_{m\to\infty} |y_m - u|^2 \leqq (3M + 1)\varepsilon$. This completes the proof.

Remark. Let X and F be as in Theorem 2.4. Then we have the following. 1°) The infinitesimal generator of S is densely defined in X_0. 2°) In general, $J_{r,h}x$ does not converge as $h \to 0+$. 3°) In addition, suppose that X is smooth. Then for $x \in X_0$ and $r>0$ $J_{r,h}x$ converges as $h \to 0+$.

REFERENCES

[1] Baillon, J. B., Générateurs et semi-groupes dan les espaces de Banach uniformément lisses, J. Funct. Anal., 29 (1978), 199 - 213.

[2] Crandall, M. and Liggett, G., Generation of semi-groups of nonlinear transformations on general Banach spaces, Amer. J. Math., 93 (1971), 265 - 293.

[3] Crandall, M. and Pazy, A., Semigroups of nonlinear contractions and dissipative sets, J. Funct. Anal., 3 (1969), 376 - 418.

[4] Diestel, J., Geometry of Banach spaces, Lecture Notes in Mathematics, Springer, 1975.

[5] Fan, K. and Glicksberg, I., Some geometric properties of the spheres in a normed linear space, Duke Math. J., 25 (1958), 553 - 568.

[6] Kato, T., Note on differentiability of nonlinear semigroups, Global Anal., Proc Symp. Pure Math., 16 (1970), 91 - 94.

[7] Kōmura, Y., Differentiability of nonlinear semi-groups, J. Math. Soc. Japan, 21 (1969), 375 - 402.

[8] Miyadera, I., Some remarks on semi-groups of nonlinear operators, Tôhoku Math. J., 23 (1971), 245 - 258.

[9] Miyadera, I., On the infinitesimal generators and the asymptotic behavior of nonlinear contraction semi-groups, Proc. Japan Acad., 58 (1982), 1 - 4.

[10] Plant, A., The differentiability of nonlinear semigroups in uniformly convex spaces, Israel J. Math., 38 (1981), 257 - 268.

[11] Reich, S., Product formulas, nonlinear semigroups, and accretive operators, J. Funct. Anal., 36 (1980), 147 - 168.

[12] Reich, S., A note on the asymptotic behavior of nonlinear semigroups and the range of accretive operators, MRC Technical Summary Report, #2198, 1981.

SET VALUED EXTENSIONS OF INTEGRAL INEQUALITIES

R. E. Moore

Department of Mathematics
The University of Texas at Arlington
Arlington, Texas

1. INTRODUCTION

Integral inequalities are useful for constructing upper and lower bounds on solutions and sets of solutions to differential and integral equations. We can extend the theory of integral inequalities to set valued mappings by basing the analysis on the set inclusion relation $\subseteq$ instead of the usual partial order relation $\leq$. In this way, we can obtain both upper and lower bounds in a single set inclusion relation instead of a pair of inequalities. A natural starting point for such an investigation is <u>interval analysis</u>. There is already a well developed body of theory and computational techniques for interval valued functions [7], [2]. Furthermore, it is easy to establish a connection between inclusion of intervals and classical inequalities via: $[a,b] \subseteq [c,d]$ iff $c \leq a \leq b \leq d$. With interval analysis, we no longer need to assume monotonicity with respect to the usual order relation $\leq$ in order to obtain set inclusion relations. Thus, if $f(Y) = \{f(y): y \in Y\}$, we have $Z \subseteq Y$ implies $f(Z) \subseteq f(Y)$ for an <u>arbitrary</u> mapping f. Finally, <u>all</u> interval valued functions (and, thus, all real valued functions) have interval integrals [2]. Interval integration preserves set inclusion; thus, $Z(t) \subseteq Y(t)$, $0 \leq t \leq x$, implies ([2],[7])

$$\int_0^x Z(t)\,dt \subseteq \int_0^x Y(t)dt.$$

If $Y(t) = [\underline{Y}(t), \overline{Y}(t)]$, then the interval integral of $Y(t)$ is the interval between the lower Darboux integral of $\underline{Y}(t)$ and the upper Darboux integral of $\overline{Y}(t)$; see [2]. With the definition of integration, we do not even need to assume continuity or measureability of $Y(t)$.

With this approach, we can obtain integral inclusions which allow the integrands to be interval valued. Thus, we can allow sets of parameter values (initial values, boundary values, coefficient values). We can approximate sets of solutions to operator equations both from inside and from outside.

The reader is referred to [7] for basic definitions, concepts and techniques of interval analysis.

2. SET VALUED EXTENSIONS OF INTEGRAL INEQUALITIES

Previously [7], [8] it has been shown that if P is an operator mapping interval (or interval vector) valued functions from a space X into the same space, then, for $B \subseteq X$, $P(B) \subseteq B$ implies that $P^i(B) \subseteq B$ for all i and $\lim_{i\to\infty} P^i(B)$ exists pointwise and componentwise, since nested sequences of intervals converge.

In this paper, we add to this result a similar result in the opposite direction: if $U \subseteq P(U)$ and P is a Volterra type integral operator, then $U \subseteq P^j(U)$ for all j and $\lim_{j\to\infty} P^j(U)$ exists pointwise and componentwise. More precisely we have the following

Theorem 1. Let $BI_n[x_0,x_0+a]$ denote the set of all bounded n-dimensional interval vector valued functions on $[x_0,x_0+a]$. Suppose that $a > 0$ and $U, H \in BI_n[x_0,x_0+a]$ with

$$\underline{B} < b_1 \le U(x),\ H(x) \le b_2 < \overline{B} \quad \text{for all } x \text{ in } [x_0,x_0+a] \tag{1}$$

where $B = [\underline{B},\overline{B}]$ is an n-dimensional interval vector. Let Λ be a q-dimensional interval vector of parameter values. Suppose that $F\colon [x_0,x_0+a]^2 \times B \times \Lambda \to E^n$ is bounded real vector valued function and M is an n-dimensional interval vector such that

$$F([x_0,x_0+a],\ [x_0,x_0+a],\ B,\Lambda) \subseteq M. \tag{2}$$

Define

$$P_{h,\lambda}[u](x) = h(x) + \int_{x_0}^{x} F(x,t,u(t),\lambda)dt \tag{3}$$

where $h(x) \in H(x)$, $\lambda \in \Lambda$, $u(x) \in B$, $x \in [x_0,x_0+a]$

and

$$P[Y](x) = H(x) + \int_{x_0}^{x} F(x,t,Y(t),\Lambda)dt \tag{4}$$

for

$$Y(x) \subseteq B,\ x \in [x_0,x_0+a].$$

Suppose, finally, that

$$U(x) \subseteq P[U](x) \quad \text{for all } x \text{ in } [x_0, x_0+a]. \tag{5}$$

Then there exists $\alpha > 0$ such that

$$P[B](x) \subseteq B \quad \text{for all } x \text{ in } [x_0, x_0+\alpha]. \tag{6}$$

For brevity, we will write $Y \subseteq Z$ if $Y(x) \subseteq Z(x)$ for all x in $[x_0, x_0+\alpha]$. Define $P^i[Y]$ recursively by $P^0[Y] = Y$, and $P^i[Y] = P[P^{i-1}[Y]]$ for $i = 1,2,\ldots$. Then we have

$$P^i[U] \subseteq P^j[B] \quad \text{for all } i, j = 0,1,2,\ldots \tag{7}$$

and furthermore U^* and U^{**} exist in $BI_n[x_0, x_0+\alpha]$ with

$$U^*(x) = \lim_{i\to\infty} P^i[B](x), \quad \text{(see [7, theorem 5.7])} \tag{8}$$

and

$$U^{**}(x) = \lim_{i\to\infty} P^i[U](x), \tag{9}$$

and

$$U^{**} \subseteq U^*. \tag{10}$$

Finally, U^* contains the set of all fixed points such that $u(x) \in B$ for $x \in [x_0, x_0+\alpha]$ of (3); and if the set of all fixed points of (3) in $[x_0, x_0+\alpha]$ contains U and is an element of $BI_n[x_0, x_0+\alpha]$, then

$$P^i[U] \subseteq U^{**} \subseteq \{\text{all fixed points in } B\} \subseteq U^* \subseteq P^j[B] \tag{11}$$

for all $i, j = 0,1,2,\ldots$.

<u>Proof</u>. The existence of a positive α such that (6) holds follows from

$$P[B](x) \subseteq H(x) + M(x-x_0). \tag{12}$$

From (1) and (12), we will have (6) if

$$H(x) + M(x-x_0) \subseteq B, \quad x_0 \leq x \leq x_0 + \alpha. \tag{13}$$

Thus, for α we can choose the supremum of x such that

$$(b_1)_k - (\underline{B})_k \geq -\underline{M}_k(x-x_0)$$

and (14)

$$\overline{B}_k - (b_2)_k \geq \overline{M}_k(x-x_0)$$

for all $k = 1,2,\ldots,n$.

The relations given in (7) follow from the set inclusion property: $Y \subseteq Z$ implies $P[Y] \subseteq P[Z]$. Using (6) and (5), we have

$$P[B] \subseteq B \quad \text{implies} \quad P^i[B] \subseteq P^j[B] \quad \text{for} \quad i \geq j,$$
$$U \subseteq P[U] \quad \text{implies} \quad P^j[U] \subseteq P^i[U] \quad \text{for} \quad i \geq j.$$

Thus, since $U \subseteq B$,

$$P^j[U] \subseteq P^i[U] \subseteq P^i[B] \subseteq P^j[B] \tag{15}$$

and (7) is proved.

The pointwise limits implied by (8) and (9) exist since, componentwise, these represent pairs of sequences of real numbers. Each of these is either a monotonic non-decreasing sequence of real numbers bounded above or else a monotone non-increasing sequence of real numbers bounded below. In either case there is a unique limiting real number which is an endpoint of one of the intervals involved. See theorem 5.7 of [7]. The relation (1) follows from (7), (8), and (9).

If $u(x)$ is a fixed point of (3) for some $h \in H$ and some $\lambda \in \Lambda$, and if $u(x) \in B$ for all x in $[x_0, x_0+\alpha]$, then it follows from $u = p_{h,\lambda}[u]$ that $u \in P[B]$. Furthermore, $u = p^i_{h,\lambda}[u] \in P^i[B]$ for all $i = 0,1,2,\ldots$. Therefore, $u \in U^*$.

If the set of all such fixed points is an element of $BI_n[x_0, x_0+\alpha]$ and if it contains U, then it also contains $P^i[U]$ and U^{**} as well by inclusion arguments like those above. We can summarize these results by the relations in (11) and thus the theorem is proved.

Remarks. We can obtain as special cases of theorem 1, the lemmas of Gronwall [4], Gronwall-Bellman [1], [3], and various nonlinear generalizations of these such as a theorem of Viswanatham [12] and others discussed in [3]. Theorem 1 also extends the result of Azbelyev and Tsaljuk [6, theorem 1.1.8] and the method of Lakshmikantham for the solution of integral inequalities [6, sec. 1.2]. Without the additional assumptions of continuity of F and the monotonicity of F with respect to u, the limiting interval vector valued frunction U^* is not necessarily the interval between the minimal and maximal fixed points. In fact, these may not exist. However, it is neverthe less true that any fixed point of (3) which is contained in B is also contained in $P^i(B)$ for all i and in U^* and in $P(U^*)$ as long as B satisfies (6). Such a B always exists as long as (1) and (2) can be satisfied.

3. EXTENSIONS OF OTHER INTEGRAL INEQUALITIES

Numerous other integral inequalities can be extended to interval valued functions or other set valued functions. Haddad [5] has studied the application of set valued mappings with non-empty compact convex values to the approximation of sets of solutions to functional differential equations (of delay type). He studies functional differential inclusions and proves, in particular, that the set of solutions is connected.

Wildenauer [13] discusses the construction of intervals containing all solutions of operator equations in ordered groups with generating cones. He shows how to apply his results to nonlinear elliptic partial differential equations. With suitable growth conditions, he is able to prove the existence of minimal and maximal solutions.

Using a somewhat different approach, Nickel [9], [10], [11] has extended lemmas of Muller, Nagumo, and Westphal to find upper and lower bounds for the set of solutions of functional differential equations with Volterra type functionals and to find upper and lower bounds for the solutions of strongly coupled systems of parabolic functional differential equations with Volterra type functionals. His approach combines ideas from interval analysis with methods from the theory of differential and integral inequalities.

REFERENCES

[1] R. Bellman, Stability Theory of Differential Equations, McGraw-Hill, London, 1953.

[2] O. Caprani, K. Madsen, and L. B. Rall, Integration of interval functions, SIAM J. Math. Anal., Vol. 12, No. 3, May 1981, pp. 321-341.

[3] J. Chandra and B. A. Fleishman, On a generalization of the Gronwall-Bellman lemma in partially ordered Banach spaces, J. Math. Anal. & Appl., 31, 1970, 668-681.

[4] T. H. Gronwall, Note on the derivatives with respect to a parameter of the solutions of a system of differential equations, Ann. Math. 20, 1918-1919, 292-296.

[5] G. Haddad, Topological properties of the sets of solutions for functional differential inclusions, Nonlinear Analysis TMA, Vol. 5, No. 2, 1981, 1349-1366.

[6] A. A. Martunyuk and P. Gutovski, Integralnuye neravyenctva y ustoychyvost dvizhennia izdatelstvo "Naukova Dymka," KIEV-1979.

[7] R. E. Moore, Methods and Applications of Interval Analysis, SIAM studies in Applied Mathematics 2, Society for Industrial and Applied Mathematics, Philadelphia, 1979.

[8] R. E. Moore, A generalization of the method of upper and lower solutions for integral equations, Nonlinear Analysis TMA, Vol. 6, No. 8, 829-831, 1982.

[9] K. Nickel, Bounds for the set of solutions of functional-differential equations, Mathematics Research Center Technical Report #1782, University of Wisconsin-Madison, 1977.

[10] K. Nickel, The lemma of Max Muller-Nagumo-Westphal for strongly coupled systems of parabolic functional-differential equations, Mathematics Research Center Technical Report #1800, University of Wisconsin-Madison, 1977.

[11] K. Nickel, Schranken fur die Losungsmengen von Funktional-Differential-Gleichungen, Freiburger Intervall-Berichte, 79/4, Institut fur Angewandte Mathematik, Universitat Freiburg i. Br. 1979.

[12] B. Viswanatham, A generalization of Bellman's lemma, Proc. Amer. Math. Soc. 14, 1963, 15-18.

[13 P. Wildenauer, Construction of domains with all solutions and the existence of extreme solutions, SIAM J. Numer. Anal. 18, No. 5, 1981, 801-807.

EQUATIONS WITH NONLINEARITIES DEFINED ONLY ON SUBSETS

R. Kent Nagle

University of South Florida
Tampa, Florida

ABSTRACT

In this paper we discuss the existence of solutions to boundary value problems of the form

$$Lu - \lambda_1 u = g(x,u) + F(x) \quad \text{in} \quad \Omega,$$

$$Bu = 0 \quad \text{on} \quad \partial\Omega,$$

where $g:\bar{\Omega}\times I \to R$ is a smooth function and I is some interval of R. For example, $g(u)=\tan u$ or $g(u)=\arcsin u$.

In many applications one is lead to a mathematical model which makes sense only for $g(x,u)$ defined for u in some interval. For example, the standard model for a simple pendulum hanging by a wire gives rise to the problem of finding 2π-periodic solutions to the equation

$$\theta'' + \text{Sin}\ \theta = p(t) ,$$

where θ is the angle the wire makes with the vertical. Clearly this model only makes sense when $|\theta| \leq \pi/2$! These problems are currently receiving little attention in the literature with most of the work now being done for the case where g is defined on $\Omega\times R$.

In order to study these problems where the nonlinearity is defined only on a subset, one is often forced to consider the pointwise behavior of the solution. This requires one to use a sup norm or have some a priori estimate on the possible values of a solution. Hence those methods which involve Hilbert space norms such as the L_2 norm are very hard to use.

In this paper we will show how the method of lower and upper solutions can be used to study these problems for elliptic equations. The success of this approach is a result of the existence of sequences of lower and upper solutions which converge monotonically to the solution, thus giving a pointwise estimate for the solution. The approach given here is similar to that of Amann [1], Kazden and Warner [3], and Sattinger [5]. Recently Lakshmikantham, Kannan, and Nieto [4] showed that the method of lower and upper solutions is related to the results of Cesari and Kannan [2].

Consider a nonlinear elliptic boundary value problem of the form

$$\begin{aligned} Lu - \lambda_1 u &= g(x,u) + F(x) \quad \text{in } \Omega, \\ Bu &= 0 \quad \text{on } \partial\Omega, \end{aligned} \tag{1}$$

where B is either a Dirichlet or Neumann boundary operator, $\Omega \subset R^n$ is a bounded domain, and $F \in C(\overline{\Omega})$. We assume that $g: \overline{\Omega} \times I \to R$ is a smooth mapping where I is an interval which may be either open or closed.

We assume L is a uniformly elliptic operator of order two of the form

$$Lu = \sum_{i,j=1}^{n} a_{ij}(x) \frac{\partial^2 u}{\partial x_i \partial x_j} + \sum_{j=1}^{n} b_j(x) \frac{\partial u}{\partial x_j} + c(x)u ,$$

where $a_{ij}(x)$'s are real, smooth, positive definite, and $a_{ij} = a_{ji}$.

Let $\phi(x)$ be an eigenfunction associated with the first eigenvalue λ_1 of L. That is,

$$L\phi = \lambda_1 \phi \quad \text{in } \Omega , \quad B\phi = 0 \quad \text{on } \partial\Omega .$$

It is well known that we can choose $\phi(x) > 0$ for x in Ω.

A smooth function u_0 is called an <u>upper solution</u> of (1) if

$$Lu_0 - \lambda_1 u_0 \leq g(x,u_0) + F(x) \quad \text{in } \Omega$$

$$Bu_0 \geq 0 \quad \text{on } \partial\Omega .$$

We say v_0 is a <u>lower solution</u> of (1) if

$$Lv_0 - \lambda_1 v_0 \geq g(x,v_0) + F(x) \quad \text{in } \Omega$$

$$Bv_0 \leq 0 \quad \text{on } \partial\Omega .$$

By a slight modification of the proof in Sattinger [5, Theorem 2.1] of a theorem of Amann [1], we have the following theorem.

<u>Theorem 1</u>. Let $u_0(x)$ and $v_0(x)$ be respectively upper and lower solutions of (1) satisfying for all x in Ω

(i) $v_0(x) \le u_0(x)$

(ii) $u_0(x) \in I$

(iii) $v_0(x) \in I$.

Then the boundary value problem (1) has a solution.

In most applications one chooses the upper and lower solutions to be of the form $s\phi(x)$ where $\phi(x)$ is the eigenfunction corresponding to the lowest eigenvalue λ_1 chocsen so that $\phi(x)>0$ in Ω. As an immediate corollary to Theorem 1 we have:

Corollary 1. If there exists constants $s_-<s_+$ such that

(i) $s_+\phi$ is an upper solution satisfying
$s_+\phi(x)\in I$ for all $x\in\Omega$, and

(ii) $s_-\phi$ is a lower solution satisfying
$s_-\phi(x)\in I$ for all $x\in\Omega$,

then the boundary value problem (1) has a solution.

When $c(x) \equiv 0$, $\lambda_1 = 0$, and 0 is an interior point of I, then we can use constant functions as upper and lower solutions. This gives us the following "asymptotic" conditions on g for existence.

Theorem 2. Let $c(x) \equiv 0$, $\lambda_1 = 0$, and $I = (a,b)$ with 0 in I. Assume

$$\lim_{s\to a^+} g(x,s) = g_a(x) \quad \text{and} \quad \lim_{s\to b^-} g(x,s) = g_b(x) , \tag{2}$$

where the convergence is uniform in x on Ω. If

$$g_a(x) < -F(x) < g_b(x) , \tag{3}$$

for all x in Ω, then the boundary value problem has a solution.

Proof. It follows from conditions (2) and (3) that there exists constants $a<s_-<0<s_+<b$ such that $u_0 = s_+$ and $v_0 = s_-$ are upper and lower solutions to (1). Theorem 2 now follows from Theorem 1.

Example 1. Let $G(x,u)$ be a smooth mapping from $\bar{\Omega}\times[-\pi/2, \pi/2]$ into R. Then the boundary value problem

$$\begin{aligned} \Delta u + u &= \tan u + G(x,u) + F(x) \quad \text{in } \Omega, \\ u &= 0 \quad \text{on } \partial\Omega , \end{aligned} \tag{4}$$

has a solution for any $F \varepsilon C(\bar{\Omega})$.

Since

$$\lim_{s \to -\pi/2^+} \tan s = -\infty , \qquad \lim_{s \to \pi/2^-} \tan s = +\infty ,$$

and $G(x,s) + F(x)$ is bounded on $\bar{\Omega} \times [-\pi/2, \pi/2]$, then for any $F \varepsilon C(\bar{\Omega})$ we can choose constants s_+ and s_- with $-\pi/2 < s_- < 0 < s_+ < \pi/2$ such that for all x in Ω

$$s_+ < \tan s_+ + G(x,s_+) + F(x)$$

and

$$\tan s_- + G(x,s_-) + F(x) < s_- .$$

It is now easy to show that $u_0 = s_+$ and $v_0 = s_-$ are upper and lower solutions and hence (4) has a solution.

<u>Example 2</u>. The boundary value problem

$$\begin{aligned} \Delta u &= \arcsin u + F(x) \quad \text{in } \Omega \\ \frac{\partial u}{\partial n} &= 0 \quad \text{on } \partial\Omega \end{aligned} \tag{5}$$

has a solution for $F \varepsilon C(\bar{\Omega})$ provided $|F(x)| < \pi/2$.

Again since

$$\lim_{s \to -1^+} \arcsin s = -\pi/2 \quad \text{and} \quad \lim_{s \to 1^-} \arcsin s = \pi/2 ,$$

we can choose constants s_+ and s_- with $-1 < s_- < 0 < s_+ < 1$ such that for all x in Ω

$$\arcsin s_- + F(x) < 0 < \arcsin s_+ + F(x) .$$

We again have $u_0 = s_+$ and $v_0 = s_-$ as upper and lower solutions and hence (5) has a solution.

The condition on $F(x)$ in Example 2 is too restrictive. It is easy to find examples where $F(x)$ has arbitrarily large sup norm, yet the boundary value problem (5) <u>may or may not</u> have a solution. We can improve the result given in Example 2 by first decomposing $F(x)$ into a component in the kernel of the linear operator plus a component in the orthogonal complement of the kernel. In this example we express $F(x) = H(x)+t$, where t is a constant and H has mean value zero.

Since $H(x)$ has mean value zero, there exists a unique function $h(x)$

with mean value zero such that

$$\Delta h(x) = H(x) \quad \text{in } \Omega$$

$$\frac{\partial h}{\partial n} = 0 \quad \text{on } \partial\Omega .$$

We can now prove the following theorem which is similar to Proposition 2.15 in [3].

Theorem 3. Let $h(x)$ be as defined above. If $\|h\|_\infty < 1/2$, then there exists constants $t_- < 0 < t_+$ depending on $H(x)$ such that the boundary value problem

$$\begin{aligned} \Delta u &= \arcsin u + H(x) + t \quad \text{in } \Omega \\ \frac{\partial u}{\partial n} &= 0 \quad \text{on } \partial\Omega \end{aligned} \tag{6}$$

has a solution if $t_- < t < t_+$, but not if $t < t_-$ or $t > t_+$. Moreover, $|t_\pm| \leq \pi/2$ meas Ω.

Proof. Let $u = v+h$. Solving (6) is equivalent to solving

$$\begin{aligned} \Delta v &= \arcsin (v+h) + t \quad \text{in } \Omega \\ \frac{\partial v}{\partial n} &= 0 \quad \text{on } \partial\Omega. \end{aligned} \tag{7}$$

Since $\|h\|_\infty < 1/2$, we can choose constants s_- and s_+ such that $s_- < 0 < s_+$ and

$$-1 < s_- + h(x) < 0 < s_+ + h(x) < 1.$$

Now choose T_- and T_+ such that

$$s_- + h(x) \leq T_- < 0 < T_+ \leq s_+ + h(x).$$

Again, $u_0 = s_+$ and $v_0 = s_-$ are lower and upper solutions for all t, $T_- \leq t \leq T_+$.

The rest of the proof is essentially the same as the proof given by Kazden and Warner for Proposition 2.15 in [3].

Theorem 3 does not completely answer the question of existence of solutions to problem (6). We did show that $\|h\|_\infty < 1/2$ is sufficient for the existence of solutions to (6) for certain values of t. But, does (6) have a solution for some value of t if $\|h\|_\infty < 1$? A more general question is how does t_- and t_+ depend on $H(x)$?

REFERENCES

1. H. Amman, "On the existence of positive solutions of nonlinear elliptic boundary value problems", Indiana Univ. Math. J. 21 (1971), 125-146.

2. L. Cesari and R. Kannan, "An abstract theorem at resonance", Proc. Amer. Math. Soc. 63 (1977), 221-225.

3. J. Kazdan and F. Warner, "Remarks on some quasilinear elliptic equations", Comm. Pure Appl. Math. 28 (1975), 567-597.

4. V. Lakshmikantham, R. Kannan, and J. Nieto, "Sufficient conditions for existence of solutions of nonlinear boundary value problems at resonance", to appear.

5. D. Sattinger, "Monotone methods in nonlinear elliptic and parabolic boundary value problems", Indiana Univ. Math. J. 21 (1972), 979-1000.

OPERATOR PARTS AND GENERALIZED INVERSES OF LINEAR MANIFOLDS WITH APPLICATIONS

Z. Nashed

Department of Mathematical Sciences
University of Delaware
Newark, Delaware

ABSTRACT

We describe some results of a comprehensive theory of (topological) operator parts and generalized inverses of linear manifolds in Banach spaces, developed recently by S. J. Lee and the author. Let M be a closed linear manifold in the direct sum of Banach spaces X, Y. Following E. A. Coddington, a linear manifold $R \subset X \oplus Y$ is called an operator part of M if M is the (topological) direct sum of R and $\{0\} \oplus M(0)$. We give a structural characterization of all operator parts of M. We also characterize generalized inverses of a multi-valued linear mapping M and relate them to operator parts of M^{-1}, the inverse of M considered as a linear relation. We indicate an application to least-squares solutions of the inclusion relation $y \in M(x)$. The theory hinges upon the notion of an adjoint subspace and suitable operational properties of linear relations, which are given in Section 1.

1. LINEAR RELATIONS AND ADJOINT SUBSPACES

Let X and Y be vector spaces, and let A be a linear manifold (algebraic subspace) of $X \oplus Y$. We call A a *linear relation* to stress our interest in those properties and problems which generalize (or markedly differ from) those when A is (the graph of) a single-valued operator. To begin with, we note that a subspace $A \subset X \oplus Y$ may be viewed as a multi-valued mapping from X into Y by defining $A(x) := \{y \in Y | \{x,y\} \in A\}$. The domain, range, and null space of A are defined, respectively, by

$$\mathcal{D}(A) := \{x \in X \mid \{x,y\} \in A \text{ for some } y \in Y\} ,$$
$$\mathcal{R}(A) := \{y \in Y \mid \{x,y\} \in A \text{ for some } x \in X\} ,$$
$$\mathcal{N}(A) := \{x \in X \mid \{x,0\} \in A \text{ for some } x \in X\} .$$

Also,

$$\alpha A := \{\{x,\alpha y\} \mid \{x,y\} \in A\} , \quad \alpha \in \mathbb{C} .$$

The inverse of a linear relation A, denoted by A^{-1}, is the linear relation

$$A^{-1} := \{\{y,x\} \mid \{x,y\} \in A\} ,$$

so $\mathcal{N}(A) = A^{-1}(0)$. The sum and algebraic sum of $A,B \subset X \oplus Y$ are defined by

$$A \dotplus B := \{a + b \mid a \in A,\ b \in B\}$$
$$A + B := \{\{x,y+z\} \mid \{x,y\} \in A,\ \{x,z\} \in B\} .$$

$A \dotplus B$ is a direct algebraic sum if $A \cap B = \{0\}$. For $A \subset X \oplus Y$ and $C \subset Z \oplus X$, we define

$$AC := \{\{x,y\} \in Z \oplus X \mid \{x,z\} \in C,\ \{z,y\} \in A\} .$$

By an "operator" we shall always mean a single-valued mapping, whereas a "mapping" need not necessarily be single-valued. A subspace $A \subset X \oplus Y$ is (the graph of) an operator if and only if $A(0) = \{0\}$.

An alternate and equivalent approach to linear relations is via mappings. A mapping T defined over a subset $\mathcal{D} \subset X$ into Y is called *linear* if its graph

$$G(T) := \{\{x,T(x)\} \mid x \in \mathcal{D},\ T(x) \in \mathcal{R}(T)\}$$

is a linear manifold. If T is linear, then the domain and range of T are linear manifolds. It is easy to show that a necessary and sufficient condition for a linear mapping to be single-valued is that $T(0)$ have the unique value 0. By a similar argument, it may be shown that if a linear relation T has two different values at some element in the domain of T, then T has infinitely many values at each element in $\mathcal{D}$. Linearity is characterized by

$$\alpha T(x_1) + \beta T(x_2) \subset T(\alpha x_1 + \beta x_2), \quad \alpha,\beta \in ¢ , \ x_1,x_2 \in \mathcal{D} .$$

From here on, we identify a linear mapping by its graph, which in turn is identified as a subspace of a direct sum of two vector spaces. Thus, we say that a subspace A is an operator if and only if $A(0) = \{0\}$.

Several (multi-valued) linear relations arise naturally in connection with (single-valued) linear operators.

I. If a linear operator A has a nontrivial null space, then A^{-1} obviously does not exist as an operator; however, A^{-1} always exists as a linear relation. Various generalized inverses (when restricted to the range of the operator) may be viewed as particular selections of the inverse relation A^{-1}. The notions of algebraic operator part and topological operator part of a subspace (to be discussed later) also yield particular single-valued selections of A^{-1}.

II. Let A be a linear operator on a domain $\mathcal{D}(A)$ in a normed space X. While A need not be closable as an operator, the closure A^c always exists as a relation.

III. If A is a linear operator in a Hilbert space H, say into H, whose domain $\mathcal{D}(A)$ is not dense in H, its adjoint is not a well-defined operator. On the other hand, the set of all pairs $\{g,h\} \in H^2 := H \oplus H$ such that $\langle Af,g\rangle = \langle f,h\rangle$ for all $f \in \mathcal{D}(A)$ is a linear manifold in H^2 which can be considered as the *adjoint* linear manifold (or *adjoint subspace*) to the graph of A in H^2. More generally, we can define the adjoint A^* of a linear relation (subspace) in the Banach space $X \oplus Y$ as the *linear relation*

$$A^* = \{\{f^+,g^+\} \in Y^* \oplus X^* | (g,f^+) = (f,g^+) \quad \text{for all} \quad \{f,g\} \in A\} ,$$

where X^* denotes the conjugate space to X. The conjugate space of a normed space W is the space of all continuous conjugate linear functionals on W. The natural pairing of W and W^* is denoted by $(\cdot,\cdot)$ and defined by

$$(w,f) = \overline{f(w)}, \quad w \in W, \ f \in W^* .$$

If A, A^+ are linear manifolds in the Banach spaces W, W^*, respectively, we define the annihilator of A and the preannihilator of A^+, respectively, as the linear manifolds:

$$A^\perp = \{f \in W^* | (w,f) = 0, \ w \in A\}$$
$$^\perp(A^+) = \{w \in W | (w,f^+) = 0, \ f^+ \in A^+\} .$$

Let J be the formal operator defined by $J\{x,y\} = \{y,-x\}$. Then the adjoint (subspace) A^* can be conveniently written in the form

$$A^* = (JA)^\perp = J(A^\perp) = \{\{y,-x\} | \{x,y\} \in A^\perp\} .$$

Clearly J is an isometry of $X \oplus Y$ onto $Y \oplus X$. Similarly, the *preadjoint* of $A^+ \subset Y^* \oplus X^*$ is defined by

$$*(A^+) := \{\{y,-x\} | \{x,y\} \in {}^\perp(A^+)\} .$$

The adjoint A^* is a weak-star closed linear manifold in $W^+ := Y^* \oplus X^*$ and the preadjoint $*(A^+)$ is a closed subspace of $W := X \oplus Y$. They also satisfy the following properties: $*(A^*) = A^c$, the closure of A in W; $(*(A^+))^* = {}^c(A^+)$, the w*-closure of A^+ in W^+. Thus $*(A^*) = A$ if and only if A is closed. Hence, an important property of adjoints of densely defined linear operators is carried over to the case of adjoint subspaces of a linear operator with a nondense domain or, more generally, of a linear relation.

We remark, however, that linear relations can exhibit a markedly different behavior from linear operators. For example, a self-adjoint linear relation can have an unbounded spectrum in finite-dimensional spaces, a phenomenon which linear operators exhibit only in infinite-dimensional spaces.

The theory of adjoint subspaces in Banach spaces and their applications to ordinary differential subspaces has been recently developed by E. A. Coddington, A. Dijksma and H. S. V. de Snoo in a series of substantial papers (see [4], [5], [6], [7]).

We devote the rest of this section to some historical remarks and brief indications of earlier applications of linear relations and their adjoint subspaces in the context of differential equations.

Elementary aspects of the theory of multi-valued linear mappings were considered in the early development of functional analysis. Von Neumann (see, e.g., [25]) considered such mappings and defined the adjoint of a nondensely-defined linear operator as a linear relation in an appropriate Cartesian product space. Operational properties of linear relations were more substantially developed in a seminal paper by Arens [1]. In particular, he indicated how a self-adjoint subspace $T \subset H \oplus H$, where H is a Hilbert space, could be written as the orthogonal sum of a purely multi-valued part and the graph of a certain self-adjoint operator on a subspace of H. This provided the starting point of several generalizations and applications.

The fact that the adjoint of a nondensely-defined linear operator does not exist as a single-valued mapping presents some problems in describing all normal extensions $\tilde{T}$ of a given formally normal densely defined operator T in a Hilbert space H_1, where $\tilde{T}$ is normal in a larger Hilbert space $H := H_1 \oplus H_2$. Because of this difficulty, in the initial approach to this problem (see [2]) attention was restricted to those $\tilde{T}$ such that $\mathcal{D}(\tilde{T}) \cap H_2$ is dense in H_2. This restriction was removed later in [3]. Coddington recognized the importance of adjoint subspaces and used them to develop an extension theory for formally normal linear manifolds (which are not necessarily the graphs of operators). A subspace $N \subset H^2 := H \oplus H$ is said to be formally normal if $\mathcal{D}(N) \subset \mathcal{D}(N^*)$ and there is an isometry V of N onto N^* of the form $V\{f,g\} = \{f,h\}$ such that $||g|| = ||h||$. A *normal subspace* is a formally normal subspace N with $\mathcal{D}(N) = \mathcal{D}(N^*)$ and isometry V of N onto N^*. Here N^* is the adjoint subspace of N as defined earlier. These definitions extend the notion of formally normal and normal operators. Coddington [3] characterized the formally normal and normal extensions in H^2 of a given formally normal subspace N in H^2, as well as the symmetric and self-adjoint extensions of a given symmetric subspace S. He also identified those extensions which are (single-valued) operators and showed that the Friedrichs extension of a semibounded symmetric subspace exists. This theory has, for example, applications to the spectral theory of non-densely defined ordinary differential operators.

In the classical development of the spectral theory for the linear differential equation $Ly = \lambda y$, where L is an ordinary differential operator which is regular on a finite interval (a,b), one considers the operator L and its adjoint in the Hilbert space $H = L^2[a,b]$. It is sometimes convenient to introduce the space $H \oplus H$ and view the operator in terms of its graph. There are problems in differential equations and other areas of analysis where the use of such direct sums is rather necessary as the basic setting for developing the theory, for example, in the case of regular boundary value problems associated with pairs of ordinary differential expressions. The recent monograph of Coddington and de Snoo [6] develops the basic spectral theory associated with the differential equation $Ly = \lambda My$, where L and M are ordinary differential operators which are regular on a finite interval and M is symmetric. Fundamental questions in boundary problems are related to certain minimal, maximal and intermediate subspaces of H^2, and their adjoint subspaces. Not all these spaces need be graphs of (single-valued) operators. Boundary conditions induce subspaces of H^2; conversely, various subspaces of H^2 can be characterized by abstract boundary conditions, and in certain cases, it is possible to describe these in concrete forms.

Another example where adjoint subspaces play a pivotal role is when the boundary conditions on an ordinary differential expression of order n are sufficiently restrictive, for instance, when multipoint conditions or Stieltjes boundary conditions are imposed. Then the natural minimal operators, say on $L^2[a,b]$, associated with this problem will not be densely defined, and hence their adjoint will be multi-valued. This leads naturally to a study of certain subspaces and their extensions.

2. OPERATOR PARTS OF SUBSPACES

The central problem which motivates many of the concepts and results of this paper is the problem of characterization and construction of all solutions of a linear inclusion: *Let* $M \subset X_1 \oplus X_2$ *be a given linear manifold in the (complex) vector space* $X_1 \oplus X_2$, *and let* $g \in X_2$ *be a given vector. Find all* $y \in \mathcal{D}(M)$ *such that* $g \in M(y)$, *if a solution exists.*

If the inclusion problem $g \in M(y)$ has no solution, a natural question is to formulate an appropriate notion of a least-squares solution and then to address the problem in the context of least-squares solvability. This will be done in Section 4, assuming X_1 and X_2 are Hilbert spaces.

A vector space R in $X_2 \oplus X_1$ is called an *algebraic operator part* of the inverse relation M^{-1} if R is the graph of a linear operator and M^{-1} is the algebraic direct sum of R and $\{0\} \oplus N(M)$. The notion of algebraic operator part makes sense, of course, for any subspace; however, we are mainly interested in operator parts of inverse relations.

The algebraic aspects do not suffice for problems where the inclusion relation $g \in M(y)$ arises from analysis. The linear manifold M has in such settings some topological properties and we are then interested in solutions with certain topological and analytic properties. A similar situation arises in generalized inverse operator theory: the algebraic theory of generalized inverses does not suffice for problems of analysis in Banach spaces; it is necessary to develop notions of generalized inverses which induce continuous projectors.

Suppose X_1 and X_2 are Banach spaces and M is a closed subspace in $X_1 \oplus X_2$. A vector space R in $X_2 \oplus X_1$ is called a (topological) *operator part* of M^{-1} if R is the graph of a *closed* linear operator and M^{-1} is the algebraic sum of R and

$\{0\} \oplus N(M)$. Similarly, a vector space R^+ in $X_1^* \oplus X_2^*$ is called a topological operator part of $(M^+)^{-1}$, where M^+ is a given w*-closed subspace of $X_2^* \oplus X_1^*$, if R^+ is the graph of a w*-closed linear operator and

$$(M^+)^{-1} = R^+ \dot{+} (\{0\} \oplus N(M^+)) .$$

It is well-known (and easy to see) that an algebraic operator part of M^{-1} always exist, but this is not always true for a (topological) operator part. The latter exists if and only if $\{0\} \oplus N(M)$ is *topologically complemented* in M^{-1}. Similarly, an operator part of $(M^+)^{-1}$ exists if and only if $\{0\} \oplus N(M^+)$ is w*-complemented in $(M^+)^{-1}$.

Remark. Let R be an algebraic operator of M. Then $\mathcal{D}(R) = R(M)$, $R(R) \subset \mathcal{D}(M)$. Moreover, $\{y,g\} \in M$ if and only if $g \in R(M)$ and $y = R(g) + u$ for some $u \in N(M)$. Also R is one-to-one if and only if M is an operator.

The following theorem characterizes all possible operator parts of M^{-1} provided that $N(M)$ is complemented in X_1.

THEOREM 2.1. *Assume that* $N(M)$ *is topologically complemented in* X_1 *and let* P_0 *be a continuous projector from* X_1 *onto* $N(M)$. *Let*

$$R_0 := \{\{g, y-P_0(y)\} \mid \{y,g\} \in M\} .$$

Then

(1) R_0 *is an operator part of* M^{-1}.

(2) R *is an operator part of* M^{-1} *if and only if*

$$R := \{\{g, R_0(g) - A(\{g, R_0(g)\}) \mid g \in R(M)\}$$

for some continuous linear operator A *from* $X_2 \oplus X_1$ *into* $\{0\} \oplus N(M)$ *such that*

$$A(\{0,y\}) = \{0,0\}, \quad \textit{for all} \quad y \in N(M) .$$

In the case when $N(M)$ is finite dimensional, (2) of Theorem 2.1 contains Theorems 3.1, 3.4 of [7] as a special case. It should be noted that (ii) of Theorem 3.4 in [7] is *not* a necessary and sufficient condition.

Aside from some technical aspects, the proof makes use of the following lemma due to Sobczyk [23] about a characterization of all possible projectors defined on a given vector space whose range is a prescribed subspace (see also [20], Proposition 1.7).

LEMMA 2.2. *Let* $L(V)$ *denote the set of all linear operators on* V *into* V, *and let* S *be a prescribed subspace of* V. *Let* P_S *be the set of all linear idempotents on* V *with range* S, *i.e.,*

$$P_S := \{P \in L(V):\ P^2 = P,\ R(P) = S\} \quad .$$

Then P_S *is an affine manifold and the subspace parallel to* P_S *is*

$$A_S := \{A \in L(V):\ R(A) \subset S \subset N(A)\} \quad ,$$

so if $P, P' \in P_S$, *then* $P - P' \in A_S$.

The notion of an "operator part" of a subspace was first introduced by Coddington for a general Banach space case (see [7] and references therein). A structural theory of an operator part was also studied in a special case. The structural theory of an operator part in a general Banach space setting was recently developed by Lee and Nashed [17] using abstract adjoint subspaces, the coordinatized adjoint theory, and the theory of projectors and topological complementation. Various problems in analysis and applications call for the use of different operator parts in the same way that different types of generalized inverses have been developed to suit different purposes.

3. GENERALIZED INVERSES OF SUBSPACES

Let X and Y be Banach spaces and let L be a bounded linear operator on X into Y. Assume that $N(L)$ and $\overline{R(L)}$ are topologically complemented in X and Y respectively, say

$$X = N(L) \oplus M \ , \quad Y = \overline{R(L)} \oplus S \ .$$

Let P be the projector of X onto $N(L)$ along M, and Q be

the projector of Y onto $\overline{R(L)}$ along S. The generalized inverse of L, relative to these decompositions of X and Y, is defined by $L^{\dagger}_{P,Q} := (L|M)^{-1}Q$, where $L|M$ is the restriction of L to M. The induced projectors $LL^{\dagger}$ and $L^{\dagger}L$ in this case are topological projectors (i.e., they are continuous). An alternate formula for the generalized inverse is

$$L^{\dagger}_{P,Q} = (I-P)L^{(-1)}Q$$

where $L^{(-1)}$ is the "inverse" (of L) considered as a multi-valued mapping. This motivates the following definition of a generalized inverse of a subspace.

DEFINITION 3.1. *Let* X_1, X_2 *be Banach spaces and let* M *be a linear manifold in* $X_1 \oplus X_2$. *Assume that* $N(M)$ *is topologically complemented in* X_1 *and* $N(M^*)$ *is topologically complemented in* X_2^*. *Then a vector space* $M^{\#}$ *in* $X_2 \oplus X_1$ *is called a generalized inverse of* M *if it has the form:*

$$M^{\#} := (\text{graph}(I-P))M^{-1}(\text{graph}(I-{}^*P^+)) ,$$

where P *is a (continuous) projector from* X_1 *onto* $N(M)$, *and* P^+ *is a* w*-*continuous projector from* X_2^* *onto* $N(M^*)$.

Thus

$$M^{\#} = \{\{g,(I-P)(y)\} | \{g,y\} \in X_2 \oplus X_1 , \text{ and } \{y,(I-{}^*P^+)(y)\} \in M\} .$$

It is easy to verify that $D(M^{\#}) = R(M) \dotplus N(M^*)$, and $R(M^{\#}) = (I-P)R(M)$.

THEOREM 3.1. *Let* $M^{\#}$ *be as above. Then*

$$D(M^{\#}) = R(M) \dotplus R({}^*P^+) , \quad \textit{direct sum,}$$

and

$$M^{\#} = (\text{graph}(I-P))M^{-1} \dotplus (R({}^*P^+) \oplus \{0\}), \quad \textit{direct sum} .$$

Since $(\text{graph}(I-P))M^{-1}$ is an operator part of M^{-1}, $M^{\#}$ can be viewed as a linear extension of an operator part (by enlarging its domain).

We describe next how to obtain *all* possible generalized inverses of a linear manifold via a pair of known of projectors $\{P_0, P_0^+\}$. The derivation uses Sobczyk's lemma as in the case of operator parts.

THEOREM 3.2. *Let* P_0 *be a continuous projector from* X_1 *onto* $N(M)$ *and let* P_0^+ *be a* w*-*continuous projector from* X_2^* *onto* $N(M^*)$ *(thus, we are assuming the existence of such projectors, or equivalently, the existence of certain topological complements as in Definition* 3.1*). Then* $M^\#$ *is a generalized inverse of* M *if and only if*

$$M^\# := (\text{graph}(I-P_0-A))M^{-1}(\text{graph}(I-{}^*P_0^+-{}^*A^+))$$

for some continuous linear operator A *from* X_1 *into* $N(M)$ *such that* $A(y) = 0$ *for all* y *in* $N(M)$ *and a* w*-*continuous linear operator* A^+ *from* X_2^* *into* $N(M^*)$ *such that* $A^+(y) = 0$ *for all* $y \in N(M^*)$.

Note that a generalized inverse of M depends on two projectors, whereas an operator part of M requires only one projector.

4. LEAST-SQUARES SOLUTIONS OF LINEAR RELATIONS IN HILBERT SPACES

Let H_1, H_2 be Hilbert spaces, M an arbitrary given vector space in $H_1 \oplus H_2$, and $y \in H_2$. A vector $u \in H_1$ is called a *least-squares solution* (LSS) of the inclusion $y \in M(x)$ if $u \in \mathcal{D}(M)$ and $d(y,\mathcal{R}(M)) = ||y-z||$ for some $z \in M(u)$, where $d(y,\mathcal{R}(M))$ is the distance between y and $\mathcal{R}(M)$.

Another generalization of the notion of a least-squares solution to the case of a multi-valued operator that seems natural is the following: An element $u \in \mathcal{D}(M)$ is called an *almost least-squares solution* of $y \in M(x)$ if $d(y,\mathcal{R}(M)) = d(y,M(u))$. Note that if M is an operator, then these two notions coincide and reduce to the usual definition of a least-squares solution of an operator equation. Here we shall be interested only in the notion of LSS; for properties of almost least-squares solutions, see [16].

It is easy to see that the inclusion $y \in M(x)$ has a LSS if and only if $y \in \mathcal{N}(M^*) \dotplus \mathcal{R}(M)$ or, equivalently, $(I-P^+)(y) \in \mathcal{R}(M)$, where P^+ is the orthogonal projector from H_2 onto $\mathcal{N}(M^*)$.

PROPOSITION 4.1. *Let* P^+ *be as above, and let* P *be the orthogonal projector of* H_1 *onto* $\mathcal{N}(M^c)$, *where* M^c *is the closure of* M. *Let* $y \in H_2$. *Then, the following statements are equivalent:*

(i) $u \in H_1$ *is a LSS of* $y \in M(x)$.
(ii) $u \in \mathcal{D}(M)$ *and* $(I-P^+)(y) \in M(u)$.
(iii) $u \in \mathcal{D}(M)$ *and* $y \in M(u) \dotplus \mathcal{N}(M^*)$.
(iv) $u \in \mathcal{D}(M)$ *and* $M(u) \subset y \dotplus \mathcal{R}(M)^{\perp}$.

Similarly, for the adjoint relation: Let $y \in H_1$ *be given. Then the following statements are equivalent.*

(i) $u \in H_2$ *and* u *is a LSS of* $y \in M^*(x)$.
(ii) $u \in \mathcal{D}(M^*)$ *and* $(I-P)(y) \in M^*(u)$.
(iii) $u \in \mathcal{D}(M^*)$ *and* $y \in M^*(u) \dotplus \mathcal{N}(M^c)$.
(iv) $u \in \mathcal{D}(M^*)$ *and* $M^*(u) \subset y \dotplus \mathcal{R}(M^*)^{\perp}$.

Proof. We prove the first part; the second part is then immediate from properties of adjoint subspaces. Assume (i). Then $d(y,\mathcal{R}(M)) = ||y-z||$ for some $z \in M(x)$ and hence $y - z \in P^+(y)$. Thus, $(I-P^+)(y) \in M(u)$ and so (i) implies (ii). Assume (ii) holds. Since $d(y,\mathcal{R}(M)) = ||y-(I-P^+)(y)||$ and $(I-P^+)(y) \in M(u)$, it follows that u is a LSS. Thus, (ii) implies (i). Clearly, (ii) implies (iii). Also (iii) implies (iv), since $\mathcal{R}(M)^{\perp} = \mathcal{N}(M^*)$. Finally, to show that (iv) implies (i), let $w = z - y$ for some $k \in \mathcal{R}(M)^{\perp}$. Then $d(y,\mathcal{R}(M)) = ||P^+(y)|| = ||P^+(z-w)|| = ||P^+(w)|| = ||(y-z)||$. This shows that u is a LSS and completes the proof.

Note that if M is an operator with $\mathcal{D}(M^*) = H_2$, then (iii) of Proposition 4.1 can be rewritten as follows: u is a LSS of $Mx = y$ if and only if $M^*Mu = M^*y$, which is the usual "normal equation" characterization for a least-squares solution of, say, a bounded linear operator equation in a Hilbert space. If $\mathcal{D}(M^*) \neq H_2$, this characterization is false and one can only state that $Mu - y \in N(M^*)$.

Suppose that $M(0)$ is *closed* in H_2 and let P denote the orthogonal projector from H_2 onto $M(0)$. Then we define the orthogonal operator part of M by

$$S_M := [\text{graph}(I-P)]M = \{\{g,(I-P)(y)\} \mid \{g,y\} \in M\} .$$

It is easy to check that S_M is an algebraic operator part of M such that S_M is orthogonal to $\{0\} \oplus M(0)$; also $\mathcal{D}(S_M) = \mathcal{D}(M)$ and $R(S_M) = R(M) \cap M(0)^\perp$. Note that this is a special case of an operator part corresponding to an orthogonal direct sum.

Let $M \subset H_1 \oplus H_2$ be a vector space such that $N(M)$ is closed. Let P and P^+ be as in Proposition 4.1. The subspace

$$M^\# := [\text{graph}(I-P)]M^{-1}[\text{graph}(I-P^+)]$$

is called the *orthogonal generalized inverse* of M. (If M is the graph of a closed densely defined linear operator, then $M^\#$ is precisely the *graph* of the Moore-Penrose inverse of that operator.)

The orthogonal generalized inverse of M is an extension of the orthogonal operator part of M^{-1}. Indeed, it is easy to show that the following properties hold: $M^\#$ is a linear *operator* such that

$$M^\# = S_{M^{-1}} \dot{+} (N(M^*) \oplus \{0\}), \quad \text{direct sum},$$

$$\mathcal{D}(M^\#) = R(M) \dot{+} N(M^*), \quad R(M^\#) = R\left(S_{M^{-1}}\right) .$$

Also,

$$MM^\# = \{\{x,(I-P^+)(x)+s \mid s \in M(0),\ x \in \mathcal{D}(M^\#)\}$$
$$M^\# M = \{\{x,(I-P)(x)\} \mid x \in \mathcal{D}(M)\} .$$

These properties should be contrasted with those in the case of an operator, see [19], [20]; in particular, it should be noted that in the case when M is a multi-valued mapping, $MM^{\#}$ is not a single-valued orthogonal projector (whereas $M^{\#}M$ is).

If M is closed, then $M^{\#}$ is closed and $(M^{\#})^*$ is the orthogonal generalized inverse of M^*. Moreover, when M is closed, $M^{\#}$ is continuous if and only if $\mathcal{R}(M)$ is closed.

We now characterize the set of all least-squares solutions of a linear relation in $H_1 \oplus H_2$ in terms of algebraic operator parts and the orthogonal generalized inverse. For a proof and related properties, see [16].

THEOREM 4.2. *Assume* $y \in \mathcal{R}(M) \dot{+} \mathcal{N}(M^*)$. *Then we have the following:*

(1) For any algebraic part R *of* M^{-1}, *the coset*

$$R(I-P^+)(y) \dot{+} \mathcal{N}(M)$$

is the set of all least-squares solutions of $y \in M(x)$.

(2) If $\mathcal{N}(M)$ *is closed, then* $M^{\#}(y) \dot{+} \mathcal{N}(M)$ *is the set of all least-squares solutions of* $y \in M(x)$, *and* $M^{\#}(y)$ *is the LSS of minimal norm.*

(3) If $M(0)$ *is closed, then*

$$d(y,\mathcal{R}(M)) = ||y-S_M M^{\#}(y)-Q(I-P^+)(y)|| \quad ,$$

where Q *is the orthogonal projector from* H_2 *onto* $M(0)$. *Moreover, the map*

$$y \to S_M M^{\#}(y) + Q(I-P^+)(y)$$

on $\mathcal{D}(M^{\#})$ *into* H_2 *is continuous.*

The theory of operator parts and generalized inverses can also be extended to the context of constrained minimization problems involving multi-valued linear operators; this will be discussed elsewhere.

5. GLIMPSES AT SOME APPLICATIONS

We indicate briefly some applications of our results in the context of operator equations involving nondensely defined unbounded linear operators; see also the paper by S. J. Lee [13] for applications of operator parts to ordinary differential subspaces. Proof and extensions will appear elsewhere.

5.1 Graph Topology Adjoint. Let M be a closed subspace in the Hilbert space $H_1 \oplus H_2$ and let $\mathcal{P}$ be (the graph of) the orthogonal projector of H_2 onto $M(0)$. Define an inner product on $\mathcal{D}(M)$ by

$$\langle x,y\rangle_M := \langle x,y\rangle + \langle S_M(x), S_M(y)\rangle \quad ,$$

where $S_M := (I-\mathcal{P})M$. It follows easily that under this inner product $\mathcal{D}(M)$ becomes a Hilbert space. The induced topology is called the graph topology (of S_M) on $\mathcal{D}(M)$.

The graph topology adjoint plays a crucial role in the characterization and construction of least-squares solutions of $M(x) = y$ in the case when M is a nondensely defined unbounded closed linear operator from H_1 into H_2. Since M^* in this case is a multi-valued mapping, the usual normal equation characterization of least-squares solutions is meaningless. However, we can recapture the normal equation using a graph topology. Let X be the Hilbert space $\mathcal{D}(M)$, endowed with the graph topology of M. Let $M_* \subset H_2 \oplus X$ be the (graph topology) adjoint of M taken in $X \oplus H_2$. Then M_* is a single-valued bounded operator on H_2 into X; thus u is a least-squares solution of $M(x) = y$ if and only if $u \in \mathcal{D}(M)$ and $M_*M(u) = M_*(y)$. This form is not useful unless M_* is explicitly described. In [15], it is shown that M_* can be explicitly related to the orthogonal operator part S_{M^*} of the *adjoint subspace* M^*; more precisely,

$$M_* = S_{M^*}(MS_{M^*}+I)^{-1} \quad .$$

5.2 Steepest Descent Method. Using the graph topology adjoint, it is possible to extend the iterative method of steepest descent to least-squares solutions of $M(x) = y$, where M is an unbounded closed nondensely defined linear operator in H_1. The algorithm takes the form $x_{n+1} = x_n - \alpha_n r_n$, where

$$r_n := S_{M^*}(MS_{M^*}+I)^{-1}(Mx_n - y) \quad ,$$

and

$$\alpha_n := \frac{||r_n||^2+||M(r_n)||^2}{||M(r_n)||^2} \quad .$$

Convergence results are given in [15].

5.3 Regularization of Ill-Posed Problems. A standard method for approximating solutions of an ill-posed problem of the form $Af = g$ involves Tikhonov's regularization. A regularized approximation is a minimizer of the functional

$$J_\alpha(f) := ||Af-g||^2 + \alpha||Lf||^2 , \qquad \alpha > 0 \quad .$$

This is equivalent to solving the operator equation $(A^*A+\alpha L^*L)f = A^*f$. However, in the case when L is not densely defined, this is not true since the latter equation does not make sense (L^* being multi-valued). Additional difficulties arise when A itself is nondensely defined or multi-valued. The theory of regularization schemes in such settings has been recently investigated by S. J. Lee and the author. This includes regularization with the aid of a "smoothing" differential operator L, subject to subsidiary boundary conditions other than two-point boundary conditions (e.g., general interface problems with finitely many jumps, conditions of the Stieltjes type, etc.).

5.4 Nonlinear Problems with a Noninvertible Linear Part. The theory and notions presented in this paper have applications to nonlinear problems of the form $Fx = Mx$ or $MFx = y$ where F is nonlinear operator and M is a nondensely defined unbounded linear operator (which is not necessarily invertible).

Finally, we remark that while most physical actions are represented by operators which are (at least) densely defined in an appropriate Hilbert or Banach space, in many cases the mathematical analysis of such problems leads naturally to operators whose domain is no longer dense. For some typical examples of problems of mathematical physics which involve operators with nondense domains, see [21] and [22].

REFERENCES

[1] R. Arens, Operational properties of linear relations, Pacific J. Math. 11 (1961), 9-23.

[2] G. Biriuk and E. A. Coddington, Normal extensions of unbounded formally normal operators, J. Math. Mech. 13 (1964), 617-638.

[3] E. A. Coddington, Extension theory of formally normal and symmetric subspaces, Memoir Amer. Math. Soc., No. 134 (1973).

[4] E. A. Coddington, Self-adjoint subspace extension of non-densely defined symmetric operators, Advances in Math. 14 (1974), 309-332.

[5] E. A. Coddington and H. S. V. de Snoo, Differential subspaces associated with pairs of ordinary differential expressions, J. Differential Equations 35 (1980), 129-182.

[6] E. A. Coddington and H. S. V. de Snoo, Regular boundary value problems associated with pairs of ordinary differential expressions, Lecture Notes in Mathematics, Vol. 858, Springer-Verlag, Berlin-Heidelberg-New York, 1981.

[7] E. A. Coddington and A. Dijksma, Adjoint subspaces in Banach spaces, with applications to ordinary differential subspaces, Annal. Mat. Pura Appl. 118 (1978), 1-118.

[8] A. Dijksma, The generalized Green's functions for regular ordinary differential subspaces in $L^2[a,b] \oplus L^2[a,b]$, Differential Equations and Applications (W. Eckhaus and E. M. de Jager, eds.), North-Holland, Amsterdam, 1978.

[9] H. W. Engl and M. Z. Nashed, New extremal characterizations of generalized inverses of linear operators, J. Math. Anal. Appl. 82 (1981), 566-586.

[10] A. M. Krall, The development of general differential and general differential-boundary systems, Rocky Mountain J. Math. 5 (1975), 493-542.

[11] S. J. Lee, Coordinatized adjoint subspaces in Hilbert spaces, with application to ordinary differential operators, Proc. London Math. Soc. 3 (40) (1980), 138-160.

[12] S. J. Lee, Boundary conditions for linear manifolds, I., J. Math. Anal. Appl. 73 (1980), 366-380.

[13] S. J. Lee, Operator parts of ordinary differential subspaces with application to nonlinear problems, this volume.

[14] S. J. Lee and M. Z. Nashed, Generalized inverses for linear manifolds and applications to boundary value problems in Banach spaces, C. R. Math. Rep. Acad. Sci. Canada 4 (1982), 347-352.

[15] S. J. Lee and M. Z. Nashed, Gradient method for nondensely defined closed unbounded linear operators, Proc. Amer. Math. Soc. (in press).

[16] S. J. Lee and M. Z. Nashed, Least-squares solutions for multi-valued linear operators, J. Approx. Theory (in press).

[17] S. J. Lee and M. Z. Nashed, Operator parts and generalized inverses of multi-valued operators, with application to ordinary differential subspaces (to appear).

[18] M. Z. Nashed, ed., Generalized Inverses and Applications, Academic Press, New York, 1976.

[19] M. Z. Nashed and G. F. Votruba, A unified approach to generalized inverses of linear operators: I. algebraic, topological, and projectional properties, Bull. Amer. Math. Soc. 80 (1974), 825-830.

[20] M. Z. Nashed and G. F. Votruba, A unified operator theory of generalized inverses, in Generalized Inverses and Applications (M. Z. Nashed, ed.), pp. 1-109, Academic Press, New York, 1976.

[21] P. Pfeifer, Adjoints of nondensely defined Hilbert space operators, J. Math. Phys. 22 (1981), 1619-1622.

[22] B. Simon, A canonical decomposition for quadratic forms with applications to monotone convergence theorems, J. Functional Analysis 28 (1978), 377-385.

[23] A. Sobczyk, Projections in Minkowski and Banach spaces, Duke Math. J. 8 (1941), 78-106.

[24] J. von Neumann, Über adjungierte Funktionaloperatoren, Ann. Math. 33 (1932), 294-310.

[25] J. von Neumann, Functional Operators, Vol. II: The Geometry of Orthogonal Spaces, Annals of Math. Studies, No. 22, Princeton University Press, Princeton, N.J., 1950.

[26] J. von Neumann, Allgemeine Eigenwerttheorie Hermitescher Junktional-operatoren, Math. Ann. 102 (1930), 49-131.

SOME GLOBAL STEEPEST DESCENT RESULTS FOR NONLINEAR SYSTEMS

John W. Neuberger

Department of Mathematics
North Texas State University
Denton, Texas

ABSTRACT

This paper gives some global steepest descent theorems which are applicable to the solution of a variety of boundary value problems for ordinary and partial differential equations.

Steepest descent methods have been used for many years to solve linear and nonlinear systems. According to [5], the idea goes back to Cauchy [4] in 1824. One may see [6], [9], [11] for some history as well as many additional references on the subject.

In this note there are given several abstract steepest descent results which relate to the study of boundary value problems for ordinary and partial differential equations. Our theorems do not require convexity nor are they local in nature. They grew out of attempts to 'explain' observed convergence of a family of numerical schemes. We refer to [7], [8], [9] for examples of concrete settings to which our present results may apply.

Many boundary value problems for nonlinear systems of ordinary or partial differential equations may be simply represented as the problem of finding x so that

$$F(x) = 0$$

where F is a function from a real Hilbert space H to a real Hilbert space K. For many such problems, F may be taken to be $C^{(2)}$ provided H is a Sobolev space of sufficiently high order (cf [1] for a general reference on Sobolev spaces). Frequently such systems are derived from a variational principle, i.e., there is a function α on an appropriate Sobolev space such that the condition

$$(\nabla\alpha)(x) = 0$$

is equivalent to the condition that $F(x) = 0$. Such variational principles are highly prized both numerically and theoretically but unfortunately they are not always available for systems of scientific interest. In this note we use a simple alternative to such variational principles:

$$\phi(x) \equiv \|F(x)\|^2/2, \quad x \in H.$$

We attempt to use steepest descent to find zeros of ϕ.

Recalling that F is required to be $C^{(2)}$, we have that the Fréchet derivative ϕ' of ϕ is given by

$$\phi'(x)h = \langle F'(x)h, F(x)\rangle = \langle h, F'(x)^*F(x)\rangle, \quad x,h \in H$$

and so

$$(\nabla\phi)(x) = F'(x)^*F(x), \quad x \in H.$$

Theorem 1. If $x \in H$, there is a unique function z: $[0,\infty) \to H$ such that

$$(*) \qquad z(0) = x, \; z'(t) = -(\nabla\phi)(z(t)), \quad t \geq 0.$$

Although the global existence and uniqueness is well-known we mention that a proof is given in [8].

In light of Theorem 1, there is a semigroup S associated with (*), i.e., there is a function S from $[0,\infty)$ with range in the space of transformations from H to H so that if $x \in H$ and $r \geq 0$, then

$$S(r)x = z(r)$$

where z satisfies (*) in Theorem 1. S has the usual one-parameter semigroup properties:

(a) $S(0) = I$, the identity on H, and

(b) $S(t)S(r) = S(t+r), \quad t,r \geq 0$.

We remark that S typically is <u>not</u> non-expansive (cf [3]) since ϕ is typically not convex for nonlinear problems. Hence results on asymptotic limits for non-expansive semigroups (cf [2], [10]) do not seem to apply directly to the present setting. However, it may be that semigroup theory will be extended to cover such cases.

For comparison purposes there is the following from [9] concerning linear problems:

Theorem 2. Suppose that T is a continuous linear transformation from H to K, $g \in K$ and

$$\phi(x) \equiv \|Tx - g\|^2/2, \quad x \in K.$$

Suppose also that there is $w \in H$ such that $Tw = g$. If $x \in H$ and z satisfies (*) of Theorem 1, then

$$u \equiv \lim_{t\to\infty} z(t)$$

exists and $Tu = g$.

Theorem 3. Suppose ϕ is a $C^{(2)}$ function from H to $[0,\infty)$ and there is a unique $u \in H$ such that $\phi(u) = 0$. In addition suppose that

(i) If $r > 0$, there is $c > 0$ such that if $\|x - u\| \leq r$, then $\|(\nabla\phi)(x)\| \geq c\phi(x)$, and

(ii) If $\varepsilon > 0$, there is $\delta > 0$ such that if $\phi(x) \leq \delta$, then $\|x - u\| \leq \varepsilon$.

If $z:[0,\infty) \to H$, $z'(t) = -(\nabla\phi)(z(t))$, $t \geq 0$ and the range of z is bounded, then

$$u = \lim_{t\to\infty} z(t).$$

Proof of Theorem 3. Suppose that $z'(t) = -(\nabla\phi)(z(t))$, $t \geq 0$ and the range of z is bounded. Denote by r a positive number such that $\|z(t)\| \leq r$, $t \geq 0$, and denote by c a positive number so that

$$\|(\nabla\phi)(x)\| \geq c\phi(x), \quad \|x\| \leq r.$$

Then, $\phi(z)'(t) = \phi'(z(t))z'(t) = -\langle(\nabla\phi)(z(t)), z'(t)\rangle$
$= -\|(\nabla\phi)(z(t))\|^2 \leq -c^2\phi(z(t))^2, \quad t \geq 0.$

Suppose there is $t_o \geq 0$ so that $\phi(z(t_o)) = 0$. Then $z(t_o) = u$ and $\phi(z)'(t_o) = 0$ since otherwise $\phi(z)(t) < 0$ for some $t > t_o$. Hence $\|z'(t_o)\| = \|(\nabla\phi)(z(t_o))\| = 0$ and by uniqueness, $z(t) = u$, $t > t_o$. So in this case $\lim_{t\to\infty} z(t) = u$.

Suppose that $\phi(z(t)) > 0$ for all $t \geq 0$. Then

$$\frac{\phi(z)'(t)}{\phi(z(t))^2} \leq -c^2 \quad \text{and hence}$$

$$-\frac{1}{\phi(z(t))} + \frac{1}{\phi(z(0))} \leq -c^2 t, \quad \text{i.e.,}$$

$$\phi(z(t)) \leq \phi(z(0))/(1+c^2\phi(z(0))t), \quad t \geq 0.$$

Hence $\lim_{t\to\infty} \phi(z(t)) = 0$ and so, using (ii), $\lim_{t\to\infty} z(t) = u$.

Theorem 4. Suppose F is a $C^{(2)}$ function from H to K and $\phi(x) \equiv \|F(x)\|^2/2$, $x \in H$. Suppose also that if $r > 0$, there is $c > 0$ such that $\|(\nabla\phi)(x)\| \geq c\|F(x)\|$ if $\|x\| \leq r$. If $x \in H$, $z\colon [0,\infty) \to H$, $z(0) = x$, $z'(t) = -(\nabla\phi)(z(t))$, $t \geq 0$ and the range of z is bounded, then

$$u \equiv \lim_{t\to\infty} z(t)$$

exists and $F(u) = 0$.

Proof of Theorem 4. As in the argument for the preceding theorem, if there is $t_0 \geq 0$ such that $\phi(z(t_0)) = 0$, then $z(t) = z(t_0)$, $t \geq t_0$ and so $u \equiv \lim_{t\to\infty} z(t)$ exists and $F(u) = 0$.

Suppose that $\phi(z(t)) > 0$ for all $t \geq 0$. Then $\phi(z)'(t) = -\|(\nabla\phi)(z(t))\|^2 \leq -c^2\|F(z(t))\|^2 = -2c^2\phi(z(t))$, $t \geq 0$. Hence

$$\frac{\phi(z)'(t)}{\phi(z(t))} \leq -2c^2,$$

$$\ln(\phi(z(t))/\phi(z(0))) \leq -2c^2 t \text{ and so}$$

$$\phi(z(t)) \leq e^{-2c^2 t}, \quad t \geq 0.$$

It follows that $\lim_{t\to\infty} \phi(z(t)) = 0$.

Since $\phi(z)'(t) = -\|z'(t)\|^2$, $t \geq 0$,

$$\phi(z(t)) - \phi(z(t+1)) = \int_t^{t+1} \|z'\|^2 \quad \text{and hence}$$

$$\left(\int_t^{t+1} \|z'\|\right)^2 \leq \int_t^{t+1} \|z'\|^2 \leq \phi(z(t)) \leq \phi(z(0))e^{-2c^2 t}, \quad t \geq 0.$$

Therefore $\int_t^{t+1} \|z'\| \leq \sqrt{\phi(z(0))}\, e^{-c^2 t}$, $t \geq 0$. Hence $\int_0^\infty \|z'\|$ exists and consequently, $u \equiv \lim_{t\to\infty} z(t)$ exists. Moreover, $\phi(u) = \lim_{t\to\infty} \phi(z(t)) = 0$. Therefore $F(u) = 0$.

The following is proved in [8]:

Theorem 5. Suppose that ϕ is a $C^{(2)}$ function from H to $[0,\infty)$, $z\colon [0,\infty) \to H$, $z'(t) = -(\nabla\phi)(z(t))$, $t \geq 0$. If there is $t_0 \geq 0$ and $\delta > 0$

such that

$$\Sigma_{n=1}^{\infty} (\phi(z(t_o+n\delta)) - \phi(z(t_o+(n+1)\delta)))^{\frac{1}{2}}$$

converges, then $u \equiv \lim_{t\to\infty} z(t)$ exists and $(\nabla\phi)(u) = 0$.

Corollary. If $\lambda \in (0,1)$, $z_n \equiv \phi(z(t_o+n\delta))$, $n = 0,1,2,\ldots$ and

$$(z_n - z_{n+1})/(z_{n-1} - z_n) \leq \lambda, \quad n = 0,1,2,\ldots,$$

then $\lim_{t\to\infty} z(t)$ exists and $(\nabla\phi)(u) = 0$.

For the linear case, Theorem 2, the situation is simple in that existence of a solution implies that a solution is reached by steepest descent regardless of starting point.

In Theorem 3 a unique solution is required in addition to two other conditions, the second of which asserts essentially that if u satisfies $F(u) = 0$, then any solution to a small perturbation of the equation has a solution near u. This second requirement seems reasonable from a physical point of view and is widely satisfied for boundary value problems in differential equations. Condition (i) is technical but also seems widely satisfied in applications. The condition of boundedness of the range of z is not particularly satisfactory. It can be shown to hold if the starting value x is near enough to us or if ϕ is convex. Although it has been observed by this writer in all numerical implementation of steepest of the present development (<u>for partial and ordinary differential equations</u>) more work is indicated in order to 'explain' this phenomena.

In Theorem 4, the condition that $\|(\nabla\phi)(x)\| \geq c\|F(x)\|$ is generally stronger than the condition $\|(\nabla\phi)(x)\| \geq c\phi(x) = c\|F(x)\|^2/2$. This stronger condition permits a conclusion like the one in Theorem 3 without requirements of existence and uniqueness or a condition analagous to (ii) of Theorem 3.

The condition in the corollary to Theorem 5 is frequently observed in calculations for λ substantially less than one. It seems to be a rather reliable practical check on convergence of steepest descent. Observe that it holds under the hypothesis of Theorem 4.

REFERENCES

[1] R.A. Adams, Sobolev Spaces, Academic Press (1975).

[2] J.B. Baillon, R.E. Bruck and S. Reich, On the Asymptotic Behavior of Nonexpansive Mappings and Semigroups in Banach Spaces, Houston J. Math. 4 (1978), 1-9.

[3] H. Brezis, Opérateurs Maximaux Monotones et Semigroups de Contractions dans les Espaces de Hilbert, North-Holland Publishing Co., Amsterdam (1973).

[4] A.L. Cauchy, Méthode générale pour las Resolution des Systemes d'Équations Simultanées, Comptes rendus, Ac. Sci. Paris 25 (1847), 536-538.

[5] H.B. Curry, The Method of Steepest Descent for Nonlinear Minimization Problems, Quarterly App. Math. 2 (1944), 258-261.

[6] M. Hestenes, Conjugate Direction Methods in Optimization, Springer-Verlag (1980).

[7] J.W. Neuberger, A. Type-independent Method for Systems of Nonlinear Partial Differential Equations: Application to the Problem of Transonic Flow, Computers and Math. Appl. 6 (1980), 67-78.

[8] ______, Steepest Descent for Systems of Nonlinear Partial Differential Equations, Oak Ridge National Laboratory CSD TM-161 (1981).

[9] ______, Steepest Descent for General Systems of Linear Differential Equations in Hilbert Space, Springer Lecture Notes for 1982 Dundee Symposium (to appear).

[10] A. Pazy, The Asymptotic Behaviour of Semigroups of Nonlinear Contractions Having Large Sets of Fixed Points, Proc. Royal Soc. Edinburgh 80A (1978), 261-271.

[11] J.M. Ortega and W.C. Rheinboldt, Iterative Solution of Nonlinear Equations in Several Variables, Academic Press (1970).

NONLINEAR PROBLEMS AT RESONANCE VIA LYAPUNOV-LIKE FUNCTIONS

J. J. Nieto

Facultad de Matematicas
Universidad de Santiago
Spain

1. INTRODUCTION

One of the well known techniques in the theory of nonlinear boundary value problems is the method of differential inequalities. The method of "Alternative Problems," a global variant of the Lyapunov-Schmidt method, has been used in the study of problems at resonance. The investigation of periodic boundary value problems, PBVP , forms an important subclass in the study of problems at resonance. Recently [7,8,10] an attempt is made successfully to combine these techniques to investigate the existence of periodic solutions of differential equations. When we wish to extend such results for systems of differential equations there are several possibilities to follow. One is to utilize the concept of Lyapunov-like function and the theory of differential inequalities. In this paper we shall discuss the existence of solutions of PBVP's for second order differential systems.

2. COMPARISON PRINCIPLES

Given $g \in C[[0,2\pi]\times\mathbb{R}\times\mathbb{R},\mathbb{R}]$ and $v \in C[[0,2\pi],\mathbb{R}]$, we say that the function $\tilde{g}(t,u,u')$ is a modified function relative to v, if

$$\tilde{g}(t,u,u') = g(t,p(t,u),u') + \frac{u-p(t,u)}{1+u^2} \tag{2.1}$$

where $$p(t,u) = \max\{v(t),u\}.$$

Let us consider the periodic boundary value problem

$$u'' = g(t,u,u') \quad u(0) = u(2\pi),\ u'(0) = u'(2\pi). \tag{2.2}$$

We then have the following comparison result for this PBVP.

Theorem 2.1. Assume that

(a) $m \in C^2[[0,2\pi],\mathbb{R}]$ and $m''(t) \geq g(t,m(t),m'(t))$ for all $t \in [0,2\pi]$ where $g \in C[[0,2\pi]\times\mathbb{R}\times\mathbb{R},\mathbb{R}]$.

(b) $m(0) = m(2\pi)$ and $m'(0) \geq m'(2\pi)$.

(c) The PBVP (2.2) has a maximal solution $r(t)$.

(d) For every lower solution v, the modified problem

$$u'' = \tilde{g}(t,u,u') \qquad u(0) = u(2\pi),\ u'(0) = u'(2\pi) \tag{2.3}$$

where $\tilde{g}(t,u,u')$ is defined by (2.1), has a solution $u(t)$. Then $m(t) \leq r(t)$ on $[0,2\pi]$

Proof: Let us consider the modified PBVP (2.3) relative to $m(t)$ and let $u(t)$ be a solution which exists by (d). If $m(t) \leq u(t)$ on $[0,2\pi]$ is not true, there exists a $t_0 \in [0,2\pi]$ and $\varepsilon > 0$ such that $m(t_0) = u(t_0) + \varepsilon$ and $m(t) \leq u(t) + \varepsilon$ for all $t \in [0,2\pi]$. If $t_0 \in (0,2\pi)$, we get $m'(t_0) = u'(t_0)$ and $p(t_0,u(t_0)) = m(t_0)$ and we have

$$m''(t_0) - u''(t_0) = \frac{m(t_0)-u(t_0)}{1 + u^2(t_0)} > 0$$

which is a contradiction. If $t_0 = 0$, we have $m'(0) \leq u'(0)$ and from (b) it follows that $m'(0) = u'(0) = 0$. Hence we obtain as before $m''(0) - u''(0) > 0$ which again is a contradiction. A similar contradiction holds at $t_0 = 2\pi$. Thus we have established that $m(t) \leq u(t)$ on $[0,2\pi]$ and we then have $m(t) \leq r(t)$ on $[0,2\pi]$.

Using similar reasonings one can prove the following comparison result

Theorem 2.2. Let $g \in C[[0,2\pi]\times\mathbb{R}\times\mathbb{R},\mathbb{R}]$ and suppose that (a) and (b) of theorem 2.1 hold, and there exists $\beta \in C^2[[0,2\pi],\mathbb{R}]$ such that $\beta''(t) \leq g(t,\beta(t),\beta'(t))$ for all $t \in [0,2\pi]$ and $\beta(0) = \beta(2\pi)$, $\beta'(0) \leq \beta'(2\pi)$. If, in addition, we have that $g(t,u,u')$ is increasing in u for each $(t,u') \in [0,2\pi]\times\mathbb{R}$, then $m(t) \leq \beta(t)$ on $[0,2\pi]$

3. THE MODIFIED PROBLEM

Consider the boundary value problem

$$u'' = f(t,u,u') \qquad u(0) = u(2\pi),\ u'(0) = u'(2\pi) \tag{3.1}$$

where $f \in C[[0,2\pi]\times\mathbb{R}^n\times\mathbb{R}^n,\mathbb{R}^n]$.

Let $\delta \in C[\mathbb{R}^+\times\mathbb{R}^+,\mathbb{R}^+]$ be such that $\delta(t,s) \leq 1$ for all $(t,s) \in \mathbb{R}^+\times\mathbb{R}^+$

and

$$\delta(t,s) = \begin{cases} 1 & \text{if} \quad t \leq M_1 \quad \text{and} \quad s \leq M_2 \\ 0 & \text{if} \quad t \geq M_1 + 1 \quad \text{or} \quad s \geq M_2 + 1 \end{cases}$$

where M_1 and M_2 are positive constants. We define the modified function F by

$$F(t,u,u') = \delta(\|u\|,\|u'\|)f(t,u,u') \tag{3.2}$$

and consider the modified problem

$$u'' = F(t,u,u') \qquad u(0) = u(2\pi),\ u'(0) = u'(2\pi). \tag{3.3}$$

Clearly F is continuous and bounded on $[0,2\pi]\times\mathbb{R}^n\times\mathbb{R}^n$. We let $E = (L_2[0,2\pi])^n$, $Lu = u''$ and N be the Nemytskii operator generator by F. The linear operator L has domain $D(L) = \{u \in E\colon u,u'$ are absolutely continuous, $u'' \in E$ and $u(0) = u(2\pi)$, $u'(0) = u'(2\pi)\}$. Then, the PBVP (3.3) is equivalent to the operator equation

$$Lu = Nu. \tag{3.4}$$

Note that $E_0 = \ker L$ consists of constant functions and hence E_1, where $E = E_0 \oplus E_1$, is the class of all functions whose average is zero. It is easily seen that the partial inverse operator H also exists and it is compact. Because the function F is bounded, any solution u_1 of the auxiliary equation satisfies $\|u_1(t)\| \leq A$ for all $t \in [0,2\pi]$, where A is a constant independent of u_1.

Then, $\left\langle N(u_0+u_1),u_0\right\rangle = \sum_{i=1}^{n} u_0^i \int_0^{2\pi} F_i(t,u_0+u_1(t),u_1'(t))dt =$

$$\sum_{i=1}^{n} u_0^i \int_0^{2\pi} \delta(\|u_0+u_1(t)\|,\|u_1'(t)\|) \cdot f_i(t,u_0+u_1(t),u_1'(t))dt.$$

Choosing $\|u_0\| = R_0$ with R_0 large, $\|u_0+u_1(t)\| \geq M_1 + 1$ for all $t \in [0,2\pi]$ and then $\delta(\|u_0+u_1(t)\|,\|u_1'(t)\|) = 0$, and it turns out that $\left\langle N(u_0+u_1),u_0\right\rangle = 0$ for $\|u_0\| = R_0$, $u_0 \in E_0$ and u_1 solution of the auxiliary equation. Thus, by the result in [4], the operator equation (3.4) has at least one solution (see Th. 2.1 in [11]).

4. EXISTENCE

Now, we shall prove the existence of solution of the PBVP (3.1).

Relative to the PBVP (3.1), let us list the following assumptions for convenience:

(i) There exists $V \in C^2[[0,2\pi]\times\mathbb{R}^n,\mathbb{R}^+]$ and $g \in C[[0,2\pi]\times\mathbb{R}^+\times\mathbb{R}^+,\mathbb{R}]$ such that for $0 \leq \lambda \leq 1$

$$V''_{\lambda f}(t,u) \geq g(t,v(t,u, v'(t,u))$$

where $V''_{\lambda f}(t,u) = U(t,u,u') + \lambda V_u(t,u)f(t,u,u')$

and $U(t,u,u') = V_{tt}(t,u) + 2V_{tu}(t,u)\cdot u' + V_{uu}(t,u)\cdot u'\cdot u'$

(ii) $V(t,u) \longrightarrow \infty$ as $\|u\| \longrightarrow \infty$ uniformly in $t \in [0,2\pi]$

(iii) $V(0,x) = V(2\pi,x)$, $V_t(0,x) \geq V_t(2\pi,x)$ and $V_u(0,x) = V_u(2\pi,x)$ for all $x \in \mathbb{R}^n$.

(iv) There exists a positive nondecreasing continuous function h such that $\lim_{s\to\infty} \frac{s^2}{h(s)} = \infty$ and

$$\|f(t,u,u')\| \leq h(\|u'\|) \quad \text{for all} \quad (t,u) \in [0,2\pi]\times\mathbb{R}^n.$$

(v) The PBVP $u'' = g(t,u,u')$, $u(0) = u(2\pi)$, $u'(0) = u'(2\pi)$ has a maximal solution $r(t)$, and for every lower solution v, the modified PBVP $u'' = \tilde{g}(t,u,u')$, where $\tilde{g}$ is defined by (2.1), has a solution.

We are now in a position to prove the following result:

Theorem 4.1. Assume that (i) - (v) hold. Then there exists at least one solution of the PBVP (3.1).

Proof: Consider the modified problem (3.3) of (3.1) and let $u(t)$ be a solution. Defining $m(t) = V(t,u(t))$ and using the assumption (i) we get

$$m''(t) \geq g(t,m(t),m'(t))$$

and by (iii) we have

$$m(0) = m(2\pi) \quad \text{and} \quad m'(0) \geq m'(2\pi)$$

Hence, by comparison Theorem 2.1, it follows that

$$m(t) \leq r(t) \quad \text{on} \quad [0,2\pi].$$

This implies that there exists K constant such that $m(t) \leq K$ for all $t \in [0,2\pi]$. Now, the assumption (ii) implies that there exists an $M_1 > 0$ such that $\|u(t)\| \leq M_1$ on $[0,2\pi]$ and, by (iv), $\|u'(t)\| \leq M_2$ on $[0,2\pi]$ for some positive constant M_2 (see [11] and [12]). These are the constant M_1 and M_2 that are used in the definition of δ. In view of the definition of F, it follows that $u(t)$ is actually a solution of the PBVP (3.1) and the proof is complete.

It is possible to prove the same result using the comparison Theorem 2.2.

Theorem 4.2. Suppose that the conditions (i) - (iv) hold and there exists $\beta \in C^2[[0,2\pi],\mathbb{R}]$ such that $\beta''(t) \leq g(t,\beta(t),\beta'(t))$ for all $t \in [0,2\pi]$ and $\beta(0) = \beta(2\pi)$, $\beta'(0) \leq \beta'(2\pi)$ and g is increasing in u for each $(t,u') \in [0,2\pi]\times\mathbb{R}^n$. Then there exists a solution to the PBVP (3.1).

Proof: As in Theorem 4.1, using the comparison Theorem 2.2, one obtains $m(t) \leq \beta(t)$ $t \in [0,2\pi]$ and from this the result follows.

Remark: There are other types of boundary conditions for which it is possible to apply this method. For instance, the second order differential system

$$u'' = f(t,u,u')$$

with Neumann conditions

$$u'(0) = u'(2\pi) = 0$$

(see [11] for further details) or the first order differential system $u' = f(t,u)$ with periodic condition $u(0) = u(2\pi)$ (see [10,11]).

REFERENCES

[1] S. Bernfeld, G. Ladde and V. Lakshmikantham, "Existence of solutions of two point boundary value problems for nonlinear systems," J. Diff. Eqns., 18 (1975), pp. 103-110.

[2] S. Bernfeld and V. Lakshmikantham, An Introduction to Nonlinear Boundary Value Problems, Academic Press, New York, 1974.

[3] L. Cesari, "Functional analysis, nonlinear differential equations, and the alternative method," Proc. of Conf. on Nonlinear Functional Analysis and Differential Equations (Eds. Cesari, Kannan, and Schuur), Marcel Dekker, Inc., New York, 1976, 1-197.

[4] L. Cesari and R. Kannan, "An abstract theorem at resonance," Proc. Amer. Math. Soc., 63 (1977), pp. 221-225.

[5] J. Chandra, V. Lakshmikantham and S. Leela, "Comparison principle and theory of Nonlinear Boundary Value Problems," The University of Texas at Arlington, Technical Report #137 (1980).

[6] P. Hartman, "On two point boundary value problems for nonlinear second order systems," SIAM J. Math. Analysis, 5 (1974), pp. 172-177.

[7] R. Kannan and V. Lakshmikantham, "Existence of Periodic Solutions of Nonlinear Boundary Value Problems and the Method of Upper and Lower Solutions," The University of Texas at Arlington Technical Report #173 (1981).

[8] R. Kannan and V. Lakshmikantham, "Periodic solution of nonlinear boundary value problems," Proc. Inter. Conf. on Recent Methods in Nonlinear Analysis and Applications, Naples, 1980, Liquori Editore, Naples (1981), pp. 99-127.

[9] V. Lakshmikantham and S. Leela, Differential and Integral Inequalities, Vol. I, Academic Press, New York, 1969.

[10] V. Lakshmikantham and S. Leela, "Existence and monotone method for periodic solutions of first order differential equations," to appear in Jour. Math. Anal. Appl.

[11] J. J. Nieto, "Problems at resonance for first and second order differential equations via Lyapunov-like functions," to appear in J. Nonlinear Analysis, TMA.

[12] K. Schmitt and R. Thompson, "Boundary value problems for infinite systems of second order differential equations," J. Diff. Eqns., 8 (1975), pp. 277-295.

SOME PROBLEMS CONCERNING NONLINEAR OSCILLATORS IN NEURAL NETWORKS

M. Namik Oğuztöreli

University of Alberta
Edmonton, Alberta, Canada

ABSTRACT

The objective of the present work is to formulate certain mathematical problems arising in the study of a class of rhythmic oscillations in neural networks.

I. INTRODUCTION

The structure and the functional organization of the brain have been investigated extensively in the literature, and a very advanced technology has been developed to study the neural activities in the living nervous systems and created artificial intelligent beings. There is an ever growing momentum in the experimental and theoretical approaches to understand how the brain really works in the local and global levels, the relationships that connect the high brain functions, such as learning, thinking, and memory, with what goes on at the most basic levels of neural activities in the local neuronal circuits. For the synaptic organization of the brain and for the local circuitries we refer to the books [35],[36].

Modelling, analysis and simulation of neural networks have attracted the attention of a number of researchers (cf. [1]-[3],[5]-[14],[18],[30]-[33],[37],[40],[41]). In a series of papers ([15],[16],[22],[26],[38],[39]) a discrete neural model has been investigated from a mathematical, computational and physiological point of view. In these articles we have investigated the existence and uniqueness of the solutions of the model equations and stability of the solutions around the steady-state solutions, the effects of the time-lags on the oscillations which are intrinsic to the neural model, dependence on the structural parameters such as external inputs, interaction coefficients. The numerical solution of the neural model, the computation of the steady-state solutions and the natural modes of the oscillations around the steady-state solutions are described. Employing the results of these works we have studied the neural activities in vertebrate retina ([27]-[29]). Further, in [23],[24],[25] the discrete model has been extended to one-and-two dimensional continuous networks.

The discrete neural model containing only finitely many neurons is described by a system of nonlinear ordinary integro-differential difference equations of the form:

$$a_{i0}^{-1}\frac{dx_i(t)}{dt} + x_i(t) = \mathcal{S}\left\{ f_i + \sum_{j=1}^{n} c_{ij}x_j(t-\tau_{ij}) + \sum_{k=1}^{m} b_{ik}\int_0^t x_i(\tau)e^{-a_{ik}(t-\tau)}d\tau \right\} \qquad (1)$$

for $t > 0$, where

t denotes time;

n is the number of the neurons in the network;

$m+1$ is the maximal order of the neurons;

$x_i = x_i(t)$ is the normalized firing rate of the i-th neuron at time t;

$f_i = f_i(t)$ denotes the external input to the i-th neuron at the time t;

a_{i0} is the rate constant characterized by the fact that a step change in input to the i-th neuron produces an exponential approach from the initial value $x_i(0)$ to a steady-state firing rate x_i^* with the rate constant a_{i0};

b_{ik} is an adaptation or self-inhibition factor of the i-th neuron in the case $b_{ik} < 0$ and is self-excitation factor in the case $b_{ik} > 0$ with the rate constant $a_{ik} (> 0)$;

c_{ij} is the interaction coefficient denoting the influence of the j-th neuron on the i-th neuron representing an inhibition in the case $c_{ij} < 0$ and an excitation in the case $c_{ij} > 0$; if the j-th neuron is not connected directly to i-th neuron we have $c_{ij} = 0$; further, we always assume that $c_{ii} = 0$ for $i = 1,2,\ldots,n$ since the self-inhibition and self-excitation in the i-th neuron are characterized by the parameters b_{ik} and a_{ik};

σ_{ij} (≥ 0) is the time-lag occurring in the transfer of the activity of the j-th neuron to the i-th neuron; we assume that $\sigma_{ii} = 0$ for $i = 1,2,\ldots,n$; and finally

$$\mathcal{S}\{u\} = \frac{1}{1+e^{-u}} . \qquad (2)$$

For the physiological interpretation of the system (1) and of the above parameters we refer to [34]. An enzymatic interpretation of this system is given in [14]. For the existence and uniqueness of periodic solutions we refer to [4],[14],[38].

The objective of the present work is to formulate certain mathematical problems arising in the study of a class of rhythmic oscillations in neural networks described by a system of integro-differential difference equations of the form (1). The physiological background of such oscillations can be found in [11],[36].

II. CHARACTERIZATION OF THE NEURONS AND NEURAL NETWORKS

Consider a neural network consisting of n neurons. Let Ⓘ be the i-th neuron in the network, $i = 1,\ldots,n$, and let $a_{i0}, a_{i1},\ldots,a_{im}, b_{i1},\ldots, b_{im}$ be the parameters appearing in Eqs. (1). To simplify our presentation we assume that m=1 or m=2, and $a_{i0}=100, a_{i1}=10$, and $a_{i2}=30$, which are physiologically feasible, [39]. The structure and processing of the neuron Ⓘ is then completely characterized by the parameters b_{i1} and b_{i2} ($b_{i2}=0$ in the case m=1). When neuron Ⓘ is isolated, then neural activites of Ⓘ under the influence of an external force f_i is determined by the equations

$$\begin{cases} 0.01\dot{x}_i + x_i = \mathcal{S}\{f_i(t) + \sum_{k=1}^{2} b_{ik} \int_0^t e^{-a_{ik}(t-\tau)} x_i(\tau)d\tau\} \\ x_i(0) = x_i^0 \end{cases} \quad (3)$$

In our simulations which will be presented in the next sections, we have indicated the cases $b_{i1} = -b_i, b_{i2} = 0$, and $b_{i1} = -b_i, b_{i2} = b_i, b_i > 0$, by the expressions

$$\text{(i)} \sim 1N \therefore \{b_i\} : f \quad , \quad \text{(i)} \sim 1N \therefore \{b_i, b_i\} : f \quad (4)$$

respectively. In these cases Eq.(2) is of the form

$$0.01\dot{x}_i + x_i = \begin{cases} \mathcal{S}\{ f_i(t) - b_i \int_0^t e^{-10(t-\tau)} x_i(\tau)d\tau \} & \text{, if } m=1 \\ \mathcal{S}\{ f_i(t) - b_i \int_0^t [e^{-10)t-\tau)} - e^{-30(t-\tau)}] x_i(\tau)d\tau\} & \text{, if } m=2 \end{cases} \quad (5)$$

with $x_i(0) = x_i^0$. In general we assume that

$$x_i(t) = 0 \quad \text{for} \quad t \leq 0. \quad (6)$$

Now, let (1) and (2) be two interconnected neurons of the above form. If these neurons are structurally identical, i.e., $a_{10} = a_{20} = 100$, $a_{i1} = a_{21} = 10$, $a_{12} = a_{22} = 30$, $b_{11} = b_{21} = -b$, $b_{12} = b_{22} = 0$ if m=1, $b_{21} = b_{22} = b$ if m=2, and they inhibit each other with equal strength, $c_{12} = c_{21} = -i$, and with equal time-lags, $\sigma_{12} = \sigma_{21} = \sigma$, and if they are subjected to external forces with equal strength, $f_1 = f_2 = f$, then we have

$$\begin{cases} 0.01x_1 + x_1 = \mathcal{S}\{f - ix_2(t-\sigma) - b\int_0^t e^{-10(t-\tau)}x_1(\tau)d\tau\} \\ 0.01x_2 + x_2 = \mathcal{S}\{f - ix_1(t-\sigma) - b\int_0^t e^{-10(t-\tau)}x_2(\tau)d\tau\} \end{cases} \tag{7}$$

in the case m=1, and

$$\begin{cases} 0.01x_1 + x_1 = \mathcal{S}\{f - ix_2(t-\sigma) - b\int_0^t [e^{-10(t-\tau)}-e^{-30(t-\tau)}]x_1(\tau)d\tau\} \\ 0.01x_2 + x_2 = \mathcal{S}\{f - ix_1(t-\sigma) - b\int_0^t [e^{-10(t-\tau)}-e^{-30(t-\tau)}]x_2(\tau)d\tau\} \end{cases} \tag{8}$$

in the case m=2. These equations will symbolly be indicated by

$$\begin{cases} 2N \therefore \{b\} : b\ ,\ f\ ,\ i\ ,\ \sigma & (\text{ if } m=1\ ) \\ 2N \therefore \{b,b\} : b\ ,\ f\ ,\ i\quad \sigma & (\text{ if } m=2\ ) \end{cases} \tag{9}$$

respectively. Although it is not difficult to describe the networks with three or more mutually inhibiting identical neurons, we will consider here mutually inhibiting networks with only two identical cells as shown above.

III. RHYTHMIC OSCILLATIONS IN NEURAL ACTIVITIES

It is well known that autonomous rhythms such as heartbeat and breathing, voluntary rhythms such as walking and swimming are driven by central nervous oscillators. It is also well known that sensory inputs play a role in the generation of rhythmic movements, and, when a central oscillator is subjected to the influence of sensory input, the amplitude and frequency of the basic rhythm generated by the central oscillator is modulated. Neurons of the central nervous system which can produce rhythmic activities fall into two general classes: endogenous polarization rhythms that depend on special oscillatory properties of individual neurons , and network oscillations that depend on oscillatory properties that arise from the interconnections linking a set of neurons with endogeneously stable polarization levels (cf.[11]). Note that cells of type (b_i,b_i) may produce endogenous polarization rhythms under constant inputs (cf.[4],[14],[26],[38],[39]), and, neurons of type (b_i) cannot produce endogenous rhythmic oscillations (cf.[26]). However, both types of neurons can produce network type rhythmic oscillations [26],[38],[39]. Network type rhythmic oscillations are observed particularly in mutual inhibition networks. The activities of the neurons are usually entrained by the action of the external forces. In our earlier studies we employed external forces of the rectangular, triangular, trapezoidal, piecewise constant, and trigonometrical form, as accustomed in the experimental investigations. For the effects of periodic inputs on the behavior of the neuronal activities we refer to [33],[34]. In our simulations presented here we employed rectangular inputs of the form

$$f_i(t) = \begin{cases} 0 & \text{for } 0 \le t < 0.2 \quad , \quad 1.2 < t \le 1.5 \\ f & \text{for } 0.2 \le t \le 1.2 \qquad (f\text{: const.}) \end{cases} \tag{9}$$

and trigonometric inputs of the form

$$f_i(t) = \frac{\rho}{2}\,[1 + \sin(2\pi\Omega t)] \tag{10}$$

IV. SIMULATIONS

We made a large number of simulations of the neural model (1), particularly Eqs.(5),(7), and (8), by using the AMDAHL 470 V/8 computing system of the University of Alberta. We present here a few cases to clarify the essences of the open problems formulated in Section V.

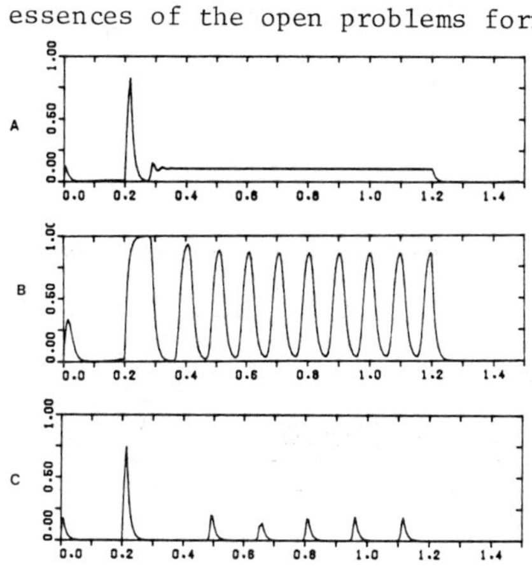

Figure 1

As we have remarked in Section II, neurons of type $1N \therefore \{b\}: f$ cannot produce rhythmic activities under constant inputs f, while neurons of type $2N \therefore \{b,b\}: f$, can produce rhythmic oscillations if b is sufficiently large. In Fig.1 (A)-(C), the responses to rectangular inputs of the form (9) of the neurons $1N \therefore \{5000\}$: f=50, $1N \therefore \{2000,2000\}$: f=50, and $1N \therefore \{50000,50000\}$: f=50 are exhibited respectively. In Fig.2(A)-(C) the entrainment and modulations of the activities of the cell in Fig.1(B) under the trigonometric inputs of the form (10) with ρ=50, Ω= 1,2,20 are shown. Modulations are more apparent for smaller frequencies. Fig.3-Fig.6 are devoted to the performances of the mutually inhibiting cells. In Fig.3(A)-(D) are presented the activity histograms of the cells in the networks

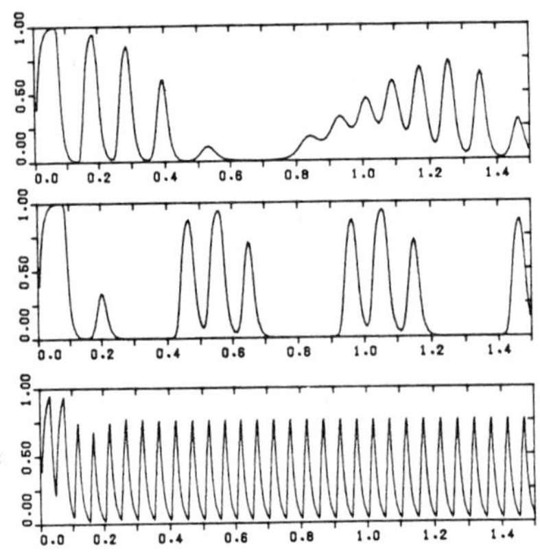

Figure 2

$2N$ ∴ {b}: b=750; f=0,10,40; i=25, σ=0 respectively, for rectangular inputs of the form (9). It is apparent that higher inputs produce higher frequencies and steady-state positions. In Fig.4(A) are exhibited the activities in the network $2N$ ∴ {b}: b=10; f=0; i=100; σ=0. It is interesting to note the generation of rhythmic oscillations in the absence of external forces. In Fig.4(B)-(D) the cases $2N$ ∴ {b}: b=1000,1500,2000; f=50 (Rectangular); i=25; σ=0 are shown. The competition between dominating modes are particularly evident in Fig.4(D). In Fig.5(A)-(D) are exhibited the histograms of mutually inhibiting networks $2N$ ∴ {b}: b=500; f=10,25, 30,50,100 (Rectangular); i=25; σ=0. The competition of the dominating modes in Fig.5(C) and (D) and the frequency variation are to be noted. Finally, the effects of the time-lags are shown in Fig. 6 for the mutually inhibiting network $2N$ ∴ {b}:b=500; f=50(Rectangular); i=25; σ=5,10,15,20,250 msec. Let us note that the frequencies decrease for increasing σ up to some limit thereafter the oscillations become more sluggish as seen in Fig. 6(E).

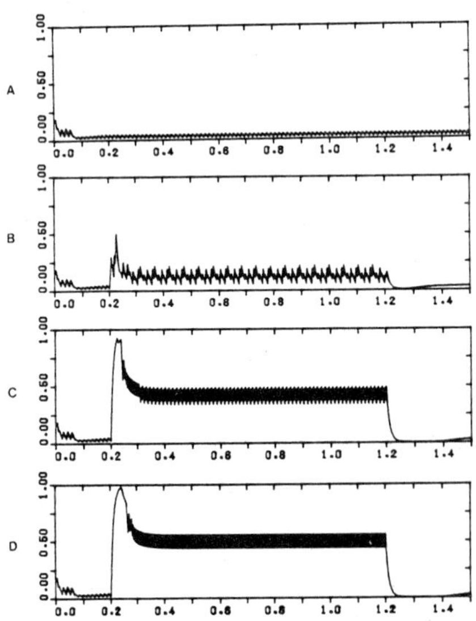

Figure 3

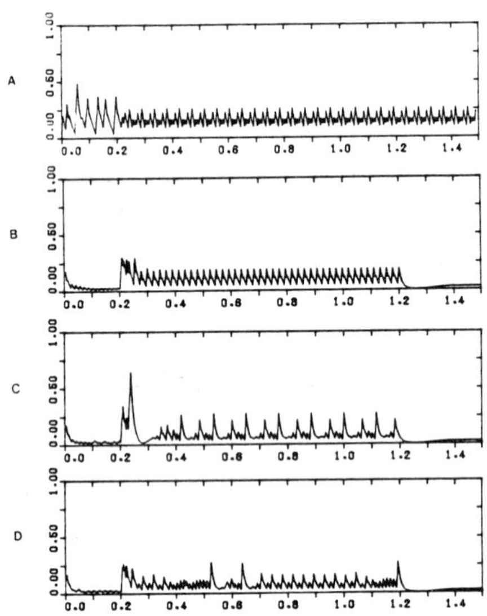

Figure 4

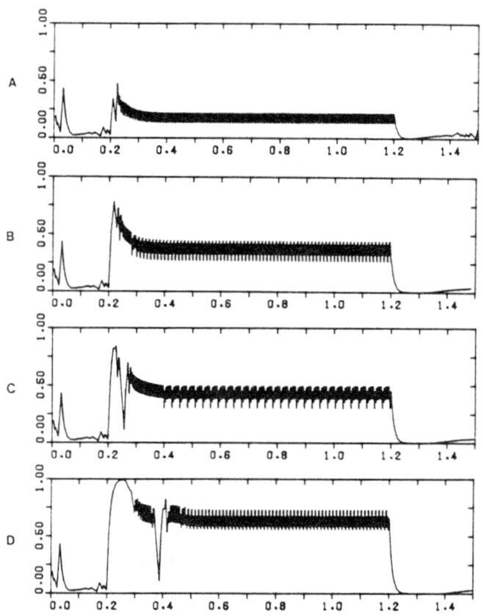

Figure 5

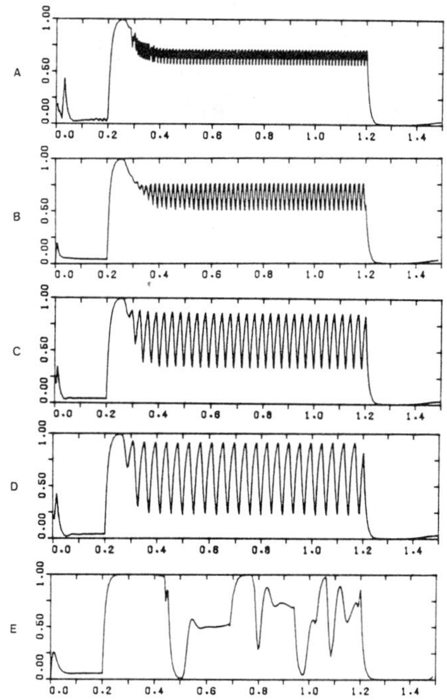

Figure 6

V. FORMULATION OF THE OPEN PROBLEMS

It is already known that a single cell of type (b_i) cannot produce periodic activities under constant external forces, but they can be entrained by periodic external forces. Further, a single cell of type $\{b_i, b_i\}$ can produce periodic activities under constant inputs, and can be modulated by periodic inputs, as well as can be entrained by any input. Furthermore, networks containing two or more identical neurons which inhibit each other with equal strength and equal time-lags can produce periodic oscillations under no external forces as well as under constant external forces with equal strength. The effects of time-lags appear to be reducing the frequency of the oscillations and producing more sluggish behaviors. It is also appearing that, under certain parametric configuration, the dominating modes compete with each other: one mode dominate the performance of the network until to some time instant, after which another mode takes the power to drive the network activities: that is, there is a quenching-like behavior in the neural activities (cf. [19],[20]).

In spite of the above mentioned progresses achieved since 1972 for the model described by Eq.(1), the following questions are still open:

Problem 1 : *Let* $\mathcal{P}$: $\{a_{i0};a_{ij};b_{ij};c_{ij};\sigma_{ij} \mid i=1,\ldots,n;j=1,\ldots,m\}$ *be an admissible parametric configuration. Then find a direct method to determine the dependency of the oscillations in the network on each parameter appearing in* $\mathcal{P}$.

Problem 2 : *Find conditions under which the network has a unique stable periodic solution when time-lags are present.*

Problem 3 : *Find exact times at which the modes of oscillations change their dominance.*

REFERENCES

[1] Aizawa, Y., Kobatake, Y.: A theory of oscillatory instability in two-state systems in relation to nerve excitation. Math. Bull. 38, 1976, 351-358.

[2] Amari, S.: Dynamics of pattern formation in lateral-inhibition type neural fields. Biol. Cybernetics. 27, 1979, 77-87.

[3] Amari, S.: A mathematical approach to neural systems. In: Systems Neuroscience. J. Metzler (Ed.), New York, London: Academic Press, 1977, 67-117.

[4] Bingxi Li: Uniqueness and stability of a limit cycle for a third order dynamical system arising in neuron modelling. Nonlinear. Anal., Theor. Meth. Appl. 5, 1981, 13-19.

[5] Braines, S.N., Napalkov, A.V., Svenchinskii, V.B.: Problems of Neurocybernetics. Akad. Med. Nauk SSR, Inst. Psikhiatrii, Moscow-USSR, 1959.

[6] Conrad, M., Guttinger, W., Cin, M.D. (Eds.): Physics and Mathematics of the Nervous Systems. Berlin, Heidelberg, New York: Springer-Verlag, 1974.

[7] Cowan, J.D.: A Statistical mechanics of nervous activity. In: Lectures on Mathematics in the Life Sciences: Vol. 2: Some Mathematical Questions in Biology. 1-57, Providence, R.I.: A.M.S., 1970.

[8] Culbertson, J.T.: The Minds of Robots. Urbana: Univ. of Illinois Press, 1963.

[9] Deimling, K.: Differential Equations in Banach Spaces. Berlin, Heidelberg, New York: Springer Verlag, 1977.

[10] Deutsch, S.: Models of the Nervous Systems. New York, London, Sydney: John Wiley and Sons, 1967.

[11] Friesen, W.O., Stent, G.S.: Neural circuits for generating rhythmic movements. Ann. Rev. Biophys. Gioeng. 7, 1978, 37-61.

[12] Green, H.S., Triffet, T.: Mathematical modelling of nervous systems. Math. Modelling 1, 1981, 41-61.

[13] Heiden, U. an der: Analysis of Neural Networks. Berlin, Heidelberg, New York: Springer-Verlag, 1980.

[14] Heiden, U an der: Existence of periodic solutions of a nerve equation Biol. Cybern. 21, 1976, 37-39.

[15] Leung, K.V., Mangeron, D., Oğuztöreli, M.N., Stein, R.B.: On a class of nonlinear integro-differential equations. III. Bull. Acad. R. Sci. Belgique, Ser. 5, 59, 1973, 492-499.

[16] Leung,K.V., Mangeron, D., Oğuztöreli, M.N., Stein, R.B.: On the stability and numerical solutions of two neural models. Utilitas Mathematika, 5, 1974, 167-212.

[17] Levin, B. Ja.: Distribution of Zeros of Entire Functions. Amer. Math. Soc. Translation Series, Vol. 5, 1964.

[18] MacGregor, R.J.: Computer simulation in FORTRAN of large nerve networks. I. The system. J. Theoret. Neurobiol. 1, 1981, 82-104.

[19] Nayfeh, A.N.: Perturbation Methods. Wiley-Interscience, New York, London, Sydney, Toronto, 1973.

[20] Nayfeh, A.H.: Mook, D.T.: Nonlinear Oscillations. Wiley-Interscience. New York, Chichester, Brisbane, Toronto, 1979.

[21] Oğuztöreli, M.N.: Time-lag Control Systems. New York, London: Academic Press, 1966.

[22] Oğuztöreli, M.N.: On the neural equations of Cowan and Stein (I). Utilitas Mathematica 2, 1972, 305-317.

[23] Oğuztöreli, M.N.: On the activities in a continuous neural network. Biol. Cybern 21, 1975, 41-48.

[24] Oğuztöreli, M.N.: Activities in a one-dimensional continuous neural network. Arch. Math. 3, Scripta Fac. Sci. Nat. Ujep Brunensis 14, 1978, 161-169.

[25] Oğuztöreli, M.N.: Activity propagation in a continuous neural network. In: "Jubilee Volume" of the Institute of Mathematics of the Soviet Academy of Sciences, Moscow-USSR, 1978, 450-458.

[26] Oğuztöreli, M.N.: Activity analysis of neural networks. Biol. Cybern. 34, 1979, 159-169.

[27] Oğuztöreli, M.N.: Modelling and simulation of vertebrate retina. Biol. Cybern. 37, 1980, 53-61.

[28] Oğuztöreli, M.N., O'Mara, K.S.: Modelling and simulation of vertebrate retina: extended network. Biol. Cybern. 38, 1980, 9-17.

[29] Oğuztöreli, M.N.: Response analysis of vertebrate analysis. Biol. Cybern. 44, 1982, 1-8.

[30] Pozin, N.V.: Modelling of Neural Structures. Ed. NAUKA, Sect. Phys. Math. Literature, Moscow-USSR, 1970.

[31] Ratliff, F.: Mach Bands: Qualitative Studied on Neural Networks in the Retina. San Franscisco, London, Amsterdam: Holden-Day, 1965.

[32] Reichardt, W.E., Poggio, T. (Eds.): Theoretical Approaches in Neurobiology. The MIT Press, 1981.

[33] Rescigno, A., Stein, R.B., Purple, R.L., Poppele, R.E.: A neuronal model for the discharge patterns produced by cyclic inputs. Bull. Math. Biophysics. 32, 1970, 337-352.

[34] Scharstein, H.: Input-output relationship of the leaky-integrator neuron model. J. Math. Biol. 8, 1979, 403-420.

[35] Schmitt, F., Worden, F.G.(Eds.): The Neurosciences: Fourth Study Program. MIT Press, 1980.

[36] Shepherd, G.M.: The Synaptic Organiztion of the Brain. London, Toronto: Oxford Univ. Press, 1979.

[37] Stein, R.B.: Nerve and Muscle: Membranes, Cells, and Systems. New York and London: Plenum Press, 1980.

[38] Stein, R.B., Leung, K.V., Oğuztöreli, M.N., Williams, D.W.: Properties small networks. Kybernetik 14, 1974, 223-230.

[39] Stein, R.B., Leung, K.V., Mangeron, D., Oğuztöreli, M.N.: Improved neuronal models for studying neural networks. Kybernetik 15, 1974, 1-9.

[40] Ward, D.S.: Stochastic Relationship for Neurons and Neuron Pair Networks. Ph.D. Dessertation, Univ. of Calif., Los Angeles, 1975. - Appl. Math. Comp. 3, 1977, 227-245.

[41] Wilson, H.R., Cowan, J.D.: A mathematical theory of the functional dynamics of cortical and thalamic nervous tissue. Cybernetik 13, 1972, 55-80.

AN INCLUSION PRINCIPLE FOR HEREDITARY SYSTEMS

Y. Ohta

Fukui University
Fukui-shi, Japan

D. D. Šiljak

University of Santa Clara
Santa Clara, California

ABSTRACT

Given two hereditary dynamic systems having different dimensions, the conditions are provided under which a part of the motion of the larger system is reproduced by the smaller system, that is, the larger system "includes" the smaller one. The conditions for inclusion are useful in applying the concept of vector Liapunov functions to stability analysis of systems composed of overlapping subsystems. By expanding a system into a larger space the overlapping subsystems appear as disjoint and standard methods can be used to conclude stability of the expanded system. Under the inclusion conditions, stability of the expansion implies stability of the original system.

1. INTRODUCTION

In mathematical modeling of physical systems, one of the basic tasks is to compare models having different dimensions and subsequently choose one which contains all of the pertinent information about the system behavior, yet simple enough to be tractable by known methods. This problem is broad and there are many unanswered questions. A specific but important aspect of the problem is to formulate conditions under which a dynamic system can reproduce precisely a part of the motion of a larger dynamic system, that is, be "included" in a system of higher dimension. For this purpose, an Inclusion Principle has been introduced in the context of linear systems [1] and later extended to nonlinear systems [2, 3] and systems for decentralized control with overlapping information sets [4]. An intuitively obvious need for the Inclusion Principle has been demonstrated in [5], where it has been used for validation of reduced-order models for complex physical systems.

Less intuitive, yet equally effective, has been the use of the Principle in the stability analysis of dynamic systems where an actual expansion of a system has been proposed to gain conceptual simplifications of the stability problems. When a system is composed of overlapping subsystems, an expansion of the system into a larger space can make the subsystems, appear as disjoint. Then, standard methods [6] of the vector Liapunov functions [7-9] can be used to determine stability of the expanded system, and applying the Inclusion Principle, conclude stability of the original system. It has been shown explicitly that the use of generalized decompositions in this context can establish stability where the standard disjoint decompositions fail [1-3].

The objective of this paper is to extend the Inclusion Principle to hereditary dynamic systems. This opens up a possibility to introduce generalized decompositions in the stability analysis of time-delay systems via vector Liapunov functions, and broaden considerably the decomposition-aggregation methods [10-17] in this context. By using a simple example of a linear system with time-delay, we can again demonstrate the superiority of generalized decompositions when used in the framework of the Inclusion Principle. Moreover, a new set of conditions are derived for stability of interconnected hereditary systems, which can further reduce the conservativeness of the decomposition-aggregation methods.

Due to the limitation in space, all proofs and the above mentioned example are omitted, but can be obtained upon request from the authors.

2. INCLUSION PRINCIPLE

Let us consider two functional differential equations representing a pair of dynamic systems

$$\mathcal{S}: \dot{x}(t) = f(t, x_t) \tag{2.1}$$

and

$$\tilde{\mathcal{S}}: \dot{\tilde{x}}(t) = \tilde{f}(t, \tilde{x}_t) \tag{2.2}$$

where $x(t) \in R^n$, $\tilde{x}(t) \in R^{\tilde{n}}$, and $n \leq \tilde{n}$. In (2.1) and (2.2), $f: R_+ \times C^n \to R^n$ and $\tilde{f}: R_+ \times C^{\tilde{n}} \to R^{\tilde{n}}$ are sufficiently smooth so that for any initial functions ϕ_0 and $\tilde{\phi}_0$ at $t = t_0$, equations (2.1) and (2.2) have unique solutions $x(t_0, \phi_0)$ and $\tilde{x}(t_0, \tilde{\phi}_0)$, respectively. $C^n = C\{[-\tau, 0] \to R^n\}$ denotes the space of continuous functions with domain $[-\tau, 0]$ and range in R^n. For $\phi \in C^n$, we define $||\phi||_0 = \max_{-\tau \leq s \leq 0} ||\phi(s)||$ where $||\cdot||$ is any convenient norm in R^n. Suppose that $x \in C\{[-\tau, \infty) \to R^n\}$. For $t \in R_+$, let x_t denote a translation of the restriction of x to the interval $[t-\tau, t]$; more specifically x_t is an element of C^n defined by $x_t(s) = x(t+s)$, $-\tau \leq s \leq 0$. We denote the value of $x(t_0, \phi_0)$ at $t \geq t_0 - \tau$ by $x(t; t_0, \phi_0)$, or simply by $x(t)$. I_n is the $n \times n$ identity matrix and

T denotes transpose.

Now, we formulate the notion of inclusion for a pair of hereditary dynamic systems $\tilde{S}$ and S by the following:

(2.3) DEFINITION. *We say that the system* $\tilde{S}$ *includes the system* S *if there exist two matrices* $U \in R^{n\times\tilde{n}}$ *and* $V \in R^{\tilde{n}\times n}$ *such that* $UV = I_n$ *and*

$$x(t_0, \phi_0) = U\tilde{x}(t_0, V\phi_0) \quad \forall(t_0, \phi_0) \in R_+ \times C^n . \tag{2.4}$$

If $\tilde{S}$ includes S, it is called an *expansion* of S, and S is referred to as a *contraction* of $\tilde{S}$. Sufficient conditions for inclusion are given by the following:

(2.5) THEOREM. $\tilde{S}$ *includes* S *if either*

(i) *there exists a matrix* $V \in R^{\tilde{n}\times n}$ *with full column rank such that*

$$\tilde{f}(t, V\phi) = Vf(t, \phi) \quad \forall(t, \phi) \in R_+ \times C^n , \tag{2.6}$$

or

(ii) *there exists a matrix* $U \in R^{n\times\tilde{n}}$ *with full row rank such that*

$$U\tilde{f}(t, \tilde{\phi}) = f(t, U\tilde{\phi}) \quad \forall(t, \tilde{\phi}) \in R_+ \times C^{\tilde{n}} . \tag{2.7}$$

The conditions of Theorem (2.5) do not imply each other. If for some matrix V with full column rank, (2.6) holds, then S is called a *restriction* of $\tilde{S}$. When (2.7) holds for a matrix U with full row rank, S is termed an *aggregation* of $\tilde{S}$.

When S and $\tilde{S}$ are linear time-delay systems, we strengthen Theorem (2.5) to get necessary and sufficient conditions for inclusion. To show this, let us consider the system S and $\tilde{S}$ described as

$$S\colon \dot{x}(t) = Ax(t) + \sum_{k=1}^{\ell} D_k x(t-\tau_k) \tag{2.8}$$

and

$$\tilde{S}\colon \dot{\tilde{x}}(t) = \tilde{A}\tilde{x}(t) + \sum_{k=1}^{\ell} \tilde{D}_k \tilde{x}(t-\tau_k) \tag{2.9}$$

where again $x(t) \in R^n$, $\tilde{x}(t) \in R^{\tilde{n}}$, and $n \leq \tilde{n}$. In (2.8) and (2.9), $A, \tilde{A}, D_k, \tilde{D}_k$ are constant matrices of appropriate dimension and time-delays are ordered as $0 < \tau_1 < \tau_2 < \dots < \tau_\ell = \tau$.

(2.10) THEOREM. $\tilde{S}$ *includes* S *if and only if there exist two matrices* $U \in R^{n\times\tilde{n}}$ *and* $V \in R^{\tilde{n}\times n}$ *such that* $UV = I$ *and*

$$U\tilde{A}^i V = A^i, \; U\tilde{A}^{i-1}\tilde{D}_k = A^{i-1}D_k U , \quad \forall i = 1, 2, \dots, \tilde{n} , \; \forall k = 1,], \dots, \ell . \tag{2.11}$$

3. STABILITY

The concept of inclusion as defined in Definition (2.3), has a natural interpretation in the stability context. In fact, for systems without hereditary effects, it has been established [1] that stability of the expansion implies stability of the original system. Our immediate objective is to state that the same is true for hereditary systems provided the equilibrium x^* of the original system S is preserved in the expansion $\tilde{S}$ under the linear transformation

$$\tilde{x} = Vx , \tag{3.1}$$

where V is the matrix specified in Definition (2.3). More precisely, if $f(t, x^*) = 0$ for all t where x^* is an equilibrium of S, then

$$\tilde{f}(t, Vx^*) = 0 \quad \forall t \in R_+ , \tag{3.2}$$

so that $\tilde{x}^* = Vx^*$ is the corresponding equilibrium of $\tilde{S}$. The condition (3.2) is implied by (i), but not by (ii) of Theorem (2.5).

(3.3) THEOREM. *Suppose that* $\tilde{S}$ *includes* S*, and* $\tilde{x}^* = Vx^*$. *Then, stability of* $\tilde{x}^*$ *in* $\tilde{S}$ *implies stability of* x^* *in* S^*.

We note that the converse of Theorem (3.3) is not true in general. When the two systems S and $\tilde{S}$ are specialized to linear constant systems with time-delays as in (2.8) and (2.9) it is possible to establish a relationship between their characteristic equations and provide a basis for the converse.

(3.4) THEOREM. *If* $\tilde{S}$ *includes* S*, then*

$$\det(\lambda I_{\tilde{n}} - \tilde{A} - \sum_{k=1}^{\ell} \tilde{D}_k e^{-\lambda\tau_k}) = \lambda^{-n} \det(\lambda I_n - A - \sum_{k=1}^{\ell} D_k e^{-\lambda\tau_k}) \det(\lambda I_n - M - \sum_{k=1}^{\ell} L_k e^{-\lambda\tau_k}) . \tag{3.5}$$

4. GENERALIZED DECOMPOSITION

Decomposition of dynamic systems into interconnected subsystems have been performed either to reduce the computational requirements in carrying out the necessary calculations, or to take advantage of special structural properties of the systems and come up with simple criteria for robust stability of dynamic systems [6]. Recent results have shown [1-4] that in certain practical situations it is advantageous to apply an unorthodox decomposition scheme whereby the resulting subsystems share common parts. An interesting result in this development has been the fact that the overlapping and generalized decompositions can be used in the context of vector Liapunov functions to show stability where disjoint decompositions fail. The aim of this section is to establish this fact for systems described by differential-difference equations.

Let us consider again the system S of (2.1). Guided by considerations of a given structure of S, which will be described later, a linear transformation (3.1) is used to produce an expansion $\tilde{S}$ of S specified by (2.2). A simple case of overlapping decomposition in this context would be to choose the transformation matrix V so that the state equations of S, which describe common parts of the physical subsystems, are repeated until an expansion $\tilde{S}$ is formulated in such a way that all the subsystems appear as disjoint. The system $\tilde{S}$ can then be decomposed as

$$\tilde{S}\colon \dot{\tilde{x}}_i(t) = \tilde{g}_i[t, \tilde{x}_i(t)] + \tilde{h}_i[t, \tilde{x}(t), \tilde{x}(t-\tau_1), \ldots, \tilde{x}(t-\tau_m)] \qquad i = 1, 2, \ldots, s \tag{4.1}$$

where $\tilde{x}_i(t) \in R^{\tilde{n}_i}$, $R^{\tilde{n}} = R^{\tilde{n}_1} \times R^{\tilde{n}_2} \times \ldots \times R^{\tilde{n}_s}$, $\tilde{g}_i\colon R_+ \times R^{\tilde{n}_i} \to R^{\tilde{n}_i}$, $\tilde{h}_i\colon R_+ \times R^{\tilde{n}} \times R^{\tilde{n}m} \to R^{\tilde{n}_i}$, and $\tau_1, \tau_2, \ldots, \tau_m$ are positive numbers. We assume that $\tilde{g}_i(t, 0) = 0$, $\tilde{h}_i(t, 0, 0, \ldots, 0) = 0$ for all t and i, so that $\tilde{x} = 0$ is an equilibrium of $\tilde{S}$.

(4.2) THEOREM. *Assume that:*

(i) *For each* $i = 1, 2, \ldots, s$, *there exists a continuous function* $v_i\colon R_+ \times R^{\tilde{n}_i} \to R_+$ *with the following properties:* $v_i(t, x_i)$ *is locally Lipschitzian in* $\tilde{x}_i$, $v_i(t, 0) \equiv 0$, *and*

$$\begin{aligned} &\psi_{i1}(\|\tilde{x}_i\|) \leq v_i(t, \tilde{x}_i) \leq \psi_{i2}(\|\tilde{x}_i\|) \\ &D^+v_i(t, \tilde{x}_i)_{(4.4)} \leq -\delta_i(t)w_i(t, \tilde{x}_i) - w_i'(t, \tilde{x}_i) \\ &\qquad \forall (t, x_i) \in R_+ \quad R^{\tilde{n}_i} \end{aligned} \tag{4.3}$$

where $\psi_{i1}, \psi_{i2} \in K_\infty$ *which is a class of comparison functions* [18]. *Furthermore,* $D^+v_i(t, \tilde{x}_i)_{(4.4)}$ *is the derivative computed with respect to the isolated subsystem*

$$\tilde{S}_i\colon \dot{\tilde{x}}_i(t) = \tilde{g}_i[t, \tilde{x}_i(t)], \tag{4.4}$$

$\delta_i\colon R_+ \to R_+$, w_i, $w_i'\colon R_+ \times R^{\tilde{n}_i} \to R_+$ *are continuous, and the functions* w_i, w'_i *satisfy either*

$$v_i(t, \tilde{x}_i) \leq \gamma_i w(t, \tilde{x}_i) \quad \forall (t, \tilde{x}_i) \in R_+ \times R^{\tilde{n}_i} \tag{4.5}$$

or

$$v_i(t, \tilde{x}_i) \leq \gamma_i' w_i'(t, \tilde{x}_i) \quad \forall (t, \tilde{x}_i) \in R_+ \times R^{\tilde{n}_i} \tag{4.6}$$

for some positive numbers γ_i *and* γ_i' . *Moreover, there exists* $\psi_{i3} \in K$ *such that*

$$w_i(t, \tilde{x}_i) \leq \psi_{i3}(||\tilde{x}||) \quad \forall(t, \tilde{x}_i) \in R_+ \times R^{\tilde{n}_i} \quad . \tag{4.7}$$

(ii) *For each* $i = 1, 2, \ldots, s$, *there exist continuous functions* $\beta_{ij}^k : [-\tau_k, +\infty) \to R_+$, $j = 1, 2, \ldots, s$; $k = 0, 1, \ldots, m$, *such that*

$$|v_i[t, \tilde{y}_i + \Delta\tilde{h}_i(t, \tilde{x}^0, \tilde{x}^1, \ldots, \tilde{x}^m)] - v_i(t, \tilde{y}_i)|$$
$$\leq \Delta \sum_{j=1}^{s} \sum_{k=0}^{m} \beta_{ij}^k(t-\tau_k)w_j(t-\tau_k, \tilde{x}_j^k) + o(\Delta)$$
$$\forall t \in R_+, \tilde{y}_i \in R^{\tilde{n}_i}, \tilde{x}^k \in R^{\tilde{n}} \tag{4.8}$$

for sufficiently small $\Delta > 0$, *where* $\tau_0 = 0$.

(iii) *There exist positive numbers* $d_1, d_2, \ldots, d_s$, θ *and* μ *such that for all* $j = 1, 2, \ldots, s$, *we have*

$$d_j\delta_j(t) - \sum_{i=1}^{s} d_i \sum_{k=0}^{m} \beta_{ij}^k(t) \geq \theta, \quad \delta_j(t) \leq \mu \quad \forall t \in R_+ \quad . \tag{4.9}$$

Then, the equilibrium $\tilde{x}^* = 0$ *of* $\tilde{S}$ *is (uniformly) asymptotically stable in the large.*

If, in addition

(iv) *For each* $i = 1, 2, \ldots, s$ *there exist positive numbers* η_{i1}, η_{i2} , *and* η_{i3} *such that for all* $j = 1, 2, 3$, *we have*

$$\psi_{ij}(||\tilde{x}_i||) = \eta_{ij}||\tilde{x}_i|| \quad \forall \tilde{x}_i \in R^{\tilde{n}_i} , \tag{4.10}$$

then the equilibrium $\tilde{x}^*$ *of* $\tilde{S}$ *is exponentially stable in the large.*

Theorem (4.2) is motivated by the results of [19] and, in particular, of [20], in which w_i is specified as v_i and $w_i' \equiv 0$ for all $i = 1, 2, \ldots, s$.

In the case of linear time-invariant differential-difference systems, we have the following corollary to Theorem (4.2), which is useful in applications. We consider the system $\tilde{S}$ of (4.1) specified as

$$\tilde{S}: \dot{\tilde{x}}_i(t) = \tilde{A}_i\tilde{x}_i(t) + \sum_{j=1}^{s} \sum_{k=0}^{m} \tilde{A}_{ij}^k\tilde{x}_j(t-\tau_k), \; i = 1, 2, \ldots, s \tag{4.11}$$

where $\tilde{A}_i$, $\tilde{A}_{ij}^k$ are constant matrices of appropriate dimensions, and state the following:

(4.12) COROLLARY. *Assume that:*

(i) *For each* $i = 1, 2, \ldots, s$, *there exists a symmetric and positive definite matrix* $\tilde{H}_i$ *such that*

$$\tilde{A}_i^T\tilde{H}_i + \tilde{H}_i\tilde{A}_i = -\tilde{G}_i \,, \tag{4.13}$$

where $\tilde{G}_i$ *is a symmetric and positive definite matrix.*

(ii) *The* $s \times s$ *aggregate matrix* $W = (w_{ij})$ *defined as*

$$w_{ij} = \begin{cases} \frac{1}{2}\lambda_m^{\frac{1}{2}}(\tilde{G}_i\tilde{H}_i^{-1}) - \sum_{k=0}^{m} ||\tilde{H}_i^{\frac{1}{2}}\tilde{A}_{ii}^k\tilde{G}_i^{-\frac{1}{2}}||_2 \,, & i = j \\ -\sum_{k=0}^{m} ||\tilde{H}_i^{\frac{1}{2}}\tilde{A}_{ij}^k\tilde{G}_j^{-\frac{1}{2}}||_2 \,, & i \neq j \end{cases} \tag{4.14}$$

is an M*-matric, where* $\lambda_m(\cdot)$ *denotes the smallest eigenvalue of the indicated matrix, and* $||\cdot||_2$ *denotes the matrix norm induced by* ℓ_2 *norm.*

Then, the equilibrium $\tilde{x}^* = 0$ *of the linear system* $\tilde{S}$ *is exponentially stable in the large.*

ACKNOWLEDGEMENTS

The research reported herein has been supported by the National Science Foundation under the Grant ECS-8011210. During the course of research, Y. Ohta was on leave at the Department of Electrical Engineering and Computer Science, University of Santa Clara, Santa Clara, California.

REFERENCES

[1] Ikeda, M., and Šiljak, D. D., Overlapping decompositions, expansions, and contractions of dynamic systems, Large Scale Systems, 1 (1980), 29-38.

[2] Ikeda, M., and Šiljak, D. D., Lotka-Volterra equations: Decomposition, stability, and structure, J. Math. Biol., 9 (1980), 65-83.

[3] Ikeda, M., and Šiljak, D. D., Generalized decompositions of dynamic systems and vector Lyapunov functions. IEEE Trans. Automatic Control, AC-26 (1981), 1118-1125.

[4] Ikeda, M., Šiljak, D. D., and White, D. E., Decentralized control with overlapping information sets. J. Optimization Theory Appl., 34 (1981), 279-310.

[5] Sezer, M. E., and Šiljak, D. D., Validation of reduced-order models for control system design. Proc. IEEE Conf. Decision Control, San Diego, California, 1981, 652-659.

[6] Šiljak, D. D., Large-Scale Dynamic Systems: Stability and Structure. North-Holland, New York, 1978.

[7] Lakshmikantham, V., and Leela, S., Differential and Integral Inequalities: Theory and Applications. Academic Press, New York, 1969.

[8] Martynyuk, A. A., Stability of Motion of Complex Systems (in Russian). Naukova Dumka, Kiev, U.S.S.R., 1975.

[9] Matrosov, V. M. (ed.), Algorithms for Proofs of Theorems in the Method of Vector Liapunov Functions (in Russian). Nauka, Novosibirsk, U.S.S.R., 1981.

[10] Ladde, G. S., Systems of functional differential inequalities and functional differential systems. Pacific J. Math., 66 (1976), 161-171.

[11] Ladde, G. S., Stability of large-scale hereditary systems under structural perturbations, IFAC Symposium on Large-Scale Systems Theory and Appl., (1976), 215-226.

[12] Ladde, G. S., Competitive processes-I. Stability of Hereditary systems. Nonlinear Analysis, 1 (1977), 607-631.

[13] Tokumaru, H., Adachi, N., and Amemiya, T., Macroscopic stability of interconnected systems, Proc. 6th IFAC World Congr., It. ID, 44.4 (1975).

[14] Araki, M., and Mori, T., On the stability criteria for composite systems including time delays (in Japanese), Trans. Soc. Instrum. & Contr., 15 (1979), 267-268.

[15] Amemiya, T., Macroscopic asymptotic stability of nonlinear interconnected systems with delays (in Japanese), Preprint 23rd JAACE Meeting on Syst. of Contr. (1979), 19-20; Trans. Soc. Instrum. and Contr., 17 (1981), 374-380.

[16] Ohta, Y., Douseki, T., and Matsumoto, T., Stability analysis of nonlinear composite systems including time delays. Proc. 8th IFAC World Congr., 2 (1981), 41-46.

[17] Amemiya, T., On the stability of non-linear interconnected systems, Int. J. Contr., 34 (1981), 513-527.

[18] Hahn, W., Stability of Motion, Springer-Verlag, New York, 1967.

[19] Anderson, B.D.O., Time delays in large-scale systems. Proc. IEEE Conf. Decision Control, Fort Lauderdale, Florida, 1979, 655-660.

[20] Lewis, R. M., and Anderson, B.D.O., Insensitivity of a class of nonlinear compartmental systems to the introduction of arbitrary time delays. IEEE Trans. Circuit Syst., CAS-27 (1980), 604-612.

ON CRITICAL CASE OF LIAPUNOV'S STABILITY OF ONE-DEGREE OF FREEDOM SYSTEM

Andrzej Olas *

Institute of Fundamental Technological Research
Warsaw, Poland

ABSTRACT

The critical case of Liapunov's stability of a trivial solution of one-degree of freedom system is considered. Using the theorem dealing with the integral from the derivative of the Liapunov's function along the solution of the system the conditions of stability consistent with the mechanical interpretation of the equation are obtained. Particularly it is shown that the only nonlinear terms influencing stability are those interpreted as nonlinear damping. The extension of the procedure for more general cases is proposed.

STATEMENT OF THE PROBLEM

Consider the Liapunov's stability of the trivial solution of the equation

$$\ddot{x} + x + a_1x^3 + a_2x^2\dot{x} + a_3x\dot{x}^2 + a_4\dot{x}^3 = 0$$

The problem of the stability of this solution is not to be solved with use of Liapunov's theorem on the stability in the first approximation since the real parts of the roots of the characteristic equation of the corresponding first order system are equal zero. Such a case is called the critical case of Liapunov's stability and the entire nonlinear system must be then investigated. No satisfying theory was developed; for instance when using the Liapunov's direct method one obtains the stability conditions being not consistent with mechanical interpretation. The goal of the present paper is to apply proposed at the paper [1] modification of the Liapunov's direct method giving the results consistent with mechanical interpretation.

We shall apply the theorem proved in [1]. Transforming the above equation to the form

*Tempoarily at Tuskegee Institute, Tuskegee, Alabama.

$$\left.\begin{aligned} \dot{y}_1 &= y_2 \\ \dot{y}_2 &= -y_1 - a_1y_1^3 - a_2y_1^2y_2 - a_3y_1y_2^2 - a_4y_2^3 \end{aligned}\right\} \quad (1)$$

that is

$$\dot{y} = Ay + P(y) \ , \ y \varepsilon R^2 \quad (2)$$

where

$$A = \begin{bmatrix} 0 & 1 \\ -1 & 0 \end{bmatrix} \ , \ P(y) = \text{col}(0, -a_1y_1^3 - a_2y_1^2y_2 - a_3y_1y_2^2 - a_4y_2^3)$$

We denote by $y(t, y_0)$ the solution of the equation. Applied to equation (2) the theorem states:

Th. If there exist:

(a) Positive-definite scalar function $V(y) \varepsilon C_y^1(R^2)$

(b) A positive number T

(c) A neighborhood U of the point $0 \varepsilon R^2$

such that the relation

$$\int_0^T \dot{V}(y\ (\tau, y_0))d\tau \leq 0$$

holds for every $y_0 \varepsilon U$ then the trivial solution of the equation is stable in the sense of Liapunov.

The above theorem introduces the investigation of the sign of the expression

$$\int_0^T \dot{V}(y(\tau, y_0))d\tau \quad (3)$$

instead of the sign of the derivative

$$\dot{V}(t, y)$$

It is easy to see that the assumptions of this theorem are weaker than those of Liapunov's theorem on stability. One can observe that the important weakness of this theorem is the fact that it assumes the knowledge of the solution $y(t, y_0)$ for a finite interval $[0, T]$. However in our case we can easily estimate the value of the expression (3) with help of approximate solution. Denote by $z(t, y_0)$ are the solution of the equation

$$\dot{z} = Az$$

Then the function

$$u(t, y_0) = y(t, y_0) - z(t, y_0)$$

is the solution of the equation

$$\dot{u} = Au + P(u - z(t, y_0))$$

fulfilling the initial condition $u(0, y_0) = 0$. It is easy to show that

$$\frac{||u(t, y_0)||}{||y_0||} \to 0$$

when $||y_0|| \to 0$. Thus we can write

$$y(t, y_0) = z(t, y_0) + 0(||y_0||)$$

SOLUTION OF THE PROBLEM

We choose the following function

$$V(y) = 1/2(y_1^2 + y_2^2)$$

obtaining

$$\dot{V}(y) = - y_2(a_1y_1^3 + a_2y_1^2y_2 + a_3y_1y_2^2 + a_4y_2^3)$$

Notice that it is difficult to draw any conclusion on stability from this from of the derivative. The function $z(t, y_0)$ is of the form

$$\left.\begin{aligned} z_1(t, y_0) &= y_{10} \cos t + y_{20} \sin t \\ z_2(t, y_0) &= - y_{10} \sin t + y_{20} \cos t \end{aligned}\right\} \quad (4)$$

with period equal 2π .

We consider the expression

$$\int_0^T \dot{V}(y(\tau, y_0))\, d\tau$$

taking as T the period of the function (4).

It is easy to see that the function $\dot{V}(y(t, y_0)$ can be written as

$$\dot{V}(y(t, y_0)) = \dot{V}(z(t, y_0)) + R(t, y_0)$$

where

$R(t, y_0) = 0(||y_o||)$. Thus we can write

$$\int_0^{2\pi} \dot{V}(y(t, y_0)) = \int_0^{2\pi} \dot{V}(z(t, y_0))\, dt + 0(||y_o||) \quad (5)$$

Using (4) we obtain

$$\int_0^{2\pi} \dot{V}(z(t, y_0))dt = A_1 \int_0^{2\pi} \sin^4 t\, dt + A_2 \int_0^{2\pi} \sin^3 t \cos t\, dt +$$

$$A_3 \int_0^{2\pi} \sin^2 t \cos^2 t\, dt + A_4 \int_0^{2\pi} \sin t \cos^3 t\, dt + A_5 \int_0^{2\pi} \cos^4 t\, dt$$

where coefficients A_1, A_2, A_3, A_4, A_5 are the homogeneous polynoms of the fourth power of the initial values y_{10}, y_{20} . Taking into consideration fact that

$$\int_0^{2\pi} \sin^4 t\, dt = \int_0^{2\pi} \cos^4 t\, dt = 3/4\pi$$

$$\int_0^{2\pi} \sin t \cos^3 t\, dt = \int_0^{2\pi} \sin^3 t \cos t\, dt = 0$$

$$\int_0^{2\pi} \sin^2 t \cos^2 t \, dt = \pi/4$$

we have

$$\int_0^{2\pi} \dot{V}(z(t,y_0))dt = 3/4\pi A_1 + \pi/4A_3 + 3/4\pi A_5$$

where

$$A_1 = -(a_1 y_{10} y_{20}^3 + a_2 y_{10}^2 y_{20}^2 - a_3 y_{10}^3 y_{20} + a_4 y_{10}^4)$$

$$A_3 = -[a_2 y_{10}^4 - 3(a_1 - a_3) y_{10}^3 y_{20} - (4a_2 - 6a_4) y_{10}^2 y_{20}^2$$

$$+ 3(a_1 - a_3) y_{10} y_{20}^3 + a_2 y_{20}^4]$$

$$A_4 = -(1a_3 y_{10} y_{20}^3 + a_2 y_{10}^2 y_{20}^2 + a_1 y_{10}^3 y_{20} + a_4 y_{20}^4)$$

Finally we have

$$\int_0^{2\pi} \dot{V}(2(t,y_0))dt = -(a_2+3a_4)y_{10}^4 - 2(a_2+3a_4)y_{10}^2 y_{20}^2 - (a_2+3a_4)y_{20}^4$$

$$= -(a_2+3a_4)(y_{10}^2+y_{20}^2)^2$$

Thus we conclude that if

$$a_2 + 3a_4 > 0$$

then $\int_0^{2\pi} \dot{V}(2(t,y_0))dt$ is the negative-definite function. On the basis of (5) it follows that for sufficiently small values of y_0 the relation

$$\int_0^{2\pi} \dot{V}(y(t,y_0))dt < 0$$

holds, what is sufficient for the stability of the trivial solution of the equation. We conclude thus that if

$$a_2 + 3a_4 > 0$$

then the trivial solution of the equation is stable.

One can observe that the coefficients a_1 and a_3 expressing the nonlinear elasticity have no influence on the stability. Moreover it is interesting to observe that stability of the trivial solution of the equation is obtained even when one of the coefficients a_2, a_4 is negative but (6) is fulfilled.

The extension of the procedure for more general cases, beginning from two-degree of freedom system should not meet principal difficulties. However the procedure may not be useful when there is the Liapunov's stability but not asymptotic one.

REFERENCES

[1] A. Olas, A Criterion of Stability of the Trivial Solution . . . , Bull. del'Acad. Polon. des Sci., Ser. des Sci. Tech., in printing.

SOME COMPUTATIONAL PROCEDURES FOR NONLINEAR PROBLEMS IN R^n

M. C. Pandian

Department of Mathematics
The University of Texas at Arlington
Arlington, Texas

1. INTRODUCTION

Consider the following nonlinear system

$$f(x) = 0, \ f: \ D \subset R^n \to R^n, \ D \subset R^n, \tag{1.1}$$

D is a bounded domain; f possesses a derivative f' which is continuous on D. Let $A \in L(R^n)$ be any nonsingular matrix. Then, we can change (1.1) into a fixed point problem $s(x) = x$, $s(x) = x - A^{-1}f(x)$. Before proceeding to compute a solution of this problem, one must know the existence of a solution in a trial region $X \subset D$. For existence theory, we have Brouwder's fixed point theorem [4]. Miranda [8] gives a different version of this theorem which is also considered as generalized intermediate value theorem. In this respect, known computational theorems are the existence theorems of Kantorovich [6] and Moore [3]. In [3] and [11], the authors provide an interval test for convergence of interval algorithms and for Newton type schemes. A new computational test is given in [5] for convergence of Newton type methods. This test is weaker than the test in [3,11]. Unlike the previous one, this test can be carried out by pointwise computations. We give (Theorem 2.3) a simplified version of the interval arithmetic theorem of Moore [3]. Also, we obtain componentwise error bounds. In the monotone method, in order to have both sided approximations converge to the unique solution, a contraction condition is assumed in [1]. In Section 3, we give a different contraction condition (3.1) which is weaker than that assumed in [1]. Here, we summarize these results. For further details as well as numerical examples, see [5].

We write, for $x = (x_i) \in R^n$, $x \geq 0$ or $x > 0$ if $x_i \geq 0$ or $x_i > 0$, respectively, for $i = 1,2,\ldots,n$. Similarly, for $A \in L(R^n)$, we write $A \geq 0$ if $a_{ij} \geq 0$, and $A > 0$ if $a_{ij} > 0$ for all $i,j = 1,2,\ldots,n$. We let $|x| = (|x_i|)$ denote the vector of absolute values, and $|A| = (|a_{ij}|)$. Let X be an n-dimensional rectangle; that is $X = \{x \in R^n | \underline{x}_i \leq x_i \leq \overline{x}_i;\ i = 1,2,\ldots,n\}$, where $\underline{x}, \overline{x} \in R^n$. Define $\rho_i = \frac{1}{2}(\overline{x}_i - \underline{x}_i)$ for $i = 1,2,\ldots,n$ and refer the vector $\rho = (\rho_i)$ as the radius vector of X. Let $\tilde{x}$ be the midpoint of X and it is given by $\tilde{x} = (\tilde{x}_i)$, $\tilde{x}_i = \frac{1}{2}(\overline{x}_i + \underline{x}_i)$ for $i = 1,2,\ldots,n$.

Definition 1.1 [2] A matrix $A \in L(R^n)$ is said to be quasi-monotone (decreasing) if $a_{ij} \leq 0$ for $i \neq j$, $i,j = 1,2,\ldots,n$.

Theorem 1.1: [7] A quasi-monotone matrix A is inverse positive if and only if there exists a test vector v such that $v \geq 0$, $Av \geq 0$, and $\sum_{j=1}^{i} a_{ij}v_j > 0$ for $i = 1,2,\ldots,n$.

Theorem 1.2 [7] Let $A,B \in L(R^n)$, $\lambda \in R$. If $A^{-1} \geq 0$, and $A^{-1}B \geq 0$, then $T = \lambda A - B$ is inverse positive if and only if $\lambda > \sigma(A^{-1}B)$, the spectral radius of $A^{-1}B$.

2. TESTS FOR EXISTENCE, UNIQUENESS, AND CONVERGENCE

In this section, we give tests for existence and uniqueness of a solution for (1.1) in an n-dimensional rectangle. We provide convergence results for modified Newton scheme, and obtain componentwise error estimates.

We assume the following Lipschitz condition on X.

$$L(x_1 - x_2) \leq f(x_1) - f(x_2) \leq M(x_1 - x_2);\ x_1, x_2 \in X,\ x_1 > x_2.$$

Remark 2.1 Here, L and M are $n \times n$ matrices, called Lipschitz matrices, and they depend on X. For example, we can choose

$$L = (\ell_{ij}),\ \ell_{ij} = \min_{x \in X}\left[\frac{\partial f_i}{\partial x_j}(x)\right], \text{ and}$$

$$M = (m_{ij}),\ m_{ij} = \max_{x \in X}\left[\frac{\partial f_i}{\partial x_j}(x)\right].$$

Recall that $s(x) = x - A^{-1}f(x)$. Let $x^{(1)} = s(\tilde{x})$, $\eta = |x^{(1)} - \tilde{x}|$, $C = (c_{ij})$; $c_{ij} = \max\{|a_{ij} - m_{ij}|, |a_{ij} - \ell_{ij}|\}$, $v = C\rho$, $u = |A^{-1}|v$.

Theorem 2.1 If the test

$$\eta_i + u_i \leq \rho_i \quad (i=1,2,\ldots,n) \tag{2.1}$$

is satisfied, then there exists a solution x^* for (1.1) in X. Further, the error estimates

$$|x_i^{(1)} - x_i^*| \leq u_i \quad (i = 1,2,\ldots,n)$$

hold.

Theorem 2.2 In addition to (2.1), if

$$u_i < \rho_i \quad (i = 1,2,\ldots,n) \tag{2.2}$$

holds, then there exists a unique solution in X. Further, the iterative scheme,

$$(*) \qquad x^{(k)} = x^{(k-1)} - A^{-1}f(x^{(k-1)}), \quad k = 1,2,\ldots,$$

converges to x^* from any starting point in X.

By defining $u^{(1)} = u$, $v^{(k)} = Cu^{(k)}$, $u^{(k+1)} = |A^{-1}|v^{(k)}$, we obtain the error estimates

$$|x_i^{(k+1)} - x_i^*| \leq u_i^{(k+1)} \quad (k = 1,2,\ldots) \tag{2.3}$$

for the scheme with starting point $x^{(0)} = \tilde{x}$.

We also have the following alternative tests. Define M and L as in Remark 2.1. Let $A = (a_{ij})$; $a_{ij} = \frac{m_{ij}+\ell_{ij}}{2}$, $R = (\gamma_{ij})$; $\gamma_{ij} = \frac{m_{ij}-\ell_{ij}}{2}$, $P = (p_{ij}) = |A^{-1}|R$, and $z = P\rho$, where ρ is the radius vector of X.

Theorem 2.3 If the test

$$\eta_i + z_i \leq \rho_i \quad (i = 1,2,\ldots,n) \tag{2.4}$$

holds, then there exists a solution for (1.1) in X. Further, the error estimates

$$|x_i^{(1)} - x_i^*| \leq z_i \quad (i = 1,2,\ldots,n)$$

hold.

Theorem 2.4 In addition to (2.4), if

$$\sum_{j=1}^{i} p_{ij}\rho_j < \rho_i \quad (i = 1,2,\ldots,n) \tag{2.5}$$

holds, then there exists a unique solution x^* in X. Further, the iterative scheme (*) converges to x^* for any starting point in X. Also, the following error estimates hold for the scheme (*) with $x^{(0)} = \tilde{x}$.

$$|x_i^{(k+1)} - x_i^*| \le z_i^{(k+1)} \quad (i=1,2,\ldots,n), \tag{2.6}$$

where

$$z^{(k+1)} = Pz^{(k)}; \quad z^{(1)} = z.$$

Remark 2.3 In general, tests (2.4) and (2.5) are easier to satisfy than the tests (2.1) and (2.2). However, in some special cases, tests (2.1) and (2.2) will be very useful. For example, in the monotone method (discussed in the next section), we use the test (2.2).

We can combine the above tests with Newton's method to compute error estimates. Let $y^{(n-1)}$, $y^{(n)}$ be the points generated in $(n-1)^{th}$ and n^{th} Newton step. When the correction $\delta^{(n)} = |y^{(n)} - y^{(n-1)}|$ becomes small, we can check the test (2.4) by defining an n-dimensional rectangle X with mid point $\tilde{x} = y^{(n)}$, and radius vector $\rho = c\delta^{(n)}$; $c = 2$ or 3, $(A = f^1(y^{(n-1)})$ for the test (2.1)). Then, one computes the error estimates using Theorem 2.3 (or Theorem 2.2). Similar procedure also holds for an application of Kantorovich's Theorem. Also, if desired, one can use methods of interval analysis to compute L and M (defined in Remark 2.1). Let $[\alpha_{ij}, \beta_{ij}]$ be an interval extension of $f'_{i,j}$ on X. Then, $L = (\ell_{ij})$; $\ell_{ij} = \alpha_{ij}$ and $M = (m_{ij})$; $m_{ij} = \beta_{ij}$. For this purpose, computer packages are available, for example, see [9,10].

3. MONOTONE METHOD

In this section, as a special case of the scheme (*), we will study a monotone iterative method.

Suppose there exist $\underline{x}$ and $\bar{x} \in R^n$ such that $f(\underline{x}) \le 0 \le f(\bar{x})$. We refer $\underline{x}$ and $\bar{x}$ as lower and upper solutions, respectively. Let $X = \{x \in R^n | \underline{x}_i \le x_i \le \bar{x}_i,\ i = 1,2,\ldots,n\}$, and M is defined as before with respect to this X. If $M^{-1} \ge 0$, setting $A = M$ in the iterative scheme (*), one gets that the iterates $\bar{x}^{(k)} = s(\bar{x}^{(k-1)})$, $\bar{x}^{(0)} = \bar{x}$ are monotonicaly decreasing to the maximal solution $\bar{x}^*$, and the iterates $\underline{x}^{(k)} = s(\underline{x}^{(k-1)})$ are monotonically increasing to the minimal solution $\underline{x}^*$.

In general, $\underline{x}^* \neq \overline{x}^*$. In order to have these two-sided approximations converge to the same solution in X, we need a contraction condition. For this, define

$$X^{(k)} = \{x \in R^n | \underline{x}_i^{(k)} \leq x_i \leq \overline{x}_i^{(k)};\ i - 1,2,\ldots,n\},$$

$$m_{ij}^{(k)} = \max_{x \in X^{(k)}} [\frac{\partial f_i}{\partial x_j}(x)],\ \ell_{ij}^{(k)} = \min_{x \in X} [\frac{\partial f_i}{\partial x_j}(x)],$$

$$c_{ij}^{(k)} = \max[|m_{ij} - m_{ij}^{(k)}|, |m_{ij} - \ell_{ij}^{(k)}|],$$

$$C^{(k)} = (c_{ij}^{(k)}),\ M^{-1} = (y_{ij}),\ v = C^{(k_0)} z, z \in R^n$$

We assume that there is a test vector $z \geq 0$ such that, for some k_0,

$$\sum_{j=1}^{n} y_{ij} v_j < z_i, \tag{3.1}$$

$(i=1,2,\ldots,n)$.

Theorem 3.1 If M is inverse positive, and (3.1) holds for some k_0, then the two sided approximations

$$\overline{x}^{(k)} = s(\overline{x}^{(k-1)});\ \overline{x}^{(0)} = \overline{x},\ \text{and}\ \underline{x}^{(k)} = s(\underline{x}^{(k-1)});\ \underline{x}^{(0)} = \underline{x},$$

converge to the unique solution x^* in X.

The proof follows essentially by the fact that the operator $P = M^{-1}C^{(k_0)}$, under the assumption of (3.1), has spectral radius less than one. In this respect, we utilize Theorem 1.1 and Theorem 1.2. For details, see [5].

Remark 3.1 In the above theorem, condition (3.1) can be replaced by the condition,

$$\sum_{j=1}^{n} (m_{ij} - c_{ij}^{(k_0)}) z_i > 0 \quad (i = 1,2,\ldots,n) \tag{3.2}$$

for some k_0, and for some vector $z \geq 0$.

We note that (3.2) is a stronger condition than (3.1). On the other hand, (3.2) is simpler to verify than (3.1). However, if one has stored M^{-1} in the memory to generate the iterates (*), then the condition (3.1) is also easier to verify. Either condition (3.1) or (3.2) can be verified in the course of iterations, and a choice for z is the radius vector of $X^{(k_0)}$.

REFERENCES

[1] E. Adams and W. F. Ames, On contracting interval iterations for non-linear problems in R^n, Part I: theory, Nonlinear Analysis, TMA 3 (1979) pp. 773-794.

[2] V. Lakshmikantham and S. Leela, Differential and Integral Inequalities, Theory and Applications, Vol. I, Academic Press, NY, London (1969).

[3] R. E. Moore, Methods and Applications of Interval Analysis, SIAM Studies 2, Philadelphia, (1979).

[4] J. M. Ortega and W. C. Rehinboldt, Iterative Solution of Nonlinear Equations in Several Variables, Academic Press, NY, 1970.

[5] M. C. Pandian, Convergence Tests and Componentwise Error Estimates for Newton-Type Methods, (submitted).

[6] L. B. Rall, Computational Solution of Nonlinear Operator Equations, John Wiley, New York, (1969).

[7] Schroder, Johann, Operator inequalities, Academic Press, London (1980).

[8] C. Miranda, Un osservatione su un teorema di Brower, Bol. Un. Nat. Ital. Series 2, No. 2, (1940) pp. 5-7.

[9] F. S. Sisser, Computer-Generated Interval Extensions of Factorable Functions and their Derivatives, Intern. J. Computer Math., 10 (1982), pp. 327-336.

[10] J. M. Yohe, The interval arithmetic package, MRC Technical Summary Report No. 1755, MRC University of Wisconsin, Madison, 1977

[11] L. Qi, A Generalization of the Krawczyk-Moore Algorithm, Interval Mathematics (K. Nickel Editor), Academic Press, New York, (1980), pp. 481-488.

MONOTONE CONVERGENCE OF TIME-DEPENDENT SOLUTIONS FOR COUPLED REACTION-DIFFUSION SYSTEMS

C. V. Pao

Department of Mathematics
North Carolina State University
Raleigh, North Carolina

Introduction

In the analysis of reaction-diffusion systems one of the most interesting problems is the assymptotic behavior of the time-dependent solution in relation to the corresponding steady-state problem. This large time behavior of the solution depends not only on the reaction function and the physical parameters of the diffusion–reaction mechanism but also on the initial functions. It is known in the case of scalar systems that for a certain class of initial functions and some mild conditions on the reaction function the corresponding time-dependent solutions converge to steady-state solutions and the convergence is monotone in time (cf. [6]). This monotone convergence property is closely related to the monotone iteration method and the associated upper and lower solutions. In fact, the class of initial functions mentioned above includes all the upper and lower solutions of the steady-state problem. It is also easy to see that for the time-dependent problem, an upper solution is necessarily an upper bound of a lower solution. An unanswered question is whether this upper-lower bound property and the monotone convergence of time-dependent solutions remain valid for coupled system of reaction diffusion equations. To answer this question in the general framework is very difficult if not impossible. The purpose of this paper is to investigate the above question for a coupled system of two equations with a certain class of reaction functions.

The equations under consideration are in the form

$$\begin{aligned} u_t - L_1 u &= f(x,u,v) \\ v_t - L_2 v &= g(x,u,v) \end{aligned} \qquad (t > 0,\ x \in \Omega) \tag{1.1}$$

where Ω is a smooth bounded domain in R^n and L_i are uniformly elliptic operators in the form

$$L_i = \sum_{k,\ell=1}^{n} a_{k\ell}^{(i)}(x)\partial^2/\partial x_k \partial x_\ell + \sum_{k=1}^{n} b_k \, \partial/\partial x_k + c_i \qquad i = 1,2$$

with $c_i(x) \le 0$ in Ω. The boundary and initial conditions are given, respectively, by

$$\begin{aligned} B_1[u] &\equiv \alpha_1(x)\partial u/\partial \nu + \beta_1(x)u = h_1(x) \\ & \qquad\qquad\qquad\qquad\qquad\qquad (t > 0,\ x \in \partial\Omega) \\ B_2[v] &\equiv \alpha_2(x)\partial v/\partial \nu + \beta_2(x)v = h_2(x) \end{aligned} \tag{1.2}$$

$$u(0,x) = u_o(x), \quad v(0,x) = v_o(x) \qquad (x \in \Omega) \tag{1.3}$$

where $\partial\Omega$ is the boundary of Ω, $\partial/\partial\nu$ is the outward normal derivative on $\partial\Omega$ and α_i, β_i are nonnegative continuous functions on $\partial\Omega$ with $\alpha_i + \beta_i > 0$, $i = 1,2$. It is assumed that the matrices $(a_{k\ell}^{(i)})$ are symmetric positive definite, the coefficients of L_i are smooth and the boundary-initial data h_1, h_2, u_o, v_o are continuous functions of their respective arguments.

In addition to the time-dependent problem (1.1)-(1.3) we also consider the corresponding steady state problem

$$\begin{aligned} -L_1 u &= f(x,u,v) \\ & \qquad\qquad\qquad (x \in \Omega) \\ -L_2 v &= g(x,u,v) \end{aligned} \tag{1.4}$$

$$B_1[u] = h_1, \qquad B_2[v] = h_2 \quad (x \in \partial\Omega) \tag{1.5}$$

Our main goal is to show that if $f_v \le 0$, $g_u \le 0$ or $f_v \ge 0$, $g_u \ge 0$ in $S(\Omega)$, where $S(\Omega)$ is given in terms of upper and lower solutions, then the solution of (1.1)-(1.3) converges monotonically to a steady-state solution of (1.4) (1.5) provided that the initial function (u_o, v_o) is taken as upper and lower solutions of (1.4) (1.5) in some suitable fashion. We also show that for this type of reaction functions an upper solution of the time-dependent problem (1.1)-(1.3) (but not the steady-state problem (1.4) (1.5)) is necessarily an upper bound of a lower solution, and each member in the maximal sequence (resp., minimal sequence) in the iteration process is an upper solution (resp., lower solution) of the problem.

2. UPPER AND LOWER SOLUTIONS

In order to describe the monotone iteration process and to identify the class of initial functions so that the corresponding time-dependent solutions converge monotonically to steady-state solutions we consider a pair of smooth functions $(\tilde{u},\tilde{v})$, $(\underset{\sim}{u},\underset{\sim}{v})$ which satisfy $(\underset{\sim}{u},\underset{\sim}{v}) \leq (\tilde{u},\tilde{v})$ and the boundary-initial inequalities

$$\left.\begin{array}{l} B_1[\tilde{u}]-h_1 \geq 0 \geq B_1[\underset{\sim}{u}]-h_1 \\ B_2[\tilde{v}]-h_2 \geq 0 \geq B_2[\underset{\sim}{v}]-h_2 \end{array}\right. \qquad (x \in \partial\Omega) \tag{2.1}$$

$$\left.\begin{array}{l} \tilde{u}(0,x)-u_o(x) \geq 0 \geq \underset{\sim}{u}(0,x) - u_o(x) \\ \tilde{v}(0,x)-v_o(x) \geq 0 \geq \underset{\sim}{v}(0,x) - v_o(x) \end{array}\right. \qquad (x \in \Omega) \tag{2.2}$$

Assume that f,g are continuously differentiable in u,v, and either $f_v \leq 0$, $g_u \leq 0$ or $f_v \geq 0$, $g_u \geq 0$ for (u,v) in S(D), where $D = [0,\infty) \times \bar{\Omega}$ and

$$S(D) = \{(u,v) \in C(D);\ (\underset{\sim}{u},\underset{\sim}{v}) \leq (u,v) \leq (\tilde{u},\tilde{v}) \text{ in } D\} \tag{2.3}$$

Then we have the following definition (cf. [4]).

Definition 2.1 Let $(\tilde{u},\tilde{v})$, $(\underset{\sim}{u},\underset{\sim}{v})$ be smooth functions in D satisfying the relations (2.1) (2.2). Then $(\tilde{u},\tilde{v})$, $(\underset{\sim}{u},\underset{\sim}{v})$ are called upper and lower solutions of (1.1)-(1.3) if

$$\begin{array}{l} \tilde{u}_t - L_1\tilde{u} - f(x,\tilde{u},\underset{\sim}{v}) \geq 0 \geq \underset{\sim}{u}_t - L_1\underset{\sim}{u} - f(x,\underset{\sim}{u},\tilde{v}) \\ \tilde{v}_t - L_2\tilde{v} - g(x,\underset{\sim}{u},\tilde{v}) \geq 0 \geq \underset{\sim}{v}_t - L_2\underset{\sim}{v} - g(x,\tilde{u},\underset{\sim}{v}) \end{array} \tag{2.4}$$

when $f_v \leq 0$, $g_u \leq 0$; and

$$\begin{array}{l} \tilde{u}_t - L_1\tilde{u} - f(x,\tilde{u},\tilde{v}) \geq 0 \geq \underset{\sim}{u}_t - L_1\underset{\sim}{u} - f(x,\underset{\sim}{u},\underset{\sim}{v}) \\ \tilde{v}_t - L_2\tilde{v} - g(x,\tilde{u},\tilde{v}) \geq 0 \geq \underset{\sim}{v}_t - L_2\underset{\sim}{v} - g(x,\underset{\sim}{u},\underset{\sim}{v}) \end{array} \tag{2.5}$$

when $f_v \geq 0$, $g_u \geq 0$.

For the steady-state problem we call the functions $(\tilde{u}_s,\tilde{v}_s)$, $(\underset{\sim}{u}_s,\underset{\sim}{v}_s)$ defined on $\bar{\Omega}$ upper and lower solutions of (1.4) (1.5) if the inequalities (2.1) (2.4) are satisfied when $f_v \leq 0$, $g_u \leq 0$; and the inequalities (2.1) (2.5) are satisfied when $f_v \geq 0$, $g_u \geq 0$. In each case, the time-derivative terms in (2.4), (2.5) are neglected.

It is clear from the above definition that if $(\tilde{u}_s,\tilde{v}_s)$, $(\underset{\sim}{u}_s,\underset{\sim}{v}_s)$ are upper and lower solutions of (1.4) (1.5) then they are also upper and lower solutions of (1.1)-(1.3) provided that $(\underset{\sim}{u}_s, \underset{\sim}{v}_s) \le (u_o, v_o) \le (\tilde{u}_s, \tilde{v}_s)$. Suppose upper and lower solutions $(\tilde{u}, \tilde{v})$, $(\underset{\sim}{u}, \underset{\sim}{v})$ of (1.1)-(1.3) exist and $(\underset{\sim}{u}, \underset{\sim}{v}) \le (\tilde{u}, \tilde{v})$. Define

$$\left.\begin{aligned} M_1 &= \sup \{f_u(x,u,v);\ x \in \Omega,\ (\underset{\sim}{u}, \underset{\sim}{v}) \le (u,v) \le (\tilde{u}, \tilde{v})\} \\ M_2 &= \sup \{g_v(x,u,v);\ x \in \Omega,\ (\underset{\sim}{u}, \underset{\sim}{v}) \le (u,v) \le (\tilde{u}, \tilde{v})\} \end{aligned}\right. \tag{2.6}$$

Then by using upper and lower solutions as initial iterations we can obtain two monotone sequences as follows:

(a) For $f_v \le 0$, $g_u \le 0$, we use the initial iteration $(\bar{u}^{(o)}, \underline{v}^{(o)}) = (\tilde{u}, \underset{\sim}{v})$ and $(\underline{u}^{(o)}, \bar{v}^{(o)}) = (\underset{\sim}{u}, \tilde{v})$ to obtain two sequences $\{\bar{u}^{(k)}, \underline{v}^{(k)}\}$, $\{\underline{u}^{(k)}, \bar{v}^{(k)}\}$ from the iteration process

$$\left.\begin{aligned} \bar{u}_t^{(k)} - L_1\bar{u}^{(k)} + M_1\bar{u}^{(k)} &= M_1\bar{u}^{(k-1)} + f(x,\bar{u}^{(k-1)}, \underline{v}^{(k-1)}) \\ \underline{v}_t^{(k)} - L_2\underline{v}^{(k)} + M_2\underline{v}^{(k)} &= M_2\underline{v}^{(k-1)} + g(x,\bar{u}^{(k-1)}, \underline{v}^{(k-1)}) \end{aligned}\right\} \tag{2.7}$$

$$\left.\begin{aligned} \underline{u}_t^{(k)} - L_1\underline{u}^{(k)} + M_1\underline{u}^{(k)} &= M_1\underline{u}^{(k-1)} + f(x,\underline{u}^{(k-1)}, \bar{v}^{(k-1)}) \\ \bar{v}_t^{(k)} - L_2\bar{v}^{(k)} + M_2\bar{v}^{(k)} &= M_2\bar{v}^{(k-1)} + g(x,\underline{u}^{(k-1)}, \bar{v}^{(k-1)}) \end{aligned}\right\} \tag{2.8}$$

(b) For $f_v \ge 0$, $g_u \ge 0$ we construct two sequences $\{\bar{u}^{(k)}, \bar{v}^{(k)}\}$, $\{\underline{u}^{(k)}, \underline{v}^{(k)}\}$ from (2.7) (2.8) by interchanging $\bar{v}$ and $\underline{v}$, where $(\bar{u}^{(0)}, \bar{v}^{(0)}) = (\tilde{u}, \tilde{v})$, $(\underline{u}^{(0)}, \underline{v}^{(0)}) = (\underset{\sim}{u}, \underset{\sim}{v})$. In both cases the boundary and initial conditions are given by

$$B_1[u^{(k)}] = h, \qquad B_2[v^{(k)}] = h_2 \qquad (x \in \partial\Omega) \tag{2.9}$$

$$u^{(k)}(0,x) = u_o(x), \qquad v^{(k)}(0,x) = v_o(x) \qquad (x \in \Omega) \tag{2.10}$$

It has been shown in [4] that for the case $f_v \ge 0$, $g_u \ge 0$ in S(D) the sequence $\{\bar{u}^{(k)}, \bar{v}^{(k)}\}$ converges monotonically from above to a unique solution (u,v) of (1.1)-(1.3) while $\{\underline{u}^{(k)}, \underline{v}^{(k)}\}$ converges monotonically from below to the same solution. When $f_v \le 0$, $g_u \le 0$ in S(D) the two sequences $\{\bar{u}^{(k)}, \underline{v}^{(k)}\}$, $\{\underline{u}^{(k)}, \bar{v}^{(k)}\}$ also converge monotonically to a unique solution (u,v) but the convergence is in mixed order in the sense that $\bar{u}^{(k)} \searrow \bar{u}$, $\underline{v}^{(k)} \nearrow \underline{v}$ and $\underline{u}^{(k)} \nearrow \underline{u}$, $\bar{v}^{(k)} \searrow \bar{v}$.

By dropping the time-derivative term and using $(\tilde{u}_s, \underset{\sim}{v}_s)$ $(\underset{\sim}{u}_s, \tilde{v}_x)$ as the initial iterations when $f_v \le 0$, $g_u \le 0$ in $S(\Omega)$, where

$$S(\Omega) = \{(u,v) \in C(\bar{\Omega});\ (\underset{\sim}{u}_s, \underset{\sim}{v}_s) \le (u,v) \le (\tilde{u}_s, \tilde{v}_s) \text{ in } \bar{\Omega}\} \tag{2.11}$$

we can also construct two sequences from (2.7) (2.9) and denote them again by $\{\bar{u}^{(k)}\ \underline{v}^{(k)}\}$ and $\{\underline{u}^{(k)}, \bar{v}^{(k)}\}$. Similarly, for $f_v \ge 0$, $g_u \ge 0$ in $S(\Omega)$ these two sequences are obtained from (2.8) (2.9) and are denoted by $\{\bar{u}^{(k)}, \bar{v}^{(k)}\}$, $\{\underline{u}^{(k)}, \underline{v}^{(k)}\}$. These two sequences possess the following properties;

(a) For $f_v \ge 0$, $g_u \ge 0$ in $S(\Omega)$, the sequences $\{\bar{u}^{(k)}, \bar{v}^{(k)}\}$ converges monotonically from above to a "maximal" solution $(\bar{u},\bar{v})$, and $\{\underline{u}^{(k)}, \underline{v}^{(k)}\}$ converges monotonically from below to a "minimal" solution $(\underline{u},\underline{v})$

(b) For $f_v \le 0$, $g_u \le 0$ in $S(\Omega)$, the sequences $\{\bar{u}^{(k)}, \underline{v}^{(k)}\}$, $\{\underline{u}^{(k)}, \bar{v}^{(k)}\}$ converge monotonically in mixed order to the respective solution $(\bar{u},\underline{v})$ and $(\underline{u},\bar{v})$. In both cases in (a), (b), the solutions satisfy the relation

$$\begin{aligned} &\underset{\sim}{u} \le \underline{u}^{(1)} \le \dots \le \underline{u} \le \bar{u} \le \dots \le \bar{u}^{(1)} \le \tilde{u} \\ &\underset{\sim}{v} \le \underline{v}^{(1)} \le \dots \le \underline{v} \le \bar{v} \le \dots \le \bar{v}^{(1)} \le \tilde{v} \end{aligned} \qquad (x \in \Omega) \tag{2.12}$$

The above conclusion implies that for both time-dependent and steady-state problems every member in the sequence $\{\bar{u}^{(k)}, \bar{v}^{(k)}\}$ (resp.,$\{\underline{u}^{(k)}\ \underline{v}^{(k)}\}$) is an upper bound (resp., lower bound) of the solution. We show in the following theorem that they are also upper and lower solutions, and for the time-dependent problem every upper solution is necessarily an upper bound of a lower solution.

<u>Theorem 2.1:</u> Let $(\tilde{u},\tilde{v})$, $(\underset{\sim}{u},\underset{\sim}{v})$ be upper and lower solutions of (1.1)-(1.3) and let either $f_v \le 0$, $g_u \le 0$ or $f_v \ge 0$, $g_u \ge 0$ for (u,v) between $(\tilde{u},\tilde{v})$ and $(\underset{\sim}{u},\underset{\sim}{v})$. Then (i) $(\tilde{u},\tilde{v}) \ge (\underset{\sim}{u},\underset{\sim}{v})$ on D, and (ii) $(\bar{u}^{(k)}, \bar{v}^{(k)})$ and $(\underline{u}^{(k)}, \underline{v}^{(k)})$ are upper and lower solutions of (1.1)-(1.3) for every $k = 1,2,\dots$.

<u>Proof:</u> Let $U = \tilde{u} - \underset{\sim}{u}$, $V = \tilde{v} - \underset{\sim}{v}$. Then by definition,

$$B_1[U] \ge 0,\quad B_2[V] \ge 0 \text{ and } U(o,x) \ge 0,\quad V(o,x) \ge 0. \tag{2.13}$$

When $f_v \le 0$, $g_u \le 0$, the relation (2.4) implies that

$$U_t - L_1 U \geq [f(x,\tilde{u},\underset{\sim}{v}) - f(x,\underset{\sim}{u},\underset{\sim}{v})] + [f(x,\underset{\sim}{u},\underset{\sim}{v}) - f(x,\underset{\sim}{u},\tilde{v})]$$

$$= f_u(x,u^*,\underset{\sim}{v})U - f_v(x,\underset{\sim}{u},v^*)V$$

$$V_t - L_2 V \geq [g(x,\underset{\sim}{u},\tilde{v}) - g(x,\tilde{u},\tilde{v})] + [g(x,\tilde{u},\tilde{v}) - g(x,\tilde{u},\underset{\sim}{v})]$$

$$= -g_u(x,u^*,\tilde{v})U + g_v(x,\tilde{u},v^*)V$$

where (u^*,v^*) is some intermediate value between $(\underset{\sim}{u},\underset{\sim}{v})$ and $(\tilde{u},\tilde{v})$. Since $f_v(x,\underset{\sim}{u},u^*) \leq 0$, $g_u(x,u^*,\tilde{v}) \leq 0$ we see from a result in [5] that $U \geq 0$, $V \geq 0$ (see also Lemma 2.1 below). This proves the conclusion in (i) when $f_v \leq 0$, $g_u \leq 0$. The proof for the case $f_v \geq 0$, $g_u \geq 0$ is similar. To show the result in(ii) it suffices to varify the inequalities in (2.4) and (2.5). Consider the case $f_v \leq 0$, $g_u \leq 0$. Then by the iteration process (2.7),

$$\bar{u}_t^{(k)} - L_1\bar{u}^{(k)} = M_1(\bar{u}^{(k-1)} - \bar{u}^{(k)}) + [f(\bar{u}^{(k-1)}, \underline{v}^{(k-1)}) - f(\underline{u}^{(k)}, \underline{v}^{(k-1)})]$$
$$+ [f(\underline{u}^{(k)}, \underline{v}^{(k-1)}) - f(\bar{u}^{(k)}, \underline{v}^{(k)})] + f(\bar{u}^{(k)}, \underline{v}^{(k)})$$

$$\underline{v}_t^{(k)} - L_2\underline{v}^{(k)} = M_2(\underline{v}^{(k-1)} - \underline{v}^{(k)}) + [g(\bar{u}^{(k-1)}, \underline{v}^{(k-1)}) - g(\bar{u}^{(k-1)}, \underline{v}^{(k)})]$$
$$+ [g(\bar{u}^{(k-1)},\underline{v}^{(k)}) - g(\bar{u}^{(k)}, \underline{v}^{(k)})] + g(\bar{u}^{(k)}, \underline{v}^{(k)})$$

Since by (2.6)

$$f(u_1,v) - f(u_2,v) \geq - M_1(u_1 - u_2) \text{ when } \tilde{u} \geq u_1 \geq u_2 \geq \underset{\sim}{u}$$

$$g(u,v_1) - g(u,v_2) \geq - M_2(v_1 - v_2) \text{ when } \tilde{v} \geq v_1 \geq v_2 \geq \underset{\sim}{v}$$

where $(u,v) \in S(D)$ we see from the relation $\bar{u}^{(k)} \leq \bar{u}^{(k-1)}$, $\underline{v}^{(k)} \geq \underline{v}^{(k-1)}$ and the monotone nonincreasing property of f,g that

$$\bar{u}_t^{(k)} - L_1\bar{u}^{(k)} \geq f(\bar{u}^{(k)}, \underline{v}^{(k)})$$

$$\underline{v}_t^{(k)} - L_2\underline{v}^{(k)} \leq g(\bar{u}^{(k)}, \underline{v}^{(k)})$$

In obtaining the above inequalities we have used the fact that $(\bar{u}^{(k)}, \bar{v}^{(k)})$ and $(\underline{u}^{(k)}, \underline{v}^{(k)})$ remain in S(D) for every k. A similar argument leads to the relation

$$\underline{u}_t^{(k)} - L_1\underline{u}^{(k)} \le f(\underline{u}^{(k)}, \bar{v}^{(k)})$$

$$\bar{v}_t^{(k)} - L_2\bar{v}^{(k)} \ge g(\underline{u}^{(k)}, \bar{v}^{(k)})$$

This shows that $(\bar{u}^{(k)}, \bar{v}^{(k)})$, $(\underline{u}^{(k)}, \underline{v}^{(k)})$ are upper and lower solutions of (1.1)-(1.3). The proof for the case $f_v \ge 0$, $g_u \ge 0$ is the same.

By neglecting the time-derivative term and the initial condition the above argument leads to a similar result for the steady-state problem (1.4) (1.5).

Corollary: Let $(\tilde{u}_s, \tilde{v}_s)$, $(\underset{\sim}{u}_s, \underset{\sim}{v}_s)$ be upper and lower solutions of (1.4) (1.5) such that $(\underset{\sim}{u}_s, \underset{\sim}{v}_s) \le (\tilde{u}_s, \tilde{v}_s)$ and let either $f_v \le 0$, $g_u \le 0$ in $S(\Omega)$ or $f_v \ge 0$, $g_u \ge 0$ in $S(\Omega)$. Then the time-independent functions $(\bar{u}^{(k)}, \bar{v}^{(k)})$, $(\underline{u}^{(k)}, \underline{v}^{(k)})$ are upper and lower solutions of (1.4) (1.5) for each k.

In the proof of Theorem 2.1 we have used a result from [5] for semi-linear parabolic system. This result can be deduced from the existence comparison theorem obtained in [4] by considering the linear functions

$$f \equiv a_1(x)u + b_1(x)v + q_1(x), \quad g = a_2(x)u + b_2(x)v + q_2(x) \tag{2.14}$$

Suppose $b_1 \ge 0$, $a_2 \ge 0$ so that $f_v \ge 0$, $g_u \ge 0$. Then the problem (1.1) - (1.3) has a unique solution (u,v) satisfying $(\underset{\sim}{u}, \underset{\sim}{v}) \le (u,v) \le (\tilde{u}, \tilde{v})$ whenever upper and lower solutions exist. Of particular interest is the case where q_i, h_i, u_o, v_o (i=1,2) are nonnegative continuous functions. In this situation a pair of upper and lower solutions may be taken as $(u,v) = (c_1 e^{\gamma t}, c_2 e^{\gamma t})$, $(\underset{\sim}{u}, \underset{\sim}{v}) = (0,0)$ for some sufficiently large constants c_1, c_2, γ. This observation leads to the following conclusion.

Lemma 2.1: Let f,g be given by (2.14) with $b_1 \ge 0$, $a_2 \ge 0$ and let q_i, h_i, u_o, v_o be nonnegative continuous functions. Then the solution (u,v) of (1.1)-(1.3) is nonnegative on D.

3. Monotone convergence of time-dependent solutions.

The results in the previous section imply that if $(\tilde{u}_s, \tilde{v}_s)$, $(\underset{\sim}{u}_s, \underset{\sim}{v}_s)$ are upper and lower solutions of (1.4) (1.5) then the set of functions $S(\Omega)$ is an invariant set of the time-dependent system (1.1) - (1.3).

Since by the corollary to Theorem 2.1, the time-independent sequences $\{\bar{u}^{(k)}, \bar{v}^{(k)}\}$, $\{\underline{u}^{(k)}, \underline{v}^{(k)}\}$ are upper and lower solutions of (1.4) (1.5); it follows that each of the sets

$$S_k(\Omega) = \{(u,v)\varepsilon\ C(\Omega);\ (\underline{u}^{(k)},\underline{v}^{(k)}) \leq (u,v) \leq (\bar{u}^{(k)},\bar{v}^{(k)})\},\quad k = 1,2,\ldots, \tag{3.1}$$

is an invariant set of (1.1)-(1.3) and $S_{k_1} \subset S_{k_2}$ for every $k_1 \geq k_2$. The aim of this section is to show that if the initial function (u_o,v_o) is taken as upper and lower solutions then the corresponding time-dependent solution not only remains in $S(\Omega)$ (or $S_k(\Omega)$) but also converges monotonically to a steady-state solution. We first establish this result for the case $f_v \leq 0$, $g_u \leq 0$ in the following

Theorem 3.1: Let $(\tilde{u}_s,\tilde{v}_s)$, $(\underset{\sim}{u}_s,\underset{\sim}{v}_s)$ be upper and lower solutions of (1.4) (1.5) with $(\underset{\sim}{u}_s,\underset{\sim}{v}_s) \leq (\tilde{u}_s,\tilde{v}_s)$ and let either $c_i(x) \not\equiv 0$ in Ω or $\beta_i(x) \not\equiv 0$ on $\partial\Omega$, $i = 1,2,$. Suppose $f_v \leq 0$, $g_u \leq 0$ on $S(\Omega)$ and denote by $(\bar{u},\underline{v})$, $(\underline{u},\bar{v})$ the solutions of (1.1)-(1.3) with $(\bar{u}(0,x),\underline{v}(0,x)) = (\tilde{u}_s,\underset{\sim}{v}_s)$, $(\underline{u}(0,x),\bar{v}(0,x)) = (\underset{\sim}{u}_s,\tilde{v}_s)$. Then (i) $\bar{u}(t,x)$, $\bar{v}(t,x)$ are monotone nonincreasing and $\underline{u}(t,x)$, $\underline{v}(t,x)$ are monotone nondecreasing in t, and

$$\underset{\sim}{u}_s(x) \leq \underline{u}(t,x) \leq \bar{u}(t,x) \leq \tilde{u}_s(x),\ \underset{\sim}{v}_s(x) \leq \underline{v}(t,x) \leq \bar{v}(t,x) \leq \tilde{v}_s(x) \tag{3.2}$$

(ii) The pointwise limits

$$\lim(\bar{u}(t,x),\ \underline{v}(t,x)) = (\bar{u}_s(x),\ \underline{v}_s(x)),\quad \lim(\underline{u}(t,x),\ \bar{v}(t,x)) = (\underline{u}_s(x),\ \bar{v}_s(x))\quad \text{as } t \to \infty \tag{3.3}$$

exist and both $(\bar{u}_s,\underline{v}_s)$ and $(\underline{u}_s,\bar{v}_s)$ are steady state solutions of (1.4) (1.5). Moreover

$$\underset{\sim}{u}_s(x) \leq \underline{u}_s(x) \leq \bar{u}_s(x) \leq \tilde{u}_s(x),\ \underset{\sim}{v}_s(x) \leq \underline{v}_s(x) \leq \bar{v}_s(x) \leq \tilde{v}_s(x). \tag{3.4}$$

(iii) Any solution (u,v) of (1.1)-(1.3) with $(\underset{\sim}{u}_s,\underset{\sim}{v}_s) \leq (u_o,v_o) \leq (\tilde{u}_s,\tilde{v}_s)$

satisfies the relation

$$\underline{u}(t,x) \le u(t,x) \le \bar{u}(t,x), \quad \underline{v}(t,x) \le v(t,x) \le \bar{v}(t,x) \tag{3.5}$$

Proof: In view of the invariant property of $S(\Omega)$ the solutions $(\bar{u},\underline{v})$, $(\underline{u},\bar{v})$ remain in $S(\Omega)$ for all $t \ge 0$. We first show the monotone property of $(\bar{u},\underline{v})$. Let $h > 0$ be fixed and let $U = \bar{u}(t,x) - \bar{u}(t+h,x)$, $V = \underline{v}(t+h,x) - \underline{v}(t,x)$. Then

$B_1[U] = B_2[V] = 0$, $U(0,x) = \tilde{u}(x) - \bar{u}(h,x) \ge 0$, $V(0,x) = \underline{v}(h,x) - \underset{\sim}{v}(x) \ge 0$. By supressing the variable x in f,g, we have

$$\begin{aligned} U_t - L_1U &= f(\bar{u}(t,x), \underline{v}(t,x)) - f(\bar{u}(t+h,x), \underline{v}(t+h,x)) \\ &= -[f(\bar{u}(t+h,x), \underline{v}(t+h,x)) - f(\bar{u}(t,x), \underline{v}(t+h,x))] \\ &\quad - [f(\bar{u}(t,x), \underline{v}(t+h,x)) - f(\bar{u}(t,x), \underline{v}(t,x))] \\ &= f_u(\bar{u}^*(t,x), \underline{v}(t+h,x))U - f_v(\bar{u}(t,x), \underline{v}^*(t,x))\, V \end{aligned}$$

$$\begin{aligned} V_t - L_2V &= g(\bar{u}(t+h,x), \underline{v}(t+h,x)) - g(\bar{u}(t,x), \underline{v}(t,x)) \\ &= -g_u(\bar{u}^*(t,x), \underline{v}(t+h,x))U + g_v(\bar{u}(t,x), \underline{v}^*(t,x))\, V \end{aligned}$$

where $\bar{u}^*$, $\underline{v}^*$ are some intermediate values between $\bar{u}(t,x)$ and $\bar{u}(t+h,x)$ and between $\underline{v}(t,x)$ and $\underline{v}(t+h,x)$, respectively. Since $(\bar{u},\underline{v}) \in S(\Omega)$ for all $t \ge 0$ and $f_v \le 0$, $g_u \le 0$ in $S(\Omega)$ we conclude from [5] (or Lemma 2.1) that $U \ge 0$, $V \ge 0$. The arbitrariness of h,t implies that $\bar{u}(t,x)$ is monotone nonincreasing and $\underline{v}(t,x)$ is monotone nondecreasing in t. The proof for the monotone property of $(\underline{u}(t,x), \bar{v}(t,x))$ is the same. To show the relation (3.2) it suffices to verify that $\underline{u} \le \bar{u}$, $\underline{v} \le \bar{v}$. Let $U = \bar{u}(t,x) - \underline{u}(t,x)$, $V = \bar{v}(t,x) - \underline{v}(t,x)$. Then U,V satisfy (2.13) and

$$\begin{aligned} U_t - L_1U &= f(\bar{u},\underline{v}) - f(\underline{u},\bar{v}) = f_u(u^*,\underline{v})U - f_v(\underline{u},v^*)V \\ V_t - L_2V &= g(\underline{u},\bar{v}) - g(\bar{u},\underline{v}) = -g_u(u^*,\bar{v})U + g_v(\underline{u},v^*)V \end{aligned} \tag{3.6}$$

for some intermediate values u^*, v^*. It follows again from $f_v \le 0$, $g_u \le 0$ in $S(\Omega)$ that $U \ge 0$, $V \ge 0$. This proves the result in (i).

In view of (3.2) and the monotone property of the solutions the pointwise limits in (3.3) exist and the relation in (3.4) holds. To justify that $(\bar{u}_s,\underline{v}_s)$, $(\underline{u}_s,\bar{v}_s)$ are steady-state solutions we consider the solution

(w_1, w_2) of the linear boundary-value problem

$$-L_1 w_1 = f(x, \bar{u}_s, \underline{v}_s), \quad B_1[w_1] = h_1(x)$$
$$\tag{3.7}$$
$$-L_2 w_2 = g(x, \bar{u}_s, \underline{v}_s), \quad B_2[w_2] = h_2(x)$$

Then the functions $U \equiv \bar{u}(t,x) - w_1(x)$, $V \equiv \underline{v}(t,x) - w_2(x)$ satisfy the relation

$$U_t - L_1 U = f(\bar{u},\underline{v}) - f(\bar{u}_s,\underline{v}_s), \quad B_1[U] = 0 \tag{3.8}$$

$$V_t - L_2 V = g(\bar{u},\underline{v}) - g(\bar{u}_s,\underline{v}_s), \quad B_2[V] = 0 \tag{3.9}$$

Clearly the functions on the right-side of (3.8), (3.9) converge to zero in $L^2(\Omega)$ as $t \to \infty$. But by the hypothesis $c_i \le 0$ and either $c_i \not\equiv 0$ or $\beta_i \not\equiv 0$, all the eigenvalues of the eigenvalue problem

$$L_i \phi + \lambda_i \phi = 0, \quad B_i[\phi] = 0$$

have positive real parts. This implies that the solutions U, V of (3.8), (3.9) converge to zero in $L^2(\Omega)$ as $t \to \infty$ (cf, [2]). But $U \to \bar{u}_s - w_1$, $V \to \underline{v}_s - w_2$ pointwise in Ω as $t \to \infty$ we conclude that $(w_1, w_2) = (\bar{u}_s, \underline{v}_s)$ a.e. in Ω. A standard argument using the embedding lemma and the schauder estimate shows that $w_1, w_2 \in C^{2+\alpha}(\Omega)$ for some $\alpha > 0$ (e.g. see [1] p. 204, [3] p. 306). Therefore $(\bar{u}_s, \underline{v}_s)$ is a classical solution of (1.4) (1.5) and the result in (ii) follows.

To show the conclusion in (iii) we observe that the functions $U = \bar{u}(t,x) - u(t,x)$, $V = v(t,x) - \underline{v}(t,x)$ satisfy the equations in (3.6) except with $\underline{u}$, $\bar{v}$ replaced by u, v respectively. Since $U(o,x) = \tilde{u} - u_o \ge 0$, $V(o,x) = v_o - \underset{\sim}{v} \ge 0$ and $B_1[U] = B_2[V] = 0$ we conclude again that $U \ge 0$, $V \ge 0$. This proves the relation $\bar{u} \ge u$, $\underline{v} \le v$. The proof for $\underline{u} \le u$, $\bar{v} \ge v$ is similar. This completes the proof of the theorem.

Using the same argument as in the proof of Theorem 3.1 we obtain the following result for the case $f_v \ge 0$, $g_u \ge 0$.

<u>Theorem 3.2.</u> Let the hypotheses in Theorem 3.1 hold except with $f_v \ge 0$, $g_u \ge 0$ in $S(\Omega)$ and with $(\bar{u}, \bar{v})$, $(\underline{u}, \underline{v})$ representing the solutions of (1.1)-(1.3) with $(\bar{u}(o,x), \bar{v}(o,x)) = (\tilde{u}_s, \tilde{v}_s)$ and $(\underline{u}(o,x), \underline{v}(o,x)) = (\underset{\sim}{u}_s, \underset{\sim}{v}_s)$. Then all the conclusions (i)-(iii) in Theorem 3.1 remain true.

References

[1] S. Agmon, A. Douglis, and L. Nirenberg, Estimates near the boundary for solutions of elliptic partial differential equations satisfying general boundary condition, I, Comm. Pure. Appl. Math., 12(1959), 623-627.

[2] A. Friedman, Partial Differential Equations of Parabolic Type, Prentice Hall, Englewood Cliffs, N.J., 1964.

[3] C. B. Morrey, Multiple Integrals in the Calculus of Variations, Springer, New York, 1966.

[4] C. V. Pao, On nonlinear reaction diffusion systems, J. Math. Anal. Appl., 87(1982), 165-198.

[5] M. H. Protter and H. F. Weinberger, Maximum principles in differential equations, Prentice Hall, 1967.

[6] D. H. Sattinger, Monotone methods in nonlinear elliptic and parabolic boundary value problems, Indiana Univ. Math. J., 21(1972), 979-1000.

A SIMPLE CONVERSION OF TWO-POINT BVP TO ONE POINT BVP

Fred R. Payne and Fung-Tai Ko

Department of Aerospace Engineering
University of Texas at Arlington
Arlington, Texas

ABSTRACT

Consider the Falkner-Skan equation which governs "similar" laminar boundary-layer flows:

$$F''' + FF'' + \beta(1 - F'^2) = 0$$

Subject to B.C.: $F'(0) = F(0) = 0$; $F'(\infty) = 1$. The simple conversion is to integrate once (or twice) formally to obtain an integro-differential equation. e.g. -

$$F' = \tau_o^+\eta - \frac{\beta\eta^2}{2} - \frac{F^2}{2} + (\beta + 1)\int_o^\eta dZ \int_o^z dwF'^2$$

where $\eta \propto y/\delta(x)$ is the similarity variable. In (1) and (2) β is known from the external, potential flow solution about the body. Solution to (2) is quickly, in about half the time of a fourth order Runge-Kutta, obtained by a "point-wise iteration" scheme applied to an initial guess of F' as linear and satisfying $F'(0) = 0$, $F''(0) = \tau_o^+$, also guessed initially. $F(0) = 0$ is automatically satisfied. The method has been successful for $\beta \, \varepsilon(-0.18, 20)$ spanning the physical range. In principle, the method should apply to any non-linear equation of the two-point BVP type. Essentially, it replaces R-Kutta integration by formal ones and simple quadratures. A serendipitous effect has been separation of cause and effect in a simulated boundary layer development as the driving pressure gradient (β) and the driven wall shear (τ_o^+) come into accommodation.

OVERVIEW

After some introductory remarks on prior techniques for solution of the Falkner-Skan equation an abbreviated derivation of the equation is given. Some typical numerical results obtained by the new method are presented and compared to fourth-order Runge-Kutta values. (Ko, F-T, MSAE Thesis, May 1982, UTA)

Application of the new technique is made to the steady Prandtl equation, which governs boundary layer flows for both laminar and turbulent modes. Some discussion is given of merits over other modellings and of extensions to unsteady, 3-D, separating and transitional flows. The closure lays claim to a broad class of applications of this simple approach.

INTRODUCTION

With modern digital computers the non-linear, third order Falkner-Skan ODE can be solved efficiently by Runge-Kutta. However, since Runge-Kutta is a one-point boundary value/initial value method, only two physical boundary conditions can be used directly; the third must be guessed in advance and "shooting" employed to match it far from the surface.

A new method, derived by formally integrating twice the Falkner-Skan equation, includes an organic iterative scheme and a scaling process without any shooting.

Computation times and number of iterations compare favorably with a 4th order Runge-Kutta method. A fundamental problem, common to all boundary-layer techniques, remains; i.e., "How far from the surface is infinity?" The current method sharpens that question via the appearance of four terms, of opposite sign, which separately diverge but whose sum is finite.

Although not exploited herein due to scope limitations this new approach to the Prandtl equations is, in effect, a flow simulator. This serendipitous result somewhat justifies faith in and bias toward the global, (integral) approach versus the local (differential) viewpoint and allows one to "follow" the development and propagation of the various mechanisms, e.g., wall shear, pressure gradient, inertial transports and geometric boundary layer growth.

DERIVATION OF NEW METHOD FOR FALKNER-SKAN SIMILARITY

Conservation of Mass

The conservation of mass ("continuity") equation in steady state is

$$\nabla \cdot \overline{V} = 0 \tag{1}$$

Conservation of Linear Momentum

The incompressible Navier-Stokes equation [1,2] is

$$\rho \frac{D\overline{V}}{Dt} - \overline{f} + \nabla p - \mu \nabla^2 \overline{V} = 0, \qquad \mu = \text{constant} \tag{2}$$

Prandtl Boundary-Layer Approximation [3]

$$\boxed{\delta << L}$$

where δ = transverse (to flow) length scale; i.e., depth of the viscous layer and L = longitudianl length scale. Within the boundary layer, the following order of magnitudes exist:

$$\begin{aligned} &u \sim U \quad \text{(Physical assumption)} && \frac{\partial}{\partial x} \sim \frac{1}{L} \quad \text{(Physical assumption)} \\ &v \sim \frac{U}{L}\,\delta \,\text{(from } u \sim U \text{ and continuity)} && \frac{\partial}{\partial y} \sim \frac{1}{\delta} \quad \text{(Physical assumption)} \end{aligned} \tag{3}$$

The Navier-Stokes equations for steady, two-dimensional flow are:

$$\begin{aligned} u\frac{\partial u}{\partial x} + v\frac{\partial u}{\partial y} &= -\frac{1}{\rho}\frac{\partial p}{\partial x} + \nu\frac{\partial^2 u}{\partial x^2} + \nu\frac{\partial^2 u}{\partial y^2} \\ u\frac{\partial v}{\partial x} + v\frac{\partial v}{\partial y} &= -\frac{1}{\rho}\frac{\partial p}{\partial y} + \nu\frac{\partial^2 v}{\partial x^2} + \nu\frac{\partial^2 v}{\partial y^2} \end{aligned} \tag{4}$$

where ν is kinematic viscosity defined by $\nu = \mu/\rho$, (3) into (4) yields: $R = \frac{UL}{\delta} >> 1$ and

$$\begin{aligned} \frac{\partial u}{\partial x} + \frac{\partial v}{\partial y} &= 0 \\ u\frac{\partial u}{\partial x} + v\frac{\partial u}{\partial y} &= -\frac{1}{\rho}\frac{\partial p}{\partial x} + \nu\frac{\partial^2 u}{\partial y^2} \end{aligned} \tag{5}$$

The outer flow is potential, and so the Bernoulli equation is valid. Hence,

$$-\frac{1}{\rho}\frac{\partial p}{\partial x} = U\frac{dU}{dx} \tag{6}$$

Hence, an alternative form of Prandtl's equation is:

$$\frac{\partial u}{\partial x} + v\frac{\partial u}{\partial y} = U\frac{dU}{dx} + \frac{\partial^2 u}{\partial y^2} \tag{7}$$

with the boundary conditions

$$\begin{aligned} y = \phi:& \quad u = \phi, \quad v = \phi \\ y = \infty:& \quad u = u(x) \end{aligned} \tag{8}$$

Falkner-Skan Equation (4)

The equation of continuity is integrated by the introduction of the stream function $\psi(x,y)$ defined by

$$u = \frac{\partial \psi}{\partial y} \qquad v = -\frac{\partial \psi}{\partial x} \tag{9}$$

Thus the equation becomes

$$\frac{\partial\psi}{\partial y}\frac{\partial^2\psi}{\partial x\,\partial y} - \frac{\partial\psi}{\partial x}\frac{\partial^2\psi}{\partial y^2} = U\frac{dU}{dx} + \nu\frac{\partial^3\psi}{\partial y^3} \tag{10}$$

with boundary condistions $\partial\psi/\partial x = 0$ and $\partial\psi/\partial y = 0$ for $y = 0$, and $\partial\psi/\partial y = U(x)$ for $y = \infty$,

Put: $\xi = \frac{x}{L}$, $\eta = \frac{y\sqrt{R}}{Lg(x)}$ (similarity assumption)

where R is Reynolds number, defined by $R = \frac{U_\infty L}{\nu}$

We obtain, under similarity, the following differential equation for $f(\xi,\eta)$

$$f''' + \alpha ff'' + \beta(1 - f'^2) = \frac{U}{U_\infty} g^2 \left(f'\frac{\partial f'}{\partial\xi} - f''\frac{\partial f}{\partial\xi}\right) \tag{11}$$

where α and β are contractions for the following functions of x:

$$\alpha = \frac{Lg}{U}\frac{d}{dx}(Ug); \quad \beta = \frac{L}{U_\infty} g^2 U' \tag{12}$$

"Similar" solutions exist if and only if f and f' do not depend on ξ;

$$\boxed{f''' + \alpha ff'' + \beta(1 - f'^2) = 0} \quad \text{(Falkner-Skan Equation)} \tag{13}$$

and for $\eta = 0$, $f = 0$, $f' = 0$; for $\eta = \infty$: $f' = 1$ (14)

The velocity distribution along a wedge is

$$\boxed{U(x) = Cx^m} \tag{15}$$

With $\alpha = 1$ in Eqn. (13), the Falkner-Skan equation becomes

$$\boxed{f''' + ff'' + \beta(1 - f'^2) = 0} \tag{16}$$

with the boundary conditions

$$\eta = \phi;\ f = \phi,\ f' = \phi;\ \eta = \infty,\ f' = 1 \tag{17}$$

New Method

Rewriting the Falkner-Skan equation

$$f''' + ff'' + \beta(1 - f'^2) = 0 \tag{16}$$

and integrating this equation once, we obtain:

$$\boxed{f'' = f''_o - ff' + (1 + \beta)\int_o^\eta f'^2\,d\eta - \beta\eta} \tag{18}$$

Integrating one more:

$$\boxed{f' = f''_o\eta - f^2/2 + (1 + \beta)\int_o^\eta\int_o^z f'^2 dz\,d\eta - \beta\eta^2/2} \tag{19}$$

where f''_o denotes the f'' value at $\eta = 0$; known is β; unknown are f''_o (the 3rd B.C.) and $f(\eta)$.

New Method

1. Assume an $f(\eta)$
2. Assume an f''_o
3. Iterate (19) until convergence obtained, varying f''_o via some scheme each k-iteration.
4. Final convergence criteria: $f'(\infty) \to 1.0$

RESULTS SPANNING FALKNER-SKAN FLOWS (See Figure 1)

Cases Calculated (F.T. Ko, MSAE Thesis, 1982)

β	Physical Flow
-0.18	Expansion corner near separation (β - .198)
0.	"Blasius" flat plate, $\nabla p = 0$
0.3	Wedge angle = $0.3\,\pi$
1.0	2-D stagnation ("Hiemenz")
5.0	Strong, favorable ∇p
10.0	Very strong, favorable ∇p
20.0	Intense, favorable ∇p

Note: Included are all physical types of f.

Conclusions (Vis-A-Vis 4th order R-Kutta)

1. Run times significantly less (for current method)
2. Accuracy seems better
3. Algorithm is far simpler

DERIVATION FOR STEADY, 2-D PRANDTL EQUATIONS

Laminar:

Integrate Prandtl once $\int_o^{y<\delta} \left[u \frac{\partial u}{\partial x} + v \frac{\partial u}{\partial y} = - \frac{1}{\rho} \frac{dp}{dx} + \nu \frac{\partial^2 u}{\partial y^2} \right] dy$

yields, using $\nabla \cdot \overline{v} = 0$, "NOSLIP" ($u(x,0) = 0$) and solid wall, $v(x,0) = 0$:

$$\nu \frac{\partial u}{\partial y} = \frac{\tau_o(x)}{\rho} + vu + \frac{1}{\rho} \frac{dP}{dx} y + 2 \frac{d}{dx} \int_o^y u^2 dy \qquad (20)$$

Integrate again

$$u = \frac{\tau_o}{\rho} y + \frac{1}{\rho} \frac{dp}{dx} y^2/2 + \int_o^y vu\, dy + 2 \frac{d}{dx} \int_o^y dy \int_o^z u^2\, dz \qquad (21)$$

FIG. 1

INTERNAL CONSISTENCY -- RATIO OF R-K ERROR TO PIT ERROR*

β	δ^*	θ
-0.18	∞	∞
0.0	∞	∞
0.3	∞	∞
1.0	∞	∞
5.0	∞	684.
10.0	∞	1.6
20.0	∞	48.2

*Error based up analytic expressions (5): $\delta^* = \eta_\infty - f(\infty)$

$$\theta = \frac{f''_o - \beta\delta^*}{1 + \beta}$$

Known: 1. $\frac{dP}{dx}$ (for any B.L. flow) and hence, U(x)

2. $v = -\frac{d}{dx}\int_o^y u\,dy$ (from continuity)

Unknown: 1. $\tau_o(x)$

2. $u(x,y)$

N.B. Structure of the problem is identical to the Falkner-Skan Case.

Method: 1. Guess $\tau_o(x)$

2. Guess $u_1(x,y)$

3. Iterate Eq. (21) with suitable τ_o updates until:

4. $u(x,\delta) \to U(x)$

Turbulent

Only change is additional $\langle uv \rangle$ term ("Reynolds' stress")(5):

$$\nu u = \frac{\tau_o}{\rho} y + \frac{1}{\rho}\frac{dP}{dx} y^2/2 + \int_o^y \left[uv + \langle uv \rangle \right] dy + 2\frac{d}{dx}\int_o^y dz \int_o^z u^2\,dz \qquad (22)$$

Where u, v are now time averaged quantities and have a 3rd unknown: $\langle uv \rangle$, Reynold's stress which must be modelled.

N.B. Usual turbulence approaches (5) must model $\langle uv \rangle$ at each x-point in a PDE,

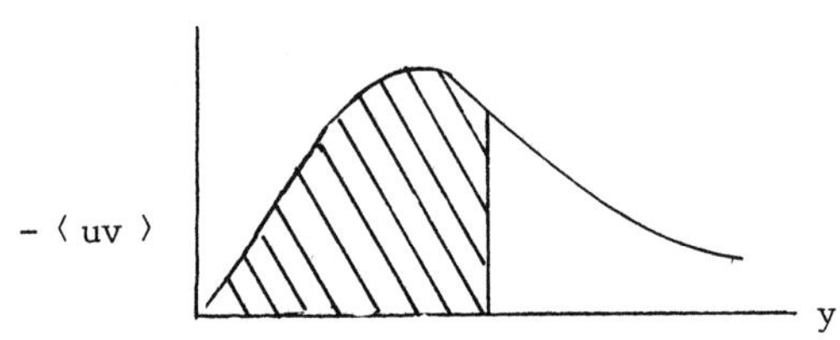

Convergence of f''_o for β = - 0.18 (Typical)

Current-Secant method

f''_o	f'max	η max(f' = max)
0.01(initial guess)	0.97862119	10.00
0.10(initial guess)	0.99238959	10.00
0.14974702	1.00844362	5.33
0.12358257	0.99857218	10.00
0.12902878	1.00013680	6.53
0.12855261	0.99997666	10.00
0.12862201	0.99999651	10.00
0.12863422	1.00000000+	8.75
0.12863422	0.99999999+	10.00

CPU Time = 1.90 Seconds

Runge-Kutta (4th Order)-Secant method

f''_o	f'max	η max(f' = max)
0.01(initial guess)	0.97862103	10.00
0.10(initial guess)	0.99238913	10.00
0.14975115	1.00844442	5.33
0.12358410	0.99857206	10.00
0.12903084	1.00013682	6.53
0.12855459	0.99997665	10.00
0.12862400	0.99999651	10.00
0.12863622	1.00000000+	8.75
0.12863622	0.99999999+	10.00

CPU Time = 7.36 Seconds

Non-dimensional Parameters

flow rate $M = \int_0^\infty f' \, d\eta$

displacement thickness $\delta^* = \int_0^\infty (1 - f') \, d\eta$

energy thickness $\delta^{**} = \int_0^\infty f'(1 - f'^2) \, d\eta$

momentum thickness $\theta = \int_0^\infty f'(1 - f') \, d\eta$

shape factor $H_{12} = \delta^*/\theta$, shape factor $H_{23} = \theta/\delta^{**}$

$\beta = -0.18$

	PIT	Runge-Kutta	Error(%)
M	8.12841550	8.12842339	-0.00010
δ^*	1.87158450	1.87157661	0.00042
θ	0.56770662	0.56770603	0.00010
δ^{**}	0.86649282	0.86649250	0.00004
H_{12}	3.29674594	3.29673547	0.00032
H_{23}	0.65517753	0.65517709	0.00007

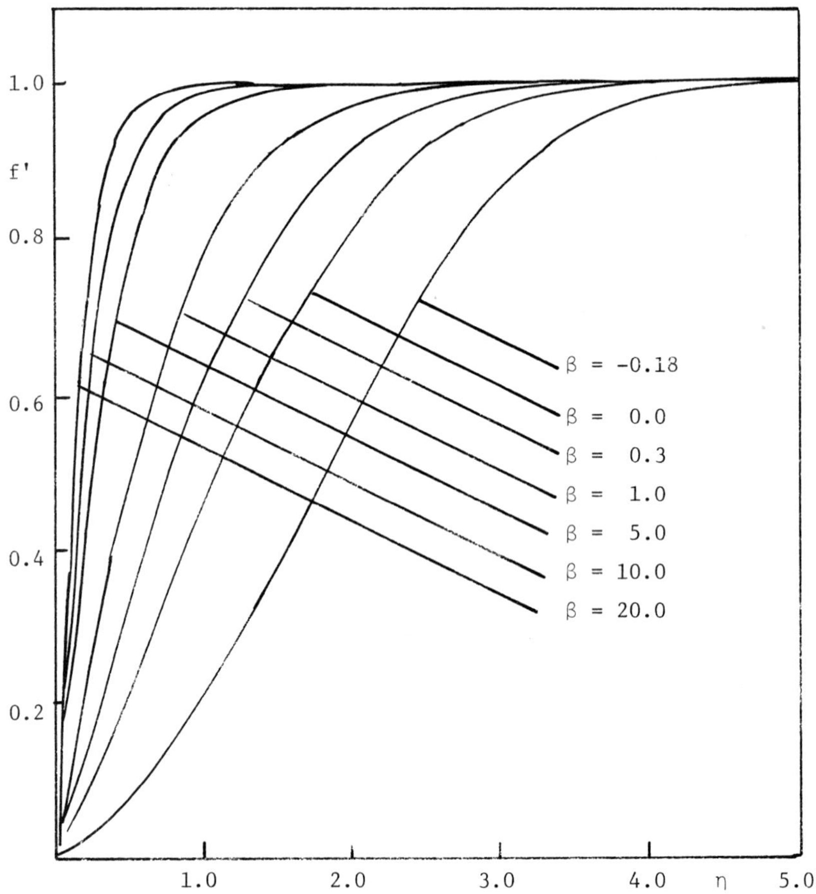

Fig. 2. Calculated Velocity Distributions

whereas current method need only model well the (variable as upper limit $y \to \delta$) shaded area under the $\langle uv \rangle$ curve: This probably yields an order of magnitude improvement (in accuracy for less computer time).

DISCUSSION

Merits

1. "Small" computer times
2. "Pre-smoothing" of results (physically appealing)
3. Simple (relatively)
4. Simpler turbulence modelling is possible

Extensions

1. See above for Prandtl, turbulence
2. Unsteady flows:

   Add a $(\frac{\partial u}{\partial t})$ term which integrates to

   $$\frac{\partial}{\partial t} \int_0^y dz \int_0^z u dz \quad \text{and}$$

   $$\frac{d}{dx} \to \frac{\partial}{\partial x} \quad \text{since} \quad f(x) \to f(x,t)$$

3. 3-D flows--2 force equations rather than one here: hence, must guess u and w and coupling obtains.
4. Separation--must use Navier-Stokes not Prandtl Eq; order of system increases from three to four.

A UNIVERSAL PANACEA?

Comment: Any ODE can be integrated formally and the iterative scheme applied.

CLOSURE

A. Method is almost trivially simple

B. Digital computers "like" to integrate but "dislike" to differentiate or take differences.

C. Fluid flow problems are "global", not local, at least in the governing dynamics.

D. Even the kinematics of turbulence, separating flows, reattaching flows subsonic flows are "global": hence, a differential approach is, at best, artificial.

E. Flow "simulation" is inherent, e.g., (eq. 21/22)

1. $\frac{\tau_o}{\rho} y$ = Linear viscous sub-layer

2. $\frac{1}{\rho}\frac{dP}{dx} y^2/2$ = Parabolic ∇P effect (Poiseuille)

3. $\int_o^y uvdy$ = inertial transport of momentum

4. $\frac{d}{dx}\int_o^y dz \int_o^z u^2 dz$ = cumulative mass transport

5. $\int_o^y \langle uv \rangle dy$ = Net effect of Reynolds's (turbulent) stress from surface to height y

i.e., cumulative total effects are readily identified as aids for "simulator/emulator".

REFERENCES

[1] M. Navier, Memoire sur les los du mouvement des fluids. Mem. de l'Acad. de Sci. 6, 386-416 (1827)

[2] G. G. Stokes, On the theories of internal friction of fluids in motion. Trans. Cambr. Phil. Soc. 8, 287-305 (1845)

[3] L. Prandtl, Under Flussigkeitsbewegung bei sehr kleiner Reibung. Prooc. Third Intern. Math. Congr. Heidelberg 1904. Reprinted in Vier Abhandlungen zun Hydro- and Aerodynamik. Gottingen, 1927; NACA TM 452 (1928); see also: Coll. Works II, 575-584 (1961)

[4] V. M. Falkner and S. W. Skan, Some approximate solutions of the boundary layer equations. Phil. Mag. 12, 865-896 (1931); ARC RM. 1314 (1930)

[5] F. M. White, Viscous Fluid Flow, McGraw-Hill, 1974 or any book on viscous flow.

CONSTRUCTIVE EXISTENCE OF SOLUTION TO A NONLINEAR EIGENVALUE PROBLEM WITH APPLICATION TO NTH ORDER REACTION DIFFUSION PROBLEMS

W. L. Perry

Department of Mathematics
Texas A&M University
College Station, Texas

ABSTRACT

The problem of determining the amount of reactant present in the steady state of a reaction diffusion problem with nth order reaction kinetics and slab geometry can be achieved by solving the boundary value problem

$$u''(x) = \phi^2 u^n(x), \qquad 0 < x < 1, \quad n \geq 1$$

$$u'(0) = 0 \qquad\qquad u(1) = 1$$

where u is a normalized (dimensionless) concentration, and ϕ is the Thiele modulus.

In this paper we consider the related nonlinear eigenvalue problem

$$y''(x) = \lambda y^n(x), \qquad 0 < x < 1$$

$$y'(0), \qquad\qquad y(1) - y'(1) = 0$$

A sequence of functions is constructed which converges to $y(x)$ in an alternating monotone way so that in numerical implementation *a posteriori* error estimates are available. In addition the solution of the original boundary value problem can be obtained from $y(x)$ and λ.

INTRODUCTION In the study of permeable catalysis the simplest governing equations for the reaction-diffusion problem in a region Ω, involving nth order reactions is [2, p. 144]

$$\Delta u(x) = \phi^2 u^n(x) \qquad x \in \Omega$$

$$c(1 - u(x)) = \frac{\partial u(x)}{\partial \nu} \qquad x \in \partial\Omega$$

where $u(x)$ is a normalized concentration, and ϕ is the Thiele modulus. If Ω is a slab and Dirichlet boundary conditions are imposed, the problem reduces to

$$u''(x) = \phi^2 u^n(x), \qquad 0 < x < 1, \quad n \geq 1$$
$$u'(0) = 0, \qquad u(1) = 1 \tag{1}$$

where width of the slab has been normalized to one. This steady state problem is of interest to biologists and engineers since it models the situation in which a slab shaped permeable catalyst or enzyme membrane has an impermeable barrier at $x = 0$.

A solution of (1) can be written in terms of hypergeometric functions along with $u(0)$ which is, of course, unknown. Regardless, the expressions can be used to gain some asymptotic information. On the other hand, these solutions are not useful in numerical calculations of concentration profiles [3].

In this paper, a constructive proof of existence of solution to the nonlinear eigenvalue problem

$$y''(x) = \lambda y^n(x)$$
$$y'(0) = 0, \qquad y(1) - y'(1) = 0$$

related to (1) is presented.

<u>THE EIGENVALUE PROBLEM</u> The problem considered is

$$y''(x) = \lambda y^n(x)$$
$$y'(0) = 0, \qquad y(1) = 1$$

If a solution (λ, y) can be found for (2) then a solution to (1) can be generated. In particular, if we make a change of variables $v(x) = ay(bx)$ then we have v satisfying

$$v''(x) = a^{1-n} b^2 \lambda v^n(x), \qquad 0 < x < b$$
$$v'(0) = 0, \qquad v\left(\frac{1}{b}\right) = a,$$

If we force $a^{1-n} b^2 \lambda = \phi^2$ and couple this with $ay(b) = 1$ we see that

$$b^2 y^{n-1}(b) = \frac{\phi^2}{\lambda} \tag{3}$$

A solution b for (3) can be found, so that $a = (\phi^2/b^2\lambda)^{1/(1-n)}$. The solution y of (2) can be extended to the right of $x = 1$ by basic continuation theorems, and we see in equation (3) that as $b \to 0$, the left side of (3) $\to 0$; as $b \to +\infty$, the left side of (3) $\to +\infty$. By continuity and the fact that $b^2y^{n-1}(b)$ is an increasing function of b there exists a unique b satisfying (3).

The solution of (2) will be achieved by an iteration scheme defined as follows: Let

$$y_0(x) = x \qquad 0 \le x \le 1$$

and define for $k = 1,2,\ldots$

$$\begin{aligned} y''_k(x) &= \lambda_k y^n_{k-1}(x), \qquad 0 < x < 1, \\ y'_k(0) &= 0, \quad y_k(1) - y'_k(1) = 0 \end{aligned} \tag{4}$$

with λ_k chosen so that $y'_k(1) = 1$. The Green's function formulation of this last problem is

$$y_k(x) = \lambda_k \int_0^1 G(x,\xi) y^n_{k-1}(\xi) d\xi, \qquad \lambda_k = \frac{1}{\int_0^1 y^n_{k-1}(\xi) d\xi} \tag{5}$$

$k = 1,2,\ldots,$ where

$$G(x,\xi) = \begin{cases} \xi & x \le \xi \\ x & x \ge \xi \end{cases}, \quad 0 \le x, \quad \xi \le 1 .$$

Now we show alternating monotonicity. Direct calculations show that $\lambda_1 > \lambda_2$, $y_1(x) > y_2(x) > y_0(x)$, $0 < x < 1$, and $y_1 - y_2$ is decreasing. We also have $y_1(x) > y_3(x) > y_2(x)$, $0 < x < 1$. Let $w(x) = y_3(x) - y_2(x)$. From (4) we have $w(1) = 0$, $w'(1) = 0$,

$$w''(x) = \lambda_3 y^n_2(x) - \lambda_2 y^n_1(x).$$

Since $\lambda_3 > \lambda_2$ we have $w''(x) > 0$ in an interval $(1-\varepsilon, 1)$. Moreover

$$\begin{aligned} w''(x) &= \lambda_3\left\{y^n_2(x) - \left(\left(\frac{\lambda_2}{\lambda_3}\right)^{1/n} y_1(x)\right)^n\right\} \\ &= \lambda_3\left\{y_2(x) - \left(\left(\frac{\lambda_2}{\lambda_3}\right)^{1/n} y_1(x)\right)\right\}\left\{y^{n-1}_2(x) + \ldots\right\} \end{aligned}$$

Since the last factor in the last equation is positive we see that $w''(x)$ can change sign at most once on $(0,1)$. Thus $y_3(x) > y_2(x)$, $0 < x < 1$ and $y_3(x) - y_2(x)$ is decreasing. A similar argument shows that

$y_1(x) > y_3(x)$, $0 < x < 1$, and $y_1(x) - y_3(x)$ is decreasing. The same argument, used inductively, leads to the

Lemma. The sequence $\{\lambda_k, y_k(x)\}$ satisfies

$$y_0(x) < y_2(x) < \cdots < y_{2n}(x) < y_{2n+1} < \cdots < y_3(x) < y_1(x)$$

$0 < x < 1$, $n = 1,2,3,\ldots$ and

$$\lambda_2 < \lambda_4 < \cdots < \lambda_{2n} < \lambda_{2n+1} < \cdots < \lambda_3 < \lambda_1 .$$

Thus we have sequences $\{y_{2j+1}(x)\}_{j=0}^{\infty}$ and $\{y_{2j}(x)\}_{j=1}^{\infty}$ which are monotone and uniformly bounded. Also from (5) we can conclude that both sequences are equicontinuous. By Ascoli's lemma there exist functions $\check{y}(x)$ and $\hat{y}(x)$, continuous on $[0,1]$ such that $y_{2j+1} \to \hat{y}$, $y_{2j} \to \check{y}$ uniformly on $[0,1]$. Clearly $\{\lambda_{2j}\}_{j=1}^{\infty}$ and $\{\lambda_{2j+1}\}_{j=0}^{\infty}$ converge; we call their limits $\check{\lambda}$ and $\hat{\lambda}$ respectively.

Utilizing the uniform convergence of the odd and even indexed sub-sequences in conjunction with (5) we have

$$\begin{aligned}
&\hat{y}'' = \hat{\lambda}\check{y}^n(x) \\
&\check{y}'' = \check{\lambda}\hat{y}^n(x) \\
&\hat{y}'(0) = \check{y}'(0) = 0, \quad \check{y}(1) - \check{y}'(1) = 0 = \hat{y}(1) - \hat{y}'(1) \\
&\hat{y}(1) = \check{y}(1) = \hat{y}'(1) = \check{y}'(1) = 1, \quad \hat{\lambda} \geq \check{\lambda}.
\end{aligned} \tag{6}$$

From the monotonicity, $\check{y}(x) \leq \hat{y}(x)$ on $[0,1]$. If $\check{y}(x) = \hat{y}(x)$ on $[0,1]$, then $\hat{\lambda} = \check{\lambda}$ and we have a solution to (2). If not we proceed as follows. We will use the terminology of Amann [1] in considering the problem

$$\begin{aligned}
L[u] &= -u''(x) = -\hat{\lambda}u^n(x) = f(u) \\
B(u) &= \begin{pmatrix} u'(0) \\ u(1) \end{pmatrix} = \begin{pmatrix} 0 \\ 1 \end{pmatrix} = g
\end{aligned} \tag{7}$$

The function $\hat{y}(x)$ is a supersolution, and a subsolution dominated by $\hat{y}$ is $c\check{y}$ where $c > 0$ is chosen so that $c^{n-1}\hat{\lambda} < \check{\lambda}$.

We can now use the Picard type iterative scheme in Theorem 9.4 of Amann [1] to obtain decreasing and increasing sequences $\{\phi_n(x)\}$ and $\{\psi_n(x)\}$ converging to a maximal and minimal solution (call them ϕ and ψ) of (7). In fact ϕ and ψ must be the same function. To see this consider $f(x) = \phi(x)\psi'(x) - \phi'(x)\psi(x)$. We have $f(0) = 0$ and $f(1) \geq 0$. Now $f' = \hat{\lambda}\phi\psi(\psi^{n-1} - \phi^{n-1})$. This implies that $f'(x) \leq 0$. From the conditions $f(0) = 0$, $f(1) \geq 0$ and $f'(x) \leq 0$, $0 < x < 1$, we conclude that $f'(x) \equiv 0$ on $(0,1)$. Thus f is a constant, and $f(x) \equiv 0$. Hence $(\phi/\psi)' = 0$. Since $(\phi/\psi) = 1$ when $x = 1$ we have $\phi \equiv \psi$ on $(0,1)$.

Theorem. A positive solution to

$$y''(x) = \lambda y^n(x) \qquad 0 < x < 1, \quad n > 1$$

$$y'(0) = 0, \quad y(1) = 1 .$$

exists and is obtainable either solely by the iterative scheme

$$y_0(x) = x$$

$$y_k''(x) = \lambda_k y_{k-1}^n(x) \qquad 0 < x < 1, \quad k = 1,2,\ldots$$

$$y_k'(0) = 0, \quad y_k(1) = 1,$$

or by this scheme in conjunction with Amann's iterative scheme

$$v_0 = \hat{y} \quad (\text{or} \quad c\check{y}, \quad 0 < c < 1, \quad c^{n-1}\hat{\lambda} < \check{\lambda})$$

$$-v_k''(x) + n\hat{\lambda}v_k(x) = -\hat{\lambda}v_{k-1}^n(x) + n\hat{\lambda}v_{k-1}(x)$$

$$v_k'(0) = 0, \quad v_k(1) = 1$$

$0 < x < 1$, $k = 1,2,\ldots$, where $\hat{\lambda}$, $\hat{y}$, $\check{y}$, and c are obtained from the first iterative scheme as described in the paper. In either instance $\hat{\lambda}$ is the eigenvalue.

Example. Consider $u''(x) = \phi^2 u^6(x)$, $0 < x < 1$, $u'(0) = 0$, $u(1) = 1$ for a Thiele modulus of 1.5. We initially solve $y''(x) = \lambda y^6(x)$, $y'(0) = 0$, $y(1) = y'(1) = 0$. In this example $\check{y}$ and $\hat{y}$ coincide, so the Amann scheme need not be used. The computed solutions y and u are tabulated in Table 1; the absolute error is $\leq .0004$. As mentioned earlier u is obtained from y by a change of variables. In this case $u(x) \doteq 1.075\, y(.93x)$. The result here agrees well with the numerical solution in [3].

TABLE 1

x	y(x)	u(x)
0	.6913	.7431
.1	.6933	.7433
.2	.6995	.7475
.3	.7102	.7559
.4	.7255	.7689
.5	.7464	.7869
.6	.7738	.8107
.7	.8093	.8415
.8	.8554	.8809
.9	.9164	.9318
1	1	1

REFERENCES

[1] H. Amann, Fixed point problems and nonlinear eigenvalue problems in ordered Banach spaces, SIAM Review, 18 (1976), pp. 620-709.

[2] R. Aris, The Mathematical Theory of Diffusion and Reaction in Permeable Catalysis, Vol. I., Clarendon Press, Oxford, 1975.

[3] J. DeSimone and J. A. Pennline, A new asymptotic analysis of the nth order reaction-diffusion problem: Analytical and Numerical Studies, Math. Biosci., 40 (1978), 303-318.

A FREQUENCY DOMAIN APPROACH FOR FINDING THE LIMIT BEHAVIOR OF RETARDED DIFFERENTIAL EQUATIONS

V.M.Popov

University of Florida
Gainesville, Florida

ABSTRACT

For the study of a class of phenomena which take place in systems described by retarded differential equations with several singular points, it is essential to find conditions which secure the following property: if a bounded solution has a bounded continuation to $-\infty$, then the corresponding α- and ω-limit points are singular points. The paper shows how such conditions can be obtained by means of a frequency-domain technique.

1. INTRODUCTION

The frequency-domain approach of control theory has many applications (see, e.g., C.Corduneanu [1], A.Halanay [2], Y.D.Landau [3], J.Moser [4], K.S.Narendra and J.H.Taylor [5], D.D.Šiljak [6], etc.) among which the case of retarded differential equations is of special interest (V.M.Popov and A. Halanay [7], Vl.Răsvan [8] : a monograph in which the subject is extensively treated). Most of the results concern the case of a unique singular point, required to be a global attractor. This paper studies retarded differential equations with several singular points. In this case, the central object of study is the "limit set" - a concept which is presented in J.Hale [9] in a very general setting.

In [10] it was recently shown that, for some systems, described by ordinary differential equations, the structure of the limit set can be completely determined. These results can be extended to retarded differential equations of the form $\dot{x}(t) = f(x_t)$ (with notations as in [9]) where f is a continuous function which maps closed bounded sets into bounded sets and the domain of f (the "state space") is, e.g., the space of continuous functions $C([-r,0],R^n)$, for some $r > 0$. We will have to study the particular, but very important solutions which can be continuated backwards to $-\infty$. The orbits corresponding to such solutions will be called "full orbits".

Let A be a bounded and closed, positively invariant subset of the state space. Then the (ω-)limit set of A is simply the union of all full orbits which lie entirely in A. A principal property of the limit set is that all the positive orbits in A approach uniformly the limit set as $t \to \infty$. Moreover, as in [10], the limit set can be characterized as the smallest nonempty, closed subset of A with such a property. It is this property that makes the limit set the appropriate tool for investigating the limit behavior of systems.

In general, limit sets cannot be determined explicitly. However, in the spirit of the qualitative theory of systems, it is possible to obtain the needed information about the behavior of the system without determining completely the limit sets. This purpose can be achieved, for instance, if the limit set consists only of "singular orbits", i.e. full orbits whose α- and ω-limit points are singular points. Then, using the index theory and other arguments, one can often show that only a limited number of singular orbits can exist and one can determine their number and the way they are connected to each other. In this case, although the limit sets are not known precisely, their "structure" is completely determined. As shown in [10], this information can often be obtained by means of a graph (the "limit-graph" of the system). Various phenomena, common in applications, can be easily visualized, understood and controlled by examining the corresponding changes of the limit graph.

In the case of retarded differential equations, the simplification achieved by introducing limit graphs is even more considerable than in the ODE-case. The limit sets lie now in $C([-r,0],R^n)$ and are hard to visualize. Nevertheless their structure is still completely determined when limit graphs are available.

In order to obtain these results, the main difficulty is to establish that the system has the following property: every bounded full orbit is singular. It is the aim of this paper to show, on a typical example (given in section 2) how a frequency domain approach (presented in sections 4-6) can fulfill this purpose. The main result is stated in section 3.

2. THE STUDIED SYSTEM

Let f_1 and f_2 be two nondecreasing, continuously differentiable functions and let $\varepsilon > 0$ be a (small) number. For some $r > 0$ we introduce the system

$$\begin{aligned} \dot{x}_1(t) &= -x_1(t-r) - \varepsilon\, x_2(t) - f_2(x_2(t)) \\ \dot{x}_2(t) &= -x_2(t-r) - \varepsilon\, x_1(t) - f_1(x_1(t)), \end{aligned} \tag{2.1}$$

assuming that for $f_1 = f_2 = 0$ every solution of the system tends to zero as $t \to \infty$. One can show that this is always the case if ε is sufficiently small and if $r \in (0, e^{-1}]$ - a condition which suffices for the theorem in section 3.

This system can be viewed as a modification of the mathematical model of the Eccles-Jordan electronic circuit, whose many variants are widely used to produce switchings and other phenomena specific to dynamical systems with several singular points. (In this case, delays can appear because of wave-propagation effects.)

3. THE MAIN RESULT AND THE IDEA OF PROOF

Theorem: Under the assumptions of section 2, if $0 < r \leqq e^{-1}$, then every bounded full orbit of the system (2.1) is singular.

Before giving the proof (sections 4-6) let us outline its main idea. We consider a bounded full orbit, fixed during the proof. To this orbit, we associate a function $J:(0,\infty)\to R$ (section 4) so chosen as to meet two conditions:

(i) J is bounded from above (section 5),

and

(ii) The boundedness from above of J implies that the orbit is singular (section 6).

The procedure of introducing a suitable function J is similar to the main approach of [11]. However the techniques of this paper are closer to those of [12].

4. CHOICE OF J

Let $x_j:R\to R$, $j=1,2$, be a bounded solution of (2.1). To avoid repetitions, we adopt the following conventions: Unless otherwise stated, every relation involving the index j alone is valid for $j=1$ and for $j=2$. Moreover, every relation involving the indices j and k is valid for $j=1$, $k=2$ and for $j=2$, $k=1$.

For every $T>0$ we define the auxiliary functions $\mu_{j,T}:[-T,\infty)\to R$ as follows:

$$\mu_{j,T}(t) = \begin{cases} f_j(x_j(t)) & \text{if} \quad -T \leqq t \leqq T \\ 0 & \text{if} \quad t > T. \end{cases} \tag{4.1}$$

We also define the functions $u_{j,T}:[-T-r,\infty)\to R$ by the initial value problem

$$\begin{aligned} \dot{u}_{j,T}(t) &= -u_{j,T}(t-r) - \varepsilon\, u_{k,T}(t) - \mu_{k,T}(t), \quad t > -T, \\ u_{j,T}(t) &= x_j(t) \quad \text{if} \quad t \in [-T-r,-T]. \end{aligned} \tag{4.2}$$

From (4.2) and (2.1) it follows that

$$u_{j,T}(t) = x_j(t) \quad \text{if} \quad t \in [-T-r,T]. \tag{4.3}$$

Next, we define the functions $v_{j,T}:[-T-r,\infty)\to R$ by

$$\begin{aligned} \dot{v}_{j,T}(t) &= -v_{j,T}(t-r) - \varepsilon\, u_{k,T}(t), \quad t > -T \\ v_{j,T}(t) &= 0 \quad \text{if} \quad t \in [-T-r,-T]. \end{aligned} \tag{4.4}$$

From the assumptions in section 2 it follows that the functions $u_{j,T}$ tend exponentially to zero as $t\to\infty$ and are uniformly bounded, i.e.

$$\sup_{T>0}\ \sup_{t>-T} |u_{j,T}(t)| < \infty. \tag{4.5}$$

Using these properties one can show that, if $r \in (0,e^{-1}]$, then $v_{j,T}$ and $\dot{v}_{j,T}$ are bounded. Hence one can define the function $J:(0,\infty)\to R$ by the (convergent) integral

$$J(T) = -\int_{-T}^{\infty}(\dot{v}_{1,T}(t)\, u_{2,T}(t) + \dot{v}_{2,T}(t)\, u_{1,T}(t))dt. \qquad (4.6)$$

5. J IS BOUNDED FROM ABOVE

We need two lemmas (proved in sections 8,9):

Lemma 1: If $r \in (0,e^{-1}]$, then the function $g:[0,\infty)\to R$, determined by the initial value problem

$$\begin{aligned} \dot{g}(t) &= -g(t-r),\ t > 0, \\ g(0) &= 1 \\ g(t) &= 0 \quad \text{if} \quad -r \leqq t < 0, \end{aligned} \qquad (5.1)$$

is monotone decreasing. Moreover, there exist strictly positive constants k_1-k_5 such that

$$e^{-k_1 t} < g(t) < k_2\, e^{-k_3 t}\ , \quad \text{if} \quad t \geqq 0, \qquad (5.2)$$

$$k_4\, e^{-k_1 t} < -\dot{g}(t) < k_5\, e^{-k_3 t}, \ \text{if} \quad t > r. \qquad (5.3)$$

Lemma 2. Let g be as in Lemma 1. Let $f:R\to R$ be a monotone increasing continuous function. Let $\xi:R\to R$ be a bounded, continuous function. For every $T > 0$, let $\mu_T:[0,\infty)\to R$ be defined as

$$\mu_T(t) = \begin{cases} f(\xi(t)) & \text{if} \quad t \in [0,2T] \\ 0 & \text{if} \quad t > 2T, \end{cases} \qquad (5.4)$$

and let $u_T:[0,\infty)\to R$ be a continuous function such that

$$u_T(t) = \xi(t) \quad \text{if} \quad t \in [0,2T]. \qquad (5.5)$$

Suppose also that $\sup_{T>0}\ \sup_{t>0} |u_T(t)| < \infty$. Then there exists $K > 0$ such that, for every $T > 0$,

$$-\int_0^{\infty} u_T(t)\, \frac{d}{dt}\, (g*\mu_T)\, dt < K, \qquad (5.6)$$

where $*$ denotes convolution.

Remark: If g satisfies (5.1), then the function $-g*\mu_T$ is identical to the function $w_T:[0,\infty)\to R$, determined by

$$\dot{w}_T(t) = -w_T(t-r) - \mu_T(t) \ , \ t \geqq 0$$
$$w_T(t) = 0 \quad \text{if} \quad t \in [-r,0],$$

and hence the conclusion (5.6) becomes $\int_0^\infty \dot{w}_T(t)\ u_T(t)\ dt < K.$

It is in this form that we shall use below this conclusion.

We next define the functions $w_{j,T}:[-T-r,\infty)\to R$ by

$$\begin{aligned} \dot{w}_{j,T}(t) &= -w_{j,T}(t-r) - \mu_{k,T}(t) \ , \ t \geqq -T \\ w_{j,T}(t) &= 0 \quad \text{if} \quad t \in [-T-r,-T]. \end{aligned} \tag{5.7}$$

Now one can apply Lemma 2 and the Remark, with $\xi(t) = x_k(t-T)$, $\mu_T(t) = \mu_{k,T}(t-T)$, $w_T(t) = w_{j,T}(t-T)$ and $u_T(t) = u_{k,T}(t-T)$. (The conditions of Lemma 2 are satisfied because of (4.1), (4.3) and (4.5).) Hence there exists $K > 0$ such that

$$\int_{-T}^{\infty} (\dot{w}_{1,T}u_{2,T} + \dot{w}_{2,T}u_{1,T})\ dt < K \ , \text{ for every } T > 0. \tag{5.8}$$

Finally, we define $z_{j,T}:[-T-r,\infty)\to R$ by

$$\begin{aligned} \dot{z}_{j,T}(t) &= -z_{j,T}(t-r), \ t > -T \\ z_{j,T}(t) &= x_j(t) \ , \ t \in [-T-r,-T]. \end{aligned} \tag{5.9}$$

From (4.2),(4.4),(5.7) and (5.9) it follows that

$$v_{j,T}(t) = u_{j,T}(t) - z_{j,T}(t) - w_{j,T}(t) \ , \ t \geqq -T.$$

Hence (4.6) can be written as $J = J_1+J_2+J_3$, where

$$J_1(T) = -\int_{-T}^{\infty} \frac{d}{dt}(u_{1,T}u_{2,T})\ dt$$

$$J_2(T) = \int_{-T}^{\infty} (\dot{z}_{1,T}u_{2,T} + \dot{z}_{2,T}u_{1,T})\ dt$$

and $J_3(T)$ is the left hand side of (5.8). Since J_3 is bounded from above, it suffices to show that J_1 and J_2 are bounded. For J_1, this follows from the boundedness of x_j, from (4.3) and from the fact that $\lim_{t\to\infty} u_{j,T}(t) = 0$ (see section 4). Finally, the boundedness of J_2 follows from (4.5), since

$$|\dot{z}_{j,T}(t)| < k_6\ e^{-k_7(t+T)}, \quad t > -T, \tag{5.10}$$

where k_6 and k_7 are strictly positive constants, independent of T. Inequality (5.10) easily follows from (5.9), by writing $z_{j,T}$ in the integral form (e.g. as in [9], Theorem

6.1, p.22) and by using the boundedness of x_j and (5.2).

6. THE ORBIT IS SINGULAR

We use another lemma (proved in section 10):

Lemma 3: Suppose that there exists $K > 0$ such that, for every $T > 0$, the functions $v_{j,T}$ (section 4) satisfy

$$\int_{-T}^{\infty} (|\dot{v}_{1,T}(t)|^2 + |\dot{v}_{2,T}(t)|^2)\, dt < K. \tag{6.1}$$

Then the orbit described by x_j is singular.

It remains to show that the conditions of this lemma are satisfied. The functions $v_{j,T}$ are in L^1 and in L^2 and one can apply the Fourier transform

$$\hat{v}_{j,T}(\omega) = \int_{-T}^{\infty} e^{-i\omega t} v_{j,T}(t)\, dt,$$

and similarly for $\hat{u}_{j,T}(\omega)$. Using (4.4) and Parseval's identity, in the standard way, we obtain

$$J(T) = \frac{\varepsilon}{2\pi} \sum_{j=1}^{2} \int_{-\infty}^{\infty} \mathrm{Re} \frac{i\omega}{i\omega + e^{-i\omega r}} |\hat{u}_{j,T}(\omega)|^2 d\omega$$

Observe now that, if $r \in (0,e^{-1}]$ (and even if $r \in (0,1)$), there exists $k > 0$ such that

$$\mathrm{Re} \frac{i\omega}{i\omega + e^{-i\omega r}} \geqq k \left| \frac{i\omega}{i\omega + e^{-i\omega r}} \right|^2, \quad \omega \in R.$$

Since J is bounded from above, it follows that there exists $K > 0$ such that, for every $T > 0$,

$$\frac{1}{2\pi} \sum_{j=1}^{2} \int_{-\infty}^{\infty} \left| \frac{i\omega}{i\omega + e^{-i\omega r}} \hat{u}_{j,T}(\omega) \right|^2 d\omega < K. \tag{6.2}$$

But the left hand sides of (6.2) and (6.1) are identical, as one can see by another application of Parseval's identity. By virtue of Lemma 3, this proves the theorem.

7. COMMENTS

Similar results can be obtained for much more general

systems, including - and extending - those considered in [12]. But even the simple case treated above constitutes a new challenge from the viewpoint of finding the best possible upper limit for r in the theorem of section 3. Apparently, if $r > e^{-1}$, one loses the monotonicity of g (which is used in the proof that J is bounded from above). Nevertheless the frequency-domain argument of section 6 holds for $r \in (0,1)$. Whether or not the theorem remains valid, in the same form, for larger values of r is unknown.

8. PROOF OF LEMMA 1

It is easy to see that, if $r \in (0,e^{-1}]$, then the equation $e^{\alpha r} = \alpha$ has a solution $\alpha \geqq e$. We first prove that

$$\alpha g(t) > g(t-r), \text{for every } t \geqq 0. \tag{8.1}$$

If (8.1) is not true, then one can find $t_0 > r$ such that

$$\alpha g(t_0) = g(t_0-r) \tag{8.2}$$

and $\alpha g(t) > g(t-r)$, for every $t \in [0,t_0)$. But then (5.1) implies that

$$-\dot{g}(t) < \alpha g(t), \text{ for every } t \in [0,t_0) \tag{8.3}$$

which contradicts (8.2) and proves (8.1). Hence (8.3) remains valid for all $t \geqq 0$. This leads to the first inequality in (5.2).

Now from (5.1) it follows that $\dot{g}(t) < 0$, for every $t > r$. Therefore g is monotone decreasing. Hence, from (8.1) (with t replaced by $t-r$) one finds that $-g(t-r) < -\alpha^{-1}g(t-2r) < -\alpha^{-1}g(t)$, for every $t > 2r$. Now from (5.1) it follows that $\dot{g}(t) < -\alpha^{-1}g(t)$, for every $t > 2r$. This leads to the second inequality in (5.2). Finally, (5.3) follows now from (5.1).

9. PROOF OF LEMMA 2

The proof is adapted from [12]. By changing the order of integration and performing other simple operations, the integral in (5.6) can be written as

$$\int_0^\infty \int_0^\infty (u_T(t) - u_T(t+\tau))\mu_T(t)\ \dot{g}(\tau)\ dt\ d\tau$$

and also (using (5.4) and (5.5))

$$\int_0^\infty \int_0^T (u_T(t) - u_T(t+\tau))\ f(u_T(t))\ \dot{g}(\tau)\ dt\ d\tau. \tag{9.1}$$

Since f is monotone increasing, one can prove that

$$-(a-b)\ f(a) \leqq V(b) - V(a)\ ,\ \text{for every } a,b \in R,$$

where $V(x) = \int_0^x f(s)\ ds$. Hence (9.1) is majorized by

$$\int_0^\infty \int_0^T (V(u_T(t+\tau)) - V(u_T(t)))\ (-\dot{g}(\tau))\ dt\ d\tau. \tag{9.2}$$

Since $\sup_{T>0} \sup_{t>0} |u_T(t)| < \infty$, one can find $K_1 > 0$ such that $|V(u_T(t))| < K_1$, for every $t > 0$ and every $T > 0$. Moreover, the inner integral in (9.2) is of the form $E(\tau)(-\dot{g}(\tau))$, where

$$E(\tau) = \int_T^{T+\tau} V(u_T(t))\ dt - \int_0^\tau V(u_T(t))\ dt.$$

One can see that $E(\tau) \leqq 2K_1\tau$. Hence (9.2) is majorized by $2K_1 \int_0^\infty (-\dot{g}(\tau))\tau\ d\tau$. From (5.3) it follows that the integral converges. This proves lemma 2.

10. PROOF OF LEMMA 3

Proposition: Let $\delta>0$ and $r>0$. Then there exists $K_1>0$ such that, for every $t\in R$ and every real, continuously differentiable function γ - defined on $[t-\delta-r, t+\delta]$ - there exists $\tilde{t} \in [t-\delta, t+\delta]$ with the property

$$\int_{t-\delta}^{t+\delta} |\rho(\tau) - \rho(\tilde{t})|^2 d\tau \leqq K_1 \int_{t-\delta-r}^{t+\delta} |\dot{\gamma}(\tau)|^2 d\tau \tag{10.1}$$

where

$$\rho(\tau) = \dot{\gamma}(\tau) + \gamma(\tau-r). \tag{10.2}$$

Proof: One can obviously choose $\tilde{t} \in [t-\delta, t+\delta]$ such that

$$|\dot{\gamma}(\tilde{t})|^2 \leqq (2\delta)^{-1} \int_{t-\delta}^{t+\delta} |\dot{\gamma}(\tau)|^2 d\tau.$$

The left hand side of (10.1) is majorized by

$$3\int_{t-\delta}^{t+\delta} |\dot{\gamma}(\tau)|^2 d\tau + 3|\dot{\gamma}(\tilde{t})|^2 2\delta + 3\int_{t-\delta}^{t+\delta} \Big| \int_{\tilde{t}-r}^{\tau-r} \dot{\gamma}(\sigma)d\sigma\Big|^2 d\tau.$$

To obtain (10.1), it suffices to estimate $\int\dot{\gamma}(\sigma)\, d\sigma$ by means of the inequality of Cauchy-Bunyakovskiĭ-Schwarz.

Now lemma 3 will be proved by contradiction. Since the orbit is bounded on $(-\infty,\infty)$, α-limit points of the orbit exist. Suppose that the orbit has an α-limit point $\lambda_j:[-r,0]\to R, j=1,2,$ which is not a singular point. (The case of an ω-limit point is similar, but simpler.) Then one can show that at least one of $\dot{\lambda}_1$ and $\dot{\lambda}_2$ is not identically zero. It follows that there exists a number $\varepsilon_0>0$ and a sequence t_n such that $\lim_{n\to\infty} t_n = -\infty$ and $|\dot{x}_j(t_n)| > \varepsilon_0$, for $n=1,2,\ldots$ and for at least one of the values $j=1$ and $j=2$. In the following, j will stand for such a value. One can also assume that $t_1+\delta < 0$ and

$$t_{n+1} < t_n - 3r, \; n=1,2,\ldots \tag{10.3}$$

Moreover, by (2.1) and since x_1 and x_2 are bounded, there exists $\delta>0$ (which will be chosen in the interval $(0,r)$) such that, for every $\tau \in [-\delta,\delta]$ and $n=1,2,\ldots$, $|\dot{x}_j(t_n+\tau)| > \varepsilon_0/2$. Hence $\dot{x}_j$ (which, by (2.1), is also continuous) does not change sign on $(t_n-\delta,t_n+\delta)$. It is easy to see that therefore there exists $\varepsilon_1>0$ such that, for every $\tilde{t}_n \in [t_n-\delta,t_n+\delta]$,

$$\int_{t_n-\delta}^{t_n+\delta} |x_j(t) - x_j(\tilde{t}_n)|^2 dt > \varepsilon_1, \; n=1,2,\ldots \tag{10.4}$$

Let now N be a positive integer such that

$$N > K K_1 \varepsilon^{-2}\varepsilon_1^{-1}. \tag{10.5}$$

Choose $T > -t_N+\delta+r$. The integral in (10.4) can be written as in the left hand side of (10.1) with ρ replaced by x_j. From (4.4) and (4.3) it follows that (10.2) holds in the interval $[-T,T]$ if ρ is replaced by x_j and γ by $-\varepsilon^{-1}v_{j,T}$. Therefore, from the above proposition it follows that, for $n=1,2,\ldots,N$, there exist $\tilde{t}_n \in [t_n-\delta,t_n+\delta]$ such that

$$\int_{t_n-\delta}^{t_n+\delta} |x_j(t) - x_j(\tilde{t}_n)|^2 dt \leqq \varepsilon^{-2}K_1 \int_{t_n-\delta-r}^{t_n+\delta} |\dot{v}_{j,T}(\tau)|^2 d\tau.$$

Hence, from (6.1),(10.3) and (10.5), since $\delta \in (0,r)$, it follows that

$$\sum_{n=1}^{N} \int_{t_n-\delta}^{t_n+\delta} |x_j(t) - x_j(\tilde{t}_n)|^2 dt < \varepsilon_1 N. \qquad (10.6)$$

On the other hand, (10.4) implies that the left hand side of (10.6) is $> \varepsilon_1 N$. The contradiction proves lemma 3.

REFERENCES

[1] C.Corduneanu, Integral Equations and Stability of Feedback Systems, Academic Press, New York, 1973.

[2] A.Halanay, Differential Equations, Academic Press, New York, 1966.

[3] Y.D.Landau, Adaptive Control, Marcel Dekker, Inc., New York, 1979.

[4] J.Moser, On nonoscillating networks, Quart. Appl. Math. XXV,1967,1-9.

[5] K.S.Narendra and J.H.Taylor, Frequency Domain Criteria for Absolute Stability, Academic Press, New York, 1973.

[6] D.Šiljak, Large-Scale Dynamic Systems, North Holland New York, 1978.

[7] V.M.Popov and A.Halanay, On the stability of nonlinear control systems with delay, Automat. Remote Control, 23,1962,849-851.

[8] Vl.Răsvan, Absolute Stability of Retarded Control Systems (in Romanian) Ed.Acad.RSR,Bucharest,1975.

[9] J.Hale, Theory of Functional Differential Equations, Springer-Verlag, New York,1977.

10 V.M.Popov, On the phase-portrait-control, Proc. Colloque National Modèles Mathématiques, Belle Isle, Sept. 1982, France.

[11] V.M.Popov, Hyperstability of Control Systems, Springer-Verlag, New York,1973.

[12] V.M.Popov, Monotonicity and Mutability, J.Differential Equations, 31,1979,337-358.

LABELED GRAPHS AND A CLASS OF GENERALIZED VOLTERRA SYSTEMS

Ray Redheffer

University of California
Los Angeles, California

Nature first made things in her own way, and then made human reason skillful enough to understand, but only by hard work, some part of her secrets. --- GALILEO

Whenever you can, count. --- GALTON

THE BEGINNING

Suppose we have ξ bunnies and η foxes, with plenty of clover for the bunnies, but bunnies as sole source of food for the foxes. Without foxes $\dot{\xi} = a\xi$, $a > 0$, and without bunnies $\dot{\eta} = -b\eta$, $b > 0$. If the number of encounters is proportional to $\xi\eta$, and if each encounter results in a certain proportion of bunnies being eaten, the equations become

$$\dot{\xi} = a\xi - c\xi\eta, \quad \dot{\eta} = -b\eta + d\xi\eta.$$

More realistic than $\dot{\xi} = a\xi$ in the absence of interaction is $\dot{\xi} = a\xi - e\xi^2$ where $e\xi^2$ is a self-limiting term preventing the population from becoming infinite. Introducing a term of this kind in both equations and changing notation, one is led to the case $m = 2$ of the Lotka-Volterra system

$$\dot{u}_i = u_i\left(e_i + \sum_{j=1}^{m} p_{ij}u_j\right), \quad u_i(0) > 0.$$

This system and generalizations thereof form the topic of the following exposition. We assume existence of a positive stationary point $q = (q_i)$, namely, a point such that

$$e_i + \sum_{j=1}^{m} p_{ij}q_j = 0, \quad q_i > 0$$

but we do not impose any condition on the sign of p_{ij} for $i \neq j$. Thus, the theory applies not only to the prey-predator interaction with which we began, but also to cooperative interaction, or to various combinations of interactions. Among questions of ecological interest are: Does $\lim u(t)$ exist as $t \longrightarrow \infty$? If so, does the limit depend on the initial conditions? How can this be decided by inspection of the matrix (p_{ij})?

Problems of this kind have a long history, starting cheifly with Verhulst in the last century. Most relevant to the exposition given here is Volterra's book (1931), LaSalle's extension of Liapunov's theorem (1960, 1968), Krikorian's use of the latter to study the case $m = 3$ (1979) and further development along these lines in joint work of Walter and Zhou with the author (1980,1981).

GRAPHICAL REPRESENTATION

The graph $G(p)$ of the matrix $p = (p_{ij})$ is constructed by joining vertices i and j if $p_{ij} \neq 0$, and by putting 0 at i if $p_{ii} = 0$, ● at i if $p_{ii} \neq 0$. For example, the graph on the left below corresponds to the equations on the right:

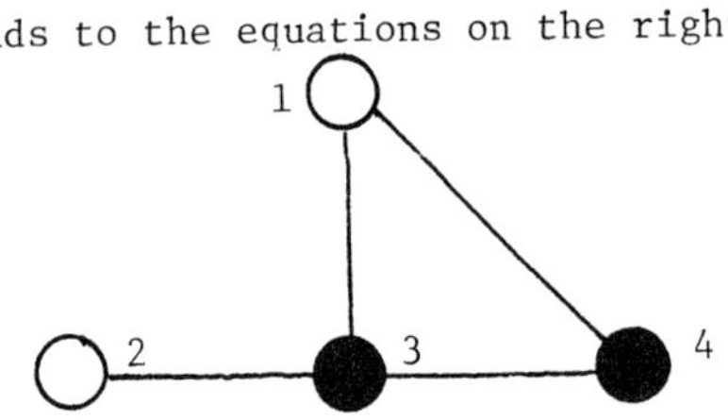

$$\dot{u}_1 = u_1(e_1+p_{13}u_3+p_{14}u_4)$$
$$\dot{u}_2 = u_2(e_2+p_{23}u_3)$$
$$\dot{u}_3 = u_3(e_3+p_{31}u_1+p_{32}u_2+p_{33}u_3+p_{34}u_4)$$
$$\dot{u}_4 = u_4(e_4+p_{41}u_1+p_{43}u_3+p_{44}u_4).$$

Although the graph is much simpler than the system itself, its complexity increases rapidly with the number of unknowns. The following three lines give, respectively, the number m of unknowns, the number of connected graphs with m nodes and a crude estimate for the number of labeled connected graphs:

2	3	4	5	6	...	10
1	2	6	21	112	...	11,716,571
3	10	48	336	3582	...	large

STABLY ADMISSIBLE MATRICES

We write $p \in A_0$ if there exist multipliers $a > 0$ such that, in the ordering provided by quadratic forms, $(a_i p_{ij}) \leq 0$. A perturbation of p is a matrix $\tilde{p}$ such that $\tilde{p}_{ij} = 0$ if, and only if, $p_{ij} = 0$. Thus, p and $\tilde{p}$ have the same labeled graph. The perturbation is small if $|p-\tilde{p}|$ is small. It is said that p is admissible, and we write $p \in A$, if every sufficiently small perturbation $\tilde{p}$ satisfies $\tilde{p} \in A_o$. This definition is motivated by the fact that, in realistic situations, the coefficients p_{ij} are known only with limited precision.

It is readily checked that, if $p \in A$, then $p_{ij}p_{ji} < p_{ii}p_{jj}$. Also one can choose the multipliers a_i in such a way that

$$\sum_{i,j=1}^{m} a_i p_{ij} w_i w_j = 0 \Rightarrow p_{ii} w_i = 0, \quad i = 1,2,\ldots,m.$$

A third property of the class A is topological, namely if $p \in A$, then every loop in $G(p)$ must have at least one pair of adjacent black dots $\bullet$. These properties are used in the sequel, but none of them follows from the weaker hypothesis $p \in A_o$. It is introduction of the class A rather than A_o which gives the theory its distinctive character.

THE VOLTERRA FUNCTION

Let $u_i = q_i e^{y_i}$, $w_i = u_i - q_i$, and, with a_i as above, let

$$V = \sum_{i=1}^{m} a_i q_i (e^{y_i} - y_i - 1).$$

Then by a short calculation

$$\dot{V} = \sum_{i,j=1}^{m} a_i p_{ij} w_i w_j.$$

Thus $p \in A_o$ implies $\dot{V} \leq 0$, hence V is bounded, and this shows that both u_i and $1/u_i$ are bounded. Introduction of the function V with $p \in A_o$ is due to Volterra. If $p \in A$ we have the further condition

$$\dot{V} = 0 \Rightarrow p_{ii} w_i = 0$$

which plays an important role in the sequel.

LASALLE'S THEOREM

For $\dot{u} = f(u)$, $u(0) = u_o$ the <u>orbit</u> is $\Lambda = \{u(t): t \geq 0\}$ and the ω

<u>limit set</u> is

$$\Lambda^+ = \{u: \lim u(t_j) = u \text{ for some sequence } t_j \to \infty\}.$$

For $V: R^m \longrightarrow R$ we have $\dot{V} = (d/dt)V(u(t)) = f(u)\cdot(\text{grad } V)(u)$. LaSalle's theorem in its simplest form then reads as follows: Let $f \in \text{Lip}(1)$, $\dot{V} \leq 0$ on Λ. Then Λ^+ is contained in the union of those orbits for which simultaneously $\dot{u} = f(u)$, $\dot{V} = 0$.

The proof involves three steps. First, if $u(t_j) \longrightarrow u^*$ then by continuity $V(u(t_j)) \longrightarrow V(u^*) = c$, where c is defined by this equation. Second, the fact that $\dot{V} \leq 0$ shows that $V(u(t))$ is decreasing, hence $\lim V(u(t)) = c$. Thus the same limit is obtained on every sequence $\{t_j\}$, $t_j \longrightarrow \infty$, and hence $V(u) = c$ on Λ^+. The third and most important step uses the fact that Λ^+ is an invariant set to deduce $\dot{V} = 0$.

It should be emphasized that the gist of the first two steps was known prior to LaSalle's work, and that the chief content of the theorem lies in the third step. For the case at hand the proof of invariance is easy. When LaSalle's theorem is extended to parabolic equations, however, the invariance is less obvious and requires discussion.

THE REDUCED GRAPH

Henceforward we consider solutions that satisfy the LaSalle condition $\dot{V} = 0$. At a black dot the equation $p_{ii}w_i = 0$ implies $w_i = 0$; conversely, if $w_i = 0$, then $p_{ii}w_i = 0$. Guided by this, we extend the symbolism so that ● at vertex i means either $w_i = 0$ or $p_{ii} > 0$. We also introduce a symbol ⊕ intermediate between 0 and ● to indicate that $w_i = \text{const}$.

Study of the differential equation together with the conditions $p_{ii}w_i = 0$ leads to simple rules for propagation of these symbols along the graph G(p). These are summarized here in the form of three <u>rules of reduction</u>:

(i) Suppose there is ● or ⊕ at i, and ● at all vertices adjacent to i except for a single vertex j adjacent to i. Then we can put ● at j.

(ii) Suppose there is ● or ⊕ at i, and ● or ⊕ at each vertex adjacent to i except for a single vertex j adjacent to i. Then we can put ⊕ at j.

(iii) Suppose there is 0 at i, and ● or ⊕ at each vertex adjacent to i. Then we can put ⊕ at i.

The <u>reduced graph</u> R(p) is obtained from G(p), the labeled graph of p, by using the rules of reduction until there is no further change. It is not difficult to show that R(p) is uniquely determined by G(p), independently of the manner in which the reduction is carried out.

THE SIMPLEST THEOREM

It is said that R(p) is of type (●) if every vertex has ●, of type (●,⊕) if every vertex has ● or ⊕ and at least one vertex has ⊕, and of type (0) if at least one vertex has 0. Under the assumption that $p \in A$ we have the following:

<u>Theorem 1.</u> (i) If R(p) is of Type (●) then p is nonsingular, the stationary point q is unique, and every solution satisfies $\lim u(t) = q$ as $t \longrightarrow \infty$.

(ii) If R(p) is of Type (●,⊕) then p is singular, q is not unique, every solution u(t) has a limit as $t \to \infty$, and the limit depends on u(0).

(iii) If R(p) is of Type (0) then there exists a matrix $\tilde{p} \in A$, with $G(\tilde{p}) = G(p)$, so that the system associated with $\tilde{p}$ has a nonconstant periodic solution.

Theorem 1 has a number of general consequences, of which the following is typical: If the graph G(p) has at most one "tail" (free end) then Case (i) holds for R(p) and hence every solution has a limit independent of the initial conditions. These general results in turn are an aid in the analysis of special cases. For example (always under the hypothesis $p \in A$) suppose $n \leq 6$, and suppose at least one vertex has ● when $n \leq 5$ while at least two have ● when $n = 6$. Then, of the 3,000 or so possible graphs G(p), only two allow a nonconstant periodic solution. In all other cases the solutions tend to a definite nonzero limit, which is independent of the initial conditions if, and only if, p is nonsingular.

What is interesting about this result is that it is obtained in only an hour or two with pencil and paper; use of a computer is unnecessary. If a computer is used, the scope of the general results is, naturally, increased.

DECOUPLING AT INFINITY

The limiting behavior is described by the differential equations associated with the reduced graph R(p), where $u_i \equiv q_i$ if a black dot in R(p) is at vertex i. The latter system in turn breaks up into separate systems for each of which the graph is a tree. The variables associated with a given tree can be analyzed independently of those in other trees and, in particular, they do not depend on the initial conditions for those in other trees. This breakup of the original system into separate systems whose graphs have no loops represents an enormous simplification. It also gives a clue to the distinction between superficial and genuine interaction, a distinction which is far from obvious in the differential equations associated with the original graph G(p).

TWO LEMMAS The ideas underlying the proof of Theorem 1 are illustrated by two lemmas which pertain to certain types of labeling. A graph admits a special labeling (a,b) if each vertex can be labeled a or b in such a way that no two a's are adjacent and the number of a's exceeds the number of b's. The first lemma then reads as follows: Let $G(p) = H \cup K$ where H admits a special labeling (a,b) with the following two properties:

$p_{ii} = 0$ at every a-vertex i,
every vertex adjacent to an a vertex is in H.

Then p is singular.

An (A,B) labeling of a graph is a labeling in which each vertex is labeled A or B, every A is adjacent to at least one other A, and every B is adjacent to at least two A's. The second lemma then reads as follows: Let $G(p) = H \cup K$ where H is a two-colorable graph. Suppose H admits an (A,B) labeling such that:

$p_{ii} = 0$ at each A-vertex i,
every vertex adjacent to an A-vertex is in H.

Then there exists $\tilde{p}$ such that $G(\tilde{p}) = G(p)$ and such that the system associated with $\tilde{p}$ admits a nonconstant periodic solution. If $p \in A$ then $\tilde{p} \in A$.

The first lemma is proved by constructing a nonzero vector v such that $pv = 0$. The second follows by a judicious interlocking of 2 by 2

Volterra systems, using the fact that the latter are known to have periodic solutions. Theorem 1 is readily obtained from the lemmas by making the identification $\oplus = a$, $\bullet = b$, $0 = A$, $\bullet$ or $\oplus = B$.

NONLINEAR INTERACTION

Let $R^+ = \{t: t > 0\}$ and $R_m^+ = \{w \in R_m: w_i > 0\}$; if $w \in R_m^+$ we write $w > 0$. Let

$$N: R_m^+ \longrightarrow R^+, \quad f: R^+ \longrightarrow R_m^+, \quad g: R^+ \longrightarrow R_m^+$$

be continuous functions such that for $j = 1,2,\ldots,m$

$$\int_0^1 \frac{ds}{f_j(s)} = \infty, \quad \int_1^\infty \frac{ds}{f_j(s)} = \infty, \quad g_j(s) \text{ is strictly increasing.}$$

Then parts (i) and (ii) of Theorem 1 apply to the generalized system

$$\dot{u}_i = N(u)f_i(u_i)\left(e_i + \sum_{j=1}^m p_{ij}g_j(u_j)\right)$$

$$u_i(0) > 0, \quad q_j > 0, \quad e_i + \sum_{j=1}^m p_{ij}g_j(q_j) = 0.$$

The proof is similar to the proof of Theorem 1.

DIFFUSION

Let $\Omega \subset R^n$ be open, bounded and orientable, with $C^{2+\alpha}$ boundary $\partial\Omega$. Let

$$L\phi = \sum_{i,j=1}^n \frac{\partial}{\partial x_i}\left(A_{ij}(x)\frac{\partial\phi}{\partial x_j}\right)$$

where $A(x) \in C^{1+\alpha}(\bar{\Omega})$, $\xi A(x)\xi \geq \lambda|\xi|^2$, $\lambda > 0$, and let ν be the associated conormal. The initial-value function u_o is assumed to satisfy

$$u_o \in C^{2+\alpha}(\bar{\Omega}), \quad u_o > 0, \quad \frac{\partial u_o}{\partial\nu} = 0.$$

With $\dot{u}_i = \partial u_i/\partial t$ we consider the system

$$\dot{u}_i = Lu_i + N(u)f_i(u_i)\left(e_i + \sum_{j=1}^m p_{ij}q_j(u_j)\right) \quad \text{in } (0,\infty) \times \Omega,$$

$$u(0,x) = u_o(x), \quad u_\nu(t,x) = 0 \quad (t>0)$$

where N,f,g,p are as above. The orbit $\Lambda = \{u(t,\cdot): t > 0\}$ is here regarded as a path in the space $C^2(\bar{\Omega})$ with the usual metric, and the ω-limit set is

$$\Lambda^+ = \phi \in C^2(\bar{\Omega}) : \lim u(t_j,\cdot) = \phi \text{ in } C^2 \text{ for some } t_j \longrightarrow \infty\}.$$

Then Λ^+ is nonempty, Λ^+ contains constant functions only, and (i) and (ii) of Theorem 1 apply without change.

AN EXTENSION OF LASALLE'S THEOREM

Proof of the above result requires an extension of LaSalle's theorem to the class of partial differential equations there considered. The extension is not complete, however, because it reduces only to a special case of LaSalle's theorem when the functions involved are independent of x. Another special case of this sort has been studied by Hadeler (1981).

A full extension of the LaSalle theorem to parabolic equations is described next. Using the same notation as above, let $u(o,x) = u_o(x)$ in Ω, $u_\nu(t,x) = 0$ in $(0,\infty) \times \partial\Omega$ and

$$\dot{u}_i = \sum_{j=1}^{n} \frac{\partial}{\partial x_j} \sum_{k=1}^{n} A_{jk}(x,u)\frac{\partial u_i}{\partial x_k} + f_i(u) \quad \text{in} \quad (0,\infty) \times \Omega.$$

Let $V: R^m \longrightarrow R$ be smooth, with $V = V(u)$ along the trajectory u. Along u suppose further

$$\lim_{|u|\to\infty} V(u) = +\infty, \quad f(u)\cdot(\text{grad } V)(u) \leq 0, \quad H(V) \geq 0$$

where H(V) is the Hessian matrix of V. Then, modulo mild regularity hypotheses, the limit-set Λ^+ contains only constant functions. Hence, Λ^+ is contained in the union of the orbits of $\dot{u} = f(u)$, $\dot{V} = 0$, just as in the theorem of LaSalle.

The main difficulty is to show compactness of the orbit $u(t,\cdot)$. This problem has been solved by Redlinger (1982). In other respects the proof parallels the proof outlined above for the ordinary differential equation $\dot{u} = f(u)$.

REFERENCES

[1] K. P. Hadeler, "Diffusion in Fisher's population model," Rocky Mountain J. of Math., 11, (1981), 39-45.

[2] N. Krikorian, "The Volterra model for three-species predator-prey systems: Boundedness and stability," J. Math. Biol., 7, (1979), 117-132.

[3] J. P. LaSalle, "Stability theory for ordinary differential equations," J. Diff. Eqs., 4, (1968), 57-65.

[4] R. M. Redheffer and Z. Zhou, "Global asymptotic stability for a class of many-variable Volterra prey-predator systems," J. Nonlinear Math. and Appls., 5, (1981), 1309-1329.

[5] R. M. Redheffer and Z. Zhou, "A class of matrices connected with Volterra prey-predator equations," SIAD, 3, (1982), 122-134.

[6] R. M. Redheffer and W. Walter, "Solution of the stability problem for a class of generalized Volterra prey-predator systems," J. Diff. Eqs. (accepted).

[7] R. M. Redheffer and W. Walter, "On parabolic systems of the Volterra prey-predator type," J. Nonlinear Math. and Appls. (accepted).

[8] R. Redlinger, "Über die C^2-Kompaktheit der Bahn von Lösungen semilinearer parabolischer Systeme," Proc. Royal Soc. Edinburgh (accepted).

[9] V. Volterra, "Lecons sur la théorie mathematique de la lutte pour la vie," Gauthieç-Villars et Cie., Paris (1931).

THE METHOD OF QUASI UPPER AND LOWER SOLUTIONS FOR EXTREMAL SOLUTIONS OF VOLTERRA INTEGRAL EQUATIONS IN ABSTRACT CONES†

G. R. Shendge*

University of Iowa
Iowa City, Iowa

ABSTRACT

Monotone iterative technique for obtaining the extremal solutions of initial and boundary value problems via coupled quasi-solutions have been recently developed, for example see [4,6]. Coupled lower and upper quasi-solutions play a vital role in this method. Hence introducing these concepts for Volterra integral equations in abstract cones we develop the monotone method for obtaining the extremal solutions of such equations in a Banach Space. These results are only briefly covered here and for a complete discussion see [5].

PRELIMINARIES

Let E be a real Banach space with norm $\|\cdot\|$. Let $K \subset E$ be a cone, that is, a closed convex subset of E such that $\lambda K \subset K$ for every $\lambda \geq 0$ and $K \cap \{-K\} = \{0\}$. A partial order $\leq$ is defined in E relative to K by $v \leq u$ if and only if $u - v \in K$.

A cone K is said to be normal if there exists a real number $N > 0$ such that $0 \leq v \leq u$ implies $\|v\| \leq N\|u\|$, where N is independent of v and u. We assume that K is a normal cone and all inequalities in E are to be understood relative to K.

Let α denote the Kuratowski's measure of noncompactness for the properties of which we refer to [1,7].

For any $v,w \in C[I,E]$, such that, $v(t) \leq w(t)$ on I, where $I = [0,T]$, let $[v,w] = \{u \in C[I,E]: v \leq u \leq w\}$ and $\Omega = \{(t,s,u) \in I \times I \times E: 0 \leq s \leq t,\ t \in I,\ v(t) \leq u \leq w(t)\}$.

†Research partially supported by U.S. Army Research Grant #DAAG29-80-C-0060

*On study-leave from Marathwada University Aurangabad 431 004, Maharashtra, India.

Consider the Volterra integral equation

(1.1) $$x(t) = h(t) + \int_0^t H(t,s,x(s))ds, \ t \in I$$

where $h \in C[I,[v,w]]$ and $H \in C[\Omega,E]$. We assume that H admits a decomposition of the form

(1.2) $$H(t,s,x) = H_1(t,s,x) + H_2(t,s,x)$$

where $H_1, H_2 \in C[I,E]$.

Definition. Let $v,w \in C[I,E]$. Then v,w are said to be coupled lower and upper quasi-solutions of (1.1) if

(1.3) $$\begin{cases} v(t) \leq h(t) + \int_0^t [H_1(t,s,v(s)) + H_2(t,s,w(s))]ds \\ w(t) \geq h(t) + \int_0^t [H_1(t,s,w(s)) + H_2(t,s,v(s))]ds \end{cases}$$

hold on I. If, in (1.3), equalities hold, then v,w are said to be coupled quasi-solutions of (1.1). The coupled minimal and maximal quasi-solutions can now be defined in the usual way.

We need the following conditions.

(A_1) $H_1(t,s,x)$ is monotone nondecreasing and $H_2(t,s,x)$ is monotone nonincreasing in x for each (t,s) relative to K.

(A_2) H_1, H_2 are uniformly continuous on Ω.

Consider the following two schemes for defining the iterates.

(s-i) $$\begin{cases} v_n(t) = h(t) + \int_0^t [H_1(t,s,v_{n-1}(s)) + H_2(t,s,w_{n-1}(s))]ds \\ w_n(t) = h(t) + \int_0^t [H_1(t,s,w_{n-1}(s)) + H_2(t,s,v_{n-1}(s))]ds \end{cases}$$

with $v_0 = v$ and $w_0 = w$ on I and

(s-ii) $$\begin{cases} v_n(t) = h(t) + \int_0^t [H_1(t,s,w_{n-1}(s)) + H_2(t,s,v_{n-1}(s))]ds \\ w_n(t) = h(t) + \int_0^t [H_1(t,s,v_{n-1}(s)) + H_2(t,s,w_{n-1}(s))]ds \end{cases}$$

with $v_0 = v$ and $w_0 = w$ on I.

MAIN RESULTS

We can now state

Theorem 1.1. Let the cone K be normal and $v, w \in C[I,E]$ be coupled lower and upper quasi-solutions of (1.1) such that $v(t) \leq w(t)$ on I. Let the conditions (A_1), (A_2) hold. Assume further that

(A_3) H_1, H_2 satisfy

$$\alpha(H_i(t,s,B)) \leq \frac{1}{2} g(t,s,\alpha(B)), \quad i = 1,2 \tag{1.4}$$

for every bounded set $B \subset E$, where $g \in C[I\times I\times R^+, R^+]$, $g(t,s,0) = 0$, $g(t,s,u)$ is monotone nondecreasing in u for each (t,s), satisfies the super-additive condition $g(t,s,u_1) + g(t,s,u_2) \leq g(t,s,u_1+u_2)$ and the integral equation $u(t) = \int_0^t g(t,s,u(s))ds$ has the only solution $u(t) \equiv 0$ on I.

Then the sequences $\{v_n\}$, $\{w_n\}$ defined by (s-i) satisfy

$$v \leq v_1 \leq v_2 \leq \cdots \leq v_n \leq \cdots \leq w_n \leq \cdots \leq w_2 \leq w_1 \leq w \tag{1.5}$$

on I and $\lim_{n\to\infty} v_n(t) = \rho(t)$, $\lim_{n\to\infty} w_n(t) = r(t)$ uniformly on I, where ρ, r are the coupled minimal and maximal quasi-solutions of (1.1) on I. Moreover, any solution $u \in [v,w]$ of (1.1) also satisfies $\rho \leq u \leq r$ on I.

Outline of the proof. We shall only indicate the main ideas involved in the proof and for details see [5]. The proof is given in three stages.

(a) Using the condition (A_1) and the scheme (s-i) repeatedly, we can show by induction that the sequences $\{v_n\}$, $\{w_n\}$ are monotone and satisfy (1.5).

(b) These sequences are then shown to be uniformly bounded, equicontinuous and for each t, the sets $\{v_n(t)\}_{n=0}^{\infty}$ and $\{w_n(t)\}_{n=0}^{\infty}$ are relatively compact. The boundedness is a consequence of the fact that the sequences lie in the conical segment $[v,w]$ in a normal cone and equicontinuity follows from standard estimates which can be obtained by using (A_2). To show that the sets $\{v_n(t)\}_{n=0}^{\infty}$ and $\{w_n(t)\}_{n=0}^{\infty}$ are relatively compact, we set $m_1(t) = \alpha(\{v_n(t)\}_{n=0}^{\infty})$, $m_2(t) = \alpha(\{w_n(t)\}_{n=0}^{\infty})$ and obtain

$$m_1(t) + m_2(t) \leq \int_0^t g(t,s,m_1(s) + m_2(s))ds$$

by using (A_3). Since g satisfies the conditions in (A_3), the foregoing inequality yields $m_1(t) = m_2(t) = 0$ by a standard result [see Th. 5.3.1, p. 322[2]]. Consequently the sets $\{v_n(t)\}_{n=0}^{\infty}$, $\{w_n(t)\}_{n=0}^{\infty}$ are relatively compact. Now on the basis of the Ascoli-Arzela Theorem and the monotone nature of the sequences it can be shown that $\{v_n\}$, $\{w_n\}$ converge uniformly and monotonically to certain limit functions.

(c) Let $\lim_{n\to\infty} v_n(t) = \rho(t)$ and $\lim_{n\to\infty} w_n(t) = r(t)$, $t \in I$. Then, using (s-i) it is verified that ρ, r are coupled quasi-solutions of (1.1) on I. Moreover, by induction we find that if u_1, u_2 are any coupled quasi-solutions of (1.1) on I such that $v \leq u_1, u_2 \leq w$, then $\rho \leq u_1, u_2 \leq r$, which implies that ρ, r are minimal and maximal quasi-solutions of (1.1) on I. Finally, we note that since any solution $u \in [v,w]$ of (1.1) on I can be looked upon as a coupled quasi-solution u,u, it satisfies $\rho \leq u \leq r$ on I and the proof is complete.

Remarks 1. Since the decomposition (1.2) of H corresponds to a similar decomposition in the finite dimensional case [3], it contains the results in [3] to Banach spaces.

Remarks 2. If $H_2 = 0$, then $H_1 = H$, where $H = H(t,s,x)$ is now monotone nondecreasing in x and the coupled lower and upper quasi-solutions become simply the usual lower and upper solutions and ρ, r reduce to the minimal and maximal solutions of (1.1) on I.

It is interesting to note that if we use the iterates v_n, w_n defined by (s-ii) instead of (s-i), we get four alternating sequences which are all monotone and the extremal quasi-solutions of (1.1) can be obtained by choosing any two of these four sequences. This offers a greater flexibility in getting the extremal solutions. We shall only state the following theorem and for the proof refer to [5].

Theorem 1.2. Let the hypotheses of Theorem 1.1 hold. Then the sequences $\{v_n\}$ and $\{w_n\}$ defined by (s-ii) satisfy

$$v \leq w_1 \leq v_2 \leq \cdots \leq w_{2n-1} \leq v_{2n} \leq \cdots \leq w_{2n} \leq v_{2n+1} \leq \cdots \leq w_2 \leq v_1 \leq w$$

on I and $\lim_{n\to\infty} v_{2n}(t) = \lim_{n\to\infty} w_{2n-1}(t) = \rho(t)$,

$\lim_{n\to\infty} v_{2n-1}(t) = \lim_{n\to\infty} w_{2n}(t) = r(t)$ uniformly on I, where ρ, r are the coupled minimal and maximal quasi-solutions of (1.1) on I.

REFERENCES

[1] V. Lakshmikantham and S. Leela, Nonlinear Differential Equations in Abstract Spaces, Pergamon Press, New York, 1981.

[2] V. Lakshmikantham and S. Leela, Differential and Integral Inequalities, Vol. I, Academic Press, New York, 1969.

[3] G. S. Ladde, V. Lakshmikantham and B. G. Pachpatte, "The Method of Upper and Lower Solutions and Volterra Integral Equations," To appear in the J. Integral Equations.

[4] V. Lakshmikantham, S. Leela and M. N. Ogustoreli, "Quasi Solutions, Vector Lyapunov Functions and Monotone Method," IEEE Trans. Automa. Control, 26 (1981), p. 1149-1153.

[5] G. R. Shendge, "Volterra Integral Equations in Abstract cones and Monotone Iterative Technique," The Univ. of Iowa, Technical Report.

[6] A. S. Vatsala, "Existence of Coupled Minimal and Maximal Periodic Quasi-Solutions for System of First Order Periodic BVPs Via Quasi-Solutions," To appear in the Proceedings of the Vth International Conference on 'Trends in Theory and Practice of Nonlinear Differential Equations' held at Arlington, Texas in June, 1982.

[7] R. L. Vaughn, "Existence and Comparison Results for Volterra Integral Equations in a Banach Space," The Univ. of Texas at Arlington, Technical Report, No. 163, May 1977.

PERTURBED BIFURCATION THEORY FOR A CLASS OF AUTONOMOUS ORDINARY DIFFERENTIAL EQUATIONS

Ratnasingham Shivaji

Department of Mathematics
Yarmouk University
Jordan

ABSTRACT

We obtain fairly detailed information about the continua of solutions for the perturbed bifurcation problem

$$-u''(x) = \lambda(u-u^3-\varepsilon),\ x \in (0,1):\ u(0) = 0 = u(1) \qquad (*)$$

where $\lambda > 0$ and $\varepsilon > 0$.

1. INTRODUCTION

In this paper, using the quadrature technique introduced by T. W. Laetsch in [1] we study about the continua of all solutions of (*). In order to do so we must extend the quadrature technique so that it applies to solutions other than non-negative solutions. Also we found a continuum containing both non-negative solutions and solutions which change sign in (0,1) which is of interest since in the standard bifurcation theory case it is well known that solutions in the same continuum have the same nodal behaviour.

2. QUADRATURE TECHNIQUE

Consider the non-negative solutions of

$$-u''(x) = \lambda f(u(x)) \text{ in } (0,1):\ u(0) = 0 = u(1) \qquad (2.1)$$

where $\lambda \geq 0$ and f is a smooth function. Since (2.1) is autonomous, if u is any solution of (2.1) then $u(1-x)$ is a solution of (2.1) and if $u'(t) = 0$ where $t < \frac{1}{2}$ then $u(2t) = 0$. Also if u is a non-negative solution of (2.1) such that $u'(0) > 0$, then u attains its maximum at $x = \frac{1}{2}$ and is strictly increasing [decreasing] in $(0,\frac{1}{2})[(\frac{1}{2},1)]$. Now if u

is any solution of (2.1) then it follows that

$$\frac{1}{2}(u'(x))^2 + \lambda F(u(x)) = \text{constant in } [0,1] \tag{2.2}$$

where $F(u) = \int_0^u f(t)dt$. So if it is non-negative with

$$\|u\| = \sup_{x \in [0,1]} u(x) = p \text{ then for } x \in [0,1]$$

$$\frac{1}{2}(u'(x))^2 + \lambda F(u(x)) = \frac{1}{2}(u'(0))^2 = \lambda F(p). \tag{2.3}$$

Hence we must have $F(p) \geq 0$ and if $F(p) > 0$ then $u'(0) > 0$ and so the solution is as described before.

<u>Theorem 2.1. (See [1] and [2])</u> Let $f \quad C([0,r))$ for some $0 < r \leq +\infty$ be such that $S: = \{u \in (0,r)/f(u)>0$ and $F(u)>F(s)$ for all $s, 0 \leq s < u\}$ is non-empty. Then for any $p \in S, \exists$ a unique $\lambda > 0$ such that (2.1) has a non-negative solution satisfying $\|u\| = p$. Also for $p \in S$ the corresponding λ is given by

$$\{\lambda(p)\}^{1/2} = (2)^{1/2} \int_0^p (F(p)-F(s))^{-1/2} ds \tag{2.4}$$

and u is symmetric with respect to $x = 1/2$ and is given by

$$(2\lambda)^{1/2} x = \int_0^{u(x)} (F(p)-F(s))^{-1/2} ds \text{ on } [0,1/2]. \tag{2.5}$$

Further $p \dashrightarrow \lambda(p)$ is a continuous function on S.

Also, a straightforward computation and use of Lebesgue dominated convergence theorem shows that

$$\frac{d}{dp}((\lambda(p))^{1/2}) = (2)^{1/2} \int_0^1 \frac{H(p)-H(pv)}{(F(p)-F(pv))^{3/2}} dv, \tag{2.6}$$

where $H(u) = F(u) - (u/2)f(u)$. (2.5) and (2.6) deduces the nature of the bifurcation curve of non-negative solutions. We shall now describe the extension required to deal with solutions which have interior zeros. (Non-positive solutions can be dealt as similar to non-negative solutions.) Consider solutions of the form

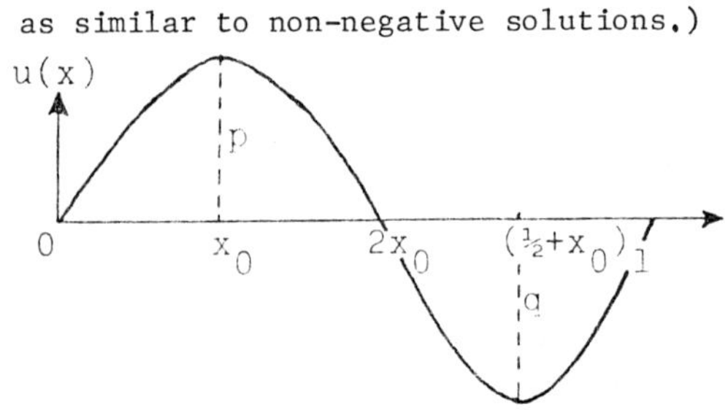

Fig. 2.1

Suppose $u'(0) > 0$ and u has an interior zero at $2x_0$. Since the equation is autonomous, it follows that u has a local maximum at x_0 and local minimum at $1/2 + x_0$. If $u(x_0) = p$ and $u(1/2+x_0) = q$, (2.2) shows that $F(p) = F(q)$. Let $f \in C((r^*,0])$ and

$$S^* = \{u \in (r^*,0)/f(u)<0 \text{ and } F(u)>F(s) \text{ for all}$$

$s, u \leq s \leq 0$ be nonempty. Then arguments similar to that used in Theorem 2.1 give the following:

<u>Theorem 2.2.</u> Let $p \in S$ and assume that $\exists$ a $q \in S^*$ such that $F(p) = F(q)$. Then $\exists$ a unique $\lambda > 0$ such that (*) has a solution as described in figure 2.1 where u is given by

$$(2\lambda)^{1/2}x = \int_0^{u(x)} (F(p)-F(s))^{-1/2}ds \quad \text{on } [0,x_0],$$

$$(2\lambda)^{1/2}(1-x) = \int_{u(x)}^{0} (F(q)-F(s))^{-1/2}ds \quad \text{on } [1/2+x_0,1]$$

and

$$(\lambda(p))^{1/2} = (2)^{1/2} p \int_0^1 (F(p)-F(pv))^{-1/2}dv -$$

$$(2)^{1/2} q \int_0^1 (F(q)-F(qv))^{-1/2}dv. \qquad (2.7)$$

Here x_0 is given by

$$x_0 = 1/2 \left(\int_0^p (F(p)-F(s))^{-1/2}ds\right)/\left(\int_q^p (F(p)-F(s))^{-1/2}ds\right).$$

(Note that u has to be symmetric about x_0 in $[0,2x_0]$ and about $1/2 + x_0$ in $[2x_0,1]$.)

Further we obtain (regarding q as a function of p)

$$\frac{d}{dp}(\lambda(p))^{1/2} = (2)^{1/2} \int_0^1 \frac{H(p)-H(pv)}{(F(p)-F(pv))^{3/2}}dv -$$

$$(2)^{1/2}\frac{dq}{dp} \int_0^1 \frac{H(q)-H(qv)}{(F(q)-F(qv))^{3/2}}dv. \qquad (2.8)$$

(2.7) and (2.8) deduce the nature of the bifurcation diagram of solutions described in Figure 2.1. In a similar fashion one can tackle all solutions which changes sign in (0,1).

Next we consider solutions which have interior zeros but which stay non-negative. First we consider solutions of the form

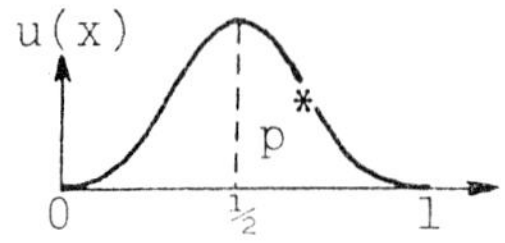

(Here $u'(0) = 0$. Such a solution is possible only if $f(0) \leq 0$.)

This is possible if $F(p^*) = 0$, $f(p^*) > 0$ and $F(p^*) > F(s)$ for all $0 \leq s < p^*$. Here $u(x)$ is given by

$$(2\lambda)^{1/2}x = \int_0^{u(x)} (-F(s))^{-1/2}\, ds \quad \text{on} \quad [0,1/2] \quad \text{and}$$

$$(\lambda(p^*))^{1/2} = (2)^{1/2}\, p^* \int_0^1 (-F(p^* v))^{-1/2}\, dv = (\mu)^{1/2} \text{ (say)}.$$

Next we consider non-negative solutions with one interior zero. It follows easily that it has to take the form below.

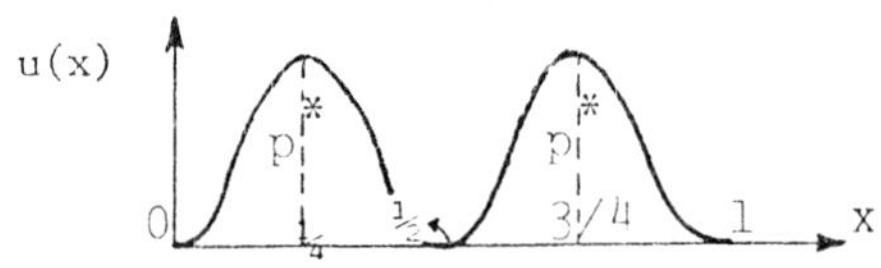

Here $\lambda = 4\mu$ and $u(x)$ is given by

$$(8\mu)^{1/2}\, x = \int_0^{u(x)} (-F(s))^{-1/2}\, ds \quad \text{on} \quad [0,1/4]\,.$$

Similarly one can deal with non-negative solutions with any number of interior zeros. (If $f(0) \geq 0$ then corresponding results follow for non-positive solutions.)

3. BIFURCATION DIAGRAM FOR (*)

Here we shall collect results on the continua of solutions for (*) obtained by the quadrature technique discussed in section 2 (for detailed proofs of these results see [3]).

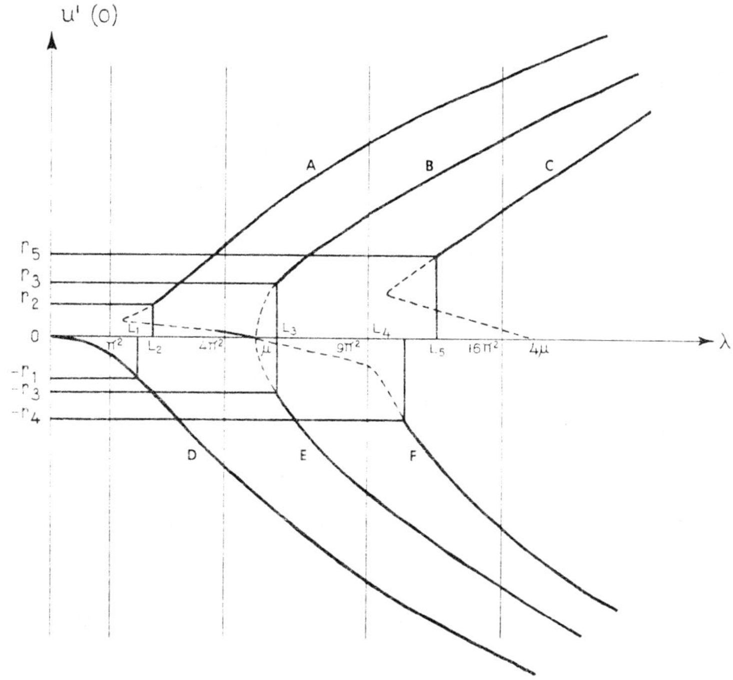

Here as $\varepsilon \longrightarrow 0$, $\lim. L_1(\varepsilon) = \lim. L_2(\varepsilon) = \pi^2$

$\lim. L_3(\varepsilon) = \lim. \mu(\varepsilon) = 4\pi^2$,

$\lim. L_4(\varepsilon) = \lim. L_5(\varepsilon) = 9\pi^2$ and

$\lim. r_s(\varepsilon) = 0$ for $s = 1,2,3,4,5$.

The solid lines represents the parts where the sign of $\frac{d\lambda}{dp}$ are known; the dotted line represents the parts where the sign of $\frac{d\lambda}{dp}$ is uncertain.

The continua A represents non-negative solutions, D represents non-positive solutions, B & E represent solutions with one interior zero and and C & F represent solutions with two interior zeros. Note that for $\lambda = \mu$ we have the nonnegative solution with $\text{Sup}\{u(x)/x\in[0,1]\} = \beta(\varepsilon)$ and $u'(0) = u'(1) = 0$ where $\beta(\varepsilon)$ is the first positive zero of $F(u)$,

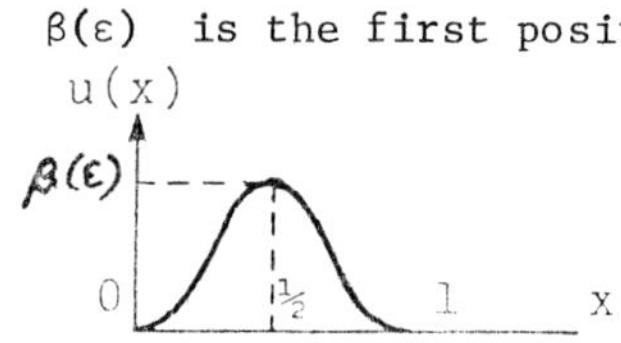

bifurcating into the following solutions.

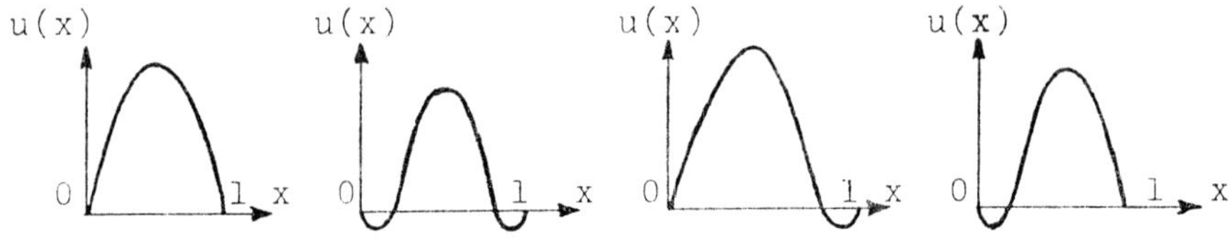

ACKNOWLEDGMENT

It is a pleasure to thank Dr. K. J. Brown for many valuable discussions on this problem.

REFERENCES

[1] T. W. Laetsch, "The number of solutions of a nonlinear two point boundary value problem," Indiana Univ. Math. J., 20 (1970/71), 1-13.

[2] K. J. Brown and H. Budin, "Multiple positive solutions for a class of nonlinear boundary value problem," J. Math. Anal., 60 No. 2 (1977), 329-338.

[3] R. Shivaji, "Solvability of certain classes of semilinear boundary value problems," Ph.D. Thesis - 1981, Heriot-Watt University, U.K.

BIFURCATION OF SUBHARMONIC SOLUTIONS FOR A PERIODIC EPIDEMIC MODEL

Hal L. Smith

Department of Mathematics
Arizona State University
Tempe, Arizona

ABSTRACT

Two methods are sketched for obtaining the existence of subharmonic solutions for a periodic epidemic model. Several stable subharmonic solutions of different periods are shown to simultaneously coexist.

In this note we outline some recent work on the existence of subharmonic solutions of an S→I→R epidemic model with annual oscillations in the transmission rate and vital dynamics. The equations describing the model are

$$
\begin{aligned}
S' &= \mu - \mu S - \beta(t)IS \\
I' &= \beta(t)IS - (\gamma+\mu)I \\
R' &= \gamma I - \mu R \qquad (1.1)\\
S + I + R &= 1 \\
\beta(t) &= \beta_0 + \beta_1 \cos 2\pi t
\end{aligned}
$$

in which S,I,R represent respectively susceptible, infectious and recovered (and immune) proportions of the population, μ is the birth and death rate, γ the recovery rate and $\beta(t)$ the transmission rate of the infection which is periodic of period one year. The equations (1.1) were first studied for periodic solutions by Dietz [2] who found apparent periodic solutions of periods 1,2,4 and 6 years from computer simulations. Later Z. Grossman, I. Gumowski, and Dietz [3] obtained in a nonrigorous and formal way, periodic solutions of period one and two years. The purpose of this article is to outline some methods from which stable subharmonic solutions of various integral periods can be rigorously obtained. A perhaps surprising result will be that several stable subharmonic solutions of different integral periods can simultaneously coexist. Full details of these methods will appear elsewhere [5,6].

For more background on the model described by (1.1) and the significance of subharmonic solutions of (1.1) the reader may consult the references [2,3,5,6].

It is clear from (1.1) that our unit of time is one year while the average life span, $1/\mu$, is 70 years and the infectious period, $P = \frac{1}{\gamma+\mu}$, one to two weeks. It is essential, then, to scale the parameters in (1.1) in order to identify the most important terms. Also, in order to obtain analytical results we need to assume small oscillations in transmission rate, $\beta_1/\beta_0 \ll 1$. For more motivation for the following scaling see [5,6]. We introduce two small parameters ε and δ and new variables x and y as follows:

$$\varepsilon^2 = \frac{\mu}{\gamma+\mu} \approx 10^{-4}, \quad \gamma+\mu = \frac{\Delta}{\varepsilon} \approx 10^2 \ , \quad \varepsilon\delta = \beta_1/\beta_0, \quad \Delta>0$$

$$QS = 1 + \varepsilon\sqrt{K}\, x \ , \quad QI = \varepsilon^2 K(1+y) \tag{1.2}$$

$$Q \equiv \beta_0 P \ , \quad K \equiv Q-1.$$

The quantity Q is the reproduction rate of the infection and in the absence of forcing ($\beta_1=0$) Q must exceed one for the existence of a stable (endemic) positive equilibrium point. We assume $Q>1$ $(K>0)$ and the change of variables to (x,y) in (1.2) reflects a translation of the origin to the endemic equilibruim point. The equations for the new variables x and y are given by

$$x' = -\upsilon[y + \frac{\varepsilon}{\sqrt{K}}\, x(K+1+Ky) + \delta\varepsilon\ (1+y)\cos 2\pi t + \delta\varepsilon^2 x\ (1+y)\cos 2\pi t]$$

(1.3)

$$y' = \upsilon[x(1+y) + \frac{\delta}{\sqrt{K}}\,(1+y)\cos 2\pi t + \varepsilon\delta x\ (1+y)\cos 2\pi t]$$

where $\upsilon = \Delta\sqrt{K}$. Setting $\varepsilon=\delta=0$ in (1.3), one obtains the autonomous system

$$x' = -\upsilon y$$

(1.4)

$$y' = \upsilon x(1+y)$$

which has a first integral $V = x^2 + 2y - 2\ln(1+y)$. The phase portrait of (1.4) is given in figure 1 below.

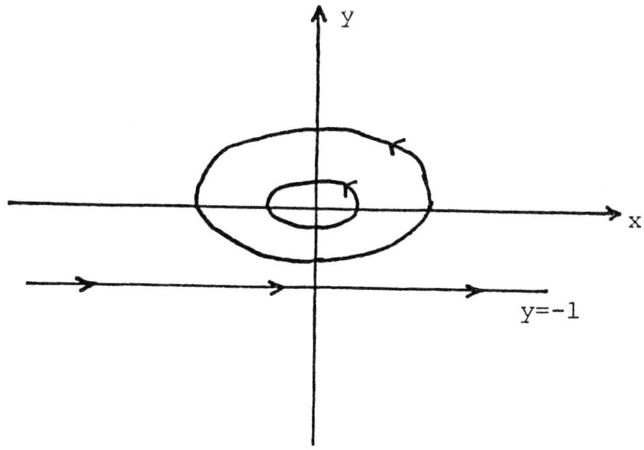

FIGURE 1

The periods of the closed orbits range from $\frac{2\pi}{\upsilon}$ to $+\infty$ at the invariant line $y=-1$.

We will describe two methods for obtaining (rigourously) subharmonic solutions of (1.3). The first method is suggested by the numerical work of Dietz [2] who found that as the amplitude of the transmission rate (β_1/β_0) is increased, a one-periodic solution of (1.1) loses stability to a two-periodic solution provided $\frac{2\pi}{\upsilon}-2$ is small, i.e., near resonance. This is also the approach used by Grossman et al [3]. Introducing a third small parameter $\eta = 1 - \upsilon/\pi$ in (1.3), one can then obtain the lowest order terms in $(\varepsilon,\delta,\eta)$ of a small-norm one-periodic solution of (1.3), (x^*,y^*), which is analytic in $(\varepsilon,\delta,\eta)$ near $(0,0,0)$ by a standard implicit function theorem argument ($(x^*,y^*) = (0,0)$ when $\varepsilon=\delta=\eta=0$). The stability of this solution is fully determined in [5]. One now introduces new variables $z_1 = x-x^*$, $z_2 = y-y^*$ and obtains the differential equations for these perturbations:

$$\text{(1.5)}\qquad \begin{aligned} z_1' &= -\pi z_2 \\ z_2' &= \pi z_1 \end{aligned} \quad + \text{ terms of higher order in } (z_1,z_2,\varepsilon,\delta,\eta)$$

See [5] for more detail. Clearly the linearization of (1.5) about $z=0$ has, for $\varepsilon=\delta=\eta=0$, two linearly independent period two solutions $u_1 = \text{col}(\cos\pi t,\sin\pi t)$ and $u_2 = \text{col}(-\sin\pi t,\cos\pi t)$. The Liapunov-Schmidt method can be applied to obtain period two solutions of (1.5) by reducing this problem to the study of the zeros of a map $G(a,\varepsilon,\delta,\eta)$ of a neighborhood of $0\varepsilon R^2 x R^3$ into R^2. Solutions of $G=0$ correspond to small norm period two solutions of (1.5), $z = a_1u_1 + a_2u_2 +$ (remainder of fourier series), and conversely. The invariance of (1.3) under the transformation $t \to t+{}^1/_2$, $\delta \to -\delta$ is inherited by (1.5) and results in the symmetry $G(a,\varepsilon,\delta,\mu) = RG(R^{-1}a,\varepsilon,-\delta,\mu)$ where R is counterclockwise rotation by 90 degrees. Applying this relation twice one sees the G is odd in a. Straightforward calculations give the following asymptotic information

$$G(a,\varepsilon,\delta,\eta) = (\varepsilon L_1+\delta L_2+\eta R)\ a + \frac{|a|^2}{24} Ra \tag{1.6}$$
$$+O(|\alpha|^2|a|+|\alpha|\,|a|^3+|a|^5)$$

where $L_1 = -\ell I$, $L_2 = m\begin{pmatrix}-1 & 0\\ 0 & 1\end{pmatrix}$, $R = \begin{pmatrix}0 & -1\\ 1 & 0\end{pmatrix}$, $m,\ell>0$, $\alpha = (\varepsilon,\delta,\eta)$. A priori estimates of solutions of $G=0$ show that no solutions are overlooked by the following scaling of variables: $\delta = \rho\cos\Theta$, $\eta = \rho\sin\Theta$, $a = \rho^{1/2}v$, $\beta_1 = {}^{\varepsilon}/_{\rho}$, $\rho>0$. With the above scaling in (1.6), solutions of $G=0$ can be obtained by an application of the implicit function theorem. We will informally describe the results by referring to figure 2 below. It should be understood that all results described below hold in a fixed small neighborhood of the origin in $(\varepsilon,\delta,\eta)$ space. The cone appearing in figure 2 is given by $\det \frac{\partial G}{\partial a}(0,\varepsilon,\delta,\eta) = 0$ and the two planes shown are $m\delta = \pm l\varepsilon$. Figure 2 gives a partitioning of $(\varepsilon,\delta,\eta)$ space into regions in which the number of solutions of $G=0$ is constant. Near $(\varepsilon,\delta,\eta)=0$ there is always the trivial solution $a=0$ which corresponds to $z=0$. Thus "inside" the cone there are two nontrivial solutions, $\pm a_1$, of $G=0$. Below the cone and between the planes a new pair of solutions, $\pm a_2$, appear in addition to $\pm a_1$. It turns out that for $\varepsilon>0$ only $\pm a_1$ correspond to stable subharmonics of period two which are translates of each other by $t\to t+1$. Asymptotic estimates of these solutions are available in [5].

The method of obtaining subharmonics of period two described above could, of course, be used to study the existence of subharmonics of period three and other integral periods. Rather than pursue this, we now describe how to take advantage of the conservative nature of (1.4) and recent work of J.Hale and P.Taboas [4] and S.Chow, J. Mallet-Paret and Hale [1]. In these works the authors consider the perturbation of the

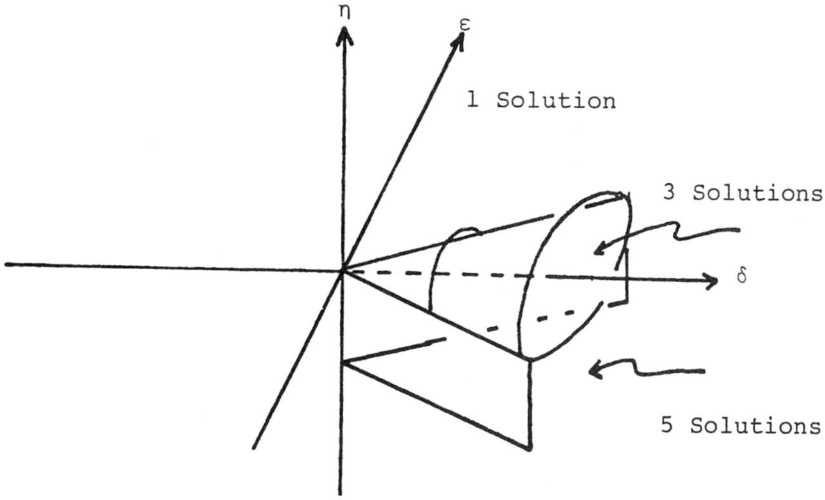

Figure 2

second order equation $x''+g(x)=0$ by small damping ($O(\varepsilon)$) and small periodic forcing ($O(\delta)$) terms. Assuming the periodic forcing term is period one, the focus is on the continuation of an n-periodic solution $p(t)$ (n a positive integer) of $x''+g(x)=0$ to an n-periodic solution of the perturbed equation. Given that certain generic conditions hold, a typical result in [1,4] is summarized in figure 3 below.

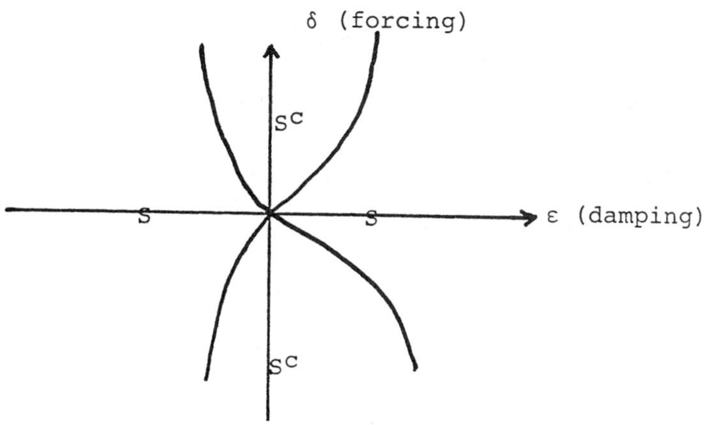

Figure 3

For (ε,δ) in S, no n-periodic solution near p(t) exist; for (ε,δ) in S^c, 2n solutions of period n exist near p(t), n of which are of saddle type, $u_1(t+k)$, k=0,1,...,n-1, and n are nodes, $u_2(t+k)$, k=0,1,...,n-1. For $\varepsilon>0$, the nodes are stable while for $\varepsilon<0$ the nodes are unstable. The result described above is only one of many possibilities described in [1,4], it is however the one which applies to our equation (1.3) (see[6]). In fact we show that there is an infinite subset P of the natural numbers containing 1,2 and 3 such that for each $n\varepsilon P$ and each υ in an open dense set (depending on n) contained in $\{\upsilon:\upsilon>\frac{2\pi}{n}\}$, there is a region S_n^c as in figure 3 near $(\varepsilon,\delta) = (0,0)$ in which (1.3) has, for $\varepsilon>0$, a stable subharmonic solution of least period n (here we are identifying the n translates of this solution). The set P can, in principle, be determined by a Poincaré-Lindstedt perturbation series. Hence (1.3) has many simultaneously stable subharmonic solutions of different integral periods (although the S_n^c may shrink with increasing n).

REFERENCES

[1] S.N. Chow, J.K. Hale and J. Mallet-Paret, An example of bifurcation to homoclinic orbits, J.D.E., 37, 1980, pp 351-373

[2] K. Dietz, The incidence of infectious diseases under the influence of seasonal fluctuations, Lecture Notes in Biomath. 11, pp 1-5, 1976.

[3] Z. Grossman, I. Gumowski and K. Dietz, the incidence of infectious diseases under the influence of seasonal fluctuations - analytic approach, "Nonlinear Systems and Applications to Life Sciences", Academic Press, New York, 1977, pp 525-546.

[4] J. Hale and P. Táboas, Interaction of Damping and Forcing in a second order equation, Nonlinear Analysis T.,M.,A., 2, 1978, pp 77-84.

[5] H.L. Smith, Subharmonic bifurcation in an S-I-R model, submitted.

H.L. Smith, Multiple stable subharmonics for a periodic epidemic model, submitted.

SOME COMPARISON AND CONVERGENCE THEOREMS IN $\lambda(n,k)$-PARAMETER FAMILIES

S. Umamaheswaram

School of Mathematics
University of Hyderabad
Hyderabad, India

ABSTRACT

Let $\lambda(n,k) = (n(1),\ldots,n(k))$ be an ordered k-tuple of positive integers satisfying $n(1)+\ldots+n(k) = n$. If F is a $\lambda(n,k)$-parameter family on an interval I, i.e., a family having the property of unique solvability for a function f in F satisfying $n(1),\ldots,n(k)$ boundary conditions at k points in I then $\lambda(n,k)$-convex (concave) functions have some properties analogous to convex (concave) functions. In this paper, solutions of $\lambda(n,k)$-BVP's in F are compared with $\lambda(n,k)$-convex (concave) functions. Also continuity with respect to some boundary conditions of solutions in F and in particular of solutions of

$$y^{(n)} = f(x,y) \tag{1}$$

satisfying some additional hypotheses is investigated.

REFERENCES

[1] P. Hartman, "On N-parameter families and interpolation problems for nonlinear ordinary differential equations," Trans. Amer. Math. Soc., 154, (1971), 201-226.

[2] P. Hartman, "Unrestricted n-parameter families," Rend. Circ. Mat. Palermo (2), 7, (1958), 123-142.

[3] R. M. Mathsen, "λ(n)-convex functions," Rocky Mtn. Jour. Math., 2, (1972), 31-43.

[4] S. Umamaheswaram, "λ(n,k)-parameter families and associated convex functions," Rocky Mtn. Jour. Math., 8, (1978), 491-501.

[5] S. Umamaheswaram, "Boundary value problems for higher order differential equations," J. Dif. Equa., 18, (1975), 188-201, MR 51, 8525.

OBSTRUCTIONS TO THE EXISTENCE OF TRANSVERSE VOLUME ELEMENTS OF FOLIATIONS

I. Vaisman

Department of Mathematics
University of Haifa
Haifa, Israel

ABSTRACT

The paper contains a full discussion of the cohomological and geometrical obstructions to the existence of a foliate (projectable) transverse volume element of a differentiable foliation.

1. Introduction. In this exposition, we shall discuss a specific problem which belongs to the global analysis of completely integrable systems of Pfaff equations. Locally i.e., on some domain in $\mathbb{R}^n$, such a system is of the form

$$\omega^a = \sum_{i=1}^{n} \alpha_i^a(x)dx^i = 0 \quad (a = 1,\ldots,q), \tag{1.1}$$

where $\operatorname{rank}(\alpha_i^a) = q$ everywhere, and the complete integrability conditions

$$d\omega^a = 0 \quad (\text{modulo } \omega^a = 0) \tag{1.2}$$

are satisfied. In (1.2), d is the exterior derivative. Hence, the global corresponding object defined, generally, on an n-dimensional differentiable manifold M^n (our convention is that differentiable is always C^∞), will be the result of glueing up local systems (1.1) given on a family of coordinate domains which cover M. Since at every point x of M, (1.1) defines a $p = (n-q)$-dimensional plane in the tangent space T_xM, we see that the global object we want is a differentiable field F of tangent p-planes on M. Such a field is called in geometry a distribution, and if (1.2) holds for the local equations of the field the distribution is completely integrable, and is called a foliation of dimension p and codimension q on M.

The integrability condition (1.2) is equivalent to the fact that whenever X,Y are vector fields on M belonging to $\mathcal{F}$, the Lie bracket $[X,Y]$ belongs also to $\mathcal{F}$. By the classical Frobenius' Theorem, this condition ensures that for every $x \in M$ there is a p-dimensional integral submanifold of $\mathcal{F}$ through x, i.e., a p-dimensional submanifold N^p of M^n such that $x \in N$, and the tangent spaces of N belong to $\mathcal{F}$. One proves that this integral submanifold extends to a maximal connected integral submanifold $L(x)$ $(x \in M)$, which is called the leaf through x. (Generally, $L(x)$ is not embedded in M i.e., it may have a manifold topology different from the one induced by M.) Thereby, M is decomposed as the union of the leaves of $\mathcal{F}$, which explains the name of a foliation.

Hence, a global completely integrable Pfaff system is a foliation, and the global solutions of the system are the leaves. The study of the behaviour of the leaves which is a main object in foliation theory is global qualitative theory of completely integrable Pfaff systems. It is but natural to be also interested in the study of the space of leaves, as part of this qualitative theory, and there is the place of the problem which we want to discuss. Roughly speaking our problem is that of the existence of elementary volumes on the space of leaves of a foliation $\mathcal{F}$.

In order to give a more precise formulation, we shall introduce some more notation. First, it is possible to use on M local coordinates $\{U_\alpha; x^a_\alpha, x^u_\alpha\}_{\alpha\in A}$, $\cup U_\alpha = M$ (where $a,b,\ldots = 1,\ldots,q$; $u,v,\ldots = q+1,\ldots,n$ will be our general index convention) such that the leaves would be given by the local canonical equations

$$x^a = \text{const.} \tag{1.3}$$

The corresponding coordinate changes are locally of the form

$$x^a_\beta = x^a_\beta(x^b_\alpha), \quad x^u_\beta = x^u_\beta(x^b_\alpha, x^v_\alpha), \tag{1.4}$$

and (x^a) can be thought of as local coordinates on the space of leaves of $\mathcal{F}$, which, however, is not a differentiable manifold, generally.

Now, the objects related to the space of leaves, like functions, differential forms, tensors, vector bundles, etc. are those objects on M which are constant along the leaves of $\mathcal{F}$, therefore whose local coordinate expressions depend locally on the coordinates x^a only. We shall use for such objects the name foliate.

In particular, a foliate differential form α on M with respect to $\mathcal{F}$ is one which expresses locally as *)

* Here and in the sequel we always use the Einstein summation convention.

(1.5) $$\alpha = \frac{1}{h!}\,\alpha_{a_1 \ldots a_h}(x^b)\,dx^{a_1} \wedge \ldots \wedge dx^{a_h},$$

and, from the classical viewpoint this is an integral invariant in the sense of E. Cartan.

Furthermore, the field F can be seen as a vector subbundle of the tangent bundle TM, and the quotient bundle $Q = TM/F$ is called the transverse bundle of the foliation F. Its transition functions (in natural bases of the above coordinates) are $\left(\frac{\partial x^a_\beta}{\partial x^b_\alpha}\right)$, which are foliate functions. Hence, Q is a foliate vector bundle and it plays the role of the "tangent bundle of the space of leaves". Therefore an elementary volume of the space of leaves has to be defined as a foliate volume on Q, and assuming that Q is an orientable bundle, such a volume has to be defined by a foliate and nowhere vanishing q-form of the type (1.5) with q instead of h. If Q is orientable, the foliation F is said to be transversally orientable, and the volume form mentioned above is called a transverse volume element of F. A transverse volume element is a particular type of an integral invariant.

Now, we can finally formulate in a precise manner our problem, and this is to characterize the foliations which posses a transverse volume element. We shall not insist here on the interest of this problem since I hope that the importance of the notion of a volume and of the integral invariants is enough of a motivation. In exchange, I shall describe a solution of the problem in terms of cohomological and geometrical obstructions. Such an obstruction was given by H. Shulman and D. Tischler [3] in the particular case where F has a trivial transverse bundle Q. In [1], C. C. Liang establishes several properties of the foliations with a transverse volume element, and gives a geometrico-analytical characterization of these foliations. Our present solution (which, in fact, generalizes [3]) uses sheaf cohomology, and it is simple and complete; it is based on an approach started in [6]. Let us also indicate that the problem could be included in the general theory of characteristic classes of foliated bundles (see, for instance, Kamber-Tondeur [2]), but it is much simpler to give it a direct approach.

2. Cohomologies Related to Foliations.

Let (M,F) be a foliated manifold, i.e., a manifold endowed with a foliation, for which we shall use the notation of Section 1. It is natural to define on M the sheaves $\Phi^h(M,F)$ $(h = 0,1,\ldots,q)$ of germs of foliate h-forms, and (with the classical model of the de Rham and Dolbeault Theorems in mind) to compute their cohomology by means of differential forms of M. This was done in our paper [4] as follows.

We chose a Riemannian metric g on M, and identify the transverse bundle Q with the normal bundle of F with respect to g. Then TM has local bases

$$X_a = \frac{\partial}{\partial x^a} - t_a^u \frac{\partial}{\partial x^u} \in Q, \; X_u = \frac{\partial}{\partial x^u} \in F, \tag{2.1}$$

and the dual cobases

$$dx^a, \; \theta^u = dx^u + t_a^u dx^a, \tag{2.2}$$

and we shall express all the elements of M by means of these bases and cobases. In particular a differential form whose local coordinate expressions contain h factors dx^a and k factors θ^u will be called of <u>type (h,k)</u>, and this is an invariant property.

Accordingly, the operator d of the exterior derivative will have a decomposition

$$d = d' + d'' + \partial \tag{2.3}$$

where d' increases the type by $(1,0)$, d'' by $(0,1)$ and ∂ by $(2,-1)$. d'' is the most simple part since it is just exterior derivative along the leaves of F. Moreover, a form ϕ is foliate if ϕ is of type $(h,0)$ and $d''\phi = 0$.

Now, let us denote by A^{hk} the space of forms of type (h,k) and by $\mathcal{A}^{hk}$ the sheaf of germs of such forms. This yields the sequence of sheaves

$$0 \to \Phi^h \xrightarrow{\subset} \mathcal{A}^{h0} \xrightarrow{d''} \mathcal{A}^{h1} \xrightarrow{d''} \cdots \xrightarrow{d''} \mathcal{A}^{hp} \to 0. \tag{2.4}$$

Furthermore, a parameter dependent version of the classical Poincaré lemma can be proven, to the effect that d'' satisfies a Poincaré lemma and, therefore, (2.4) is a fine resolution of Φ^h. This provides the following de Rham-Dolbeault type theorem

$$H^i(M,\Phi^h) = \frac{\ker(d'': A^{hi} \to A^{h,i+1})}{\operatorname{im}(d'': A^{h,i-1} \to A^{hi})}. \tag{2.5}$$

For $h = 0$, we set $\Phi^0 = \Phi$ - the sheaf of germs of foliate functions of M.

Let us also exhibit an interesting exact cohomology sequence of a foliated manifold [5]. Let $\Psi(M,F)$ denote the sheaf of germs of forms of type (1,0), which are d-closed (i.e., belong to ker d). The operator d' induces a sheaf homomorphism $d'\colon \Phi \to \Psi$ by sending each germ of a foliate function f to the germ of d'f, and (since in this case d'f = df) the kernel of this homomorphism is $\mathbb{R}$. Moreover, in view of the Poincaré lemma, d' is onto. Hence, we get the exact sequence of sheaves

$$0 \to \mathbb{R} \overset{\subset}{\to} \Phi \xrightarrow{d'} \Psi \to 0, \tag{2.6}$$

and the corresponding exact cohomology sequence

$$\dots \to H^{i-1}(M,\Psi) \to H^{i}(M,\mathbb{R}) \to H^{i}(M,\Phi) \to H^{i}(M,\Psi) \to \dots, \tag{2.7}$$

which relates the real cohomology (hence the topology) of M, and the foliation F.

The cohomology of Ψ can also be computed by means of differential forms as follows. Consider the spaces of differential forms defined by

$$C_1^r = \bigoplus_{\substack{h+k=r \\ h\geq 1}} A^{hk}(M), \tag{2.8}$$

and let $\mathcal{C}_1^r$ be the corresponding sheaves of germs. Then, one can prove [5] that

$$0 \to \Psi \xrightarrow{\subset} \mathcal{C}_1^1 \xrightarrow{d} \mathcal{C}_1^2 \xrightarrow{d} \dots \tag{2.9}$$

is a fine resolution of Ψ, therefore,

$$H^i(M,\Psi) = \frac{\ker(d\colon C_1^{i+1} \to C_1^{i+2})}{\operatorname{im}(d\colon C_1^{i} \to C_1^{i+1})} . \tag{2.10}$$

3. Obstructions to volume elements. Let F be a transversally orientable foliation on M. By the definition of Section 1, a transverse volume element of F is given locally by

$$\theta_\alpha = f_\alpha(x^a)dx_\alpha^1 \wedge \dots \wedge dx_\alpha^q, \tag{3.1}$$

where $\{U_\alpha; x_\alpha^a, x_\alpha^u\}_{\alpha\in A}$ is the atlas of Section 1, and $f_\alpha > 0$ are some locally defined foliate functions on M. Since (3.1) glue up to a global form one has

$$f_\beta = f_\alpha \det\left(\frac{\partial x_\alpha^a}{\partial x_\beta^b}\right) , \tag{3.2}$$

which is equivalent to

$$\ell n\, f_\beta = \ell n\, f_\alpha + \ell n\, \det\left(\frac{\partial x^a_\alpha}{\partial x^b_\beta}\right). \tag{3.3}$$

Now, let us denote

$$\nu_{\alpha\beta} = \ell n\, \det\left(\frac{\partial x^a_\alpha}{\partial x^b_\beta}\right). \tag{3.4}$$

This is obviously a 1-cocycle with values in the sheaf Φ of germs of foliate functions on M, whose cohomology class $[\nu]_\Phi(F)$ is well defined by F, and which exists without reference to the transverse volume problem.

Proposition 1. The transversally orientable foliation F has a transverse volume element iff $[\nu]_\Phi(F) = 0$.

Indeed, if the volume (3.1) exists, then formula (3.3) yields $[\nu]_\Phi(F) = 0$. Conversely, if $[\nu]_\Phi(F) = 0$, we can determine the foliate local functions $f_\alpha > 0$ which satisfy (3.3), then define the volume by (3.1), q.e.d.

We can also find in a simple way a representative form of type (0,1) for $[\nu]_\Phi$ in agreement with Formula (2.5). The method is known from general sheaf theory (e.g. [5]). By a partition of unity argument, we can always get a globally defined and nowhere zero (q,0)-form, given locally as

$$\phi_\alpha = \ell_\alpha(x^a, x^u)\, dx^1_\alpha \wedge \ldots \wedge dx^q_\alpha. \tag{3.5}$$

(Or, alternatively, just take $\ell_\alpha = \sqrt{\det(g_{ab})}$ for the metric g of Section 2.) Then, like for (3.3), we have

$$\nu_{\alpha\beta} = \ell n\, \ell_\beta - \ell n\, \ell_\alpha, \tag{3.6}$$

whence $\xi|_{U_\alpha} = d''\ell n\, \ell_\alpha$, yields a d"-closed (0,1)-form globally defined on M. This form ξ represents $[\nu]_\Phi(F)$ via (2.5), and F has a transverse volume element iff ξ is also d"-exact. From the definition of ξ, it follows that our cohomological obstruction reduces to the one given by Schulman-Tischler [3] if the transverse bundle Q is trivial.

Now, we shall give an ampler discussion of the obtained obstruction in the cohomological context of Section 2.

Let us consider the following portion of the exact cohomology sequence (2.7):

$$(3.7)\qquad \dots \to H^1(M,\mathbb{R}) \xrightarrow{\alpha} H^1(M,\Phi) \xrightarrow{\beta} H^1(M,\Psi) \xrightarrow{\gamma} H^2(M,\mathbb{R}) \to \dots,$$

where α, β, γ denote the homomorphisms entering in (2.7). Then, the cohomology class

$$(3.8)\qquad [\mu]_\Psi(F) = \beta([\nu]_\Phi(F))$$

is also an obstruction to the existence of transversal volumes in the sense that if $[\mu]_\Psi \neq 0$, F has no transverse volume element. This is interesting because of the fact that we shall be able to give a clear geometrical meaning of $[\mu]_\Psi(F)$.

First, let us give the representation of $[\mu]_\Psi(F)$ by means of a differential form according to (2.10). Namely, $[\nu]_\Phi$ is represented by $\{\nu_{\alpha\beta}\}$ of (3.4), hence $[\mu]_\Psi$ is represented by $\{d'\nu_{\alpha\beta}\}$, where d' is given by (2.3). From (3.6), we have

$$(3.9)\qquad d'\nu_{\alpha\beta} = d'\ell n\, \ell_\beta - d'\ell n\, \ell_\alpha,$$

whence it follows that $[\mu]_\Psi$ is represented by μ given by

$$(3.10)\qquad \mu|_{U_\alpha} = dd'\ell n\, \ell_\alpha = (d'+d''+\partial)d'\ell n\, \ell_\alpha = -d\xi|_{U_\alpha},$$

where the final result is obtained by using $d''\xi = 0$, and the following consequences of $d^2 = 0$:

$$(3.11)\qquad \begin{gathered} d''^2 = 0,\ \partial^2 = 0,\ d'^2 + d''\partial + \partial d'' = 0, \\ d'd''+d''d' = 0,\quad d'\partial+\partial d' = 0. \end{gathered}$$

Hence, we have $\mu = -d\xi$, which is a form belonging to $\ker(d\colon C_1^2 \to C_1^3)$. Of course, μ is exact but in general it is not the differential of a form in C_1^1 hence, generally, $[\mu]_\Psi \neq 0$.

Now, we shall prove

Proposition 2. *Let F be a transversally orientable foliation. Then $[\mu]_\Psi(F) = 0$ iff the line bundle $K(F) = \wedge^q Q$ has a flat basic connection.*

First, a flat connection is one which has a vanishing curvature. Second, the bundle $K(F)$ (the qth exterior power of Q) is clearly foliate (i.e., it has foliate transition functions), and a basic connection on a foliate bundle is defined as one whose connection forms are of the type (1,0).

Now, if the volume (3.5) is taken arbitrarily for F, we have

$$\ell n\, \ell_\beta = \ell n\, \ell_\alpha + \ell n \det\left(\frac{\partial x^a_\alpha}{\partial x^b_\beta}\right),$$

whence

$$d'\ln \ell_\beta = d'\ln \ell_\alpha + d \ln \det\left(\frac{\partial x^a_\alpha}{\partial x^b_\beta}\right), \tag{3.12}$$

and this proves that $\{d'\ln \ell_\alpha\}$ are the connection forms of a basic connection on $K(F)$. By (3.10), the form μ is precisely the curvature of this connection.

Furthermore, $[\mu]_\Psi = 0$ means $\mu = d\gamma$ for some (1,0)-form γ. If this happens, $\{d'\ln \ell_\alpha - \gamma|_{U_\alpha}\}$ yields a new connection on $K(F)$ which, obviously, has a vanishing curvature.

Conversely, let us assume that $\{\omega_\alpha\}$ defines a basic flat connection on $K(F)$. Then ω_α has the type (1,0), and

$$\omega_\beta = \omega_\alpha + d\ln \det\left(\frac{\partial x^a_\alpha}{\partial x^b_\beta}\right), \quad d\omega_\alpha = 0. \tag{3.13}$$

Hence $\omega_\alpha = d\ln f_\alpha$, where $f_\alpha > 0$ are local foliate functions, and

$$d\{\ln f_\beta - \ln f_\alpha - \ln \det\left(\frac{\partial x^a_\alpha}{\partial x^b_\beta}\right)\} = 0. \tag{3.14}$$

This means that the bracket above is a real 1-cocycle, and it can be put under the form $\ln \psi_\alpha - \ln \psi_\beta$ (= const.), where $\psi_\alpha > 0$ are local differentiable (but not necessarily foliate or constant) functions. Thereby, we obtain that $\{(f_\alpha \psi_\alpha) dx^1_\alpha \wedge \ldots \wedge dx^q_\alpha\}$ yields a global volume on Q i.e., we can take $\ell_\alpha = f_\alpha \psi_\alpha$.

Now, (3.10) and (3.11) will give $\mu = dd'\ln \psi_\alpha$, while $\ln \psi_\alpha - \ln \psi_\beta =$ const. proves that $\{d'\ln \psi_\alpha\}$ is a global (1,0)-form. The conclusion is therefore $[\mu]_\Psi(F) = 0$. Q.e.d.

Corollary. *If F is a transversally orientable foliation whose bundle $K(F)$ has no flat basic connection, F has no transverse volume element.*

(Note that this can also be seen straightforwardly.)

Now, let F be a transversally orientable foliation with $[\mu]_\Psi(F) = 0$. Then, by the exact sequence (3.7), there is a cohomology class $[\nu']_{\mathbb{R}}(F) \in H^1(M,\mathbb{R})$ such that $[\nu]_\Phi = \alpha([\nu']_{\mathbb{R}})$. In order to obtain such a class, we consider a flat basic connection $\{\omega_\alpha\}$ on $K(F)$. Then we have (3.13), (3.14), whence

$$\nu'_{\alpha\beta} \overset{\text{def}}{\equiv} -\ln f_\beta + \ln f_\alpha + \ln \det\left(\frac{\partial x^a_\alpha}{\partial x^b_\beta}\right) = \tag{3.15}$$

$$= \nu_{\alpha\beta} + \ln f_\alpha - \ln f_\beta = \text{const.}$$

This formula shows that $\{\nu'_{\alpha\beta}\}$ is a real 1-cocycle, which defines the class $[\nu']_{\mathbb{R}}(F)$ needed.

Of course, there is no reason for $[\nu']_{\mathbb{R}}$ to be unique, and different connections $\{\omega_\alpha\}$ can yield different classes $[\nu']_{\mathbb{R}}$ (but the same class $[\nu]_\Phi$) . The situation is easily seen to be as follows. If $\{\omega_\alpha\}$, $\{\omega'_\alpha\}$ are two basic flat connections of $K(F)$ then $\omega'_\alpha - \omega_\alpha = \tau_\alpha$ where $\tau = \{\tau_\alpha\}$ yields a global closed (1,0)-form. We shall say that these connections are <u>cohomologous</u> if τ is an exact form. The two connections define the same class $[\nu']_{\mathbb{R}}$ iff they are cohomologous. If we call $[\nu']_{\mathbb{R}}$ the <u>characteristic class</u> of the connection $\{\omega_\alpha\}$, this means that there is a bijection between cohomology classes of basic flat connections and characteristic classes.*)

Now, of course, $[\nu']_{\mathbb{R}} = 0$ implies $[\nu]_\Phi = \alpha([\nu']_{\mathbb{R}}) = 0$, where the existence of a transverse volume element follows. Conversely, if transverse volume element exists, it follows from (3.3) that there is a basic flat connection with a vanishing characteristic class.

This finally yields our main result.

<u>Theorem</u>. <u>Let F be a transversally orientable foliation with transverse bundle Q. Then F has a transverse volume element iff $K(F)$ has a basic connection with vanishing curvature and vanishing characteristic class.</u>

<u>Remark</u>. If $H^1(M,\mathbb{R}) = 0$, the characteristic class always vanishes, and the only remaining condition for the existence of the transverse volume element is the existence of a basic flat connection i.e. $[\mu]_\Psi(F) = 0$.

---

*) Of course, $K(F)$ is differentiably isomorphic to a locally constant bundle, and, on the latter, the local connection forms unify themselves up to a global closed 1-form. The introduced characteristic class is precisely the 1-dimensional cohomology class represented by this form.

REFERENCES

[1] C. C. Liang, "On the volume preserving foliations," Math. Ann., 223, (1976), 13-17.

[2] F. W. Kamber and Ph. Tondeur, "Foliated bundles and characteristic classes," Lecture Notes in Math. 493, Springer-Verlag, New York, 1975.

[3] H. Shulman and D. Tischler, "Leaf invariants for foliations and the Van Est isomorphism," J. Diff. Geometry, 11, (1976), 535-546.

[4] I. Vaisman, "Variétés riemanniennes feuilletées," Czechosl. Math. J., 21, (1971), 46-75.

[5] I. Vaisman, Cohomology and differential forms, M. Dekker, Inc., New York, 1973.

[6] I. Vaisman, "Differential Geometry of Projective Foliations," Univ. of Haifa, Math. Publication Series, Report #16, 1981.

THE MONOTONE METHOD FOR A SECOND-ORDER DIFFERENTIAL EQUATION

C. Vargas

Department of Mathematics
University of Texas at Arlington
Arlington, Texas

1. INTRODUCTION

The monotone method has been successfully applied to find solutions of nonlinear boundary problems, see [1-4] and references therein.

In this talk we are interested in obtaining periodic solutions for the equation $-u'' = f(t,u(t))$, under the assumption of existence of lower and upper solutions. We show that the method gives existence and a constructive way to find a solution. These results can be generalized to partially ordered spaces and will appear elsewhere.

2. PROBLEM AND ASSUMPTIONS

We want to find $u \in C^2_{[0,2\pi]}$, such that

2.1)
$$\begin{aligned} -u'' &= f(t,u), \quad f \in C^0_{[0,2\pi]\times R} \\ u(0) &= u(2\pi), \\ u'(0) &= u'(2\pi). \end{aligned}$$

We assume the existence of $\alpha(t)$, $\beta(t)$ lower and upper solutions, i.e.,

2.2)
$$\begin{aligned} -\alpha''(t) &\leq f(t,\alpha(t)) \\ -\beta''(t) &\geq f(t,\beta(t)), \quad \alpha_1\beta \in C^2_{[0,2\pi]} \end{aligned}$$

and

2.3)
$$\alpha(t) \leq \beta(t), \quad t \in [0,2\pi]$$

It is also assumed that:

$$\begin{aligned} &\alpha(0) = \alpha(2\pi),\ \alpha'(0) \geq \alpha'(2\pi) \\ 2.4)\qquad &\beta(0) = \beta(2\pi),\ \beta'(0) \leq \beta'(2\pi). \end{aligned}$$

3. AUXILIARY LEMMA

Before we proceed with our problem, let us establish a lemma upon which we shall base all of our results.

Lemma. If $\phi \in C^2_{[0,2\pi]}$ and

$$\begin{aligned} &\phi''(t) \geq M\phi(t),\ M > 0,\ t \in [0,2\pi] \\ 3.1)\qquad &\phi(0) = \phi(2\pi), \\ &\phi'(0) \geq \phi'(2\pi), \end{aligned}$$

then, $\phi(t) \leq 0$ for $t \in [0,2\pi]$.

Proof: We deny the claim, therefore there exists a $t_0 \in [0,2\pi]$ for which $\phi(t_0) > 0$, and we can say that t_0 is a maximum of ϕ. It can be proved immediately that t_0 can not be an interior point of $[0,2\pi]$, therefore $t_0 = 0$ or $t_0 = 2\pi$. Again expansion in Taylor series and noticing that $\phi'(0) \leq 0$ and $\phi'(2\pi) \geq 0$; together with 3.1) gives us a contradiction.

4. THE MONOTONE METHOD

We begin defining an operator

$$A:\ C^2_{[0,2\pi]} \longrightarrow C^2_{[0,2\pi]} \quad \text{in the following manner:}$$

$$A\eta = u$$

where u is the solution of

$$\begin{aligned} &- u''(t) = f(t,\eta(t)) - M(u(t)-\eta(t)) \\ 4.1)\qquad &u(0) = u(2\pi),\ u'(0) = u'(2\pi). \end{aligned}$$

This differential equation can be rewritten as

$$- u'' + Mu = f(t,\eta) + M\eta$$

$$u(0) = u(2\pi),\ u'(0) = u'(2\pi),$$

and it is easily shown that this system has a unique solution.

The next step is to define a sequence of functions $\alpha_n(t)$ by means of the operator A, that is,

$$4.2)\qquad \alpha_n(t) = A\alpha_{n-1}(t)$$

and we prove that beginning with $\alpha_0 \equiv \alpha$ we obtain an increasing sequence.

This can be done by induction and the first step proceeds as follows:

$$\alpha_1 = A\alpha_0 \quad \text{satisfies}$$

4.3)
$$-\alpha_1'' + M\alpha_1 = f(t,\alpha(t)) + M\alpha(t)$$
$$\alpha(0) = \alpha_1(2\pi), \; \alpha_1'(0) = \alpha_1'(2\pi).$$

Taking $\phi = \alpha_0 - \alpha_1$, the use of 2.2) and 4.3) enable us to see that ϕ satisfies the conditions of our lemma, therefore $\alpha_0(t) \leq \alpha_1(t)$. A repetition of the argument, proves the induction step and we can conclude that $\{\alpha_n\}$ is an increasing sequence.

In the same way is shown that

$$\alpha_n(t) \leq \beta(t), \; t \in [0,2\pi]$$

therefore

4.4)
$$\min_{[0,2\pi]} \alpha(t) \leq \alpha_n(t) \leq \max_{[0,2\pi]} \beta(t).$$

Thus the sequence α_n is uniformly bounded and increasing and it must have a pointwise limit, say $a(t)$. And we shall prove that $a(t)$ is a solution of the original problem.

5. a(t) is a solution

First we realize that the sequence $\{\alpha_n''\}$ is uniformly bounded, since:

$$\alpha_n''(t) = M\alpha_n(t) - f(t,\alpha_{n-1}(t)) - M\alpha_{n-1}(t)$$

implies

$$\|\alpha_n''(t)\| \leq 2M\|\alpha_n\| + \|f\| \leq 2M\|\beta\| + \|f\|,$$

as a consequence $\{\alpha_n'\}$ is equicontinuous. Using the fact that α_n' has a zero in $(0,2\pi)$ and the uniform boundedness of $\{\alpha_n''(t)\}$ it is possible to establish that α_n' is pointwise bounded.

By use of the Arzela-Ascoli theorem we know that $\{\alpha_n'\}$ is uniformly bounded. Repeating the argument, $\{\alpha_n\}$ is equicontinuous and 4.4) gives more than pointwise boundedness, therefore, there exists a subsequence $\{\alpha_{n_k}\}$ which is uniformly convergent to $a(t)$. By the monotonicity of $\{\alpha_n\}$, the whole sequence $\{\alpha_n\}$ converges to $a(t)$. To prove that $a(t)$ is a solution of 2.1), it is enough to prove that $a(t)$ is a solution of the equivalent integral equation.

5.1)
$$u(t) = u(0) + \frac{t}{2\pi}\int_0^{2\pi} dr \int_0^r f(s,u(s))ds - \int_0^t dr \int_0^r f(s,u(s))ds$$

with
$$\int_0^{2\pi} f(s,u(s))ds = 0.$$

Let us recall that α_n satisfies:

5.2)
$$- \alpha_n''(t) + M\alpha_n = f(t,\alpha_{n-1}(t)) + M\alpha_{n-1}(t)$$
$$\alpha_n(0) = \alpha_n(2\pi); \; \alpha_n'(0) = \alpha_n'(2\pi)$$

Integrating twice and using the boundary conditions we obtain

$$\alpha_n(t) = \alpha_n(0) + \frac{t}{2\pi}\int_0^{2\pi} dr \int_0^r f(s,\alpha_{n-1}(s))ds -$$

$$\frac{Mt}{2\pi}\int_0^{2\pi} dr \int_0^r (\alpha_n(s)-\alpha_{n-1}(s))ds +$$

$$M\int_0^t dr \int_0^r (\alpha_n(s)-\alpha_{n-1}(s))ds - \int_0^t dr \int_0^r f(s,\alpha_{n-1}(s))ds$$

and

$$\int_0^{2\pi} f(s,\alpha_{n-1}(s))ds = M\int_0^{2\pi} (\alpha_n(s)-\alpha_{n-1}(s))ds,$$

letting $n \longrightarrow \infty$ and using the uniform convergence of α_n, we obtain 5.1).

The same type of arguments allow us to prove that if u is a solution and $\alpha \leq u \leq \beta$ then $a \leq u$; which means that $a(t)$ is the minimal solution.

All that have been said for the increasing sequence $\{\alpha_n\}$, can be said for a decreasing sequence $\{\beta_n\}$, if the monotone method is started with $\beta_0 \equiv \beta$. In this case we obtain a maximal solution.

REFERENCES

[1] H. B. Keller, "Elliptic Boundary Value Problems Suggested by Nonlinear Diffusion Processes," Arch. Rat. Mech. Anal. 35 (1969), pp. 363-381.

[2] D. H. Sattinger, "Monotone Methods in Nonlinear Elliptic and Parabolic Boundary Value Problems," Indiana Univ. Math. J. 21 (1972), pp. 979-1000.

[3] J. C. Chandra and P. W. Davis, "A Monotone Method for Quasilinear Boundary Value Problems," Arch. Rat. Mech. Anal. 54 (1974), pp. 257-266.

[4] S. R. Bernfeld and V. Lakshmikantham, "Linear Monotone Method for Nonlinear Boundary Value Problems in Banach Spaces," To appear in Rocky Mtn. J.

EXISTENCE OF COUPLED MINIMAL AND MAXIMAL PERIODIC QUASISOLUTIONS FOR SYSTEMS OF FIRST ORDER PBVP VIA QUASISOLUTIONS

A. S. Vatsala

Department of Mathematics
Bishop College
Dallas, Texas

ABSTRACT

The method of upper and lower solutions and Lyapunov-Schmidt method have been used to prove the existence of periodic solutions for first and second order equations in [1,2]. Also monotone method has been developed for first order scalar equations in [2]. In this paper we develop monotone iteraitve technique which yields coupled minimal and maximal periodic quasisolutions for systems of first order equations. Further under a uniqueness assumption, we obtain a unique periodic solution for first order systems. See [3] for details.

MONOTONE ITERATIVE TECHNIQUE

Consider the periodic boundary value problem (PBVP for short)

$$u' = f(t,u),\ u(0) = u(2\pi) \tag{2.1}$$

where $f \in C[[0,2\pi]\times R^n, R^n]$. We shall develop monotone iterative technique which offers a constructive method to obtain coupled maximal and minimal periodic quasisolutions of (2.1). In the special case when f satisfies uniqueness conditions, we can show that coupled maximal and minimal quasisolutions coincide and gives the unique solution of the system (2.1).

To define quasisolutions of (2.1) we fix for each $i \in \{1,2,,,n\}$ two nonnegative integers p_i, q_i such that $p_i + q_i = n - 1$ and split $u \in R^n$ into $u = (u_i, [u]p_i, [u]q_i)$. Then the system (2.1) becomes

$$u_i' = f_i(t,u_i,[u]p_i,\ [u]q_i),\ u_i(0) = u_i(2\pi). \tag{2.2}$$

We assume that $i = 1,2,\ldots n$ and all inequalities between vectors are componentawise.

The functions $u,v \in C'[[0,2\pi],R^n]$ are said to be coupled periodic quasisolutions of (2.2) or (2.1), if

$$u_i' = f_i(t,u_i,[u]p_i,[v]q_i), \quad u_i(0) = u_i(2\pi)$$

$$v_i' = f_i(t,v_i,[v]p_i,[u]q_i), \quad v_i(0) = v_i(2\pi).$$

The function f is said to possess a mixed quasimonotone property (mqmp for short), if for each i, $f_i(t,u_i,[u]p_i,[u]q_i)$ is monotone nondecreasing in $[u]p_i$ and monotone nonincreasing in $[u]q_i$.

In order to develop monotone method for system (2.1) we need the following assumptions which we list below.

(A_1) $\alpha, \beta \in C[[0,2\pi],R^n]$, $\alpha(t) \leq \beta(t)$ on $[0,2\pi]$ and

$$D_-\alpha_i(t) \leq f_i(t,\alpha_i,[\alpha]p_i,[\beta]q_i) \qquad t \in (0,2\pi)$$

$$D_-\beta_i(t) \geq f_i(t,\beta_i,[\beta]p_i,[\alpha]q_i).$$

(A_2)(i) $\alpha_i(0) \leq \alpha_i(2\pi)$, $\alpha \in C'[(0,2\pi],R^n]$

(ii) $\beta_i(0) \geq \beta_i(2\pi)$, $\beta \in C'[(0,2\pi],R^n]$.

(A_3)(i)(a) $D^+\alpha_i(0) > f_i(0,\alpha_i(0),[\alpha(0)]p_i,[\beta(0)]q_i) + M_i(\alpha_i(0)-\underline{\alpha}_{i,0})$

and $\underline{\alpha}_{i,0} = \min_{t\in[0,2\pi]} \alpha_i(t)$

(i)(b) $\displaystyle\int_0^{2\pi} e^{M_i s} f_i(s,\alpha_i(s),[\alpha(s)]p_i,[\beta(s)]q_i)ds \geq 0$

(ii)(a) $D^+\beta_i(0) < f_i(0,\beta_i(0),[\beta(0)]p_i,[\alpha(0)]q_i) - M_i(\overline{\beta}_{i,0}-\beta_i(0))$

and $\overline{\beta}_{i,0} = \max_{t\in[0,2\pi]} \beta_i(t)$

(ii)(b) $\displaystyle\int_0^{2\pi} e^{M_i s} f_i(s,\beta_i(s),[\beta(s)]p_i,[\alpha(s)]q_i)ds \leq 0$

(A_4) $f_i(t,u_i,[u]p_i,[u]q_i) - f_i(t,v_i,[u]p_i,[u]q_i) \geq - M_i(u_i-v_i)$

whenever $\alpha(t) \leq u,v \leq \beta(t)$, $t \in [0,2\pi]$.

(A_5) (A_2)(i), (A_3)(ii)(a), (b),

(A_6) (A_2)(ii), (A_3)(i), (a), (b).

To proceed further we require the following lemma on which our main discussion of this section depends on.

<u>Lemma 2.1</u>. Let $m \in C'[[0,2\pi],R]$ and $m'(t) \leq - Lm(t)$ for $0 < t \leq 2\pi$ where $L > 0$. Then (i) $m(0) \leq m(2\pi)$ implies $m(t) \leq 0$ on $[0,2\pi]$ (ii)

$m(0) > m(2\pi)$ and $m'(0) > 0$ implies $m(t) \leq 0$ on $[0,2\pi]$.

Proof: If the conclusion is false, there exists a $t_0 \in [0,2\pi]$ and an $\varepsilon > 0$ such that $m(t_0) = \varepsilon$ and $m(t) \leq \varepsilon$ on $[0,2\pi]$. Let $t_0 \in (0,2\pi]$, then $m'(t_0) \geq 0$. It then follows that

$$0 \leq m'(t_0) \leq - Lm(t_0) = - L\varepsilon < 0$$

which is a contradiction. If $t_0 = 0$, and (i) holds, then $m(2\pi) \geq m(0) = \varepsilon$ and hence $0 \leq m'(2\pi) \leq - Lm(2\pi) \leq - L\varepsilon < 0$ which is a contradiction. If (ii) holds then $m(0) = \varepsilon$, and $m'(0) \leq 0$. Therefore we get

$$0 \geq m'(0) > 0$$

which is again a contradiction. This contradiction proves the lemma.

The next theorem gives us the main result.

Theorem 2.1. Let A_1,A_4 hold. Also let f possess mqmp property. Further if any one of the condition A_2, A_3, A_5 or A_6 holds, then there exist monotone sequences $\{\alpha_n(t)\}$, $\{\beta_n(t)\}$ with $\alpha_0 = \alpha$, $\beta_0 = \beta$ such that $\lim_{n\to\infty} \alpha_n(t) \longrightarrow \rho(t)$, $\lim_{n\to\infty} \beta_n(t) \longrightarrow r(t)$ uniformly and monotonically on $[0,2\pi]$ and that (ρ,r) are coupled minimal and maximal periodic quasi-solutions of the PBVP (2.1). That is if (u_1,u_2) are any coupled periodic solutions of (2.1) such that $\alpha \leq u_1, u_2 \leq \beta$, then

$$\alpha = \alpha_0 \leq \alpha_1 \leq \ldots \leq \alpha_n \leq \rho \leq u_1,u_2 \leq r \leq \beta_n \leq \ldots \leq \beta_1 \leq \beta = \beta \text{ on } [0,2\pi]. \tag{2.3}$$

Furthermore, if u is any solution of (2.1) such that $\alpha(t) \leq u(t) \leq \beta(t)$, also satisfies the inequality (2.3). That is

$$\alpha = \alpha_0 \leq \alpha_1 \leq \ldots \leq \alpha_n \leq \rho \leq u \leq r \leq \beta_n \leq \ldots \leq \beta_1 \leq \beta_0 = \beta \quad \text{on} \quad [0,2\pi].$$

Proof: For any $\eta,\mu \in [\alpha,\beta] = [u\in C[0,2\pi],R],\alpha(t)\leq u\leq\beta(t)]$ on $[0,2\pi]$, consider the PBVP

$$u_i' = G_i(t,u), \quad u_i(0) = u_i(2\pi) \tag{2.4}$$

where $G_i(t,u) = f_i(t,\eta_i,[\eta]p_i,[\mu]q_i) - M_i(u_i-\eta_i)$.

By a direct computation it is easy to calculate the unique solution of of (2.4) as

$$u_i(t) = [u_i(0) + \int_0^t [f_i(s,\eta_i(s),[\eta(s)]p_i,[\mu(s)]q_i) + M_i\eta_i(s)]e^{M_i s}ds]e^{-M_i t} \tag{2.5}$$

where

$$(2.6)\quad u_i(0) = \frac{1}{e^{2M_i\pi}-1}\int_0^{2\pi}[f_i(s,\eta_i(s),[\eta(s)]p_i,[\mu(s)]q_i)+M_i\eta_i(s)]e^{M_is}ds.$$

We now define a mapping $A[\eta,\mu] = u$, where u is the unique solution of (2.4) and show that

(a) $\alpha \leq A[\alpha,\beta]$, $\beta \geq A[\beta,\alpha]$

(b) A possess mixed monotone property on the segment

$$[\alpha,\beta] = [u\in R^n;\ \alpha(t)\leq u\leq\beta(t),\ t\in[0,2\pi]].$$

Before proving (a) we claim that $u(0) = u(2\pi)$ given by (2.6) for each $\eta,\mu \in [\alpha,\beta]$ satisfies $\underline{\alpha}_0 \leq u(0) = u(2\pi) \leq \overline{\beta}_0$ where $\underline{\alpha}(0) = \min_{t\in[0,2\pi]} \alpha(t)$, and $\overline{\beta}_0 = \max_{t\in[0,2\pi]} \beta(t)$. We just prove $\underline{\alpha}_{0,i} \leq u_i(0) = u_i(2\pi)$ for $i = 1,2,\ldots,n$ and a similar proof holds good for $u(0) = u(2\pi) \leq \overline{\beta}_0$.

$$u_i(0) = \frac{1}{e^{2M_i\pi}-1}\int_0^{2\pi} e^{M_is}[f_i(s,\eta_i(s),[\eta(s)]p_i,[\mu(s)]q_i)+M_i\eta_i(s)]ds$$

$$\geq \frac{1}{e^{2M_i\pi}-1}\int_0^{2\pi} e^{M_is}[f_i(s,\alpha_i,[\alpha]p_i,[\beta]q_i)-M_i(\eta_i-\alpha_i)+M_i\eta_i]ds$$

$$\geq \frac{e^{2M_i\pi}}{e^{2M_i\pi}-1}\int_0^{2\pi} f_i(s,\alpha_i,[\alpha]p_i,[\beta]q_i)ds + \frac{M_i\underline{\alpha}_{0,i}}{e^{2M_i\pi}-1}\int_0^{2\pi} e^{M_is}ds \geq \underline{\alpha}_{0,i}.$$

That is $\underline{\alpha}_{0,i} \leq u_i(0) = u_i(2\pi)$.

To prove (a) set $A[\alpha,\beta] = \alpha_1$ where α_1 is the unique solution of the PBVP (2.4) with $\eta = \alpha, \mu = \beta$. Set $p_i(t) = \alpha_i - \alpha_{1,i}$, then

$$p_i' = \alpha_i' - \alpha_{1,i}' \leq f_i(t,\alpha_i,[\alpha]p_i,[\beta]q_i) - f_i(t,\alpha_i,[\alpha]p_i,[\beta]q_i) + M_i(\alpha_{1,i}-\alpha_i)$$
$$= -M_ip_i(t).$$

That is $$p_i' \leq -M_ip_i \quad 0 < t \leq 2\pi.$$

If (A_2) holds, then $p_i(0) \leq p_i(2\pi)$. If (A_3) holds, then we get

$$D^+p_i(0) = D^+\alpha_i(0) - D^+\alpha_{1,i}(0) \geq f_i(0,\alpha_i(0),[\alpha(0)]p_i,[\beta(0)]q_i) + M_i(\alpha_i(0)-\underline{\alpha}_{0,i})$$
$$- f_i(0,\alpha_i(0),[\alpha(0)]p_i,[\beta(0)]q_i) + M_i(\alpha_{1,i}(0)-\alpha_i(0))$$
$$= M_i(\alpha_{1,i}(0)-\underline{\alpha}_{0,i}) \geq 0$$

using the claim on (2.6). Hence $p_i(t)$ satisfies the hypothesis of the lemma which proves $\alpha(t) \leq \alpha_1(t) = A[\alpha,\beta]$. Similarly one can prove $\beta \geq A[\beta,\alpha]$.

To prove (b), let $\eta,\mu \in [\alpha,\beta]$ such that $\eta \leq u$. Suppose that $u_1 = A[\eta,\mu]$, $u_2 = A[\mu,\eta]$. Setting $p_i(t) = u_{1,i}(t) - u_{2,i}(t)$, we see that

$$\begin{aligned} p_i'(t) = u_{1,i}'(t) - u_{2,i}'(t) &= f_i(t,\eta_i,[\eta]p_i,[\mu]q_i) - M_i(u_{1,i}-\eta_i) \\ &\quad - f_i(t,\mu_i,[\mu]p_i,[\eta]q_i) + M_i(u_{2,i}-\mu_i) \\ &\leq M_i(\mu_i-\eta_i) - M_i(u_{1,i}-u_{2,i}) + M_i(\eta_i-\mu_i) \\ &= - M_i p_i(t), \end{aligned}$$

using mqmp of f and (A_4). Also $p_i(0) = u_{1,i}(0) - u_{2,i}(0) = u_{1,i}(2\pi) - u_{2,i}(2\pi) = p_i(2\pi)$. Now using Lemma 2.1 we get $u_{1,i}(t) \leq u_{2,i}(t)$.

It is now easy to define the sequences $\{\alpha_n(t)\}$, $\{\beta_n(t)\}$ with $\alpha = \alpha_0$, $\beta = \beta_0$ such that $\alpha_n = A[\alpha_{n-1},\beta_{n-1}]$ and $\beta_n = A[\beta_{n-1},\alpha_{n-1}]$ and we can conclude $\alpha_0 \leq \alpha_1 \leq \ldots \leq \alpha_n \leq \beta_n \leq \ldots \leq \beta_1 \leq \beta_0$ on $[0,2\pi]$. It then follows using standard arguments that $\lim_{n\to\infty} \alpha_n(t) = \rho(t)$, $\lim_{n\to\infty} \beta_n(t) = r(t)$ uniformly and monotonically on $[0,2\pi]$. This implies that

$$(2.7) \qquad \begin{aligned} \rho_i'(t) &= f_i(t,\rho_i,[\rho]p_i,[r]q_i), \ \rho_i(0) = \rho_i(2\pi) \\ r_i'(t) &= f_i(t,r_i,[r]p_i,[\rho]q_i), \ r_i(0) = r_i(2\pi). \end{aligned}$$

We shall show that (ρ,r) are coupled minimal and maximal quasisolutions respectively. Let (u,v) be any coupled solution of (2.1) such that $u,v \in [\alpha,\beta]$. Let us assume that for some integer $k > 0$, $\alpha_{k-1} \leq u, v \leq \beta_{k-1}$ on $[0,2\pi]$. Then setting $p_i = \alpha_{k,i} - u_i$ and employing mqmp property of f and (A_4) we arrive at

$$\begin{aligned} p_i' &= f_i(t,\alpha_{k-1,i},[\alpha_{k-1}]p_i,[\beta_{k-1}]q_i) - M_i(\alpha_{k,i}-\alpha_{k-1,i}) \\ &\quad - f_i(t,u_i,[u]p_i,[v]q_i) \\ &\leq M_i(u_i-\alpha_{k-1,i}) - M_i(\alpha_{k,i}-\alpha_{k-1,i}) = - M_i(\alpha_{k,i}-u_i) = - M_i p_i. \end{aligned}$$

Since $p_i(0) = p_i(2\pi)$, this implies $\alpha_{k,i} \leq u_i$ using Lemma 2.1. Using similar arguments $\alpha_k \leq u,v \leq \beta_k$ on $[0,2\pi]$. Since $\alpha = \alpha_0 \leq u,v \leq \beta_0$ on $[0,2\pi]$, it follows by induction $\alpha_n \leq u,v \leq \beta_n$ on $[0,2\pi]$ for all n. Hence we have $\rho \leq u,v \leq r$ on $[0,2\pi]$ proving (ρ,r) are coupled minimal and maximal periodic quasisolutions of (2.1). Since any solution u of (2.1) such that $u \in [\alpha,\beta]$ can be considered as (u,u) coupled periodic quasisolution of 2.1, we also have $\rho \leq u \leq r$ on $[0,2\pi]$. This completes the proof.

Remark 2.1. If $q_i = 0$, then ρ,r are minimal and maximal periodic quasisolutions respectively of (2.1).

Theorem 2.2. Let the hypothesis of Theorem 2.1 hold. Further let f satisfy the following condition

$$\sum_{i=1}^{n} (u_i-v_i)(f_i(t,u_i,[u]p_i,[v]q_i)-f_i(t,v_i,[v]p_i,[u]q_i) \leq - L\|u-v\|^2 \tag{2.8}$$

whenever $\alpha(t) \leq u(t), v(t) \leq \beta(t)$. Then there exists a unique periodic solutions u of (2.1).

Proof. It is enough to prove $\rho = r$. Consider $p = \sum_{i=1}^{n} |\rho_i - r_i|^2$. Then it follows using (2.8), that

$$p' = 2\sum_{i=1}^{n} (\rho_i-r_i)(\rho_i'-r_i') \leq - 2L\|\rho-r\|^2 = - 2Lp.$$

Also $p(0) = \|\rho(0)-r(0)\|^2 = \|\rho(2\pi)-r(2\pi)\|^2 = p(2\pi)$. Now using Lemma 2.1, it follows that $p \leq 0$ which implies $p(t) \equiv 0$. Hence we have $\rho \equiv r = u$ in (2.7) which concludes the theorem.

REFERENCES

[1] R. Kannan and V. Lakshmikantham, "Existence of periodic solutions of nonlinear boundary value problems and the method of upper and lower solutions," To appear, Technical report #173, The University of Texas at Arlington.

[2] V. Lakshmikantham and S. Leela, "Existence and monotone method for periodic solutions of first order differential equations," To appear in J. Math. Anal. and Appl.

[3] A. S. Vatsala, "On the existence of periodic quasisolutions for first order systems," To appear in J. Nonlinear Analysis.

DIFFERENTIAL EQUATIONS WITH PIECEWISE CONSTANT DELAYS

Joseph Wiener
Pan American University
Edinburg, Texas

ABSTRACT

Differential equations in Banach space with delays $[t]$ and $t - [t]$ are considered. The initial value problem is posed at $t = 0$, and the solution is sought for $t > 0$. Existence and uniqueness theorems are established for equations with bounded and unbounded operators. Some results are stated also for the asymptotic behavior of solutions as $t \to \infty$. Equations with unbounded delay arise in cases of several argument deviations. In such problems the initial function is prescribed on $(-\infty, 0]$ and the solution is considered for $t > 0$.

Profound and close links exist between functional and functional differential equations. Thus, the study of the first often enables one to predict properties of differential equations of neutral type. On the other hand, some methods for the latter in the special case when the deviation of the argument vanishes at individual points has been used to investigate functional equations [1]. Functional equations are directly connected with difference equations of a discrete (for example, integer-valued) argument, the theory of which has been very intensively developed in the book [2] and in numerous subsequent papers. Bordering on difference equations are the impulse functional differential equations with impacts and switching, loaded equations (that is, those including values of the unknown solution for given constant values of the argument), equations

$$x'(t) = f(t, x(h(t)))$$

with lagging arguments of the form $h(t) = [t]$, that is, having intervals of constancy, etc. A substantial theory of such equations is virtually undeveloped [3].

In this note we study linear differential equations in Banach space with arguments $h(t) = [t]$ and $h(t) = t - [t]$. The initial value problem is posed at $t = 0$, and the solution is sought for $t > 0$. The properties of solutions to equations with bounded operators are similar to those of solutions to systems of ordinary differential equations which can be viewed as equations in a finite-dimensional Banach space. In particular, it is not difficult to find the closed form for the solutions of linear equations with constant bounded operators. Some nonlinear equations are also investigated. Equations with unbounded delay [4] arise in cases of several argument deviations. In such problems the initial function is prescribed on $(-\infty, 0]$ and the solution is considered for $t > 0$.

Consider in Banach space E the differential equation

$$x'(t) = Ax(t) + A_0x([t]) + A_1x'([t]) \tag{1}$$

with linear constant operators A, A_0, A_1: $D \longrightarrow E$ having the same domain D dense in E. Here $[t]$ denotes the greatest-integer function. We introduce the definition of a solution to (1) that modifies the given in [5] for the equation

$$x'(t) = Ax(t). \tag{2}$$

DEFINITION 1. A solution of Eq. (1) on the segment $[0, T]$ is a function $x(t)$ satisfying the conditions:

(i) The function $x(t)$ is continuous on $[0, T]$ and its values lie in the domain D for all $t \in [0, T]$.

(ii) At each point $t \in [0, T]$ there exists a strong derivative $x'(t)$ with the possible exception of the points $[t] \in [0, T]$ where one-sided derivatives exist.

(iii) Eq. (1) is satisfied on each interval $[n, n + 1) \subset [0, T]$ with integral ends and for $[t] \leq t \leq T$, if T is finite.

The Cauchy problem on $[0, T]$ is to find a solution of the equation on $[0, T]$ satisfying the initial condition

$$x(0) = c_0 \in D \tag{3}$$

Let $L(E, E)$ be the space of bounded linear operators from E into E. In this case we consider a more general equation

$$x'(t) = Ax(t) + \sum_{k=0}^{n} A_kx^{(k)}([t]). \tag{4}$$

THEOREM 1. If $A, A_k \in L(E, E)$ and the operators A and $B = I - \sum_{k=1}^{n} A_kA^{k-1}$ have bounded inverses, then problem (3) - (4) on $[0, \infty)$ has a unique solution

$$x(t) = \Phi(t - n)\, \Phi^{n}(1)c_0, \quad (n \leq t \leq n + 1), \quad n \geq 0 \tag{5}$$

where

$$\Phi(t) = I + (e^{At} - I)A^{-1}B^{-1}(A + A_0).$$

COROLLARY. In the conditions of Theorem 1 each solution of the Cauchy problem for Eq. (4) grows to infinity no faster than exponentially, and the exponential types of all solutions are bounded above.

THEOREM 2. If in addition to the hypotheses of Theorem 1, there exists a bounded inverse of the operator $\Phi(1)$, then solution (5) has a unique backward continuation on $(-\infty, 0]$:

$$x(t) = \Phi(t + n)\Phi^{-n}(1)c_0, \quad (-n \leq t \leq -n + 1).$$

THEOREM 3. Suppose that Eq. (1) with linear constant operators A, A_0, A_1 satisfies the following hypotheses:

(i) The operators are closed, the domain D is dense in E.

(ii) The weakened Cauchy problem for the homogeneous equation (2) is correct on D and the operator A has at least one regular point.

(iii) The operator $B = I - A_1$ has a bounded inverse and $P = A_0 + A_1B^{-1}(A + A_0)\colon\ D \longrightarrow D$.

Then on $[0, \infty)$ problem (1) - (3) has a unique solution

$$x(t) = \left(U(t - n) + \int_n^t U(t - s)P\,ds\right) \prod_{i=n}^{1} \left(U(1) + \int_{i-1}^{i} U(i - s)P\,ds\right)c_0,$$

$$(n \leq t \leq n + 1)$$

where $U(t)$ is the semigroup operator generated by (2) - (3).

THEOREM 4. If in the equation

$$x'(t) = A(t)x(t) + B(t)x([t]) \tag{6}$$

the operators $A(t)$, $B(t) \in L(E, E)$ and are strongly continuous on $0 \leq t \leq [T] + 1$, then there exists a unique solution of problem (3) - (6) given by the formula

$$x(t) = U(t)\left(U^{-1}(n) + \int_n^t U^{-1}(s)B(s)\,ds\right)c_n, \quad (n \leq t \leq n + 1) \tag{7}$$

where

$$c_n = \prod_{i=n}^{1} U(i) \left(U^{-1}(i - 1) + \int_{i-1}^{i} U^{-1}(s)B(s)ds\right)c_0.$$

THEOREM 5. If $A(t)$, $B(t) \in L(E, E)$ and are strongly continuous on $0 \le t \le \infty$, then solution (7) satisfies the estimate

$$\| x(t) \| \le e^{(t+1)a(t)} c^{t+1}(t) \| c_0 \|,$$

where

$$a(t) = \max_{0\le s\le t} \| A(s) \| > 0, \quad b(t) = \max_{0\le s\le t} \| B(s) \|,$$

$$c(t) = 1 + \left(b(t)/a(t)\right).$$

THEOREM 6. If the closed operators $A(t)$ and $B(t)$ are strongly continuous and have the common constant domain D for all $0 \le t \le [T] + 1$ which is dense in E, the Cauchy problem for $x' = A(t)x$ is uniformly correct and $B(t)$: $D \longrightarrow D$, then (7) yields the unique solution of problem (3) - (6).

Now we consider the equation

$$x'(t) = \sum_{k=0}^{N} A_k x(t - k[t]) \tag{8}$$

with linear constant operators A_k: $D \longrightarrow E$, having the same domain D dense in E.

DEFINITION 2. The function $x(t)$ is called a solution of the initial-value problem for (8), if the following conditions are satisfied:

(i) The function $x(t)$ is continuous on $(-\infty, \infty)$ and its values lie in the domain D for all $t \in (-\infty, \infty)$.

(ii) At each point $t \in [0, \infty)$ there exists a strong derivative $x'(t)$ with the possible exception of the points $[t]$ where one-sided derivatives exist.

(iii) On $(-\infty, 0]$, $x(t)$ coincides with a given continuous function $\phi(t) \in D$ and satisfies Eq. (8) on each interval $[n, n + 1)$.

The solution $x_0(t)$ of (8) on $[0, 1)$ satisfies the equation

$$x_0'(t) = \left(\sum_{k=0}^{N} A_k\right) x_0(t), \quad x_0(0) = \phi(0). \tag{9}$$

We also employ the homogeneous equation

$$x'(t) = A_0x(t), \quad x(1) = x_0(1) \tag{10}$$

corresponding to (8) for $t \geq 1$. Denote

$$f_i(t) = A_1x_0(t - i) + \sum_{k=2}^{N} A_k\phi(t - ik), \; i = 1, 2, \ldots$$

$$(i \leq t \leq i + 1)$$

THEOREM 7. Suppose that Eq. (8) satisfies the following hypotheses:

(i) The Cauchy problem for (9) is correct.

(ii) The Cauchy problem for (10) is uniformly correct.

(iii) The values $f_i(t) \in D$, the functions $f_i(t)$ and $A_0f_i(t)$ are continuous. Then the initial-value problem for (8) has a unique solution

$$x(t) = U(t - 1)x_0(1) + \sum_{i=1}^{n-1} \int_i^{i+1} U(t - s)f_i(s)ds +$$

$$+ \int_n^t U(t - s)f_n(s)ds, \; (n \leq t \leq n + 1), \quad n \geq 1 \tag{11}$$

where U(t) is the semigroup operator generated by (10).

THEOREM 8. If in the conditions of Theorem 7 the functions $f_i(t)$ are bounded in D uniformly relative to t and i: $\| f_i(t) \| \leq K$, $(i \leq t \leq i + 1)$, $i = 1, 2, \ldots$, then solution (11) of the initial-value problem for Eq. (8) satisfies the inequality

$$\| x(t) \| \leq Me^{wt}, \; t \geq 0$$

where M and w are some positive constants.

THEOREM 9. Suppose the equation

$$x'(t) = A(t)x(t) + f(t, x(t - [t]), x(t - 2[t]), \ldots, x(t - N[t])) \tag{12}$$

satisfies the following hypotheses:

(i) The operator $A(t) \in L(E, E)$ and is strongly continuous on $R_+ = [0, \infty)$.

(ii) The operator $f\colon R_+ \times E^N \longrightarrow E$ is continuous on the direct product of R_+ and N copies of E.

(iii) The function $\phi\colon R_- \longrightarrow E$ is continuous on $R_- = (-\infty, 0]$.

(iv) The solution of the equation

$$x_0'(t) = A(t)x_0(t) + f(t, x_0(t), x_0(t), \ldots, x_0(t)), \; x_0(0) = \phi(0)$$

exists on $0 \leq t \leq 1$ and is unique.

Then the initial-value problem $x(t) = \phi(t)$, $t \in R_-$ for (8) has a unique solution

$$x(t) = U(t)\Big(U^{-1}(1)x_0(1) + \sum_{i=1}^{n-1} \int_i^{i+1} U^{-1}(s)f_i(s)ds + \int_n^t U^{-1}(s)f_n(s)ds\Big),$$

$$(n \leq t \leq n + 1), \quad n \geq 1$$

where $U(t)$ is the solution of $U'(t) = A(t)U(t)$, $U(0) = I$ and $f_i(s) = f\big(s, x_0(s - i), \phi(s - 2i), \ldots, \phi(s - Ni)\big)$

THEOREM 10. If $f(x, \lambda) \in C(R^2)$ is different from zero everywhere and the solutions of $dx/dt = f(x, \lambda)$ can be extended over $[0, \infty)$, then the problem

$$x'(t) = f\big(x(t), x([t])\big), \quad x(0) = c_0, \quad (0 \leq t < \infty) \tag{12}$$

has a unique solution. If $f(c_0, c_0) = 0$ and $\int f^{-1}(x, c_0)\, dx$ diverges when $x \longrightarrow c_0$, then $x = c_0$ is the unique solution of (12). And if this integral converges for $x \longrightarrow c_0$, then (12) has more than one solution.

REFERENCES

[1] G.P. Pelyukh and A.N. Sharkovskii, "Introduction to the Theory of Functional Equations" (Russian), Naukova Dumka, Kiev, 1968.

[2] A. Halanay and D. Wexler, "Teoria Calitativa a Sistemelor cu Impulsuri," Acad. RSR, Bucuresti, 1968.

[3] A.D. Myshkis, On certain problems in the theory of differential equations with deviating argument, Uspekhi Mat. Nauk 32:2(1977), 173-202.

[4] C. Corduneanu and V. Lakshmikantham, Equations with unbounded delay, Nonlinear Analysis, Vol. 4, No. 5, 1980, 831-877.

[5] S.G. Krein, "Linear Differential Equations in Banach Spaces", Transl. of Math. Monogr. Vol. 29, Amer. Math. Soc., Providence, 1971.

ASYMPTOTIC BEHAVIOR OF SOLUTIONS IN NONAUTONOMOUS SYSTEMS

Taro Yoshizawa

Tohoku University
Sendai, Japan

ABSTRACT

We discuss some sufficient conditions for attractivity of a closed set and show how to apply our result and the growth condition to the study of asymptotic behaviors of solutions, and we discuss also global properties of solutions of a differential equation and its limiting equations.

The classical Liapunov's theorem on asymptotic stability has been improved by many authors so that the conditions in Liapunov's theorem have been weakened. For autonomous systems, Barbashin and Krasovskii [4] generalized Liapunov's theorem to the case where the derivative along the solution of a Liapunov function is not negative definite. LaSalle [10] made a simple observation based on the invariance property of solutions. LaSalle and other many authors extended the invariance principle to nonautonomous systems and dynamical systems, see [12] for the references.

In order to discuss the invariance principle in a nonautonomous system

(1) $$x' = f(t,x),$$

where $f(t,x)$ is continuous on $I\times D$, $I = [0,\infty)$ and D is an open set in R^n, Yoshizawa [20] gave a sufficient condition for the attractivity of a closed set in D by assuming

(a) $f(t,x) = h(t,x) + g(t,x)$, where $h(t,x)$ and $g(t,x)$ are continuous on $I\times D$, $h(t,x)$ is bounded for all t when x belongs to a compact set in D and $\int_{t_0}^{\infty} |g(s,x(s))|ds < \infty$ if $x(t)$ is continuous and $x(t) \in K$ for all $t \geqq t_0$ and some compact set K in D.

LaSalle [11] extended the result by weakening the boundedness condition of $f(t,x)$. On the other hand, Burton, Rouche, Haddock and others relaxed conditions on $f(t,x)$ but required certain relations between $\dot{V}(t,x)$ of a Liapunov function $V(t,x)$ and $|f(t,x)|$, see [7,13] for the references. Recently, LaSalle [13] has made a further extension of Liapunov's second method for a broad class of nonautonomous systems and Hatvani [8] has generally discussed the attractivity for nonautonomous systems so as to improve known results. Burton [5] has discussed also the attractivity of a closed set for functional differential equations. As an application of the invariance principle, Yoshizawa [20], Onuchic and others [17,18] discussed the asymptotic stability for a second order differential equation, where the boundedness condition was assumed. In order to relax the boundedness condition, Thurston and Wong [19], Artstein and Infante [3] considered some growth condition on the vector field.

On the other hand, the study of relationships between the behaviors of solutions of a given differential equation and its limiting equation is also important and fruitful for discussing stability properties. Many efforts devoted to this topics can be found in references in [1,2,6].

Now, using Liapunov functions, we want to find a closed set E in D which attracts every bounded solution of (1). We say that $V: I\times D \to R$ is a Liapunov function for (1), if

(i) $V(t,x)$ is continuous on $I\times D$,

(ii) $V(t,x)$ is locally Lipschitzian in x,

(iii) given a compact set K in D, there exists an $m(K)$ and a continuous function $\gamma_K(t) \geqq 0$ such that $m(K) \leqq V(t,x)$ for all $(t,x) \in I\times K$, $\gamma_K \in L_1[0,\infty)$ and $\dot{V}_{(1)}(t,x) \leqq \gamma_K(t)$ for all $(t,x) \in I\times K$, where

$$\dot{V}_{(1)}(t,x) = \limsup_{\delta\to 0+} \frac{1}{\delta}\{V(t+\delta, x+\delta f(t,x)) - V(t,x)\}.$$

Let E be a nonempty closed set in the topology of D. Suppose that for any $\varepsilon > 0$ and any compact set K in D, there exist $\alpha(\varepsilon,K) > 0$ and $T(\varepsilon,K) > 0$ such that

$$\dot{V}_{(1)}(t,x) \leqq -\alpha(\varepsilon,K) + \gamma_K(t) \tag{2}$$

for $t \geqq T(\varepsilon,K)$ and $x \in K\cap N(\varepsilon,E)^c$, where $N(\varepsilon,E)$ is the ε-neighborhood of E. Let $x(t)$ be a solution of (1) such that $x(t) \in K$ for $t \geqq t_0$ and some compact set K in D. Denote by $B(\delta,p)$ the open ball of radius δ about p. Suppose a positive limit point p of $x(t)$ is not in E. Then for some $\varepsilon > 0$, either $x(t) \in B(2\varepsilon,p)\cap K$ for all $t \geqq T_1$ and some $T_1 \geqq t_0$ or $x(t)$ must enter $B(\varepsilon,p)\cap K$, then leave $B(2\varepsilon,p)\cap K$ and do this an infinite number of times. In the former case, (2) implies

(3) $$V(t,x(t)) \to -\infty \quad \text{as } t \to \infty,$$

which contradicts the boundedness of $V(t,x(t))$. In the latter case, if we have some condition which guarantees that $x(t)$ must be in $B(2\varepsilon,p) \cap K$ an infinite time, then (2) implies also (3). Thus we can conclude that $x(t)$ approaches E. One of such conditions is condition (a) mentioned before. If there exists a Liapunov function $V(t,x)$ for (1) such that $\dot{V}_{(1)}(t,x) \leqq -W(x)$, where $W(x) \geqq 0$ is continuous, and if $f(t,x)$ satisfies condition (a), then every bounded solution of (1) approaches $E = \{x \in D\ ;\ W(x) = 0\}$. A generalization of this result has been given by Onuchic [17]. He assumed that

(b) in condition (a), for every compact set K in D and every continuous function $u(t) \in K$ defined on I, $\int_\sigma^{\sigma+t} g(s,u(s))ds \to 0$ uniformly for $t \in [0,1]$ as $\sigma \to \infty$.

If $|g(t,u(t))| \in L_p[0,\infty)$, $1 \leqq p < \infty$, then g satisfies this condition.

Onuchic et al [18] considered a Liapunov function $V(t,x)$ such that $\dot{V}_{(1)}(t,x) \leqq -W(t,x)$, where $W(t,x)$ satisfies certain conditions so that $\liminf_{t\to\infty} W(t,x) = W^*(x)$ exists and is continuous on D. Then, under condition (b), every bounded solution of (1) approaches $E = \{x \in D\ ;\ W^*(x) = 0\}$. LaSalle [13] has extended his own result so that the result by Onuchic et al is included. To do this, a closed set E is defined as follows: $x \in E$ if there is a sequence (t_k,x_k) with the property that $(t_k,x_k) \to (\infty,x)$ and $\dot{V}_{(1)}(t_k,x_k) \to 0$ as $k \to \infty$. He assumed that

(c) given any compact set K in D, any continuous $u : I \to K$ and any $\alpha > 0$, there exist $T = T(\alpha,u) > 0$ and $\beta = \beta(\alpha,u) > 0$ such that $|\int_a^{a+t} f(s,u(s))ds| > \alpha$ for $a > T$ and $t > 0$ implies $t > \beta$.

In this case, we have (2) and condition (c) is just a condition that the solution travels in $B(2\varepsilon,p)$ an infinite time. Both of conditions (a) and (b) imply condition (c). The result by Onuchic et al is included also in Theorem 1 below.

Now we consider a simple equation

(4) $$x' = y, \quad y' = -(t \sin^2 t)y - x$$

and a Liapunov function $V(x,y) = (x^2+y^2)/2$. Then $\dot{V}_{(4)}(x,y) = -(t \sin^2 t)y^2$. This equation does not satisfy condition (c) and a closed set E defined above is the whole space. Thus we discuss another condition so that we can conclude that every solution of (4) approaches $E = \{(x,y);y = 0\}$. Hatvani [8] also discussed this situation in general form.

Let $p(t) \geqq 0$ be continuous on I. If $\int_J p(t)dt = \infty$ holds on every set $J = \bigcup_{m=1}^{\infty} (a_m,b_m)$ such that $0 \leqq a_1$, $a_m < b_m \leqq a_{m+1}$, $b_m - a_m \geqq \delta > 0$ (m =

$1,2,\cdots$), we say that $p(t)$ is integrally positive [8,15]. This is equivalent to saying that $p(t)$ is integrally positive if for every $\delta > 0$, $\liminf_{t\to\infty} \int_t^{t+\delta} p(s)ds > 0$.

Theorem 1. Let $V(t,x)$ be a Liapunov function for (1). Suppose that for any compact set K in D, there exist continuous function $e_K(t) \geqq 0$ and $T(K) > 0$ such that $\int_t^{t+1} e_K(s)ds \to 0$ as $t \to \infty$ and that $\dot{V}_{(1)}(t,x) \leqq -p(t)U(x)W(x) + e_K(t)$ for all $t \geqq T(K)$ and $x \in K$, where $p(t)$ is integrally positive, $U(x) \geqq 0$ and $W(x) \geqq 0$ are continuous on D. Suppose that $U(x) > 0$ for $x \in E^c$, where $E = \{x \in D;\ W(x) = 0\}$. Assume that

(d) for any $\alpha > 0$, any $\delta > 0$ and any bounded solution $x = x(t)$ of (1), there exist $T = T(\alpha, \delta, x) > 0$ and $\beta = \beta(\alpha,\delta,x) > 0$ such that if $a > T$, $t > 0$, $x(t+a) \in N(\delta,E)^c$ and $\alpha < W(x(a+t))-W(x(a))$, then $t > \beta$, or

(e) for any $\alpha > 0$, any $\delta > 0$ and any bounded solution $x = x(t)$ of (1), there exist $T = T(\alpha,\delta,x) > 0$ and $\beta = \beta(\alpha,\delta,x) > 0$ such that if $a > T$, $t > 0$, $x(t+a) \in N(\delta,E)^c$ and $W(x(a+t)) - W(x(a)) < -\alpha$, then $t > \beta$.

Then every bounded solution of (1) approaches E.

The ancestor of these conditions can be found in [11]. If there exists a continuous function $q(t) \geqq 0$ such that $\int_0^t q(s)ds$ is uniformly continuous on I and $\dot{W}(x(t)) \leqq q(t)$, then condition (d) is satisfied, where $\dot{W}(x(t)) = \limsup_{\delta\to 0^+} \delta^{-1}\{W(x(t+\delta)) - W(x(t))\}$. Here it should be noticed that condition (c) implies condition (d).

Theorem 2. Let $V(t,x)$ be a Liapunov function for (1). Suppose that for every compact set K in D, there exist continuous function $e_K(t) \geqq 0$ and $T(K) > 0$ such that $e_K \in L_1[0,\infty)$ and that $\dot{V}_{(1)}(t,x) \leqq -p(t)U(x)W(x) + e_K(t)$ for $t \geqq T(K)$ and $x \in K$, where $p(t) \geqq 0$ is continuous, $\int_0^\infty p(t)dt = \infty$, $U(x) \geqq 0$ and $W(x) \geqq 0$ are continuous on D. Suppose that $U(x) > 0$ for $x \in E^c$, where $E = \{x \in D\ ;\ W(x) = 0\}$. Moreover, suppose that for any bounded solution $x(t)$ of (1) and any $\delta > 0$, there exist $c > 0$ and continuous functions $q(t) \geqq 0$ and $r(t) \geqq 0$ such that $cq(t) \leqq p(t)$, $r \in L_1[0,\infty)$ and $\dot{W}(x(t)) \leqq q(t) + r(t)$ for all large t such that $x(t) \in N(\delta,E)^c$ or $-r(t)-q(t) \leqq \dot{W}(x(t))$ for all large t such that $x(t) \in N(\delta,E)^c$. Then every bounded solution of (1) approaches E.

In applications to the study of asymptotic stability, we assume the growth condition. Let $H : I\times R^m \to I$ be a continuous function. We say $H(t,x)$ satisfies the growth condition for $r > 0$ if for any continuous function $u : I \to K$ for some compact set K in R^m, there exist $d > 0$ and a sequence

$\{s_k\}$, $s_k > 0$, such that $s_{k+1} - s_k \geqq d$ and that $H(t,u(t)) = 0$ on $[s_k, s_k+d]$ for all k or $\sum_{k=1}^{\infty} (\int_{s_k}^{s_k+d} H(t,u(t))dt)^{-r} = \infty$. For $x'' + h(t,x,x')x' + f(x) = 0$, Thurston and Wong [19] assumed this growth condition for $h(t,x,y)$ with $r = 1$. Artstein and Infante [3] also assumed the growth condition that $T^{-2}\int_0^T h(t,x(t),y(t))dt \leqq B$ for some $B > 0$ and all $T > 0$ and $(x(t),y(t))$ in a compact set. This condition is somewhat easily checked in an application, but it satisfies actually Thurston and Wong's condition which is weaker, for example consider $h(t) = t \log t$.

<u>Lemma.</u> Let $H(t) \geqq 0$ and $z(t) \geqq 0$ be continuous on I, and let $p > 0$, $q > 0$ be constants such that $1/p + 1/q = 1$. If $\int_0^\infty H(t)z^q(t)dt < \infty$ and $H(t)$ satisfies the growth condition for q/p, then there are $d > 0$ and a sequence $\{t_k\}$ such that $t_k > 0$, $t_{k+1} - t_k \geqq d$ and $\int_{t_k}^{t_k+d} H(t)z(t)dt \to 0$ as $k \to \infty$.

Now consider the following system which originated in some problems of reactor dynamics and was discussed by Levin and Nohel [14] and recently by Murakami [16]. Letting $c_i > 0$, $i = 0,1,\cdots,n$, be constants, $x, z_0 \in R$, $z = (z_1,z_2,\cdots,z_n)$ and all functions be continuous, for the system

$$(5)\qquad \begin{aligned} x' &= -\sum_{i=0}^{n} c_i b_i(t) z_i \\ z_i' &= -h_i(t,x,z_0,z)z_i + b_i(t)f(x) + e_i(t), \quad i = 0,1,\cdots,n, \end{aligned}$$

we make the following assumptions:

(i) $e_i(t) \in L_1[0,\infty)$, $i = 0,1,\cdots,n$.

(ii) $\int_0^t |b_i(s)|ds$ is uniformly continuous on $[0,\infty)$, $i = 0,1,\cdots,n$.

(iii) $xf(x) > 0$ $(x \neq 0)$ and $F(x) = \int_0^x f(u)du \to \infty$ as $|x| \to \infty$.

(iv) There exists a continuous function $k(x,z_0,z) > 0$ for $z \neq 0$ and an integrally positive function $p(t)$ such that $h_i(t,x,z_0,z) \geqq p(t)k(x,z_0,z)$ for all $i = 1,2,\cdots,n$.

(v) $h_0(t,x,z_0,z) \equiv 0$ and $h_1(t,x,z_0,z)$ satisfies the growth condition for $r = 1$.

(vi) $\liminf_{t\to\infty} |b_0(t)| \neq 0$ and $\liminf_{t\to\infty} |b_1(t)| \neq 0$.

Then every solution of (5) approaches the origin.

To see this, consider a Liapunov function $V(t,x,z_0,z) = e^{-E(t)}\{F(x) + \sum_{i=0}^{n} c_i z_i^2/2 + 1\}$, where $E(t) = \sum_{i=0}^{n} \sqrt{c_i}\int_0^t |e_i(s)|ds$. Then

$\dot{V}_{(5)}(t,x,z_0,z) \leqq -e^*p(t)k(x,z_0,z)W(z)$, where $e^* = e^{-E(\infty)}$ and $W(z) = \sum_{i=1}^{n} c_i z_i^2$. Thus the solutions of (5) are uniformly bounded. Since we can find a continuous function $q(t) \geqq 0$ such that $\dot{W}(z(t)) \leqq q(t)$ and $\int_0^t q(s)ds$ is uniformly continuous on $[0,\infty)$, $W(z(t))$ satisfies condition (d). By Theorem 1, $z(t) \to 0$ as $t \to \infty$. Since $\dot{V}_{(5)}(t,x,z_0,z) \leqq -e^* \sum_{i=1}^{n} c_i h_i(t,x,z_0,z)z_i^2$ implies $\int_{t_0}^{\infty} h_1(s,x(s),z_0(s),z(s))z_1^2(s)ds < \infty$, applying the lemma, there exist $d > 0$ and a sequence $\{t_k\}$ such that $t_k > 0$, $t_{k+1} - t_k \geqq d$ and

$$(6) \quad \int_{t_k}^{t_k+t} h_1(s,x(s),z_0(s),z(s))z_1(s)ds \to 0 \text{ uniformly on } [0,d] \text{ as } k \to \infty.$$

Set $x_k(t) = x(t+t_k)$ and $y_k(t) = z_0(t+t_k)$ for $t \in [0,d]$. Then $\{x_k(t)\}$ and $\{y_k(t)\}$ are uniformly bounded and equicontinuous on $[0,d]$, and hence we can assume that $x_k(t)$ and $y_k(t)$ converge uniformly on $[0,d]$. Integrating z_1' and x', and using (6), (iii) and (vi), we can see that $x(t_k) \to 0$, $z_0(t_k) \to 0$ as $k \to \infty$. On the other hand, $e^* \leqq V(t,x(t),z_0(t),z(t)) \leqq V(t_k,x(t_k),z_0(t_k),z(t_k))$ for all $t \geqq t_k$. But $V(t_k,x(t_k),z_0(t_k),z(t_k)) \to e^*$ as $k \to \infty$, and thus $V(t,x(t),z_0(t),z(t)) \to e^*$, which shows that $x(t) \to 0$, $z_0(t) \to 0$, $z(t) \to 0$ as $t \to \infty$.

By the same idea as in the above, we can discuss the asymptotic stability of

$$(7) \qquad x' = y, \; y' = -h(t,x,y)|y|^{\alpha}y - f(x),$$

where $\alpha \geqq 0$, $h(t,x,y)$ is continuous on $I\times R\times R$ and $f(x)$ is continuous on R. We assume that (i) there exists an integrally positive function $p(t)$ and a continuous function $k(x,y)$ such that $h(t,x,y) \geqq p(t)k(x,y)$ and $k(x,y) > 0$ for $y \neq 0$, (ii) $xf(x) > 0$ for $x \neq 0$ and $F(x) = \int_0^x f(u)du \to \infty$ as $|x| \to \infty$, and (iii) $h(t,x,y)$ satisfies the growth condition for $r = (\alpha+1)^{-1}$. Then the zero solution of (7) is asymptotically stable. Actually, it is equiasymptotically stable, because uniform stability and attractivity imply this. In this case, consider a Liapunov function $V(x,y) = F(x) + y^2/2$.

In the case where $k(x,y) \geqq 0$ and $h(t,x,y)$ satisfies condition (c), the growth condition is automatically satisfied, and by Theorem 1, $(x(t),y(t)) \to E = \{(x,y) \,;\, k(x,y)y=0\}$ as $t \to \infty$. Thus, if we assume that $(-\infty,0)$ and $(0,\infty)$ are connected components in $E-\{(0,0)\}$, we can conclude that $y(t) \to 0$ and then every solution approaches the origin.

Finally, we shall make some remarks on global properties in limiting equations. For global properties of solutions, the concept of boundedness is very important. It was pointed out in [6] that there is a misunderstanding in "if" part in Theorem A given in [2] which is concerned with global asymptotic stability. The "only if" part in global stability property in [2] also is incomplete, because the continuation property in limiting equations is lacking. We denote by $C(I\times R^n, R^n)$ the space of continuous functions defined on $I\times R^n$ with values in R^n with compact-open topology. A function $f \in C(I\times R^n, R^n)$ is said to be positively precompact if for any sequence $\{t_k\}$ in I such that $t_k \to \infty$ as $k \to \infty$, the sequence $\{f(t+t_k, x)\}$ contains a subsequence which converges uniformly on any compact set in $I\times R^n$. We denote by $\Omega(f)$ the set of all limit functions g such that $f(t+t_k, x)$ converges to g uniformly on any compact set in $I\times R^n$ for some sequence $\{t_k\}$, $t_k \to \infty$ as $k \to \infty$.

Now consider the system

(8) $$x' = f(t,x),$$

where $f \in C(I\times R^n, R^n)$ and $f(t,x)$ is assumed to be positively precompact. A system

(9) $$x' = g(t,x)$$

is said to be a limiting equation of (8) when $g \in \Omega(f)$. We say that system (8) is regular if the solutions of every limiting equation of (8) are unique for the initial value problem. The solutions of (8) are said to be uniformly ultimately bounded with bound B, if there is a constant $B > 0$ and for any $\alpha > 0$, there is a $\sigma(\alpha) \geqq 0$ such that every solution $x(t)$ of (8) satisfies $|x(t)| \leqq B$ for all $t \geqq \tau + \sigma(\alpha)$ whenever $\tau \geqq 0$ and $|x(\tau)| \leqq \alpha$. For the case where $f(t,0) \equiv 0$, the zero solution of (8) is said to be globally uniformly asymptotically stable if it is uniformly stable and if the solutions of (8) are uniformly ultimately bounded with arbitrary small bound. This is the same as the definition given in [2]. Notice that for the usual definition, the uniform boundedness of solutions is required [21]. Then we have the following theorems.

Theorem 3. Suppose that system (8) is regular. If the solutions of (8) are uniformly bounded and uniformly ultimately bounded, then for any $g \in \Omega(f)$ the solutions of (9) are uniformly ultimately bounded.

Theorem 4. Suppose that system (8) is regular and $f(t,0) \equiv 0$. If the zero solution of (8) is globally uniformly asymptotically stable and if the solutions of (8) are uniformly bounded, then the zero solution of every limiting equation is globally uniformly asymptotically stable.

We should notice here that neither the regularity nor uniform boundedness in the assumptions can be dropped even for an almost periodic system, see [9] for the details. The following example shows that we can not drop the uniform boundedness without supplementary conditions.

Define a function $h(t,x)$ which is 4-periodic in t as follows:

$$h(t,x) = \begin{cases} -x & \text{for } 0 \leqq x \leqq 1,\ 0 \leqq t < \infty, \\ -x^2 & \text{for } 1 \leqq x,\ 0 \leqq t \leqq 1, \\ -x^2(2-t)+(x^2-2)(t-1) & \text{for } 1 \leqq x,\ 1 \leqq t \leqq 2, \\ x^2-2 & \text{for } 1 \leqq x,\ 2 \leqq t \leqq 3, \\ (x^2-2)(4-t)-x^2(t-3) & \text{for } 1 \leqq x,\ 3 \leqq t \leqq 4, \end{cases}$$

and $h(t,x) = -h(t,-x)$ for $x \leqq 0$ and $0 \leqq t < \infty$. Let $f(t,x) = h(t,x)-a(t)x^3$, where $a(t)$ is an almost periodic function due to Sibuya (refer to p.143 in [21]) with the following properties : (i) $a(t) > 0$ for all $t \geqq 0$ and (ii) there is a $b \in \Omega(a)$ such that $b(t) = 0$ on $[0,5]$. Then $f(t,x)$ is almost periodic in t uniformly for $x \in R$. Consider the equation

$$x' = f(t,x). \tag{10}$$

Clearly, equation (10) is regular since every $g \in \Omega(f)$ has a form $h(t+m,x)-b(t)x^3$ with a constant m, $0 \leqq m \leqq 4$, and a $b \in \Omega(a)$. Since $a(t) > 0$ for all $t \geqq 0$, for any compact interval $J \subset I$ we can find an $r > 0$ such that $xf(t,x) < 0$ for all $t \in J$ and $|x| \geqq r$. This guarantees that every solution of (10) is continuable up to $t = \infty$. Moreover, the region $I\times\{x ; |x| < 1\}$ is positively invariant. Since $xx' \leqq -x^2|x|$ on $[4k,4k+1]\times R$ for any integer k, every solution $x(t)$ of (10) starting at $t = 4k$ must satisfy $|x(4k+1)| < 1$. Thus we can see that every solution $x(t)$ of (10) starting at $t = \tau$ reaches the line $t = 4k$, satisfies $|x(4k+1)| < 1$, remains in $|x| < 1$ for all $t \geqq 4k + 1$ and satisfies the inequality $x(t)x'(t) \leqq -x^2(t)$ there, where k is the smallest integer such that $4k \geqq \tau$. Therefore it is not difficult to show that the zero solution of (10) is globally uniformly asymptotically stable. However, by property (ii) of $a(t)$, we can find a $g \in \Omega(f)$ such that $g(t,x) = (x^2-2)\operatorname{sgn} x$ for $0 \leqq t \leqq 1$ and $|x| \geqq 1$. Therefore every solution $x(t)$ of $x' = g(t,x)$ satisfying $|x(0)| > \sqrt{2}\,(\exp(2\sqrt{2})+1)/(\exp(2\sqrt{2})-1)$ can not be continuable up to $t = 1$.

REFERENCES

[1] Artstein, Z., Limiting equations and stability of nonautonomous ordinary differential equations, Appendix A in J. P. LaSalle, The Stability of Dynamical Systems, CBMS Regional Conference Series in Applied Math. 25, SIAM, Philadelphia 1976.

[2] Artstein, Z., Uniform asymptotic stability via the limiting equations, J. Differential Equations, 27(1978), 172-189.

[3] Artstein, Z. and E. F. Infante, On the asymptotic stability of oscillators with unbounded damping, Quart. Appl. Math., 34(1976), 195-199.

[4] Barbashin, E. A. and N. N. Krasovskii, On global stability of motion, Dokl. Akad. Nauk SSSR, 86(1952), 453-456.

[5] Burton, T. A., Stability theory for functional differential equations, Trans. Amer. Math. Soc., 255(1979), 263-275.

[6] D'Anna, A., A. Maio and V. Moauro, Global stability properties by means of limiting equations, Nonlinear Anal., 4(1980), 407-410.

[7] Haddock, J., Stability theory for non-autonomous systems, Dynamical Systems. An International Symposium (L. Cesari, J. K. Hale and J. P. LaSalle, eds.), Vol. 2, 271-274, Academic Press, New York 1976.

[8] Hatvani, L., Attractivity theorems for non-autonomous systems of differential equations, Acta Sci. Math. (Szeged), 40(1978), 271-283.

[9] Kato, J. and T. Yoshizawa, Remarks on global properties in limiting equations, Funkcial. Ekvac., 24(1981), 363-371.

[10] LaSalle, J. P., The extend of asymptotic stability, Proc. Nat. Acad. Sci. U.S.A., 46(1960), 363-365.

[11] LaSalle, J. P., Stability theory of ordinary differential equations, J. Differential Equations, 4(1968), 57-65.

[12] LaSalle, J. P., Stability theory and invariance principle, Dynamical Systems. An International Symposium (L. Cesari, J. K. Hale and J. P. LaSalle, eds.), Vol. 1, 211-222, Academic Press, New York 1976.

[13] LaSalle, J. P., Stability of nonautonomous systems, Nonlinear Anal., 1(1976), 83-91.

[14] Levin, J. J. and J. A. Nohel, Global asymptotic stability for nonlinear systems of differential equations and applications to reactor dynamics, Arch. Rational Mech. Anal., 5(1960), 194-211.

[15] Matrosov, V., On the stability of motion, Prik. Mat. Meh., 26(1962), 885-895.

[16] Murakami, S., Asymptotic behavior of solutions of ordinary differential equations, Tohoku Math. J. (to appear).

[17] Onuchic, N., Invariance properties in the theory of ordinary differential equations with applications to stability problem, SIAM J. Control, 9(1971), 97-104.

[18] Onuchic, N., L. R. Onuchic and P. Z. Taboas, Invariance properties

in the theory of stability for ordinary differential systems and applications, Applicable Anal., 5(1975), 101-107.

[19] Thurston, L. H. and J. S. W. Wong, On global asymptotic stability of certain second order differential equations with integrable forcing terms, SIAM J. Appl. Math., 24(1973), 50-61.

[20] Yoshizawa, T., Asymptotic behavior of solutions of a system of differential equations, Contrib. Differential Eqs., 1(1963), 371-387.

[21] Yoshizawa, T., Stability Theory and the Existence of Periodic Solutions and Almost Periodic Solutions, Applied Math. Sciences 14, Springer-Verlag, New York 1975.